RETURN
LIBRARY
THE STANDARD OIL CO. (OHIO)
WARRENSVILLE LAB

RETURN
LIBRARY
THE STANDARD OIL CO. (OHIO)
WARRENSVILLE LAB

Adsorption from Solution at the Solid/Liquid Interface

Adsorption from Solution at the Solid/Liquid Interface

Edited by

G. D. Parfitt

Department of Chemical Engineering,
Carnegie-Mellon University,
Pittsburgh, Pennsylvania, USA

and

C. H. Rochester

Chemistry Department,
University of Dundee,
Dundee, UK

1983

Academic Press

A Subsidiary of Harcourt Brace Jovanovich, Publishers

London New York
Paris San Diego San Francisco São Paulo
Sydney Tokyo Toronto

ACADEMIC PRESS INC. (LONDON) LTD.
24/28 Oval Road
London NW1 7DX

United States Edition published by
ACADEMIC PRESS INC.
111 Fifth Avenue
New York, New York 10003

British Library Cataloguing in Publication Data

Adsorption from solution at the solid/liquid interface.
1. Adsorption 2. Solids
I. Parfitt, G. D. II. Rochester, C. H.
541.3′453 QD506

ISBN 0-12-544980-1

LCCCN 82-73220

Printed in Great Britain by Page Bros (Norwich) Ltd.

Contributors

J. S. Clunie Procter and Gamble Ltd, Newcastle Technical Centre, P.O. Box Forest Hall No. 2, Newcastle-upon-Tyne NE12 9TS, UK

G. J. Fleer Laboratory for Physical and Colloid Chemistry, Agricultural University, De Dreijen 6, Wageningen, The Netherlands

C. H. Giles Department of Pure and Applied Chemistry, University of Strathclyde, 295 Cathedral Street, Glasgow G1 1XL, UK

F. Th. Hesselink Koninklijke Shell Exploratie en Produktie Laboratorium, Shell Research B.V., Postbus 60, Rijswijk, The Netherlands (Present address: Brunei Shell Petroleum Company Limited, Seria, State of Brunei)

D. B. Hough Unilever Research, Port Sunlight Laboratory, Quarry Road East, Bebington, Wirral, Merseyside L63 3JW, UK

B. T. Ingram Procter and Gamble Ltd, Newcastle Technical Centre, P.O. Box Forest Hall No. 2, Newcastle-upon-Tyne NE12 9TS, UK

J. E. Lane Division of Applied Organic Chemistry, C.S.I.R.O., P.O. Box 4331, Melbourne, Victoria 3001, Australia

J. Lyklema Laboratory for Physical and Colloid Chemistry, Agricultural University, De Dreijen 6, Wageningen, The Netherlands

G. D. Parfitt Department of Chemical Engineering, Carnegie-Mellon University, Pittsburgh, Pennsylvania 15213, USA

H. M. Rendall Department of Chemistry, Paisley College of Technology, High Street, Paisley, Renfrewshire PA1 2BE, UK

C. H. Rochester Department of Chemistry, University of Dundee, Dundee DD1 4HN, UK

Preface

Justification for a book concerned with adsorption at the solid/liquid interface is hardly necessary. The biological, environmental, and technological importance of adsorption phenomena at the boundaries between solids and liquids can never be in doubt. Interactions between the components of liquids or solutions and the surfaces of solids make, in a wide variety of ways, a momentous contribution to our everyday lives. To give a comprehensive description of systems in which solid/liquid interfaces are a significant feature would require a book in itself. A few brief illustrative examples must suffice.

Macroscopic solid/liquid interfaces are exemplified by the surfaces of electrodes and by interfacial phenomena involved in the processes of corrosion, lubrication and adhesion. Surface-adsorbate interactions are a key feature of such processes and are often deliberately modified by appropriate surface treatments. Modification of the surface of a solid before contact with a liquid may be carried out in order to change the wetting properties of the surface. Alternatively solute species, such as corrosion inhibitors which may be chemisorbed or surface-active agents which may be physisorbed, can alter the adsorptive characteristics of the surface of a solid *in situ* at the solid/liquid interface.

Colloidal systems particularly involving dispersions of solids in liquids often occur as paints and inks, foodstuffs, medicines, cosmetics and pharmaceutical or agricultural products. The adsorption of species at the solid/liquid interface frequently has a dominating influence on colloid stability. The addition of stabilizing agents is commonplace. Interactions between the surfaces of particles and molecules of additive provide the first step in the mechanism of the stabilization process. Conversely modification of the interfacial region in a sol may be deliberately affected to cause instability and hence flocculation. Interparticulate interactions are of importance in thixotropic systems such as drip-free paint, toothpaste and lubricating greases. The thixotropic properties are influenced by the adsorption of medium or additives onto the surfaces of the gelling agents. For example, aerosil silicas give a good gelling effect in non-hydrogen bonding liquids but aerogel silicas are better in aqueous systems.

There is an abundance of systems between those containing a macroscopic solid/liquid interface and those which are colloidal in nature. Dyes, soaps or detergents are adsorbed from solution onto the surfaces of fibres in the dyeing and washing processes. Coal cleaning, mineral extraction and soil science are areas where adsorption behaviour at the solid/liquid interface is of paramount importance. The adsorptive capacity of powdered solids is frequently used to advantage, for example in sugar decolorization, beer clarification, chromatographic separation and effluent or sewage treatment. Biological systems are attracting increasing attention and include problems associated with the adsorption of macromolecules (polysaccharides, lipids, proteins) onto artificial vessels and capillary or cell walls. Heterogeneous catalysis at the solid/liquid interface is exemplified by the use of powdered catalysts for the polymerization of alkenes in the presence of a solvent, either for the alkene or for the polymeric product.

The book is divided into two sections, the first dealing with adsorption of non-electrolytes and the second with electrolytes. In both sections the adsorption of small molecules/ions, surface-active agents and polymeric adsorbates are treated in separate chapters.

Chapter 1 considers the adsorption of small molecules that have limited solubility in the solvent, and includes partially miscible liquid systems. Critical analysis of adsorption isotherms, and data from immersional calorimetry and infrared spectroscopic analysis of the solid/liquid interface is seen to provide information on the interaction between solute and surface and on the structure of the adsorbed layer, although such a comprehensive analysis is relatively rare. The subject of adsorption from mixtures of miscible liquids is discussed in Chapter 2. Following a resumé of the basic thermodynamic relations a detailed analysis is given of the models used to describe the solid/solution interface. These models are usually based on either a lattice model of the surface and solution phases, or by assuming a homogeneous structureless surface phase in equilibrium with a structureless liquid. The results of both models are compared with experimental data. Chapter 3 deals with non-ionic surfactants, amphiphilic molecules that form micelles above a well-defined solution concentration. Examples are the alkylpolyoxyethyleneglycol monoethers, dimethyldodecyl phosphine oxide, and molecules based on the difunctional sulphinylalkanol group. The discussion is mainly concerned with dilute aqueous systems, involving carbon blacks, polymer latexes, silica, pigments, silver iodide, metals and textile fibres. Concluding the section on non-electrolytes is a review (Chapter 4) of polymer adsorption. Experimental trends are identified, and related to theory. The theoretical treatments outlined are those of Hoeve and Silberberg, and of Roe and Scheutjens-Fleer, all of which account explicitly for polymer-solvent interactions using Flory-Huggins theory for polymer solutions. Techniques leading to surface coverage, bound fraction, and layer thickness are discussed, with examples.

The section on electrolytes begins with a critical account of the current problems associated with the adsorption of small ions at the solid/liquid interface (Chapter 5). Such questions as ion binding, adsorption specificity, discreteness of charge, lyotropic sequences, and negative adsorption are considered. The subject of complex ion adsorption is not included in this book since current understanding of the factors involved is confused. The traditional principle of splitting the adsorption potential into a series of terms does not satisfactorily explain experimental findings, and no acceptable solution exists at present. Chapter 6 reviews the parameters, both electrical and non-electrical, that contribute to the adsorption of ionic surfactants, mainly from aqueous solution. The problems of interpretation are highlighted, with illustrations from experiments with carbons, polymer latexes, oxides, ionic solids, clays and silicates. Adsorption of dyes is considered in Chapter 7, with particular reference to the dyeing of textile fibres for which both the dye-fibre interactions and the diffusion of dye into the internal regions of the fibre contribute to the kinetics and thermodynamics of the process. Finally, Chapter 8 deals with the adsorption of ionized polymers (polyelectrolytes) which includes biological molecules such as proteins and nucleic acids. The behaviour of polyelectrolytes in solution is discussed, and Hesselink's train/loop model for adsorption is considered in detail. This theory unfortunately leaves many questions unanswered at present.

Many areas in which technological innovation has involved solid/liquid interfaces have been developed more through art and craft than through science. Basic understanding of the scientific principles often lagged far behind in part because the study of surfaces requires exceedingly careful experimentation if meaningful and reproducible results are to be obtained. In recent years, however, considerable effort has been increasingly directed toward closing the gap between theory and practice. Much has been achieved, in part through the availability of new techniques which probe sub-surface layers or interfacial regions. However, many reported studies have little more than qualitative value since insufficient attention has been paid to crucial variables. During the past decade we have seen significant advances in theory, following the development of more satisfactory models for bulk phase solutions. But in the cases where electrostatic and non-electrical interactions are involved, e.g. for complex ions and polyelectrolytes, there is still much to be done. It is hoped that the broad aim of this book to review the current status of knowledge concerning adsorption at the solid/liquid interface will also provide a stimulus for future work in this important area.

January 1983

G. D. Parfitt
C. H. Rochester

Contents

Non-electrolytes

1. Adsorption of Small Molecules

G. D. PARFITT and C. H. ROCHESTER

I. Introduction

This chapter deals with the adsorption of small molecules from solution at the solid/liquid interface. Electrolytes and polymers are considered in later chapters. Here we are essentially concerned with molecular solute species that have limited solubility in the solvent. This will include partially miscible liquid systems, and solutions of solids; liquids that are completely miscible over the whole range of concentration are discussed in Chapter 2.

A. THE ADSORPTION ISOTHERM

In adsorption from solution at the solid/liquid interface the experimental measurement is the change in concentration of the solution which results from adsorption. Since the solution contains more than one component an isotherm of concentration change is essentially a "composite isotherm" and for a two-component system it is described by the relation

$$n^0 \Delta x/m = n_1^s(1 - x) - n_2^s x \tag{1}$$

where Δx is the decrease in mole fraction x of component 1 when n^0 moles of original solution are brought into contact with m grams of adsorbent, and n_1^s and n_2^s are the numbers of moles of components 1 and 2 respectively adsorbed per gram of solid (Kipling, 1965). Equation (1) contains two unknown quantities, hence the individual isotherms for the two components can only be obtained with the use of supplementary information. For the systems under consideration here we may consider two situations:

(a) For very dilute solutions, and for which the limiting solubility of the solute occurs at a low value of the mole fraction, even if n_2^s is large the product $n_2^s x$ is small compared with $n^0 \Delta x/m$, and $n_1^s(1 - x)$ is approximately equal to n_1^s. Hence for this situation, e.g. solutions of stearic acid in benzene, we have

$$n^0 \Delta x/m \approx n_1^s \tag{2}$$

and the composite isotherm gives a very reasonable individual isotherm for the solute despite the fact that an appreciable amount of solvent may also be adsorbed.

(b) Cases exist for which the adsorption isotherm exhibits a maximum at relatively low values of solute mole fraction below the solubility limit. In this case, despite the relatively low values of x, the composite isotherm cannot represent that of the solute. An acceptable analysis assumes that a complete monolayer of solute is adsorbed at a relatively low concentration, and n_2^s is zero. The adsorption equation then becomes

$$n^0 \Delta x/m = n_1^s(1 - x) \tag{3}$$

and the composite isotherm falls linearly with concentration when the monolayer of solute is complete. Extrapolation of the isotherm to $x = 0$ gives the value of n_1^s for the monolayer. For this analysis to be correct, extrapolation to $x = 1$ should give a value for $n^0 \Delta x/m$ of zero; if this is not the case it may be assumed that the adsorbed layer is of constant composition after the maximum, and contains solvent. The composition of the monolayer may then be calculated using the analysis proposed by Schay and Nagy (1961).

The most common shapes of adsorption isotherm are given in Fig. 1, although a variety of other shapes have been reported. At low concentrations the data can often be fitted to an equation of the Freundlich type

$$w = \alpha c^n \tag{4}$$

where w is the weight of solute adsorbed per unit weight of solid, c is the concentration of solute in the solution at equilibrium, and α and n are

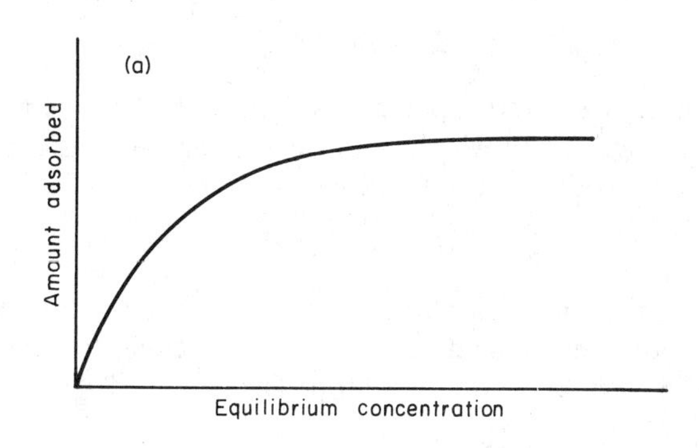

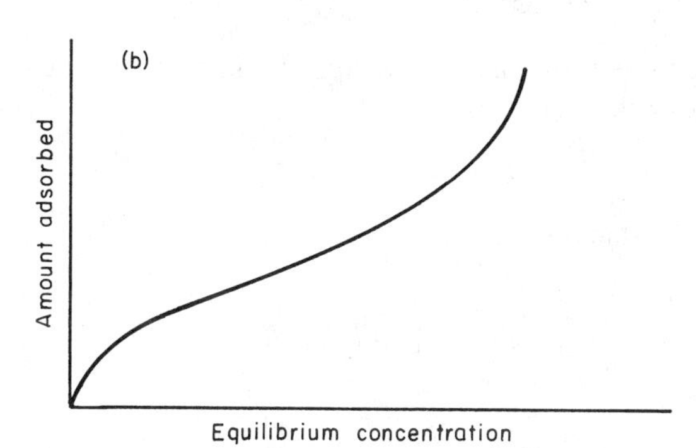

Figure 1 The most common isotherm shapes for adsorption from (a) solutions of solids, and (b) solutions of liquids of limited solubility.

constants ($n < 1$). However, this simple expression does not usually apply over the whole of the available concentration range, and its validity for any part of the isotherm may not be meaningful. The Freundlich equation results from a consideration of the heterogeneity of the surface, when applied to the adsorption of vapours on solids. If data for adsorption from solution fit the equation it is possible, but not proven, that the surface is heterogenous (Adamson, 1982).

Equations of the Langmuir type have also been applied, probably because of the similarity in shape of the isotherm shown in Fig. 1(a) to Type 1 observed for adsorption of gases on solids. The Langmuir equation can be derived for the special case of dilute solutions, as follows. For the adsorption equilibrium

$$(1)^{l} + (2)^{s} \rightleftharpoons (1)^{s} + (2)^{l} \tag{5}$$

corresponding to the exchange reaction between components of equal size, the equilibrium constant is given by

$$K = a_1^s a_2^l / a_1^l a_2^s \tag{6}$$

where a is the activity, and superscripts denote the surface (s) and liquid (l) phases. For the case in which both phases can be considered as ideal, activities are replaced by mole fractions, and

$$K = x_1^s x_2^l / x_1^l x_2^s \tag{7}$$

which is essentially the separation factor S that relates the experimental quantities. Everett rearranged equation (7) to include the experimentally determined $n^0 \Delta x_1^l / m$ [cf. equation (1)] to give

$$\frac{x_1^l x_2^l}{(n^0 \Delta x_1^l / m)} = \frac{m}{(n_1^s)_m} \left(x_1^l + \frac{1}{K-1} \right) \tag{8}$$

where $(n_1^s)_m$ is the monolayer capacity of component 1 (Everett, 1964). For the case of dilute solutions,

$$n^0 \Delta x_1^l / m \approx n_1^s \quad \text{and} \quad x_2^l \approx 1$$

and equation (8) becomes

$$\frac{x_1^l}{n_1^s} = \frac{1}{(n_1^s)_m} \left(x_1^l + \frac{1}{K-1} \right) \tag{9}$$

or

$$\frac{x_1^l}{n_1^s} = \frac{1}{(n_1^s)_m (K-1)} + \frac{x_1^l}{(n_1^s)_m} \tag{10}$$

which is the familiar form of the Langmuir equation. A plot of x_1^l / n_1^s against x_1^l is linear, and values of $(n_1^s)_m$ and K can be deduced from the slope and intercept.

The fact that Everett's thermodynamic treatment for the exchange reaction involving a binary mixture of equal sized molecules leads to an equation of the Langmuir form, for adsorption from dilute solution when the solute is strongly adsorbed ($K \gg 1$), gives some credence to the popular application of equation (10) to the analysis of adsorption from dilute solution (Brown and Everett, 1975). However, for most systems analysed in this way the assumptions implicit in the thermodynamic argument are far from being realized. Nevertheless, experimental isotherms often can be fitted to equation (10) and values obtained for the maximum capacity of the solid. This can be useful if only the early part of the isotherm is available, and the adsorption limit has not been reached, but such an equation would have no significance in terms of a model of adsorption, and the constants may have no physical meaning.

B. ADSORPTION THERMODYNAMICS

The standard free energy of adsorption is related to the equilibrium constant by

$$\Delta G^\circ = -RT \ln K \tag{11}$$

so values of ΔG° may in principle be calculated from adsorption data. The standard state for the adsorbed layer may be taken as $\theta = 0.5$, so that ΔG° corresponds to the free-energy change associated with transfer of solute from a solution of unit concentration to a surface at $\theta = 0.5$ (Crisp, 1956). ΔG° may be positive or negative and its value depends on the units chosen for the concentration. From the temperature coefficient of the equilibrium constant, ΔH° can be deduced, hence ΔS°. For dilute solutions equation (11) may be put in the approximate form

$$\Delta G^\circ = -RT \ln(x_1^{\mathrm{s}}/x_1^{\mathrm{l}}) \tag{12}$$

although the solution or surface phases are not likely to be ideal. Other standard states may be chosen, e.g. for a solute that is solid at the adsorption temperature the standard state of the surface phase is that of the solid solute and that for solution is the concentration at which the surface becomes saturated with solute (Groszek, 1975).

The differential heat of adsorption of component 1 is given by

$$\Delta \overline{H} = -RT^2 \left(\frac{\partial \ln a_1}{\partial T} \right)_{P, n_1^{\mathrm{s}}, n_2^{\mathrm{s}}} \tag{13}$$

but this can only be strictly correct if $n_1^{\mathrm{s}}/n_2^{\mathrm{s}}$ is independent of temperature, which is generally not true (Corkill *et al.*, 1966). So the use of the Clausius–Clapeyron equation to obtain heats of adsorption from adsorption isotherms which represent the interfacial excess of only one component at different temperatures can only give misleading results. A better method is to measure the heat of immersion of the solid in solutions at constant temperature, as a function of n_1^{s}.

Consider a two-component solution containing initially n_1 and n_2 moles of components 1 and 2 respectively, and that after immersion there are n_1^{s} and n_2^{s} moles of these components in unit area of the adsorbed layer. Simple addition of the enthalpies of the two solution components and the outgassed solid before and after immersion gives the total enthalpy change $\Delta_w H_s$ during the immersion process

$$\Delta_w H_s = h_{ss}^s - h_s^s - n_1^s \bar{h}_1^l - n_2^s \bar{h}_2^l - [n_1(\bar{h}_1^{l*} - \bar{h}_1^l) + n_2(\bar{h}_2^{l*} - \bar{h}_2^l)] \quad (14)$$

where h_{ss}^s is the enthalpy of the solid/solution interface, h_s^s the enthalpy of the solid/vacuum interface, and $\bar{h}_{1(2)}^{l*}$ and $\bar{h}_{1(2)}^l$ are the partial molar enthalpies of components 1(2) before and after adsorption respectively. The terms inside the square brackets represent the enthalpy change in bulk solution due to the concentration change. For a dilute solution this can be neglected, and

$$\Delta_w H_s = h_{ss}^s - h_s^s - n_1^s \bar{h}_1^l - n_2^s \bar{h}_2^l \quad (15)$$

The enthalpy of the solid/solution interface may be related to the composition of the adsorbed layer by

$$h_{ss}^s = \theta \bar{h}_1^s + (1 - \theta) \bar{h}_2^s \quad (16)$$

where $\bar{h}_{1(2)}^s$ is the partial molar enthalpy of the adsorbed components, and θ is the fractional coverage of component 1. It follows that

$$\left(\frac{\partial \Delta_w H_s}{\partial n_1^s}\right)_{P,T,n_2^s} = \bar{h}_1^s - \bar{h}_1^l = \Delta \bar{H}_1 \quad (17)$$

So the differential heat of adsorption $\Delta \bar{H}_1$ of component 1 may be obtained by measuring the heats of immersion $\Delta_w H_s$ of the adsorbent in a range of dilute solutions from which component 1 is preferentially adsorbed, together with the adsorption isotherm of component 1 (Corkill *et al.*, 1966; Armistead *et al.*, 1971b). In practice the total heat change on immersion of an evacuated solid in the solution involves both the wetting process and a dilution of the solution phase; the latter is measured separately and subtracted from the total immersional heat.

Enthalpy data may also be obtained from flow-microcalorimetry. Assuming that solvent is displaced from the solid/liquid interface by solute at a rate proportional to the concentration of the solution, and that under conditions of dynamic equilibrium solute molecules are leaving the surface at the same rate as they are being adsorbed, it has been shown (Allen and Patel, 1968/9) that

$$c/q = 1/kq_0 + c/q_0 \quad (18)$$

where q is the heat evolved when the solute concentration is c, and q_0 corresponds to monolayer coverage. Thus the integral enthalpy of displacement is related to the concentration by a Langmuir type equation, and this has been borne out in practice for a number of systems (Allen and

Patel, 1970, 1971; Groszek, 1975). From the linear plot of c/q against c a value of q_0 is obtained, which combined with adsorption data gives the integral displacement enthalpy at saturation. The differential enthalpy may also be obtained from the integral heat measurements and related adsorption data (Kern *et al.*, 1978; Liphard *et al.*, 1980).

C. MECHANISM OF ADSORPTION

Although adsorption from solution by solids is of great practical importance and a vast number of papers have been published, it has been only over the last fifteen years or so that a fundamental understanding has developed. Much of the early work was concerned with relatively dilute solutions, and interpretation of the isotherms was not unreasonably analogous to that used for adsorption from the gas phase, and experimental results were fitted to equations of the Langmuir and Freundlich types. The role of the solvent and its competition with the solute for surface sites has now become recognized as an important factor, particularly as a result of a large number of studies on the adsorption from binary liquid mixtures and their thermodynamic analysis. However, relatively few studies involving dilute solutions of molecular solutes are sufficiently comprehensive to provide an adequate mechanism for the process. In principle one needs accurate adsorption data on well defined surfaces from well defined solutions, at various temperatures, together with reliable thermodynamic data.

An increasing awareness of the effect of the chemistry of the surface on the adsorption process has followed the development of techniques that characterize both qualitatively and quantitatively the species present. The development of infrared spectroscopic methods for the study of adsorption behaviour at the solid/liquid interface has gained impetus in the last decade (Rochester, 1980). Results suggest that infrared spectroscopy provides an unrivalled method for the characterization of surface–adsorbate interactions. Adsorption of a single adsorbate on distinguishable sites or competition between several adsorbates for particular sites can be monitored. Fractional coverages of sites can be deduced from absorbance data. In contact with aqueous solutions oxide surfaces exhibit a surface charge which is dependent on the pH of the solution; all oxides have a well defined pH at which this charge is zero. The variety of surface groups on carbon blacks may be identified by chemical methods (Boehm, 1966). Most adsorbents are heterogeneous in the surface chemical sense in that they contain a wide range of high to low energy sites, both polar and non-polar, as well as impurity atoms and adsorbed material, e.g. moisture and organic contaminants arising from exposure to the environment. Mineral surfaces are often poorly defined yet the chemical species exposed might well determine the character of the adsorption, i.e. whether it is a chemical or physical interaction. Of all the surfaces studied one of the most uniform and non-polar is that of a graphitized carbon black, having basal planes of graphite exposed. Evaluation of the surface area of an adsorbent using gas adsorption is a well established procedure, and reliable information can be obtained on pores of all sizes. Porosity will obviously be very relevant to adsorption from solution.

Studies on solution properties are well documented. Of importance to adsorption behaviour is the physical state of the solute species, e.g. stearic acid is dimeric in benzene solution, tensides form micelles and the activity of the solute only increases slowly above the critical micelle concentration. Interactions in the liquid state are important, and are reflected in deviations from ideality. Gross deviations must be taken into account as they might well be relevant to the "escaping" tendency of the solution species, hence activity coefficients are required for any quantitative analysis. Adsorption from aqueous solution also reflects the extraordinary structural and solvating properties of water.

The interaction between the surface and adsorbed species may be either chemical or physical. Several types of bonding can be identified as follows:

(1) Chemical adsorption (chemisorption)—stearic acid from benzene solution on metal powders.
(2) Hydrogen bonding—long-chain alcohols from hydrocarbon solution on dry oxide surfaces.
(3) Hydrophobic bonding—association of hydrocarbon chains to "escape" from an aqueous environment, e.g. acids on polystyrene (Schneider *et al.*, 1965).
(4) van der Waals forces.

The net interaction of an adsorbate molecule with a surface might involve more than one type of interaction, depending on the chemical structure of both components.

The most favoured approach to an investigation of the adsorption mechanism is a study of the isotherm. The important aspects are: (a) the rate of adsorption; (b) the shape of the isotherm; (c) the significance of the plateau found in many isotherms; (d) the extent of solvent adsorption; (e) whether the adsorption is monomolecular or extends over several layers; (f) the orientation of the adsorbed molecules; (g) the effect of temperature; (h) the nature of the interaction between adsorbate and adsorbent. Although most isotherms for adsorption from dilute solutions have the shapes shown in Fig. 1, a variety of other shapes have been reported. They were classified by Giles *et al.* (1960), and later a theoretical basis was given to the classification adopted (Giles *et al.*, 1974a). The various isotherm shapes considered are shown in Fig. 2.

Four characteristic classes are identified, based on the form of the initial part of the isotherm; the subgroups relate to the behaviour at higher concentrations. The L (Langmuir) class is the most common and is characterized by an initial region which is concave to the concentration axis. The L2 isotherm reaches a plateau, further adsorption above this value gives the L3 isotherm, and if that reaches a second plateau it is designated

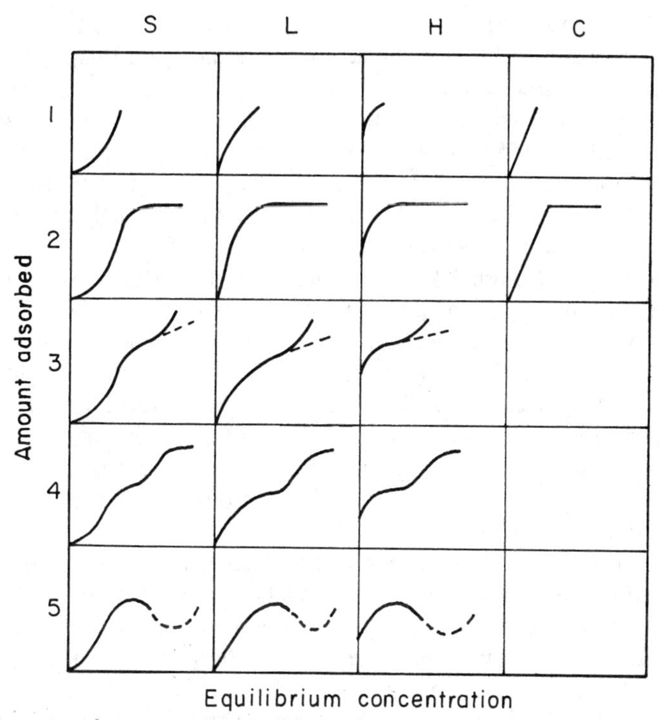

Figure 2 Classification of isotherm shapes. (Giles *et al.*, 1960, 1974a)

L4. The fifth L type shows a maximum and reflects a special set of circumstances—they are found with solutes that associate in solution (tensides and certain dyes) and contain highly surface-active impurities (a maximum is not thermodynamically possible in a pure system). A similar set of isotherms are associated with the other classes, although not all have been observed. For the S class the initial slope is convex to the concentration axis, and this is frequently followed by a point of inflection leading to an S-shaped isotherm; the H (high affinity) class results from extremely strong adsorption at very low concentrations giving an apparent intercept on the ordinate; the C (constant partition) class has an initial linear portion which indicates constant partition of the solute between solution and adsorbent, and occurs with microporous adsorbents.

In their theoretical treatment of the various isotherm shapes Giles *et al.* (1974a) have demonstrated that much useful information can be obtained on the adsorption mechanism, and this is illustrated with examples from the literature (Giles *et al.*, 1974b). The analysis is similar to that used by Langmuir for gas adsorption, and the constant in the Langmuir equation is related to the energy of activation for removal of the solute from the surface. If the interaction between adsorbed molecules is negligible, the activation energy will be independent of coverage, and this leads to an L or H isotherm. When the force of interaction is significant relative to that between solute and adsorbent, the activation energy will be higher, and cooperative adsorption occurs corresponding to the S isotherm. In this case the solute molecules tend to be packed in rows or clusters on the surface, and this situation is encouraged when the solvent is strongly adsorbed and the solute is monofunctional, e.g. the adsorption of a monohydric phenol on a polar surface (alumina) from a polar solvent (water); resorcinol gives an L type isotherm. The adsorption isotherm for *p*-nitrophenol on silica from dry benzene is L shape; saturation of the benzene with water leads to the S shape, since the water strongly competes with the solvent for the surface. Parallel orientation of solute gives L type isotherms. Some dyes form aggregates in solution, and these are adsorbed giving the S shape, e.g. certain cyanine dyes on silver halides. The H curves are associated with chemisorption or other strong interactions, e.g. stearic acid from benzene on metal powders.

The C curve is found with microporous adsorbents, and is consistent with conditions in which the number of adsorption sites remains constant throughout the concentration range. As sites are covered new sites appear, and the surface available expands proportionally with the amount of solute adsorbed. Models for this process were developed by the authors. Cases of C type linear adsorption have been found after an initial portion of L or H type, e.g. *p*-nitrophenol from benzene on microporous alumina. The inflections in subgroup 3 and the second plateau in subgroup 4 may reflect a change in orientation of the adsorbed solute, e.g. sodium dodecyl sulphate on Graphon from aqueous solution, or the formation of a second layer, e.g. long-chain ammonium salts on polystyrene latexes. Numerous examples of all these shapes are given by Kipling (1965) and in the first paper by Giles *et al.* (1960).

Usually no distinction is made between adsorption from solutions of compounds that are normally solid at room temperature and are sparingly soluble in the solvent, and adsorption from solutions of liquid solutes of limited solubility or solutions of solid solutes of unlimited solubility. However, the isotherms show characteristic differences in shape. For the former the isotherm [Fig. 1(a)] usually rises rapidly to a limiting value at low concentration followed by little or no increase in adsorption (Giles L2), whereas the isotherms [Fig. 1(b)] for adsorption of liquid solutes are often S-shaped, the amount adsorbed increasing rapidly as the solubility limit is approached (Giles L3). Little attempt has been made to analyse this difference in terms of the chemistry of the solution components. Monolayer theories do not predict the S-shaped isotherm. However, Roe's (1974) analysis, involving a multilayer theory of adsorption from polymer solutions based on Flory–Huggins statistics, predicts the shape, and he demonstrated that his theory fitted quite well to published data for the adsorption of n-octane, n-decane, and n-dodecane on Graphon from methanol solutions. Furthermore, he was able to fit his theory to adsorption from solution of solids of limited solubility, using published data for stearic acid on Spheron 6 (heated at 1000°C) from solutions in cyclohexane, benzene, and carbon tetrachloride, for which isotherms were of the shape shown in Fig. 1(a) (Roe 1975). For this case the isotherm could in principle have a point of inflection and subsequently increasing adsorption, but this may occur only at concentrations well above the solubility limit. Perhaps this approach is more justified theoretically than analysis using equations of the Freundlich and Langmuir types.

II. Experimental Methods

A. QUANTITATIVE ESTIMATION OF AMOUNT ADSORBED

The simple technique that has been used for many years by a large number of experimenters is described by Kipling (1965). Essentially, a sample of solution is added to a known weight of solid, the mixture is tumbled in a thermostat until adsorption equilibrium is reached (usually several hours), and a sample of the solution is then removed for analysis. Kipling has reviewed the techniques commonly used for the analysis (interferometry, titrimetric and gravimetric analysis, use of radiotracers, etc.). Each has its own deficiencies, which are particularly relevant when the quantities adsorbed are small and there is a need to measure very accurately small concentration changes. Accurate examination of the submonolayer region with low-area powders is not really feasible with the simple technique. Furthermore, unless extreme precautions are taken, the effect of surface contamination can lead to irreproducible and erroneous results. Efforts in this direction make the completion of a full isotherm an extremely tedious procedure, as was demonstrated by Parfitt and Thompson (1971) for the adsorption of hydrocarbon mixtures on titanium dioxide. The use of such a polar solid demands extreme care to prevent contamination of the surface by moisture which, if not a constant parameter in each individual experiment, can lead to errors, and these are particularly obvious if the adsorption data are subject to critical examination in a comparison with theoretical analysis.

Perhaps the most critical current need is to obtain accurate isotherms over a range of temperature, so that meaningful thermodynamic analysis can be carried out. Such a comprehensive analysis by the simple technique would be very tedious, if account were taken of all the experimental uncertainties. Everett and coworkers (Everett, 1978; Ash *et al.*, 1973) have developed a technique to overcome these difficulties, thereby increasing precision and speed. Their system consists essentially of two closed circuits one of which contains the adsorbent in a thermostatted cell. The liquids in the two systems are pumped through the arms of a recording differential refractometer, and the measured change in refractive index of the liquid in contact with the solid gives the amount adsorbed. The measurements are relatively rapid, and by changing the temperature of the thermostat a range of values at different temperatures can be quickly acquired. An order of magnitude improvement in precision is claimed.

Everett reports that the technique is less precise with dilute solutions and was replaced by a precision liquid chromatograph. The chromatographic technique is based on the frontal analysis procedure of Tiselius, and was used by Claesson (1946). The adsorbent is contained in a column which is initially filled with solvent. The solution is then run through the column and the concentration of eluate is measured continuously by interferometric analysis. A refined technique described by Sharma and Fort (1973) proved both useful and accurate for the analysis of adsorption on low surface area graphite fibres ($\sim 1\ m^2\ g^{-1}$) from solutions of octadecane and benzoic and stearic acids in cyclohexane. Further refinements of the technique were incorporated in the apparatus used by Everett and Podoll (Everett, 1978), to permit accurate measurements of the adsorption by Graphon (a graphitized carbon black) from solutions of dodecanol in heptane at high dilution (at mole fractions $<0.9 \times 10^{-3}$).

B. DERIVATION OF THERMODYNAMIC QUANTITIES

The standard free energy of adsorption of one component from a dilute solution may be estimated from the adsorption data by application of the Langmuir equation, and, from measurements of adsorption at different temperatures, use of an equation of the Clausius–Clapeyron type has provided enthalpy data (Crisp, 1956; Groszek, 1975). However, as pointed our earlier, this procedure can give misleading results. Direct calorimetry is the only method to obtain reliable enthalpy data.

Two types of calorimeter have been used to study the adsorption process. The first involves measurement of the heat evolved when the evacuated solid adsorbent is brought into contact with the solution in a calorimeter.

The solid is contained in a bulb which is broken beneath the surface of the solution. High precision calorimeters have been designed for this experiment and are available commercially, and several have been developed in research laboratories (Tyler *et al.*, 1971; Zettlemoyer *et al.*, 1953; Corkill *et al.*, 1964; Everett and Findenegg, 1969). Since the composition of the adsorbed layer is different from that of the bulk liquid it is normally necessary to measure the heat change associated with this dilution, and this may be achieved by breaking an evacuated bulb containing the solute into the solvent, corrections being made for the heat of bulb breaking and the evaporation of the solvent into the empty space in the bulb. The variation of the heat of immersion of the solid in the solution is determined at constant temperature, and is compared with the adsorption isotherm to give the differential enthalpy of adsorption (Corkill *et al.*, 1966; Mills and Hockey, 1975b).

The second calorimetric technique involves a flow system. A known quantity of solution is injected into a stream of solvent flowing through a column of adsorbent (~0.2 g) contained in the calorimeter cell. The heat change recorded by this procedure is essentially an enthalpy of displacement, and may be related to an adsorption isotherm to yield enthalpies of (preferential) adsorption. Commercially available flow microcalorimeters have been used for such studies (Husbands *et al.*, 1971/2; Kern *et al.*, 1978; Morimoto and Naono, 1972).

In both calorimetric procedures a careful analysis of the data is required before a meaningful interpretation of the adsorption process can be achieved.

C. INFRARED SPECTROSCOPIC EXAMINATION OF SURFACE SPECIES

The usefulness of infrared spectroscopy for the study of surface species at the solid/vapour interface is well documented (Little, 1966; Hair, 1967) but the development of infrared spectroscopic methods for the study of adsorption from solution has, until recently, been slow. The addition of a liquid component makes the detection of infrared bands due to surface species more difficult, particularly for liquids which are themselves strong absorbers of infrared radiation. For this reason many studies have involved the separation of the solid and liquid phases before spectroscopic examination of the solid plus adsorbed molecules. The solid adsorbent, which has generally been a powdered oxide, is immersed in the solution from which adsorption takes place. After equilibration the solid is removed from the solution and freed from solvent by drying if necessary. The spectrum of the solid is then recorded. The simplest method is to use a self-supporting disc of oxide compacted before immersion in the liquid phase (Erkelens and Liefkens, 1975). Spectra of discs can be recorded directly by conventional transmission spectroscopy. Alternatively self-supporting discs may be pressed from loosely powdered oxide after separation of the phases (Nechtschein and Sillion, 1971). A further method involves mixing the separated powder with potassium bromide (Sherwood and Rybicka, 1966; Raghavan and Fuerstenau, 1975) or potassium iodide (French *et al.*, 1954; Eyring and Wadsworth, 1956) before compacting a disc for spectroscopic examination. Spectra of powder dispersed in nujol or fluorolube mulls may also be recorded (Little and Ottewill, 1962; Sherwood and Rybicka, 1966; Kapler and Nekrasov, 1971). In general all these methods suffer from the disadvantage that separation of the solid and liquid phases before infrared spectra are recorded may influence the orientation of adsorbed species or the nature of the surface–adsorbate interactions. The spectroscopic results do not therefore necessarily give a true indication of adsorption behaviour *in situ* at the solid/liquid interface. Spectroscopic study of powdered adsorbents *in situ* in the liquid phase provides unambiguous information about the solid/liquid interface and is to be preferred. Separation of the solid and liquid phases before spectroscopic examination of the dried powder has been used for studying adsorption from aqueous solutions. However, Neagle and Rochester (1982) have shown that *in situ* studies are possible for systems containing aqueous solutions.

Infrared spectra of powders dispersed in liquids may be recorded by placing the dispersions in a conventional cell for the measurement of spectra of solutions (Killmann and Strasser, 1973; Rupprecht and Kindl, 1975). High surface area of the solid favours the possible appearance of absorption bands due to surface species, provided that the bands are not obscured by absorption maxima due to vibrations of the bulk solid or of molecules in solution. A blank cell containing solvent (Bascom and Timmons, 1972) or solution (Thies *et al.*, 1964; Thies, 1966, 1968; Botham and Thies, 1970) may be positioned in the reference beam of the infrared spectrometer. Solution for the reference cell may be taken from the adsorption system by decantation (Hayashi *et al.*, 1969). Equilibrium concentrations of adsorbate in solution may be evaluated from spectra of the solutions alone (Killmann *et al.*, 1972). If the added and equilibrium concentrations are sufficiently different, the weights of species adsorbed can also be deduced from the spectroscopic results (Hayashi *et al.*, 1969). Solutions may be obtained free of solid by filtration (Bascom and Timmons, 1972) or centrifugation (Fontana and Thomas, 1961) if sedimentation of the solid does not occur.

Study of compacted self-supporting discs of powdered solids has been the most generally used method for the infrared spectroscopic investigation of surface species at the solid/vapour interface (Little, 1966; Hair, 1967). The method has been adapted for study of the adsorption of small molecules onto discs of oxides immersed in liquids or solutions. Carbon tetrachloride has, for spectroscopic reasons, been the most commonly studied predominant component of the liquid phase. Suitable cells have been described by Low and Hasegawa (1968), Hasegawa and Low (1969a), Griffiths *et al.* (1974), Marshall and Rochester (1975b), and Buckland *et al.* (1980). Cells with a shorter optical path length (Rochester and Trebilco, 1977a; Buckland *et al.*, 1978; Rochester and Yong, 1980a) have been developed for the study of systems in which the main constituents of the liquid phase are hydrocarbons.

Infrared cells used in the study of compacted discs immersed in liquids are glassblown to a conventional vacuum apparatus which includes a section for the storage and mixing of liquids and solutions. The simplest cell so far devised is that of Rochester and Trebilco (1977a) in which was incorporated a pair of silica optical windows 0.5 mm apart. Aerosil silica was pressed into a disc which was placed between the windows. The cell was linked to the vacuum apparatus and evacuated with the disc heated at 1023 K by an external furnace. After removal of the furnace, liquid was admitted and the spectrum of the disc immersed in liquid was recorded. Addition of pure solvent followed by aliquots of concentrated solutions of adsorbate allows measurement in a single experiment of a series of spectra corresponding to increasing coverages of the silica surface. The short path length of the cell enabled spectra to be recorded of surface species on silica immersed in liquid hydrocarbons. However, the silica windows precluded study of the spectral region below ~2100 cm^{-1}.

Window materials transparent in the spectral region below 2100 cm^{-1} cannot be glassblown to glass of silica cell bodies. Cells have therefore been devised with optical windows fixed with adhesive to bodies made from pyrex (Buckland *et al.*, 1980) or metal (Griffiths *et al.*, 1974). The optical compartment of the cells could not be heated to high temperatures by an external furnace, so the cells consisted of two sections, an upper section surrounded by a furnace and a base section to which were attached fluorite windows. Following heat treatment in a vacuum, silica discs were lowered to a position between the windows, liquids were admitted, and the spectra recorded. Minimum optical path lengths attainable for such cells (~2.5 mm) were governed by the need to be able to lower a disc in a suitably thin holder between the windows. The path lengths allowed spectra to be recorded of species adsorbed from solution in carbon tetrachloride but were too long to allow study of systems involving liquid hydrocarbons as solvent. A further disadvantage of the cells so far described for the study of oxide discs is the absence of thermostatting or stirring of the adsorption system.

Simultaneous thermostatting and stirring is achieved by circulating the liquid phase in a cyclic system which includes the optical compartment of the infrared cell (Low and Hasegawa, 1968; Hasegawa and Low, 1969a;

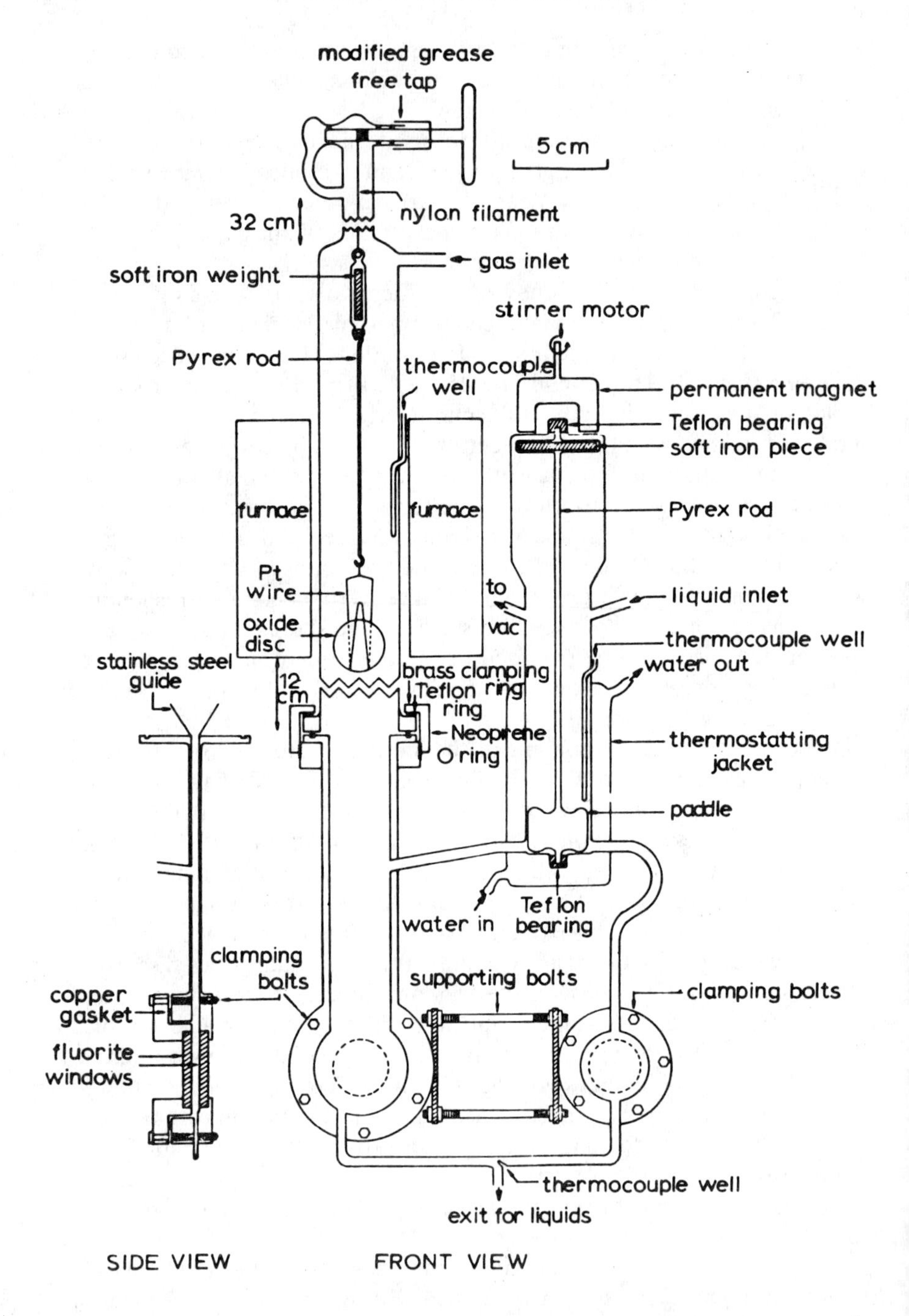

Figure 3 Infrared cell of Marshall and Rochester (1975b).

Marshall and Rochester, 1975b; Rochester and Yong, 1980b). The cell shown in Fig. 3 illustrates one type of circulation system and exemplifies the use of an upper cell section with an external furnace and a lower stainless steel section with attached fluorite windows. The subsidiary cell compartment, also fitted with fluorite windows, allowed the infrared spectrum of the circulating solutions to be independently monitored. Alternatively the cells used by Low and Hasegawa (1968; Hasegawa and Low, 1969a) in their pioneering studies of oxide discs immersed in solutions of adsorbates in carbon tetrachloride contained a subsidiary compartment which provided solvent blanking in the reference beam of the spectrometer. Circulation systems have also been incorporated in the design of cells for the study of adsorption from hydrocarbon solutions. A silica cell with silica windows did not require separate furnace and optical sections because discs could be heated *in situ* between the silica windows (Rochester and Yong, 1980b). A cell similar to that in Fig. 3 contained modifications which enabled the path lengths of the main and subsidiary compartments to be varied (Buckland *et al.*, 1978). A disc could be lowered to a position between the windows in the main compartment, and then the path length could be shortened to a minimum corresponding to the thickness of the oxide disc. An alternative method of achieving a short optical path length of liquid while spectra were recorded was described by Mills and Hockey (1975c). Their cell allowed silica discs to be immersed in liquid which was then drained into a reservoir after adsorption had taken place. The base of the disc remained in the surface of the liquid and capillary rise kept the whole disc wetted by the liquid.

The usefulness of infrared reflection spectroscopy for the study of adsorption from solution has been demonstrated by Yang *et al.* (1973). Early reflection studies had involved separation of the solid and liquid phases before spectra were recorded (Francis and Ellison, 1959; Leja *et al.*, 1963; Poling and Leja, 1963). Separation was necessary for adsorption systems containing aqueous solutions. Yang *et al* (1973) recorded reflection spectra of stearic acid adsorbed on germanium PdO-coated germanium, and the (0001) faces of α-alumina immersed in carbon tetrachloride. Advantages of the reflection method are that single crystal faces can be investigated, and the use of polarized radiation can give information about the orientation of adsorbed species. The use of a Fourier transform spectrometer allowed complete spectra to be recorded rapidly, so rates of adsorption and desorption could be determined by recording spectra as a function of time.

III. Analysis of Experimental Adsorption Data

It is a relatively simple matter to determine the amount of solute adsorbed under a particular set of circumstances, and in many practical situations that is all that is required. To determine the mechanism of adsorption and the relevance of the variety of possible variables is altogether more difficult, and there are not too many examples in the literature where all the factors have been adequately defined for a full story to emerge. For the present analysis it will be convenient to identify certain critical features and then to examine published data, hopefully to establish guiding principles. We shall examine (a) the role of the surface, (b) the nature of the solute, its chemical structure and its interaction with the solvent, (c) the influence of the solvent, (d) the nature of the interactions between surface and adsorbed solute, (e) the structure of the adsorbed layer, (f) the effect of temperature, and (g) the thermodynamics of adsorption.

A. THE ROLE OF THE SURFACE

A thorough understanding of the chemistry of the solid surface would appear to be a prerequisite. Much work has been done with ill-defined surfaces, and it is not surprising that reports from different laboratories show significant variations for what might appear to be similar materials. Few solids are completely reproducible. Contamination from exposure to the environment might be sufficient to cause significant variations. Preparative methods vary, and removal of surface impurities is usually not always treated with the attention it deserves. The graphitized carbon black may be considered one of the most reproducible materials, in terms of surface character, available to the surface chemist, and its use in solution adsorption studies has led to valuable and reliable information. Other carbons exhibit a wide variety of surface chemical behaviour related to the range of acidic and basic groups that exist on the surface. The situation is similar for oxides, e.g. silica occurs in a variety of forms, both porous and non-porous, and the type and number of surface hydroxyl groups depend on the method of preparation and the treatment given before the adsorption experiments are carried out. Titanium dioxide may be either rutile or anatase, and may originate from either the sulphate or the chloride process, both of which leave on the surface significant amounts of anionic impurities which profoundly affect the surface chemistry. Graphites, charcoals, minerals, clays, etc. are often even less well defined. So it is not surprising that in only relatively few studies can the role of the surface be explicitly defined.

Brown and Everett (1975) have reviewed the wide-ranging behaviour exhibited by carbons, from published data mainly involving binary liquid mixtures for which the adsorption of both components is assessed. The marked effect that the complexes on the surface of commercial blacks have on the competition for the surface between the two components is demonstrated. Wright's study of the adsorption of a series of dibasic acids (from oxalic to sebacic) from aqueous solution at low concentrations by Spheron 6, Spheron 6 heat-treated at 1000°C, and Graphon (Spheron 6 heated at 2700°C) shows the effect of removal of the oxygen complexes. The fraction of surface covered, determined from the limiting adsorption values, decreases in the order Graphon > Spheron 6 (1000°C) > Spheron 6, reflecting the decreasing affinity of the surface for water (Wright, 1966a). Changes in the adsorptive capacity of carbons for phenol and nitrobenzene by chemical treatment, particularly with respect to the formation and removal of acidic surface oxides, show that the amount adsorbed from dilute aqueous solutions is greatly reduced by increasing oxidation of the surface (Coughlin and Ezra, 1968). Polycyclic hydrocarbons are preferentially adsorbed on graphite from dilute solutions in cyclohexane, but the amount adsorbed may be strongly influenced by the preparation of the graphite sample (Groszek, 1975). When the surface consists predominantly of basal planes the amount of pyrene adsorbed is significantly less than that required to form a close-packed monolayer, but is very much increased when the surface contains a relatively high proportion of "polar" edge sites; the adsorption on Graphon is also low.

Benzene and naphthalene are adsorbed more strongly on Spheron 6 than on Graphon from solutions in cyclohexane, carbon tetrachloride, and cyclohexanone, but the converse is observed with anthracene (Wright and Powell, 1972; Kiselev and Khopina, 1969). However, enthalpy of immersion data indicate that all three aromatic hydrocarbons interact more strongly with Spheron 6 than with Graphon. The reversal in the case of anthracene is ascribed to an entropy effect. In all cases the entropy decreases when the aromatic hydrocarbon displaces cyclohexane from the surface, the decrease being larger for Spheron 6, but only in the case of anthracene on Spheron 6 does this unfavourable entropy change outweigh the energy term. It is proposed (Kiselev and Khopina, 1969) that the adsorption of anthracene on Spheron 6 occurs in two steps, the initial adsorption associated with the most active surface sites followed by adsorption on less active sites for which the solvent competes more effectively. A study by Kiselev and Shikalova (1970) of the adsorption of 6-methylhept-1-ene and phenanthrene from dilute solutions in n-heptane by adsorbents having a wide range of polar character also demonstrates the role of the surface. For the alkene the adsorption order is hydroxylated ($8.5\ \mu mol\ OH\ m^{-2}$) silica > dehydroxylated ($0.7\ \mu mol\ OH\ m^{-2}$) silica > oxidized carbon > graphitized carbon black, reflecting the interaction between the π-electrons of the double bond in the alkene and the surface hydroxyl and oxide groups. The order was different for phenanthrene: oxidized carbon > hydroxylated silica ≈ graphitized carbon black > dehydroxylated silica, which indicates that the aromatic rings interact more strongly with the oxidized and graphitic surfaces than does the one double bond of the alkene.

Adsorption data for polar adsorbates from hydrocarbon and carbon-tetrachloride solutions on to oxide surfaces illustrate the importance of well defined pretreatment procedures. Ganichenko *et al.* (1959) used three non-porous silica gels of different degrees of hydration (2.8, 4.2, and 7.9 μmol OH m^{-2}) to demonstrate that the adsorption of aliphatic alcohols from dilute carbon tetrachloride solutions increases with the degree of hydration of the surface. Silica surfaces normally contain two types of hydroxyl group: isolated and hydrogen-bonded interacting in pairs. Armistead *et al.* (1971a) found that n-fatty acids (C_6–C_{16}) are adsorbed strongly from benzene solutions on fully hydroxylated surfaces, but removal of the hydrogen-bonded groups reduced the adsorption to virtually nothing, a somewhat surprising result since the isolated hydroxyls are known to be the strongest adsorption centres for hydrogen-bonded molecules, but benzene also interacts with the isolated groups to form a π-electron–proton complex. However, a surface containing only isolated hydroxyls gave, for a series of n-fatty acid methyl esters from benzene solution, similar adsorption behaviour to that of the fully hydroxylated surface, which supports the thesis that the isolated groups are the adsorption centres (Mills and Hockey, 1975a).

The effect of adsorbed water on the adsorption behaviour of acids and alcohols on titanium dioxide has been reported. The adsorption of octadecanol from *p*-xylene solutions on to rutile is reduced markedly as the amount of pre-adsorbed water is increased (Day and Parfitt, 1967; Parfitt and Wiltshire, 1964). A similar effect was observed for stearic acid from benzene solution on anatase (Sherwood and Rybicka, 1966) and on commercial titanium dioxide pigments (Doorgeest, 1962; Heertjes *et al.*, 1979; Hirst and Lancaster, 1951). Husbands *et al.* (1971/2) report that the heat of displacement of heptane from the surface of α-ferric oxide by stearic acid, as measured in a flow microcalorimeter, is also very dependent on the amount of pre-adsorbed water. Pre-adsorbed water also reduces the adsorption from n-heptane solutions of stearic acid and n-octadecylamine on to silica surfaces (Brooks, 1958), but the effect is reversed for 1,1-diphenyl-2-picrylhydrazyl from benzene solutions on to rutile (Misra, 1975). Awareness of the surface moisture effect has increased over recent years—it is hardly mentioned by Kipling (1965) in his book.

The physical structure of the surface of an adsorbent is readily assessed by the well established procedures associated with the adsorption of gases and vapours (Gregg and Sing, 1982). The B.E.T. surface area is normally used for calculating the maximum capacity of a non-porous solid in adsorption from solution, in conjunction with estimated molecular dimensions for the adsorbed molecule. For porous solids the dimensions of the pore will determine the extent to which this extra area/volume will be available to the solution components. With molecules of significantly different size preferential adsorption of the smaller component may be enhanced. The chemistry of the two types of surface is also relevant; it is quite probable that the internal surface behaves in a different way from the external surface, not only because the chemistry of the surface may be different but the close proximity of the adjacent surfaces in the pores might result in stronger adsorption than at the external surface. So an analysis of adsorption from solution data for porous solids must take account of both chemical and physical aspects of the available surface, and in any particular case it may not be possible to separate the two effects. Consequently, interpretation of published data is somewhat confused because the relevance of each aspect has not been rcsolved.

What does become clear from the above discussion is that one cannot ignore the chemistry of the surface when embarking on measurements of adsorption from solution.

B. THE NATURE OF THE SOLUTE AND ITS INTERACTION WITH THE SOLVENT

Several features of solute structure are obviously relevant, e.g. chain length, ring structure, nature of polar groups, and physical state in solution. Some others (dipole moment, Hammett constant) have been considered. Solubility is also important, a factor naturally related to the others. Deviations from ideality in solution might also be expected to affect the adsorption behaviour.

A relationship between the level of adsorption and chain length has been recognized for a long time. Traube's early experiments on the surface tension of aqueous solutions of a homologous series of organic solutes, for which surface activity increased in a regular manner as the series was ascended, led to the well known Traube's rule (Traube, 1891). A similar trend has been reported for a number of different solid/liquid systems, and regarded as reflecting the same relationship as established by Traube. But Traube's rule, which was given a physical significance by Langmuir, requires that the increase is regular, i.e. a constant ratio between the molar concentrations of successive homologues leads to the same level of surface activity (surface tension lowering). The rule also only applies at very high dilution when the adsorbed films are in the two-dimensional gaseous state. This is by no means true in the cases reported of adsorption at the solid/liquid interface, and a quantitative relationship does not exist.

Nevertheless the qualitative trends are of interest. Freundlich (1926) established that on charcoal the adsorption of the lower n-fatty acids from aqueous solutions increased with chain length. Data for the adsorption of six of the lower n-fatty acids from aqueous solutions by a non-porous carbon black (Spheron 6), whilst showing the increasing effect with chain length, fell on one curve when plotted against solute activity rather than concentration (Hansen and Craig, 1954). Wright observed a similar regular increase in adsorption with chain length for dicarboxylic acids (C_2–C_{10}) from aqueous solutions on Spheron 6, Spheron 6 heated at 1000°C, and Graphon, but plots of adsorption level against relative concentration (x/x_0, where x_0 is the mole fraction at the solubility limit) show an alternation in the fraction of surface covered such that oxalic (C_2) < succinic (C_4) < malonic (C_3) < adipic (C_6) < glutaric (C_5). This reflects the alternation in the solubilities of the acids (Wright, 1966a). Nekrassov (1928) showed the opposite effect for a charcoal prepared under strongly oxidizing conditions. The reversal presumably relates to the increased affinity of the surface for the solvent. The former effect seems to be found when adsorption occurs on a relatively non-polar solid from a polar solvent, and the latter might be expected if the polarities are reversed, e.g. lower fatty acids from toluene on silica gel (Holmes and McKelvey, 1928), methyl esters of n-fatty acids (C_6, C_{10}, and C_{14}) from benzene solutions on silica (Mills and Hockey, 1975a), lower n-alcohols from carbon tetrachloride solutions on silica gel (Bonetzkaya and Krasilnikov, 1957), and n-fatty acids (C_6–C_{16}) from n-hexane solutions on silica, although for C_{12}–C_{16} from benzene solutions it was independent of chain length (Armistead *et al.* 1971a). Mills and Hockey suggested that the adsorption of the esters at 25°C is determined by the presence at the interface of an "ordered" layer of solvated hydrocarbon chains which is responsible for the limiting adsorption values being lower than that corresponding to the utilization of all the surface adsorption sites; increasing the temperature to 50°C removed the chain length dependence and increased the adsorption limit to a level close to that corresponding to the number of adsorption sites (isolated hydroxyl groups). In contrast, adsorption of the esters from carbon tetrachloride solutions was found to be independent of chain length, and the adsorption limit corresponded closely to the concentration of isolated silanol groups, indicating that the availability of these surface groups and not solvation phenomena determines the extent of adsorption. Solvation phenomena were also discounted for the fatty acid/hydrocarbon/SiO_2 system, for both solvents (Armistead *et al.*, 1971a).

The trends are not always comparable. Crisp (1956) studied the adsorption of long chain alcohols (C_{16}, C_{18}, and C_{31}) on alumina from benzene solution and found no chain length effect; the isotherms were superimposed when plotted as fraction of surface covered against equilibrium concentration, and the heats and free energies of adsorption were found to be independent of chain length. Two interesting cases of change of direction with increasing chain length have been reported, both involving titanium dioxide and both sets of data covering the range C_2–C_{18}: alcohols from *p*-xylene solutions (Parfitt and Wiltshire, 1964), and acids from n-heptane solutions (Heertjes *et al.*, 1979). A minimum occurs at C_6–C_8, when the molar volumes of solute and solvent are similar.

The effect of increasing the number of aromatic rings in the molecule was demonstrated by Kiselev and Shikalova (1970). The fraction of hydroxylated silica surface covered at a given concentration of benzene, naphthalene, and phenanthrene in n-heptane solution increased with the number of aromatic rings.

Some solutes are associated in solution. The formation of micelles in solutions of tensides is well known; their relationship to adsorption from solution is discussed in Chapter 2. Fatty acids are mainly in the dimeric form in organic media, but their configuration in an adsorbed layer depends on the relative strength of their interaction with the surface and of the association. Hence, on non-polar solids, e.g. Graphon, they are probably adsorbed as dimers with the major axis parallel to the surface (Kipling and Wright, 1962), but in perpendicular orientation on alumina and titanium dioxide (Kipling and Wright, 1964). Armistead *et al.* (1971a) relate their n-fatty acid adsorption data, from benzene and hexane solutions on silica, to adsorption of the monomer because the limiting values were determined by the upper limit of the concentration of monomeric species in solution as set by the aggregation process.

An attempt to relate adsorption to the dipole moment of the adsorbate has been made by Ibbitson and coworkers, although several previous attempts had not been successful. Using alumina as adsorbent, it was found that the magnitude of the standard free energy of adsorption for a series of substituted phenols from cyclohexane solution reflected the intensity of the O–H bond moment (Erić *et al.*, 1960) but no correlation exists for substituted azobenzenes from benzene solutions (Ibbitson *et al.*, 1960). A related study (Davis *et al.*, 1973) of the adsorption of *p*-substituted phenols from non-polar solvents on to silica gel showed that $\log K_1$ decreased linearly with increase in the Hammett σ-constant [$K_1 = (x_1^l)_{sat.}(K - 1)$, where $(x_1^l)_{sat.}$ is the mole fraction solubility of the solute]. No such relationship is found if $\log K$ is used, and the meaning of the correlation between $\log K_1$ and σ is obscure (Everett and Podoll, 1979). Using alumina as adsorbent (Davies *et al.*, 1974) the K values appeared to increase with σ, although there is significant scatter. Kagiya *et al.* (1971) reported that $\log K$ decreases approximately linearly with increase in σ for various cyclic ethers adsorbed on to silica gel from hydrocarbon solutions. Such correlation of equilibrium parameters with dipole moment, σ, etc. is obviously not yet clearly established from isotherm data.

Solubility is, of course, an important parameter. For a given solvent the less soluble solutes are generally more strongly adsorbed than the more soluble, and solubility is related to the chemical structure of the solute.

Solvent–solute interactions reflect the same phenomenon; a given solute is generally more strongly adsorbed from a poor solvent than from a good solvent, provided that other effects, e.g. specific interactions with the surface, do not dominate. These are qualitative, and perhaps obvious, observations but are nevertheless useful guiding principles.

Too few data are available for conclusions to be reached on the relevance of such parameters as dipole moment etc., and appropriate solution data are usually not available for adequate assessment of deviations from ideality and their relevance to the adsorption process. Even the chain length dependence is not readily predictable since such factors as specific site adsorption and the structure of the adsorbed layer can be so significant as to make simple chain–solvent interactions of little consequence.

C. INFLUENCE OF THE SOLVENT

We might expect the solvent to influence the adsorption behaviour in three ways associated with (a) its interaction with the solute in solution, hence contributing to solubility and departures from ideality, (b) its interaction with the adsorbent, which will depend on the chemical structure of both, and (c) its interaction with the solute in the adsorbed layer.

Much of the earlier work did not seem to recognize that the solvent is also adsorbed, resulting in faulty interpretation of experimental data particularly for cases in which the equilibrium solution concentration is sufficiently high that x in equation (1) cannot be neglected. Composite isotherms having both negative and positive regions are indicative of competitive adsorption of both components, and these are frequently obtained for adsorption from binary liquid mixtures, but not for dilute solutions. Few cases are known for which a maximum is followed by an extensive decline; one example is the adsorption on Spheron 6 from aqueous solutions of malonic acid (Cornford *et al.*, 1962). For this system $n^0 \Delta x/m$ decreases linearly after the maximum, virtually to zero at the solubility limit ($x = 0.21$), and corresponds to the formation of a mixed monolayer of constant composition over the concentration range for which the isotherm is linear. This is exceptional behaviour, and in most of the systems studied the solvent effect is negligible in relation to the calculation of the individual isotherm for the solute from experimental data, and isotherm shapes as classified by Giles are obtained (see Section I.C).

It is likely that solvent effects would be important in adsorption on to polar surfaces, especially when the solvent is polar and/or contains aromatic rings. The solvent order of adsorption for lauric acid on alumina is n-pentane > benzene > diethyl ether (de Boer *et al.*, 1972), for lauric acid on titanium dioxide it is n-heptane > benzene > acetone (Heertjes *et al.*, 1979), for *p*-substituted phenols on silica gel it is n-hexane > cyclohexane

> carbon tetrachloride > toluene > benzene [a linear relationship exists between $\log K$ (for phenol) and the solvent parameter E_T (Davis *et al.*, 1973)], for n-fatty acids on silica it is n-heptane > benzene (Armistead *et al.*, 1971a), and for methyl esters of n-fatty acids on silica it is carbon tetrachloride > benzene (Mills and Hockey, 1975a). In all cases the competition for adsorption sites by the solvent molecules is indicated, which means that surface areas obtained by this method must be viewed with caution.

With carbon blacks the polar surface sites also contribute to the solvent effect. In the adsorption of the aromatic hydrocarbons benzene and naphthalene on to Spheron 6 the solvent order is cyclohexane > carbon tetrachloride > cyclohexanone (Kiselev and Khopina, 1969). A similar result was obtained for the same system (plus anthracene) by Wright and Powell (1972). Furthermore, preferential adsorption of anthracene, at similar bulk concentrations, is greatest from cyclohexane and least from cyclohexanone but, because of the much greater solubility of anthracene in cyclohexanone, the highest adsorption value is obtained using cyclohexanone solutions (Wright and Powell, 1972).

D. THE INTERACTION BETWEEN SOLUTE AND SURFACE

1. *Adsorption Isotherms*

The strength of the solute–surface bond varies over a wide range from the weakest (van der Waals forces) to the strongest (chemical adsorption), and all types have been observed in studies of adsorption from solution. The time taken to reach equilibrium is normally indicative of the type of interaction, although this may not be completely unambiguous in the case of porous adsorbents for which pore filling might be slow. At room temperature chemisorption is usually a much slower process than physical adsorption. The rate of desorption with an appropriate solvent may also be a useful guide, together with an analysis of the desorbed material. The shape of the adsorption isotherm also provides qualitative information on the nature of the solute–surface interaction, e.g. the H type in Giles's classification (Fig. 2) indicates a very strong bond. The S and L type isotherms reflect a form of adsorption other than chemical, i.e. dispersion forces, hydrogen bonding, or hydrophobic bonding. An accurate analysis can only be carried out by investigating the chemistry of the adsorbed layer on a microscopic scale using, for example, infrared spectroscopy, a technique that has only recently been fully developed for studies of the solid/liquid interface.

Physical adsorption is very common, and the specific nature of the interaction is usually implied from the chemical nature of the materials involved. Energies of interaction provide important supplementary information, although their interpretation may be difficult since they refer to the exchange process involving both solution components, and entropy factors might be dominant. From heat of immersion and isotherm data Wright and Pratt (1974) derived free energy, enthalpy, and entropy data for the adsorption of the lower fatty acids from solutions in carbon tetrachloride on to Spheron 6 and Graphon. No significant differences were found between the two adsorbents, free energy changes were large and negative and decreased with increasing coverage, the negative adsorption enthalpies increased with coverage, and the large and (unusual) positive entropy changes increased with coverage. It appears that the dimeric acid is adsorbed on the non-polar regions of the carbon surface in a non-specific manner. In contrast, similar data for aromatic hydrocarbons adsorbed from cyclohexane on the same two carbons reflect specific (π-bond) interactions with the more polar surface (Spheron 6); the negative entropy changes decreased with coverage (increased with Graphon) (Wright and Powell, 1972). The heat of adsorption (from immersional heats) of n-fatty acids (C_6–C_{16}) from benzene solution on silica surfaces appears to be independent of surface coverage (a decrease would be expected from gas adsorption data for hydrogen bonding adsorbates), suggesting a compensating effect due to mixing in the adsorbed layer (Armistead *et al.*, 1971b).

Evidence for specific interactions between the π-electrons of the double bond in 6-methylhept-1-ene and of the aromatic rings of phenanthrene with the polar groups on silica and carbon black surfaces was reported by Kiselev and Shikalova (1970) (see Section III.A).

The nature of the specific interaction between phenol, *m*-nitrophenol, and *p*-nitrophenol and the surface of an active carbon, when adsorbed from aqueous solution, was studied by Mattson *et al.* (1969) using infrared reflection spectroscopy. The role of each of the three functional groups (OH, NO_2, and the aromatic π-electron system) was investigated, and they concluded that neither of the substituent groups was involved directly in the interaction with the surface, but that they contribute to the electron acceptor characteristics of the aromatic ring. The phenols are adsorbed on the active carbon by a donor–acceptor complex mechanism involving surface carbonyl oxygens acting as electron donors. A much stronger interaction would be expected with 2,4-dinitrophenol and this has been established experimentally using the same carbon (DiGiano and Weber, 1969).

Chemisorption of fatty acids on metal surfaces from hydrocarbon solutions is well established; the isotherm is horizontal over the whole concentration range, and the level of adsorption is independent of the solvent used and of the chain length (Smith and Fuzek, 1946). It may be combined with physical adsorption, leading to a total isotherm that increases with concentration and has a positive intercept on the ordinate well above the origin (Cook and Hackerman, 1951). Both processes are involved in the adsorption of stearic acid from n-heptane solution on α-ferric oxide (Husbands *et al.*, 1971/2), from cyclohexane solution on alumina (Kipling and Wright, 1964), and fron nitrobenzene solution on iron, nickel, chromium, and stainless steel (Timmons, 1973). In the last example the acids were desorbed and the relative concentrations of both types of adsorbed species determined; for iron 70% was in the physically adsorbed form, for chromium 40%, and for nickel and stainless steel 50–60%. Lauric and oleic acids are chemisorbed on ferric oxide from aqueous solutions (Han *et al.*, 1973).

2. *Infrared Measurements*

(a) *Characterization of surface–adsorbate interactions*. The immersion of oxides in liquids causes a perturbation of surface hydroxyl groups which can be observed by infrared spectroscopy. For example, the band at 3750 cm^{-1} due to the OH stretching vibrations of isolated silanol groups on silica was shifted to 3705 cm^{-1} when a silica disc was immersed in 2,2,4-trimethylpentane (Rochester and Trebilco, 1979). The magnitude of the shift $\Delta\bar{\nu}_{OH}$ varies for different liquids (Filimonov, 1956; Filimonov and Terenin, 1956; Rochester, 1980) and gives a measure of the strength of the interaction between surface OH groups and molecules of liquid at the solid/liquid interface. Spectra of silica heated at 423 K and cooled to ambient temperature exhibit a broad maximum at 3535 cm^{-1} which may be assigned to adjacent surface silanol groups involved in lateral hydrogen bonding interactions. The band shifted to 3490 cm^{-1} when silica was immersed in 2,2,4-trimethylpentane (Rochester and Trebilco, 1979) showing that the lateral hydrogen bonds were not broken, but were perturbed to a small extent, on immersion.

Hydrogen bonding interactions between surface hydroxyl groups on oxides and adsorbate molecules may be characterized by the shifts $\Delta\bar{\nu}_{OH}$ in the positions of the infrared maxima due to OH stretching vibrations. Typical results for the adsorption of propionitrile on silica immersed in n-heptane are shown in Fig. 4. The maximum at 3705 cm^{-1} was progressively replaced by a broader band at 3395 cm^{-1} as the surface coverage of isolated silanol group sites by propionitrile molecules increased. The absorbance values at 3705 and 3395 cm^{-1} were linearly related. The interaction involved the formation of hydrogen bonds between isolated silanol groups and the cyano groups of adsorbed propionitrile molecules. Cyano groups were also perturbed by the interaction and gave an infrared band at 2260 cm^{-1} [Fig. 4(b)–(h)] whereas the corresponding band for propionitrile in solution in n-heptane was at 2249 cm^{-1} [Fig. 4(j)]. Other adsorbates for which perturbation of surface hydroxyl groups and of functional groups in adsorbed molecules have been observed spectroscopically include substituted pyridines (Griffiths *et al.*, 1974), ketones (Cross and Rochester, 1978a), esters (Cross and Rochester, 1979), and carboxylic acids (Hasegawa and Low, 1969b; Marshall and Rochester, 1975a).

Chemisorptive interactions at the solid/liquid interface may be characterized by the observation of infrared bands due to vibrations of surface species generated by chemisorption. Typical examples are the formation of surface carboxylate groups by the adsorption of carboxylic acids on

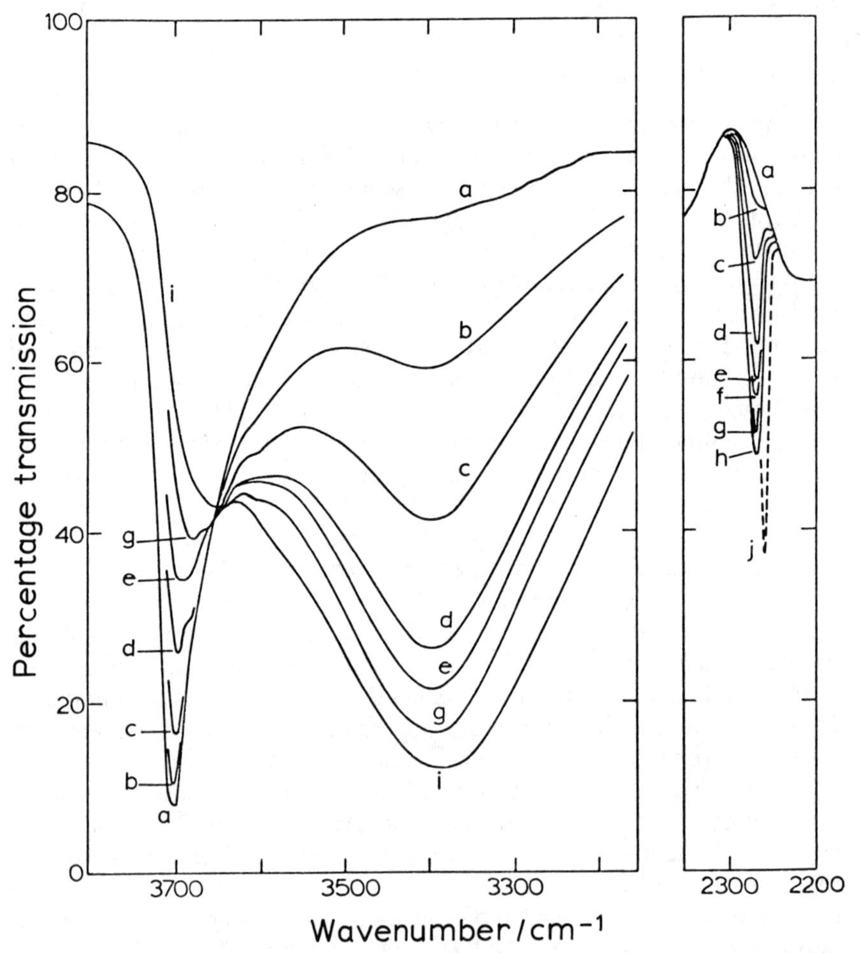

Figure 4 Infrared spectra of silica after heat treatment of 873 K and immersion in (a) n-heptane, (b)–(i) solutions of propionitrile at increasing concentration in n-heptane; (j) spectrum of propionitrile in n-heptane. (Rochester and Trebilco, 1978a)

alumina (Hayashi *et al.*, 1969; Hasegawa and Low, 1969b; Cross and Rochester, 1978b), zinc oxide (Hasegawa and Low, 1969a), magnesia (Hasegawa and Low, 1969b), goethite and haematite (Buckland *et al.*, 1980). Similarly the adsorption of aliphatic amines on silica immersed in carbon tetrachloride in part involved proton transfer from surface silanol groups to amine molecules with the resulting appearance in spectra of infrared bands characteristic of substituted ammonium cations (Rochester and Yong, 1980b). In general the infrared spectroscopic characterization of physical adsorption involves the observation of perturbations of surface groups or of adsorbed molecules. Chemisorption is recognized by the appearance of infrared bands due to vibrations of the products of adsorption.

(b) *Distinguishable modes of adsorption.* (i) *Adsorbates containing a single functional group.* Different modes of adsorption involving particular surface sites can be distinguished using infrared spectroscopy. For example, single isolated silanol groups on silica immersed in carbon tetrachloride interacted with adsorbed mono-, di-, or tri-n-butylamine molecules in one of two ways (Rochester and Yong, 1980a). The perturbation of silanol groups led to the appearance of broad maxima at ~3250 cm^{-1} which were consistent with the existence of hydrogen bonds between silanol groups and the nitrogen atoms of amine molecules. Infrared bands characteristic of substituted ammonium ions showed that some chemisorption had also occurred. In contrast two types of hydrogen bonded complex on silica were recognized from spectra of cyclohexanone adsorbed from solution in 2,2,4-trimethylpentane or 1,4-dimethylbenzene (Buckland *et al.*, 1978). The predominant mode of adsorption involved the formation of hydrogen bonds between single isolated silanol groups and the carbonyl groups of single cyclohexanone molecules. However, particularly at low coverages, pairs of isolated silanol groups were involved in the simultaneous formation of pairs of hydrogen bonds with single cyclohexanone molecules. The resulting shift in the position of the infrared band due to the C=O stretching vibration of cyclohexanone was approximately double the corresponding shift when single silanol groups, rather than pairs, constituted the adsorption sites.

The surface of silica which has been heated at temperatures below ~750 K contains not only isolated surface silanol groups but also adjacent interacting silanol groups. Infrared spectra of adsorbed ketones (Griffiths *et al.*, 1974; Marshall and Rochester, 1975c; Cross and Rochester, 1978a; Rochester and Trebilco, 1979), ethyl acetate (Cross and Rochester, 1979), and carboxylic acids (Marshall and Rochester, 1975a) have enabled adsorption onto isolated silanol group sites and adjacent interacting silanol group sites to be distinguished. The hydrogen bonds between pairs of interacting silanol groups were broken and each pair of groups formed two hydrogen bonds to the carbonyl group of a single adsorbate molecule. This mode of adsorption was recognized from the resulting double perturbation of the carbonyl groups and was more significant for silica that had been preheated at relatively low temperatures and therefore retained a high residual surface population of interacting silanol groups (Griffiths *et al.*, 1974; Rochester and Trebilco, 1979; Cross and Rochester, 1979). Adsorption involving pair sites was more favourable at low overall surface coverages than adsorption involving single isolated silanol group sites. This conclusion followed from correlations between the intensities of infrared bands not only in the 1650–1750 cm^{-1} spectral region (Rochester and Trebilco, 1979) but also due to the OH stretching vibrations of unperturbed and perturbed silanol groups (Marshall and Rochester, 1975a, c).

(ii) *Adsorbates containing two or more functional groups.* Acceptance of hydrogen bonds by carbonyl groups gives a shift $\Delta\bar{\nu}_{CO}$ in the position of the infrared band due to C=O stretching vibrations. Fontana and Thomas (1961) showed that the spectroscopic distinction between perturbed and unperturbed carbonyl groups could be used to assess the proportion of carbonyl groups, in adsorbed molecules containing several carbonyl groups, which were involved in H-bonding interactions with surface silanol groups on silica. Spectra of three diesters and one triester on silica only contained bands due to perturbed carbonyl groups (Fontana, 1966). All the ester groups in adsorbed molecules were linked to the oxide surface through hydrogen bonding interactions. Similarly hexane-2,5-dione was adsorbed on silica immersed in carbon tetrachloride with both ketone groups in each molecule involved in hydrogen bonding interactions with isolated surface silanol groups (Cross and Rochester, 1978a). In contrast, the adsorption of hexane-2,3-dione involved the formation of hydrogen bonds between single silanol groups and only one carbonyl group in each adsorbate molecule. The second carbonyl group in each adsorbed molecule remained unperturbed by the silica surface. Results of a study of the chemisorption of dicarboxylic acids on haematite and goethite also exemplify how infrared spectroscopy can be used to characterize the mode of adsorption of molecules containing more than one functional group of the same type (Buckland *et al.*, 1980).

Infrared spectroscopy provides an unrivalled method for distinguishing between surface–adsorbate interactions involving molecules containing two or more functional groups of different type. Isolated silanol groups on silica were perturbed to give two infrared bands at 3000 and 3380 cm^{-1} by the adsorption of 3-acetylpyridine from solution in carbon tetrachloride (Griffiths *et al.*, 1974). Silanol groups linked by hydrogen bonds to the nitrogen atoms of adsorbed molecules were responsible for the band at 3000 cm^{-1}. The concomitant perturbation of 3-acetylpyridine molecules resulted in the shift of a band at 1422 cm^{-1} for 3-acetylpyridine in solution to 1430 cm^{-1} for 3-acetylpyridine in the adsorbed state. The maximum at 3380 cm^{-1} was assigned to silanol groups which were perturbed by hydrogen bonding interactions with carbonyl groups in adsorbed molecules. The resulting perturbation of the carbonyl groups was also accompanied by a spectroscopic shift from 1697 to 1682 cm^{-1} in the position of the band due to the C=O stretching vibrations. Similar results were obtained for the adsorption of 4-formylpyridine.

Infrared spectra of silica immersed in solutions of substituted anilines (Low and Hasegawa, 1968; Rochester and Trebilco, 1978b), anisoles (Rochester and Trebilco, 1978c), or phenols (Rochester and Trebilco, 1978d; Rupprecht and Fuchs, 1978) have established the existence of hydrogen bonding interactions between surface silanol groups and either substituent groups in the benzene rings or π-electrons in the aromatic nuclei. Spectra generally contained two infrared bands due to hydroxyl stretching vibrations of perturbed silanol groups. For the adsorption of anisoles and phenols the band positions and their relative intensities were a function of the electron withdrawing or donating abilities of substituents in the benzene ring. The spectroscopic results provide direct characterization of the nature of surface–adsorbate interactions involving adsorption of aromatic molecules on silica immersed in liquids.

(c) *Substituent effects on surface–adsorbate interactions.* The formation of hydrogen bonds between surface silanol groups on silica and adsorbate molecules leads to shifts $\Delta\bar{\nu}_{OH}$ in the position of the infrared band due to the OH stretching vibrations of silanol groups. The magnitude of the shifts can be used as an indication of the strengths of the hydrogen bonds. Griffiths *et al.* (1974) measured spectra of silica immersed in solutions of pyridine and seven substituted pyridines in carbon tetrachloride. The observed shifts $\Delta\bar{\nu}_{OH}$ were an approximately linear function of the pK_a values of the corresponding pyridinium ions in water. The more basic was the substituted pyridine, the stronger was the hydrogen bonding interaction at the solid/liquid interface and the greater was $\Delta\bar{\nu}_{OH}$.

Shift data in Fig. 5 relate to the adsorption of benzene, seven anisoles, toluene, nitrobenzene, and nitrotoluene on isolated silanol group sites on silica immersed in n-heptane (Rochester and Trebilco, 1978c). Lines (a) and (c) in Fig. 5 were for silanol groups forming hydrogen bonds with methoxy groups and aromatic π-electrons, respectively, in adsorbate molecules. Electron withdrawing substituents weakened both types of interaction as evidenced by the reductions in $\Delta\bar{\nu}_{OH}$ with increasing σ-constant. For 4-nitroanisole only one band due to perturbed silanol groups was observed and the shift $\Delta\bar{\nu}_{OH}$ was not consistent with either line (a) or line (c) in Fig. 5 (σ-constant for 4-nitro, 1.24). However, the $\Delta\bar{\nu}_{OH}$ value was consistent with data for the adsorption of nitrobenzene and 4-nitrotoluene [Fig. 5(b)] showing that the adsorption of 4-nitroanisole involved the formation of hydrogen bonds between surface silanol groups and nitro groups in adsorbed molecules. This result illustrates how shift data can be used to diagnose surface–adsorbate interactions when several alternatives are possible.

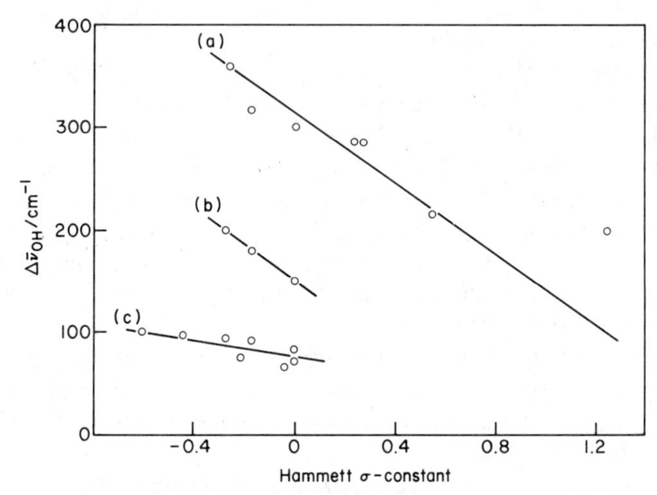

Figure 5 Correlations between $\Delta\bar{\nu}_{OH}$ and Hammett σ-constants for the adsorption of (a) substituted anisoles (b) substituted nitrobenzenes, and (c) substituted benzenes on silica in n-heptane. (Rochester and Trebilco, 1978c)

Detailed studies of the adsorption of phenols on silica (Rochester and Trebilco, 1978d; Rupprecht and Fuchs, 1978) also illustrate how infrared spectroscopy gives information about the effects of substituents on surface–adsorbate interactions. Correlations between $\Delta\bar{\nu}_{OH}$ shifts or band intensities and Hammett σ-constants have been observed. The influence of steric, as well as electronic, substituent effects on adsorption behaviour can be deduced from the infrared data.

(d) *Fractional surface coverages*. Measurement of intensities of infrared bands of surface species at the solid/liquid interface has enabled the calculation of fractional surface coverages of specific adsorption sites (Buckland *et al*., 1978; Rochester and Trebilco, 1977a, 1978a,c,d, 1979). In general, adsorption onto isolated silanol group sites has been studied. Marshall and Rochester (1975b), using the cell in Fig. 3, achieved simultaneous measurement of infrared spectra and weights of phenol adsorbed for the adsorption of phenol on silica immersed in carbon tetrachloride. Hence, with the assumption that one phenol molecule was adsorbed per isolated silanol group site, the number of silanol groups per unit area of the silica surface could be calculated.

A study of the adsorption of cyclohexanone on silica immersed in 2,2,4-trimethylpentane has illustrated how infrared spectroscopy may allow total adsorption isotherms to be dissected into the contributions from adsorption onto distinguishable surface sites (Rochester and Trebilco, 1979). Fractional surface coverages of isolated silanol group sites as a function of cyclohexanone concentration were similar for silica that had been preheated at 423 and 1073 K. The spectroscopic results established that adsorption involving isolated silanol groups was unaffected by the presence of a high population of adjacent interacting OH groups on the surface. The surface coverages of interacting OH group pair sites by cyclohexanone molecules were separately estimated. The promise of infrared spectroscopy in this context is impressive since the fractional coverage data refer to surface–adsorbate interactions which are simultaneously characterized by the spectroscopic results.

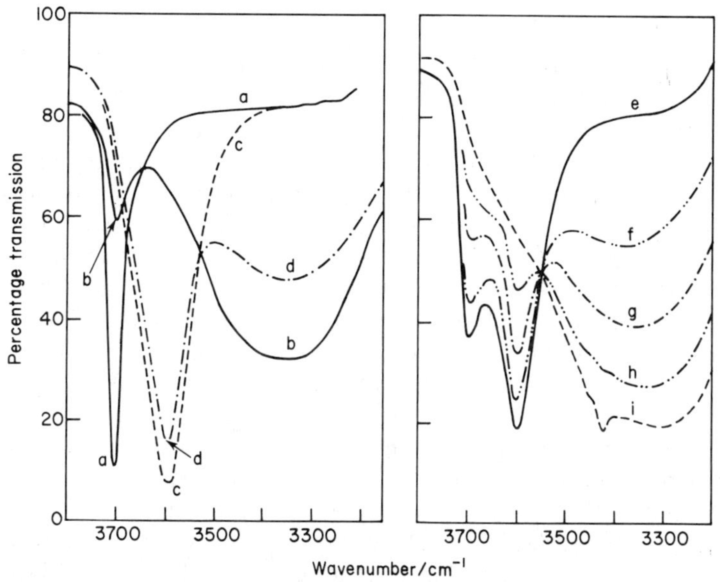

Figure 6 Spectra of silica after evacuation at 873 K and immersion in (a) n-heptane, (b) n-heptane–acetone mixture, (c) toluene, (d) toluene–acetone mixture, (e)–(i) n-heptane–toluene mixtures ($x_t:x_h = 0.2:1$) containing increasing concentrations of acetone. (Rochester and Trebilco, 1977b)

(e) *Adsorption from multicomponent solutions*. The perturbation of surface silanol groups on silica by the constituents of multicomponent solutions may be detected by infrared spectroscopy. Figure 6 illustrates typical results for silica immersed in n-heptane–toluene–acetone mixtures. Silanol groups perturbed by n-heptane, toluene, or acetone gave three infrared bands, the positions of which were independent of the composition of the mixtures. The absence of appreciable solvent effects on the spectra and the applicability of the Beer–Lambert law was further substantiated by the presence of an isosbestic point in the spectra for three-component mixtures. More detailed studies of 2,2,4-trimethylpentane–1,4-dimethylbenzene–cyclohexanone mixtures (Buckland *et al.*, 1978) and 2,2,4-trimethylpentane–toluene–propionitrile mixtures (Rochester and Yong, 1980a) have shown that the proportion of isolated silanol groups perturbed by each component of the liquid phase can be evaluated from absorbance data taken from the spectra. The extent of adsorption of each component onto specific surface sites can therefore be quantitatively determined.

E. STRUCTURE OF THE ADSORBED LAYER

1. *Adsorption Isotherms*

In an analysis of the structure of the adsorbed layer at the solid/liquid interface several features should be considered. Its thickness is a primary factor. For the majority of systems involving adsorption from dilute solutions a monolayer is postulated which can contain one or both components. Steps in adsorption isotherms reflect either a change in orientation at the surface or the formation of multimolecular layers. The specific orientation adopted depends on (a) the chemical nature of the materials involved, (b) the solution concentration, (c) the interaction between adsorbate and adsorbent, and (d) lateral interactions between adsorbed species.

It is usual to infer the orientation from values of adsorption limit, surface area, and molecular dimensions, whilst in a few cases the configuration has been established by examining the adsorbate–adsorbent interactions using infrared spectroscopy. Some examples of the two orientation extremes inferred from adsorption data are given in Table 1. Certain trends are evident although not consistent; monocarboxylic acids are adsorbed perpendicular to the surface when chemisorption takes place, e.g. on alumina from hydrocarbon solution, but in other cases the dimeric acid molecule orients parallel to the surface, as with dicarboxylic acids. Alcohols can be oriented either way depending on the strength of interaction with the surface. Hydrocarbons tend to adopt the parallel orientation in all situations.

A number of cases of stepped isotherms have been reported. Examples are (a) lauric acid from carbon tetrachloride solution on Spheron 6, the step occurring well below the monolayer value, and attributed to a phase change in the adsorbed layer (Kipling and Wright, 1963), (b) octadecanol and ethyl stearate from benzene solution on metal (Fe, Ag, Ni, Cu)

Table 1 Orientation of adsorbate molecules at the solid/liquid interface, inferred from data for adsorption from solution.

Solute	Solvent	Adsorbent	Orientation	Reference
Stearic acid	Cyclohexane	SiO_2	Parallel	Kipling and Wright, 1964
Stearic acid	Cyclohexane	Al_2O_3	Perpendicular	Kipling and Wright, 1964
Stearic acid	Cyclohexane	TiO_2	Perpendicular	Kipling and Wright, 1964
Benzoic acid	Cyclohexane	SiO_2	Parallel	Wright and Pratt, 1974
Benzoic acid	Cyclohexane	Al_2O_3	Perpendicular	Wright and Pratt, 1974
Lauric acid	Benzene–heptane	TiO_2	Perpendicular	Heertjes *et al.*, 1979
Fatty acids	Benzene	SiO_2	Perpendicular	Armistead *et al.*, 1971a
Fatty acids	n-Hexane	SiO_2	Perpendicular	Armistead *et al.*, 1971a
Stearic acid	Cyclohexane–ethanol	Graphon	Parallel	Kipling and Wright, 1962
Benzoic acid	Cyclohexane	Graphon	Parallel	Wright and Pratt, 1974
Esters	Benzene	SiO_2	Perpendicular	Mills and Hockey, 1975a
Esters	Carbon tetrachloride	SiO_2	Perpendicular	Mills and Hockey, 1975a
Octadecanol	*p*-Xylene	TiO_2	Perpendicular	Parfitt and Wiltshire, 1964
Octadecanol	Benzene	Al_2O_3	Perpendicular	Crisp, 1956
Octanol	Carbon tetrachloride	SiO_2	Parallel	Ganichenko *et al.*, 1959
Dicarboxylic acids	Water	Graphon/Spheron 6	Parallel	Wright, 1966a
Adipic acid	Water	Al_2O_3	Parallel	Wright, 1966b
Aromatic hydrocarbons	Cyclohexane	Graphon/Spheron 6	Parallel	Wright and Powell, 1972
Alkylbenzenes	n-Heptane	Graphon	Parallel	Parfitt and Willis, 1964
Docosane	n-Heptane	Graphon	Parallel	Kern *et al.*, 1978

powders, due to a change in orientation from perpendicular to parallel, the step becoming less sharply defined at higher temperature (Kipling and Wright, 1962), (c) 1,1-diphenyl-2-picrylhydrazyl from benzene solution on titanium dioxide, due to an orientation change (Misra, 1975), (d) oleic, linoleic, and linolenic acids from n-heptane solutions on a dry rutile surface, interpreted in terms of reorientation (Ottewill and Tiffany, 1967), and (e) phenol from aqueous solution on a (channel) carbon black, the first plateau being associated with parallel orientation since it changed with modification of the surface polarity (by oxidation and reduction), whilst the second was independent of the nature of the surface suggesting the phenol molecules were vertically oriented (Coughlin and Ezra, 1968). Theory based on the adsorption of dimers predicts that such a change in orientation would occur if the molecule contains one segment that interacts much more strongly with the surface than the other (Ash *et al.*, 1968).

Evidence for adsorption in excess of a monolayer is given for (a) dicarboxylic acids (C_7 and C_{10}) from aqueous solution on Graphon, for which the adsorption rises steeply after a long plateau (Wright, 1966a), (b) pyrene from cyclohexane solution on graphites containing a high proportion of polar edge sites (Groszek, 1975), (c) acetic and butyric acids from aqueous and hydrocarbon solutions on styrene–divinylbenzene copolymer containing large pores (Paleos, 1969), and (d) oleic, linoleic, and linolenic acids from n-heptane solution on dry rutile (Ottewill and Tiffany, 1967).

Evidence for structuring in adsorbed layers has been presented by Everett for the system n-hexanol/n-hexane/Graphon, from accurate measurements of adsorption over a range of temperature, and it is shown that as the temperature is raised the change in the surface phase has thermodynamic characteristics similar to those of melting (Everett, 1978). These structuring effects appear to be confined to straight chain compounds and surfaces with a well defined crystal structure. An apparent transition was also found for the adsorption of stearic acid from benzene solutions on to Graphon. Isotherms at 25–30°C showed an upward trend after the plateau value, but the step became less distinct as the temperature was raised to 40°C and disappeared at 60°C. Similar behaviour was found with the dodecanol/n-heptane/Graphon system (Everett, 1978), the n-hexadecane/n-heptane/Graphon system (Parfitt and Tideswell, 1981), and with n-docosane/n-heptane/graphitized carbon black (Kern *et al.*, 1978). The latter system was discussed in terms of strong lateral interactions between the long chain hydrocarbon molecules adsorbed on the graphite basal planes on the surface; a phase separation is indicated at lower temperatures leading to a close-packed ordered array of n-docosane molecules. Further evidence for the strength of such interactions was provided by studies from the same laboratory using 1-decanol and decanoic acid (Liphard *et al.*, 1980) and long chain n-paraffins C_{22}, C_{28}, and C_{32} (Kern and Findenegg, 1980). Lateral interactions were also invoked to explain deviations from Langmuir behaviour in the adsorption of substituted phenols from aqueous solution by styrene–divinylbenzene copolymers (Paleos, 1969).

The formation of a hydrophobic bond between the chains of the lower n-fatty acids from aqueous solution on to the surface of polystyrene was studied by Schneider *et al.* (1965) from adsorption isotherms and calorimetric data in the range 4–40°C. Approximate agreement with the theory of Nemethy and Scheraga was obtained, but not for alcohols having the same chain length as the acids, suggesting that hydrophobic bond formation is not necessarily independent of the head group. Dipole–dipole interactions in the adsorbed phase on a carbon surface, and containing *p*-nitroaniline molecules adsorbed from aqueous solution, were assumed by Koganovskii *et al.* (1972) to describe thcir adsorption data, the dipole moment being set equal to that of the amino group. Such lateral interactions as are involved in the formation of a hydrophobic bond and between adsorbed dipolar species have received relatively little attention in studies of adsorption from dilute solutions of non-electrolytes at the solid/liquid interface.

2. *Infrared Measurements*

(a) *Orientation of hydrocarbon chains.* Three methods involving infrared spectroscopy have been used to gain information about the orientation of hydrocarbon chains in molecules adsorbed at the solid/liquid interface. Long chain carboxylic acids have been studied. The predominant surface–adsorbate interactions were therefore hydrogen bonding between silanol groups and carboxylic acid groups for silica adsorbent, or chemisorption to give carboxylate groups when alumina, magnesia, or zinc oxide was the adsorbent oxide.

Hasegawa and Low (1969a,b) recorded the positions of infrared bands in the range 1150–1380 cm^{-1} for stearic acid adsorbed on alumina and zinc oxide, and for decanoic acid on magnesia. Analysis showed that the band positions for model stearates or decanoates were a function of the stacking of the hydrocarbon chains in the model compounds. Comparison of band positions for adsorbed stearic or decanoic acid and the model compounds suggested that stearate chains on alumina were arranged irregularly but that the stacking of hydrocarbon chains in decanoate anions on magnesia was fairly regular. Stearic acid was adsorbed on zinc oxide immersed in carbon tetrachloride as stearate anions with their hydrocarbon chains probably tilted at about 60°.

Yang *et al.* (1973) measured absorbance ratios $A_{\perp}/A_{\parallel}$ for a band at 2928 cm^{-1} in infrared spectra, recorded using perpendicular and parallel polarized radiation, of stearic acid adsorbed from carbon tetrachloride onto (0001) cut faces of α-alumina. Expected ratios were 1.28 and 2.13

respectively for a randomly oriented surface film and for close-packed hydrocarbon chains normal to the alumina surface. The observed absorbance ratio of 1.62 for adsorbed stearic acid suggested that the hydrocarbon chains were partially oriented on the oxide surface. Unfortunately the carbon tetrachloride had to be removed from contact with the surface before spectra could be recorded, and therefore the orientation of the hydrocarbon chains may not have been the same as when liquid was present. However, the possible usefulness of infrared reflection spectroscopy in this context has been clearly demonstrated by the results of Yang *et al.* (1973) who also made similar measurements for stearic acid adsorbed on germanium immersed in carbon tetrachloride. The better optical characteristics of the system involving germanium, rather than alumina, enabled spectra to be recorded of adsorbed films *in situ* at the solid/liquid interface. The resulting value of $A_{\perp}/A_{\parallel}$ indicated that hydrocarbon chains in adsorbed stearic acid molecules were randomly oriented on the surface of germanium.

Marshall and Rochester (1975a,c) showed that the predominant interaction between the surface of silica and oleic, linoleic, or linolenic acid molecules in carbon tetrachloride involved the formation of hydrogen bonds between silanol groups and carboxylic acid groups. However, infrared bands in spectra of the adsorbed acids due to CH stretching vibrations of –CH=CH– groups were much weaker than corresponding bands in spectra of the acids in solution. This effect was attributed to perturbation of the alkenyl segments by the silica surface (Marshall *et al.*, 1976), and it was concluded that the hydrocarbon chains were oriented parallel to the surface or in a looped configuration. A similar conclusion was consistent with spectra of linoleic acid adsorbed on alumina immersed in carbon tetrachloride (Cross and Rochester, 1978b). An infrared study has also shown that alkenyl segments in the hydrocarbon chains of substituted succinic acids promote orientations of the chains which hinder chemisorptive interactions between the carboxylic acid groups and the surfaces of haematite or goethite immersed in carbon tetrachloride (Buckland *et al.*, 1980).

(b) *Changes of orientation with coverage.* Infrared spectroscopy has proved extremely valuable in studies of the orientation of adsorbed polymers as a function of coverage on silica immersed in liquids. Polymer adsorption has been reviewed elsewhere (Rochester, 1980). For small molecules infrared spectroscopy allows different modes of adsorption as a function of total surface coverage to be characterized. For example, the adsorption of ketones at very low coverages on silica primarily involved pairs of hydroxyl groups acting as adsorption sites (Rochester and Trebilco, 1979). At higher coverages adsorption onto single OH group sites constituted the predominant adsorptive interaction. Similarly carboxylic acids were adsorbed on pair OH sites at low coverages, single OH sites at moderate coverages, and as acid dimers at high coverages (Hasegawa and Low, 1969b; Low and Lee, 1973; Marshall and Rochester, 1975a, c). Different modes of adsorption involving different sites have therefore been recognized. There is less evidence for changes in orientation or mode of adsorption as a function of coverage for small molecules adsorbed on particular sites. However, three examples involved the adsorption of 2,6-dimethylanisole (Rochester and Trebilco, 1978c), 4-methylphenol and 2,4,6-trimethylphenol (Rochester and Trebilco, 1978d) on silica immersed in n-heptane. For high surface coverages infrared spectra showed that hydrogen bonds between silanol groups and hydroxy or methoxy groups in adsorbate molecules were being formed in part at the expense of interactions between the same silanol groups and aromatic π-electrons.

F. EFFECT OF TEMPERATURE ON ADSORPTION

The level of adsorption at any particular concentration usually decreases with increase in temperature, i.e. the overall process is exothermic. If the increased solubility was the only factor, isotherms obtained at different temperatures could be superimposed when plotted against the relative concentration (x/x_0 where x_0 is the mole fraction at the solubility limit). There is not much evidence of this applying. When the solubility is reduced with increase in temperature (aqueous solutions of n-butanol) the opposite effect is observed (Bartell *et al.*, 1951). Unfortunately the number of studies carried out on adsorption from dilute solutions at different temperatures is small.

Mills and Hockey (1975a, b) related the increase in adsorption with temperature of lauric acid and methyl esters of n-fatty acids (C_{10} and C_{14}) from benzene solution on to silica, to a decrease in "ordering" of the adsorbed layer as the temperature is raised. From infrared and calorimetric data they conclude that the hydrocarbon chains do not displace adsorbed benzene, hence are oriented perpendicular to the surface, and the solvated hydrocarbon chains form an ordered layer. Structuring at the interface was also observed by Everett for the stearic acid/benzene/Graphon and dodecanol/n-heptane/Graphon systems, and by Findenegg and coworkers for the n-docosane and decanol/n-heptane/graphitized carbon black system, but in all these cases, for which parallel orientation of the long chain molecule is probable, the adsorption decreased with increase in temperature (Everett, 1978; Kern *et al.*, 1978; Liphard *et al.*, 1980). More evidence of structuring at the interface is available from studies on adsorption from binary liquid mixtures (Everett, 1978), but the data that are available for dilute solutions are too limited to permit general conclusions to be made.

G. ADSORPTION THERMODYNAMICS

The number of published thermodynamic data for adsorption of non-electrolytes from dilute solution is rather small. Free energy, enthalpy, and entropy data have been calculated from adsorption isotherms over a limited range of temperature, and enthalpy values have been obtained from calorimetric measurements.

From adsorption data at 5° and 25°C Ottewill and Tiffany (1967) calculated ΔG° for the adsorption of stearic, oleic, and linoleic acids from n-heptane solutions on dry rutile surfaces, and obtained similar values (-25 to -29 kJ mol^{-1}) for each acid, corresponding to adsorption via the carboxyl group in the initial region of the isotherm. The values were lower (17–21 kJ mol^{-1}) for the region after the change in orientation (step in the isotherm) for the unsaturated acids. The differential enthalpy and entropy values for stearic acid adsorption at 25°C decreased with coverage ($\Delta\bar{H}$ from -71 to -21 kJ mol^{-1}; $\Delta\bar{S}$ from -159 to -8 J K^{-1} mol^{-1}) as adsorption approached the saturation limit.

The differential enthalpies of adsorption for long chain alcohols from benzene solution appear not to change appreciably with coverage, and are slightly lower than the calorimetric (integral) value of about 33 kJ mol^{-1} (Crisp, 1956). Free energy changes corresponding to $\theta = 0.5$ were found to be ~17 kJ mol^{-1}.

Differential enthalpies were calculated from adsorption and calorimetric data at 27°C [equation (17)] for n-fatty acids (C_6–C_{16}) from benzene solution on silica by Armistead *et al.* (1971b) and found to be independent of surface coverage and chain length. It was postulated that the expected decrease in adsorption heat due to adsorbate–adsorbent interaction is offset by a decreasing enthalpy of mixing of the adsorbed film. Similar results were obtained for the adsorption of methyl esters of the n-fatty acids from solutions in benzene and carbon tetrachloride (Mills and Hockey 1975b). The enthalpies of adsorption from benzene are 30–35 kJ mol^{-1} and independent of temperature (25–50°C), and from carbon tetrachloride about 52 kJ mol^{-1}.

Allen and Patel (1970) derived values for the integral enthalpy of displacement of monolayer coverage of n-alcohols (C_2–C_{18}) from n-heptane solutions on α-Fe_2O_3 using flow microcalorimetry. The data followed equation (18). The enthalpies increased with chain length from 15.5 kJ mol^{-1} for ethanol to 35.9 kJ mol^{-1} for n-octanol, and thereafter remained approximately constant. A similar study with a range of fatty acids (C_2–C_{18}) led to different behaviour—for the lower acids (C_2–C_6) the enthalpy values were similar (about 9.2 kJ mol^{-1}) after which they increased up to 22 kJ mol^{-1} for stearic acid (Allen and Patel, 1971). The results were interpreted in terms of the configuration adopted by the alcohol and acid molecules at the interface.

There are several reports containing thermodynamic data for adsorption on to carbon blacks. Wright and Pratt (1974) obtained free energy (from adsorption isotherms), integral enthalpy (from heats of immersion) and entropy data for the adsorption of aromatic carboxylic acids (benzoic and 1- and 2-naphthoic) from carbon tetrachloride solutions at $x_1^l < 0.044$ on Spheron 6 and Graphon at 30°C. ΔG and ΔS decrease and ΔH increases with increasing surface coverage, there are no significant differences between the values for the two adsorbents, and the ΔG and ΔS values for the naphthoic acids are significantly higher than those for benzoic acid. All ΔS values are large and positive (100–500 J K^{-1} mol^{-1}). Using the same procedures and adsorbents, Wright and Powell (1972) derived thermodynamic data for the adsorption of benzene, naphthalene, and anthracene from cyclohexane solutions ($x_1^l < 0.15$) at 30°C. As surface coverage increases, ΔG decreases in all cases, ΔH and ΔS increase for Graphon but decrease for Spheron 6, and all three quantities increase with increasing size of the aromatic molecule that is adsorbed. ΔS is negative throughout (100–300 J K^{-1} mol^{-1}). These data are consistent with greater preferential adsorption of benzene and naphthalene on Spheron 6 than on Graphon, and with the competitive nature of the adsorption process for Spheron 6.

Groszek used flow calorimetry to determine integral heats of displacement at monolayer coverage of hydrocarbons on Graphon and graphites: pyrene from cyclohexane solutions (Groszek, 1975) and n-dotriacontane from n-heptane solutions (Groszek, 1970). The data obeyed the Langmuir type equations, and reflect the specific interaction between the hydrocarbon and the basal plane surface of graphite. The same technique was used by Findenegg and coworkers in studies of the adsorption of long chain n-paraffins from n-heptane solutions on graphitized carbon black surfaces. S-shaped isotherms were obtained for n-docosane at 25–45°C, and the increase in differential enthalpy with coverage was explained in terms of increasing lateral interaction between the adsorbed molecules (Kern *et al.*, 1978). Evidence for a similar cooperative effect was found in the submonolayer region of the isotherm for n-dotriacontane (Kern and Findenegg, 1980).

It is seen that adsorption measurements at different temperatures together with related calorimetric data can provide useful information on the dominant factors in the adsorption process. More attention to this aspect is warranted.

References

Adamson, A. W. (1982). *Physical Chemistry of Surfaces,* 4th edn, p. 373, John Wiley and Sons, New York.
Allen, T. and Patel, R. M. (1968/9). *Powder Technol.* **2**, 111.
Allen, T. and Patel, R. M. (1970). *J. Appl. Chem.* **20**, 165.
Allen, T. and Patel, R. M. (1971). *J. Colloid Interface Sci.* **35**, 647.
Armistead, C. G., Tyler, A. J. and Hockey, J. A. (1971a). *Trans. Faraday Soc.* **67**, 493.
Armistead, C. G., Tyler, A. J. and Hockey, J. A. (1971b). *Trans. Faraday Soc.* **67**, 500.
Ash, S. G., Everett, D. H. and Findenegg, G. H. (1968). *Trans. Faraday Soc.* **64**, 2645.
Ash, S. G., Bown, R. and Everett, D. H. (1973). *J. Chem. Thermodyn.* **5**, 239.
Bartell, F. E., Thomas, T. L. and Fu. Y. (1951). *J. Phys. Chem.* **55**, 1456.
Bascam, W. D. and Timmons, R. B. (1972). *J. Phys. Chem.* **76**, 3192.
Boehm, H. P. (1966). *Adv. Catal.* **16**, 179.
Bonetzkaya, A. K. and Krasilnikov, K. G. (1957). *Proc. Acad. Sci. U.S.S.R. (Phys. Chem.)* **114**, 1257.
Botham, R. and Thies, C. (1970). *J. Polym. Sci. (C)* **30**, 369.
Brooks, C. S. (1958). *J. Colloid Sci.* **13**, 522.
Brown, C. E. and Everett, D. H. (1975). In *Colloid Science,* Specialist Periodical Reports, Vol. 2, Ch. 2, Chemical Society, London.
Buckland, A. D., Rochester, C. H., Trebilco, D.-A. and Wigfield, K. (1978). *J. Chem. Soc. Faraday Trans. 1* **74**, 2393.
Buckland, A. D., Rochester, C. H. and Topham, S. A. (1980). *J. Chem. Soc. Faraday Trans. 1* **76**, 302.
Claesson, S. (1946). *Arkiv Kemi Min. Geol.* **23A**, No. 1.
Cook, E. L. and Hackerman, N. (1951). *J. Phys. Chem.* **55**, 549.
Corkill, J. M., Goodman, J. F. and Tate, J. R. (1964). *Trans. Faraday Soc.* **60**, 996.
Corkill, J. M., Goodman, J. F. and Tate, J. R. (1966). *Trans. Faraday Soc.* **62**, 939.
Cornford, P. V., Kipling, J. J. and Wright, E. H. M. (1962). *Trans. Faraday Soc.* **58**, 74.
Coughlin, R. W. and Ezra, F. S. (1968). *Environ. Sci. and Technol.* **2**, 291.
Crisp, D. J. (1956). *J. Colloid Sci.* **11**, 356.
Cross, S. N. W. and Rochester, C. H. (1978a). *J. Chem. Soc. Faraday Trans. 1* **74**, 2130
Cross, S. N. W. and Rochester, C. H. (1978b). *J. Chem. Soc. Faraday Trans. 1* **74**, 2141.
Cross, S. N. W. and Rochester, C. H. (1979). *J. Chem. Soc. Faraday Trans. 1* **75**, 2865.
Davis, K. M. C., Deuchar, J. A. and Ibbitson, D. A. (1973). *J. Chem. Soc. Faraday Trans. 1* **69**, 1117.
Davis, K. M. C., Deuchar, J. A. and Ibbitson, D. A. (1974). *J. Chem. Soc. Faraday Trans. 1* **70**, 417.
Day, R. E. and Parfitt, G. D. (1967). *Powder Technol.* **1**, 3.
deBoer, J. H., Houben, G. M. M., Lippens, B. C., Meijs, W. H. and Walrave, W. H. A. (1962). *J. Catal.* **1**, 1.

DiGiano, F. J. and Weber, W. J. (1969). Tech. Pub. T-69-1, Dept. of Civil Eng., University of Michigan, U.S.A.
Doorgeest, T. (1962). Fatipec Congress Book, 93.
Eríć, B., Goode, E. V. and Ibbitson, D. A. (1960). *J. Chem. Soc.* 55.
Erkelens, J. and Liefkens, T. J. (1975). *J. Catal.* **39**, 173.
Everett, D. H. (1964). *Trans. Faraday Soc.* **60**, 1803.
Everett, D. H. (1978). *Progr. Colloid Polym. Sci.* **65**, 103.
Everett, D. H. and Findenegg, G. H. (1969). *J. Chem. Thermodynamics* **1**, 573.
Everett, D. H. and Podoll, R. T. (1979). In *Colloid Science,* Specialist Periodical Reports, Vol. 3, Ch. 2, Chemical Society, London.
Eyring, E. M. and Wadsworth, M. E. (1956). *Mining Eng.* **8**, 531.
Filimonov, V. N. (1956). *Optika i Spekt.* **1**, 490.
Filimonov, V. N. and Terenin, A. N. (1956). *Dokl. Akad. Nauk S.S.S.R.* **109**, 982.
Fontana, B. J. (1966). *J. Phys. Chem.* **70**, 1801.
Fontana, B. J. and Thomas, J. R. (1961). *J. Phys. Chem.* **65**, 480.
Francis, S. A. and Ellison, A. H. (1959). *J. Opt. Soc. Am.* **49**, 131.
French, R. O., Wadsworth, M. E., Cook, M. A. and Cutler, I. B. (1954). *J. Phys. Chem.* **58**, 805.
Freundlich, H. (1926). *Colloid and Capillary Chemistry,* Methuen, London.
Ganichenko, L. G., Kiselev, V. F. and Krasil'nikov, K. G. (1959). *Dokl. Akad. Nauk S.S.S.R.* **125**, 1277.
Giles, C. H., MacEwan, T. H., Nakhwa, S. N. and Smith, D. (1960). *J. Chem. Soc.* 3973.
Giles, C. H., Smith, D. and Huitson, A. (1974a). *J. Colloid Interface Sci.* **47**, 755.
Giles, C. H., D'Silva, A. P. and Easton, I. A. (1974b). *J. Colloid Interface Sci.* **47**, 766.
Gregg, S. J. and Sing, K. S. W. (1982). *Adsorption, Surface Area and Porosity,* 2nd edn. Academic Press, London.
Griffiths, D. M., Marshall, K. and Rochester, C. H. (1974). *J. Chem. Soc. Faraday Trans. 1* **70**, 400.
Groszek, A. J. (1970). *Proc. Roy. Soc.* (London) *A***314**, 473.
Groszek, A. J. (1975). *Faraday Discuss. Chem. Soc.* **59**, 109.
Hair, M. L. (1967). *Infrared Spectroscopy in Surface Chemistry,* Marcel Dekker, New York.
Han, K. N., Healy, T. W. and Fuerstenau, D. W. (1973). *J. Colloid Interface Sci.* **44**, 407.
Hansen, R. S. and Craig, R. P. (1954). *J. Phys. Chem.* **59**, 211.
Hasegawa, M. and Low, M. J. D. (1969a). *J. Colloid Interface Sci.* **29**, 593.
Hasegawa, M. and Low, M. J. D. (1969b). *J. Colloid Interface Sci.* **30**, 378.
Hayashi, S., Takenaka, T. and Gotoh, R. (1969). *Bull. Inst. Res. Kyoto Univ.* **47**, 378.
Heertjes, P. M., Smits, C. I. and Vervoorn, P. M. M. (1979). *Progr. Org. Coatings* **7**, 141.
Hirst, W. and Lancaster, J. K. (1951). *Trans. Faraday Soc.* **47**, 315.
Holmes, H. N. and McKelvey, J. B. (1928). *J. Phys. Chem.* **32**, 1522.
Husbands, D. I., Tallis, W., Waldsax, J. C. R., Woodings, C. R. and Jaycock, M. J. (1971/2). *Powder Technol.* **5**, 32.
Ibbitson, D. A., Jackson, T., McCarthy, A. and Stone, C. W. (1960). *J. Chem. Soc.* 5127.
Kagiya, T., Sumida, Y. and Tachi, T. (1971). *Bull. Chem. Soc. Japan* **44**, 1219.
Kapler, R. and Nekrasov, L. I. (1971). *J. Phys. Chem.* **45**, 419.

Kern, H. E. and Findenegg, G. H. (1980). *J. Colloid Interface Sci.* **75**, 346.
Kern, H. E., Piechocki, A., Brauer, U. and Findenegg, G. H. (1978). *Progr. Colloid Polym. Sci.* **65**, 118.
Killmann, E. and Strasser, H. J. (1973). *Angew. Makromol. Chem.* **31**, 169.
Killmann, E., Strasser, H. J. and Winter, K. (1972). 6th International Congress on Surface Active Agents, Zurich, 221.
Kipling, J. J. (1965). *Adsorption from Solutions of Non-Electrolytes,* Academic Press, London.
Kipling, J. J. and Wright, E. H. M. (1962). *J. Chem. Soc.* 855.
Kipling, J. J. and Wright, E. H. M. (1963). *J. Chem. Soc.* 3382.
Kipling, J. J. and Wright, E. H. M. (1964). *J. Chem. Soc.* 3535.
Kiselev, A. V. and Khopina, V. V. (1969). *Trans. Faraday Soc.* **65**, 1936.
Kiselev, A. V. and Shikalova, I. V. (1970). *Colloid J. (U.S.S.R.)* (Engl. transl.) **32**, 588.
Koganovskii, A. M., Levchenko, T. M., Sollogubovskaya, L. I. and Kirichenko, V. A. (1972). *Dopov. Akad. Nauk Ukr. R.S.R.* (Ser. B) **34**, 724.
Leja, J., Little, L. H. and Poling, G. W. (1963). *Trans. Inst. Min. and Met.* **72**, 407.
Liphard, M., Glanz, P., Pilarski, G. and Findenegg, G. H. (1980). *Progr. Colloid Polym. Sci.* **67**, 131.
Little, L. H. (1966). *Infrared Spectra of Adsorbed Species,* Academic Press, London.
Little, L. H. and Ottewill, R. H. (1962). *Can. J. Chem.* **40**, 2110.
Low, M. J. D. and Hasegawa, M. (1968). *J. Colloid Interface Sci.* **26**, 95.
Low, M. J. D. and Lee, P. L. (1973). *J. Colloid Interface Sci.* **45**, 148.
Marshall, K. and Rochester, C. H. (1975a). *J. Chem. Soc. Faraday Trans. 1* **71**, 1754.
Marshall, K. and Rochester, C. H. (1975b). *J. Chem. Soc. Faraday Trans. 1* **71**, 2478.
Marshall, K. and Rochester, C. H. (1975c). *Faraday Discuss. Chem. Soc.* **59**, 117.
Marshall, K., Rochester, C. H., Smith, R. and Smith T. N. (1976). *Chem. and Ind.* (London) 409.
Mattson, J. S., Mark, H. B., Malbin, M. D., Weber, W. J. and Crittenden, J. C. (1969). *J. Colloid Interface Sci.* **31**, 116.
Mills, A. K. and Hockey, J. A. (1975a). *J. Chem. Soc. Faraday Trans. 1* **71**, 2384
Mills, A. K. and Hockey, J. A. (1975b). *J. Chem. Soc. Faraday Trans. 1* **71**, 2392.
Mills, A. K. and Hockey, J. A. (1975c). *J. Chem. Soc. Faraday Trans. 1* **71**, 2398.
Misra, D. N. (1975). *J. Colloid Interface Sci.* **50**, 108.
Morimoto, T. and Naono, H. (1972). *Bull. Chem. Soc. Japan* **45**, 700.
Neagle, W. and Rochester, C. H. (1982). *J. Chem. Soc. Chem. Commun.* 398.
Nechtschein, J. and Sillion, B. (1971). *Tetrahedron Lett.* 2213.
Nekrassov, B. (1928). *Z. Phys. Chem.* (Leipzig) **32**, 676.
Ottewill, R. H. and Tiffany, J. M. (1967). *J. Oil Colour Chem. Assoc.* **50**, 844.
Paleos, J. (1969). *J. Colloid Interface Sci.* **31**, 7.
Parfitt, G. D. and Thompson, P. C. (1971). *Trans. Faraday Soc.* **67**, 3372.
Parfitt, G. D. and Tideswell, M. W. (1981). *J. Colloid Interface Sci.* **79**, 518.
Parfitt, G. D. and Willis, E. (1964). *J. Phys. Chem.* **68**, 1780.
Parfitt, G. D. and Wiltshire, I. J. (1964). *J. Phys. Chem.* **68**, 3545.
Poling, G. W. and Leja, J. (1963). *J. Phys. Chem.* **67**, 2121.
Raghavan, S. and Fuerstenau, D. W. (1975). *J. Colloid Interface Sci.* **50**, 319.
Rochester, C. H. (1980) *Adv. Colloid Interface Sci.* **12**, 43.
Rochester, C. H. and Trebilco, D.-A. (1977a). *J. Chem. Soc. Faraday Trans. 1* **73**, 883.

Rochester, C. H. and Trebilco, D.-A. (1977b). *J. Chem. Soc. Chem. Commun.* 621.
Rochester, C. H. and Trebilco, D.-A. (1978a). *Chem. and Ind.* (London) 127.
Rochester, C. H. and Trebilco, D.-A. (1978b). *Chem. and Ind.* (London) 348.
Rochester, C. H. and Trebilco, D.-A. (1978c). *J. Chem. Soc. Faraday Trans. 1* **74**, 1125.
Rochester, C. H. and Trebilco, D.-A. (1978d). *J. Chem. Soc. Faraday Trans. 1* **74**, 1137.
Rochester, C. H. and Trebilco, D.-A. (1979). *J. Chem. Soc. Faraday Trans. 1* **75**, 2211.
Rochester, C. H. and Yong, G. H. (1980a). *J. Chem. Soc. Faraday Trans. 1* **76**, 1158.
Rochester, C. H. and Yong, G. H. (1980b). *J. Chem. Soc. Faraday Trans. 1* **76**, 1466.
Roe, R.-J. (1974). *J. Chem. Phys.* **60**, 4192.
Roe, R.-J. (1975). *J. Colloid Interface Sci.* **50**, 64.
Rupprecht, H. and Fuchs, G. (1978). *Pharm. Industrie.* **40**, 1174.
Rupprecht, H. and Kindl, G. (1975). *Arch. Pharm.* (Weinheim) **308**, 46.
Schay, G. and Nagy, L. G. (1961) *J. Chim. Phys.* 149.
Schneider, H., Kresheck, G. C. and Scheraga, H. A. (1965). *J. Phys. Chem.* **69**, 1310.
Sharma, S. C. and Fort, T. (1973). *J. Colloid Interface Sci.* **43**, 36.
Sherwood, A. F. and Rybicka, S. M. (1966). *J. Oil Colour Chem. Assoc.* **49**, 648.
Smith, H. A. and Fuzek, J. F. (1946). *J. Am. Chem. Soc.* **68**, 229.
Thies, C. (1966). *J. Phys. Chem.* **70**, 3783.
Thies, C. (1968). *Macromolecules* **1**, 335.
Thies, C., Peyser, P. and Ullmann, R. (1964). 4th International Congress of Surface Activity, Brussels, 1041.
Timmons, C. O. (1973). *J. Colloid Interface Sci.* **43**, 1.
Traube, J. (1891). *Annalen* **265**, 27.
Tyler, A. J., Taylor, J. A. G., Pethica, B. A. and Hockey, J. A. (1971). *Trans. Faraday Soc.* **67**, 483.
Wright, E. H. M. (1966a). *J. Chem. Soc.* (*B*) 355.
Wright, E. H. M. (1966b). *J. Chem. Soc.* (*B*) 361.
Wright, E. H. M. and Powell, A. V. (1972). *J. Chem. Soc. Faraday Trans. 1* **68**, 1908.
Wright, E. H. M. and Pratt, N. C. (1974). *J. Chem. Soc. Faraday Trans. 1* **70**, 1461.
Yang, R. T., Low, M. J. D., Haller, G. L. and Fenn, J. (1973). *J. Colloid Interface Sci.* **44**, 249.
Zettlemoyer, A. C., Young, G. J., Chessick, J. J. and Healey, F. H. (1953). *J. Phys. Chem.* **57**, 649.

2. Adsorption from Mixtures of Miscible Liquids

J. E. LANE

I. Introduction

The study of adsorption from completely miscible solutions is generally made with a view to improving our understanding of adsorption rather than for any immediate technological gain. Until very recently, there had been no convincing model of bulk phase solutions and this hindered the development of satisfactory models of the solid/solution interface. There is now ample opportunity for theoreticians to apply the new-found understanding of liquids to problems at the boundary with a solid.

There have been problems also for the experimentalist. In order to form a completely miscible solution, the two components need to have similar intermolecular potentials, and in general this will mean that each component will have a similar interaction with the solid. As a result, any preferential adsorption of one component will be small and difficult to measure. Furthermore, models of the interface generally assume that the solid boundary is homogeneous both in a physical and in a chemical sense. This homogeneity is very difficult to attain with real solids.

Notwithstanding the experimental and theoretical difficulties, we now have a qualitative understanding of adsorption from completely miscible solutions. There is reason to believe that a better quantitative understanding is not far away.

II. Thermodynamics of Adsorption and of Liquid Mixtures

It is a necessary, but not sufficient, condition that any model of an interface satisfy all the many restrictions imposed by equilibrium thermodynamics. The following is a summary of the more important thermodynamic relations. In order to avoid confusion, particularly with regard to the status of the chemical potential, a very brief account of the formal development is included. A more complete description of the formal structure for bulk phases has been given by Callen (1960) who has amplified this aspect of the pioneering work of Gibbs (1961).

We commence with the assumption that any phase α may be subdivided into ν regions, each having a well defined value for the internal energy U, the entropy S, the volume V, the number of moles n_i of component i, the temperature T, the pressure p, and chemical potential μ_i of component i. The extensive variables U, S, V, and n_i for the whole phase are obtained as the sums over the values for each region:

$$U = \sum_{k=1}^{\nu} U^k \tag{1}$$

$$S = \sum_{k=1}^{\nu} S^k \tag{2}$$

$$V = \sum_{k=1}^{\nu} V^k \tag{3}$$

$$n_i = \sum_{k=1}^{\nu} n_i^k \tag{4}$$

the superscript k indicating that the variable applies to the kth region, and Σ indicating summation. The next assumption is that the variation of U^k is given by

$$\mathrm{d}U^k = T^k\,\mathrm{d}S^k - p^k\,\mathrm{d}V^k + \sum_i \mu_i^k\,\mathrm{d}n_i^k \tag{5}$$

The condition for equilibrium in any isolated system is that the internal energy has its minimum value for a given value of the entropy. In mathematical terms this may be expressed as

$$(\mathrm{d}U)_S = 0 \qquad (\mathrm{d}^2U)_S > 0 \tag{6}$$

Allowing the interchange of entropy, volume, and matter between regions, but isolating phase α so that the total volume, entropy, and amount of each component is fixed, equilibrium is established when (6) is satisfied, or

$$\sum_{k=1}^{\nu} \mathrm{d}U^k = 0$$

and therefore,

$$T = T^1 = T^2 = \ldots = T^\nu \quad (7)$$

$$p = p^1 = p^2 = \ldots = p^\nu \quad (8)$$

$$\mu_i = \mu_i^1 = \mu_i^2 = \ldots = \mu_i^\nu \quad (9)$$

At equilibrium, the intensive variables T, p, and μ_i are uniform throughout phase α.

Equation (5) is homogeneous and first-order in the variables, so by Euler's theorem (Callen, 1960) and relations (1)–(5) and (7)–(9) we obtain

$$U = TS - pV + \sum_i \mu_i n_i \quad (10)$$

and the Gibbs–Duhem relation

$$S\,\mathrm{d}T - V\,\mathrm{d}p + \sum n_i\,\mathrm{d}\mu_i = 0 \quad (11)$$

Using a Legendre transformation (Callen, 1960) it is possible to define other energy functions for a bulk phase in which one or more of the independent variables is intensive. These include the enthalpy H,

$$H = U + pV = TS + \sum_i \mu_i n_i \quad (12)$$

$$\mathrm{d}H = T\,\mathrm{d}S + V\,\mathrm{d}p + \sum_i \mu_i\,\mathrm{d}n_i \quad (13)$$

the Helmholtz free energy A,

$$A = U - TS = -pV + \sum_i \mu_i n_i \quad (14)$$

$$\mathrm{d}A = -S\,\mathrm{d}T - p\,\mathrm{d}V + \sum_i \mu_i\,\mathrm{d}n_i \quad (15)$$

and the Gibbs function G,

$$G = U - TS + pV = \sum_i \mu_i n_i \quad (16)$$

$$\mathrm{d}G = -S\,\mathrm{d}T + V\,\mathrm{d}p + \sum_i \mu_i\,\mathrm{d}n_i \quad (17)$$

The Legendre transformation ensures that each function adopts a minimum value at equilibrium when the corresponding independent variables are fixed.

The Maxwell relations (Callen, 1960) are obtained by noting that the order of obtaining second derivatives for any of the functions U, H, A, and G is arbitrary. Examples are

$$\frac{\partial^2 U}{\partial S \partial V} = -\left(\frac{\partial p}{\partial S}\right)_{V,n_i} = \frac{\partial^2 U}{\partial V \partial S} = \left(\frac{\partial T}{\partial V}\right)_{S,n_i} \tag{18}$$

$$\frac{\partial^2 G}{\partial T \partial n_i} = \left(\frac{\partial \mu_i}{\partial T}\right)_{p,n_j \neq n_i} = \frac{\partial^2 G}{\partial n_i \partial T} = -\left(\frac{\partial S}{\partial n_i}\right)_{T,p,n_j \neq n_i} \tag{19}$$

The last derivative of (19) is of particular importance, belonging to a class known as partial molar quantities. If X is any extensive quantity, then X_i defined by

$$X_i = \left(\frac{\partial X}{\partial n_i}\right)_{T,P,n_{j \neq i}} \tag{20}$$

is a partial molar quantity, and (Guggenheim, 1957)

$$X = \sum_i n_i X_i \tag{21}$$

According to the Gibbs–Duhem relation (11), the chemical potential of each component, at a given temperature and pressure, is a function only of the phase composition. This may be stated in mathematical form as

$$\mu_i = \mu_i^0 + RT \ln f_i x_i \tag{22}$$

where μ_i^0 is the chemical potential of the pure component at the same temperature and pressure, R is the gas constant, and x_i and f_i are respectively the mole fraction and activity coefficient of component i.

The discussion of the thermodynamics of mixing is based on the concept of an ideal solution (Guggenheim, 1957; Prigogine and Defay, 1954) defined as one in which all components satisfy (at all temperatures)

$$\mu_i = \mu_i^0 + RT \ln x_i \tag{23}$$

The change in Gibbs function $\Delta_m G$ on mixing the pure components to form one mole of mixture is given by (22) and (16) (note that the chemical potential is also the partial molar Gibbs function) as

$$\Delta_m G = \sum_i x_i RT \ln f_i x_i \tag{24}$$

which for an ideal solution gives

$$\Delta_m G(\text{ideal}) = \sum_i x_i RT \ln x_i \tag{25}$$

The excess molar Gibbs function G_m^E is defined as the difference between the actual value for a solution and the value for an equivalent ideal solution, or from equations (22)–(25),

$$G_m^E = \Delta_m G - \Delta_m G(\text{ideal}) = \sum_i x_i RT \ln f_i \tag{26}$$

The excess molar entropy S_m^E is defined in a similar fashion, and from equations (17) and (26) we obtain

$$S_m^E = -\left(\frac{\partial G_m^E}{\partial T}\right)_{p,n_i} = -RT \sum_i x_i \left(\frac{\partial \ln f_i}{\partial T}\right)_{p,n_i} - R \sum_i x_i \ln f_i \tag{27}$$

Similarly the excess molar enthalpy H_m^E is obtained as

$$H_m^E = -T^2 \left(\frac{\partial (G_m^E/T)}{\partial T}\right)_{p,n_i} = -RT^2 \sum_i x_i \left(\frac{\partial \ln f_i}{\partial T}\right)_{p,n_i} \tag{28}$$

The enthalpy of mixing is zero for an ideal mixture, and for this reason H_m^E is often called the heat of mixing.

There are several ways of proceeding in order to include surfaces. The most complete is the continuum approach in which every point in the surface region has definite properties assigned to it (Melrose, 1968). It is particularly suited to molecular models in which the molecules interact via continuous potential functions, but is too complex for our purpose.

The alternative approaches generally introduce one or more fictitious dividing surfaces at which there is a discontinuity in density of mass, energy, entropy, force, and molecular composition. The original scheme of this type is due to Gibbs (1961) and is the one used in this chapter. The interested reader should consult the original work of Gibbs or a modern account such as that by Defay *et al.* (1966). The I.U.P.A.C. manual of symbols and terminology may also be helpful (Everett, 1972).

Consider the two phases α and β meeting at a plane interface of area A_s. The surface tension γ is assumed to be related to the internal energy and the area by

$$\gamma = \left(\frac{\partial U}{\partial A_s}\right)_{S,V,n_i} \tag{29}$$

The concept of a Gibbs dividing surface (denoted by the symbol σ) is now introduced. It is an imaginary surface located within the interfacial region such that at every point the constituent matter is in the same condition. Two further imaginary surfaces s_a and s_b are now placed on either side of the dividing surface and at a distance sufficient to ensure that they lie within the homogeneous regions of the system. The arrangement is shown in Fig. 1. The surfaces s_a and s_b divide the system into three parts, the regions a and b containing only phase α and β respectively, the whole of the interface lying within the central region s. If the positions of s_a, s_b, and the external boundaries are fixed (A_s is then determined), any variation

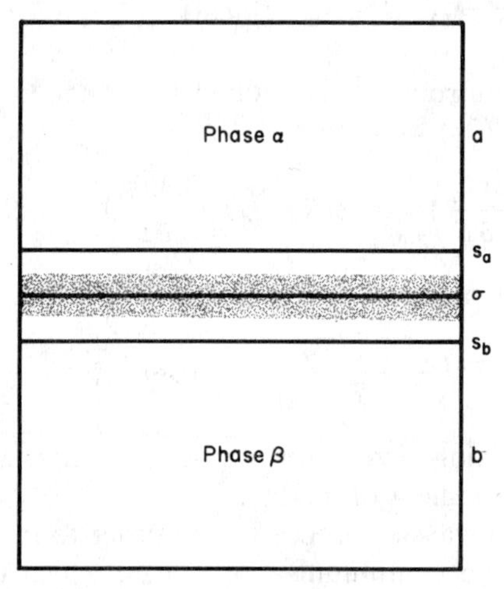

Figure 1 The Gibbs dividing surface σ, lying within the interfacial region, and the surfaces s_a and s_b, parallel to σ and lying within the two bulk phases.

in the internal energy of the three regions is given by equations (5) and (29) as

$$dU^a = T^a \, dS^a + \sum_i \mu_i^a \, dn_i^a \tag{30}$$

$$dU^b = T^b \, dS^b + \sum_i \mu_i^b \, dn_i^b \tag{31}$$

$$dU^s = T^s \, dS^s + \sum_i \mu_i^s \, dn_i^s \tag{32}$$

If the total entropy and matter for the whole system is fixed, but interchange between regions is allowed, then

$$dS^a + dS^b + dS^s = 0 \tag{33}$$

$$dn_i^a + dn_i^b + dn_i^s = 0 \tag{34}$$

The condition (6) specifies that, at equilibrium,

$$dU^a + dU^b + dU^s = 0 \tag{35}$$

and therefore the relations (30)–(34) are satisfied only if

$$T = T^{a} = T^{b} = T^{s} \tag{36}$$

$$\mu_i = \mu_i^{a} = \mu_i^{b} = \mu_i^{s} \tag{37}$$

The result (37) is particularly important in the development of models of an interface. That the chemical potential of each species in the interface is the same as it is in any of the bulk phases (of which the species is a component) follows directly from the definition of the chemical potential as

$$\mu_i = \left(\frac{\partial U}{\partial n_i}\right)_{S,V,A_s,n_j \neq n_i} \tag{38}$$

Alternative definitions of the chemical potential lead to relations different from (37).

We now finish with the auxiliary surfaces s_a and s_b and return to the discussion of the Gibbs surface σ. The definition given earlier is incomplete since there are an infinite number of imaginary surfaces that may be placed within the interfacial region and satisfy the condition that at every point the constituent matter is in the same condition.

Any choice of Gibbs surface defines two volumes V^{α} and V^{β} such that

$$V = V^{\alpha} + V^{\beta} \tag{39}$$

the volume V^{α} being on the same side of the dividing surface as the bulk phase α. The surface excess Ω^{σ} of any extensive property Ω of the system is defined by

$$\Omega^{\sigma} = \Omega - V^{\alpha}\omega^{\alpha} - V^{\beta}\omega^{\beta} \tag{40}$$

where ω^{α} and ω^{β} are the densities of Ω in the homogeneous phases α and β respectively. In particular, the surface excess of component i is given by

$$n_i^{\sigma} = n_i - V^{\alpha}c_i^{\alpha} - V^{\beta}c_i^{\beta} \tag{41}$$

where c_i^{α} and c_i^{β} are the respective concentrations (mol/vol) in phases α and β. The adsorption Γ_i^{σ} of component i is defined by

$$\Gamma_i^{\sigma} = n_i^{\sigma}/A_s \tag{42}$$

It is common practice to fix the position of the Gibbs surface by specifying the adsorption, and two common choices are

$$\Gamma_1^{\sigma} = 0 \tag{43}$$

$$\sum_i \Gamma_i^{\sigma} = 0 \tag{44}$$

For the choice (43) the adsorption of components other than 1 is called the relative adsorption (denoted by $\Gamma_i^{(1)}$), and for (44) it is termed the reduced adsorption ($\Gamma_i^{(n)}$) for all components. It is possible to evaluate the relative or reduced adsorption from the data for any arbitrary position of the Gibbs surface (Everett, 1972). For a curved interface, the dividing surface is usually chosen so that equation (29) still applies (the area of a curved surface is dependent upon the position of the dividing surface), and this fixes the surface excess quantities n_i^σ and hence the Γ_i^σ. This particular Gibbs surface is generally called the surface of tension.

A lower-case symbol is generally used to describe a surface excess quantity per unit area, i.e.

$$\omega^\sigma = \Omega^\sigma / A_s \tag{45}$$

An exception is the adsorption where the symbol Γ_i^σ is used (the extensive property already has a lower-case symbol n_i^σ).

The results (36) and (37) do not include a description of mechanical equilibrium. It is sufficient at this stage to state, for a plane interface at equilibrium, that

$$p = p^\alpha = p^\beta \tag{46}$$

By adding equations (30)–(32), and using equations (29), (36), (37), and (46), the variation in internal energy for the whole system is given as

$$\mathrm{d}U = T\,\mathrm{d}S - p\,\mathrm{d}V + \gamma\,\mathrm{d}A_s + \sum_i \mu_i\,\mathrm{d}n_i \tag{47}$$

The definition of surface excess quantities (40) leads to

$$\mathrm{d}U^\sigma = T\,\mathrm{d}S^\sigma + \gamma\,\mathrm{d}A_s + \sum_i \mu_i\,\mathrm{d}n_i^\sigma \tag{48}$$

The absence of any volume term in equation (48) arises as a result of combining (39) and (40), making $V^\sigma = 0$ for all dividing surfaces. Euler's theorem may be applied to (48), giving

$$U^\sigma = TS^\sigma + \gamma A_s + \sum_i \mu_i n_i^\sigma \tag{49}$$

Differentiation of (49), combining with (48), dividing by the area, and rearranging gives

$$\mathrm{d}\gamma = -s^\sigma\,\mathrm{d}T - \sum_i \Gamma_i^\sigma\,\mathrm{d}\mu_i \tag{50}$$

Equation (50) is known as the Gibbs adsorption equation and is the surface analogue of the Gibbs–Duhem relation (11).

If one phase is solid there is a possibility of residual stresses in the solid phase near the interface. It is then better to regard γ as an energy per unit area, and from (49) it may be written as

$$\gamma = u^\sigma - Ts^\sigma - \sum_i \mu_i \Gamma_i^\sigma \tag{51}$$

It is possible to define other surface excess energy functions. This has often been a source of controversy (Everett, 1950) and we require only the surface excess Helmholtz free energy A^σ defined (Everett, 1972) by

$$A^\sigma = U^\sigma - TS^\sigma = \gamma A_s + \sum_i \mu_i n_i^\sigma \tag{52}$$

Differentiation of (52) and combination with (48) gives

$$\mathrm{d}A^\sigma = -S^\sigma \,\mathrm{d}T + \gamma \,\mathrm{d}A_s + \sum_i \mu_i \,\mathrm{d}n_i^\sigma \tag{53}$$

Again, Maxwell relations can be obtained from second derivatives of the energy functions, e.g.

$$\frac{\partial^2 A^\sigma}{\partial A_s \partial T} = \frac{\partial \gamma}{\partial T} = \frac{\partial S^\sigma}{\partial A_s} = s^\sigma \tag{54}$$

Equations (47)–(54) apply to any choice of position of the dividing surface for a plane interface.

III. Experimental Methods

Adsorption at the solid/solution interface is measured as a change in concentration of the solution on contact with the solid. The solution phase α has n^0 moles of solution with initial mole fraction $x_i(\text{in})$ before contact with the solid, so the amount of each substance is

$$n_i = n^0 x_i(\text{in}) \tag{55}$$

After contact with the solid the final mole fractions are $x_i(\text{fin})$. A Gibbs surface is chosen so that the volume V^α attributed to the solution phase still contains n^0 moles and therefore the amount of each substance n_i^α in this hypothetical volume is

$$n_i^\alpha = n^0 x_i(\text{fin}) \tag{56}$$

Assuming that none of the solution components enters the solid phase, the surface excess is given by (40) as

$$n_i^\sigma = n^0[x_i(\text{in}) - x_i(\text{fin})] = n^0 \Delta x_i \tag{57}$$

where Δx_i is the change of concentration on contact with the solid. If the surface area is A_s the adsorption is given by (42) and (44) as

$$\Gamma_i^{(n)} = n^0 \Delta x_i / A_s \tag{58}$$

and

$$\sum_i \Gamma_i^{(n)} = n^0 \left(\sum_i \Delta x_i \right) / A_s = 0 \tag{59}$$

The surface area of the solid is not always known, so the adsorption is often reported as $n^0 \Delta x_i / m$ where m is the mass of the solid.

The traditional experimental procedure is to seal the adsorbent and solution in a clean glass tube which is then shaken or rotated in a thermostat. After equilibration, a sample of the solution is removed and analysed. In earlier experiments the sealing was accomplished by freezing the solution at one end of the tube and sealing the other end with a gas flame. This technique is not recommended since trapped vapours are released from the heated glass and condense on the frozen solution with consequent contamination. Also, the tubes may explode as a result of stresses introduced into the glass by the sealing process. It is better to use tubes with a screw cap and teflon seal.

Any accurate method of measuring the change of concentration is suitable. Refractometric methods are popular since the concentration changes are small and the solutions may not be dilute. In practice refractometry is restricted to two-component solutions although in principle very accurate measurements at several wavelengths might distinguish composition changes for more components. Traces of impurity with a significantly different refractive index from either of the major components can introduce serious errors.

If the difference in refractive index between the two pure components is large, the composition of any mixture can be obtained directly from the measurement of refractive index with a Pulfrich refractometer. It is essential that monochromatic light be used and that the temperature be the same for the unknown and calibrating solutions.

Equation (57) indicates that concentration differences are required rather than concentrations. Thus a differential refractometer is particularly useful since samples of the solution, before and after exposure to the adsorbent, can be compared directly.

Rayleigh interferometers also measure, both directly and accurately, the refractive index difference between solutions. They are not entirely suited for use with solutions in which the two components have a markedly different optical dispersion, or variation of refractive index with wavelength (Candler, 1951; Adams, 1915). Mixtures of benzene with saturated hydrocarbons provide examples of this behaviour.

There is a simple way of overcoming this dispersion problem, if the adsorbent equilibrates and settles rapidly on contact with the solution, as for example graphitized carbon black in hydrocarbon mixtures. The twin compartments of the interferometer cell are filled with the original solution of components 1 and 2. The cell is placed in the interferometer and the position of the interference band is noted. A small weighed amount of adsorbent is then added to one of the compartments (it may be necessary to add a fixed prism of rectangular cross-section to the light path to divert the beam so that it passes through the solution above the settled adsorbent). The solid will reduce the concentration of one component (say 2). Using a precision microlitre syringe, pure component 2 is added until the interference pattern is returned to its original position. If the area of added solid is A_s, and n_2^+ moles of component 2 were needed, the adsorption is calculated from (40) and (43) as

$$\Gamma_1^{(1)} = 0 \qquad \Gamma_2^{(1)} = n_2^+/A_s \tag{60}$$

The added liquid is a direct measure of the surface excess of component 2 with respect to a dividing surface placed so that the surface excess of component 1 is zero. It can be related (Everett, 1972) to the reduced adsorption, as in equation (44), by

$$\Gamma_1^{(n)} = -(n_2^+/A_s)x_1 \qquad \Gamma_2^{(n)} = (n_2^+/A_s)x_1 \tag{61}$$

This procedure also removes the necessity of calibrating the interferometer (it is merely an indicator for the end-point of a titration). The calibration of a Rayleigh interferometer can be both tedious and prone to error.

More recently, Ash *et al.* (1973) have described an apparatus capable of high precision and rapidity for measurements over a range of temperatures. It consists of two closed loops, A and B, each loop being connected to one compartment of a differential refractometer cell thermostatted at a temperature T_c. One loop (B) has a section S (which can be by-passed) that holds the solid adsorbent, and this section is thermostatted at a temperature T_s. At the start of a run loops A and B (including S) are evacuated. The section S is now isolated, the solution is introduced into loops A and B, and the zero of the differential refractometer is checked. The solid is now exposed to the solution in loop B and the change in refractive index is monitored. The solid temperature T_s may now be changed and the variation of refractive index measured (note that T_c is held constant throughout). At the conclusion of the run the differential refractometer is calibrated, by injecting into loop A small quantities of the non-preferentially adsorbed component. The solution is removed from the apparatus and the procedure is repeated with a solution of different composition. The same adsorbent sample is used for all mixtures of the two liquid components. Kurbanbekov *et al.* (1969) have described a similar apparatus.

IV. Types of Isotherm Observed Experimentally

Experimental isotherms are generally reported in the form of surface excess isotherms showing the quantity $n^0\Delta x_i/m$ as a function of liquid mole fraction x_i. Such curves are referred to as composite isotherms, although this terminology is not recommended by I.U.P.A.C. (Everett, 1972). Schay and Nagy (1966) have classified these isotherms into five types as shown in Fig. 2. This classification is more than adequate to describe the observed experimental data. Indeed two groups would be sufficient, the first three types in Fig. 2 belonging to the class in which the quantity $n^0\Delta x/m$ always has the same sign and the remaining two types belonging to the class in which the quantity changes sign.

Figure 3 shows the experimental data of Ash *et al.* (1973) for the system benzene/n-heptane/Graphon at 283.15 K and 343.15 K. Thus the same

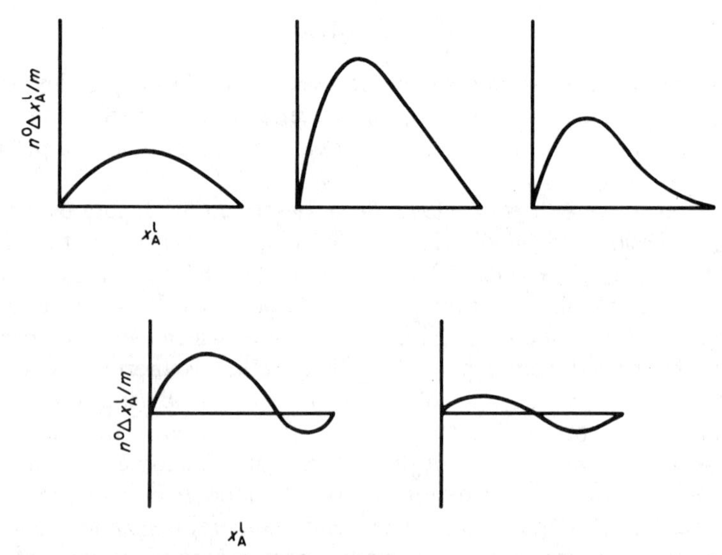

Figure 2 The Schay and Nagy (1966) classification of composite adsorption isotherms.

system, but at two different temperatures, exhibits both of the major classes of adsorption behaviour. These high quality measurements were made using the double-loop technique described in the preceding section.

For a two-component system, equation (59) indicates that

$$n_2^\sigma = n^0 \Delta x_2 = -n^0 \Delta x_1 = -n_1^\sigma \quad (62)$$

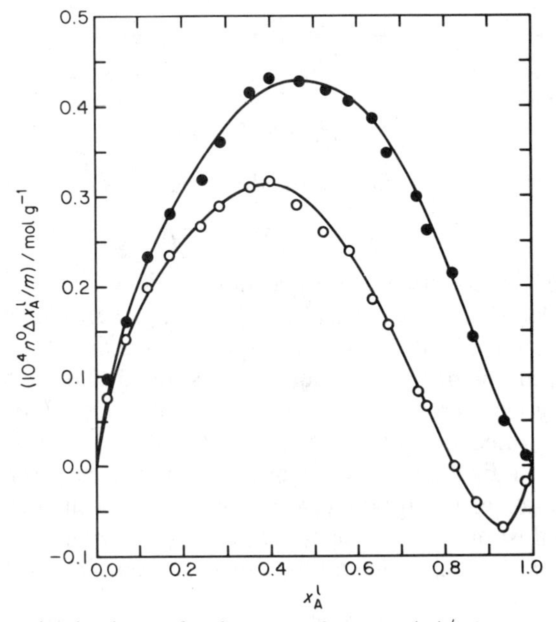

Figure 3 Adsorption isotherms for the system benzene (A)/n-heptane (B)/Graphon: ○ 283.15 K; ● 343.15 K. (Experimental data from Ash *et al.*, 1973)

so the surface excess isotherm for component 2 is just the negative of the isotherm for component 1. However, if it is assumed (i) that the adsorption is monomolecular, (ii) that the area per mole a_i of component i in the adsorption layer is independent of liquid concentration, and (iii) that the solution components are not found in the solid, it is possible to evaluate the amounts n_i^s of the two components in the adsorption layer on m grams of solid. If $(A_s)_m$ is the area of m grams of solid and $(n_i^s)_m$ is the amount of component i in the adsorption layer when the liquid consists only of i, then

$$(A_s)_m = (n_1^s)_m a_1 = (n_2^s)_m a_2 \quad (63)$$

and by assumption (ii) above,

$$(A_s)_m = n_1^s a_1 + n_2^s a_2 \tag{64}$$

or by combining (63) with (64),

$$1 = n_1^s/(n_1^s)_m + n_2^s/(n_2^s)_m \tag{65}$$

If there are n^0 moles of solution of composition x_i^0 before contact with the solid, and n^1 moles of solution (excluding the adsorbed layer) of composition x_i after contact, then by definition

$$n_1^s = (n^0/m)x_1^0 - (n^1/m)x_1 \tag{66}$$

and

$$n^1 = n^0 - mn_1^s - mn_2^s \tag{67}$$

These equations may be combined and on rearrangement give

$$n^0\Delta x_1/m = x_2 n_1^s - x_1 n_2^s \tag{68}$$

Equations (65) and (68) provide two independent relations that can be solved provided that $n^0\Delta x_1/m$, x, $(n_1^s)_m$, and $(n_2^s)_m$ are known. The latter two quantities refer to the solid/pure liquid interface and are inaccessible to direct measurement, but it is generally assumed that they are equal to corresponding values for the solid/pure vapour interface. Plots of n_i^s versus x_i are often termed individual isotherms. Because of the many approximations used, individual isotherms are not reliable but they may help in the interpretation of adsorption from solution. Figure 4 shows the composite and individual isotherms for the system ethanol/p-xylene/rutile at 25°C as calculated from the experimental data of Parfitt and Wiltshire (1964).

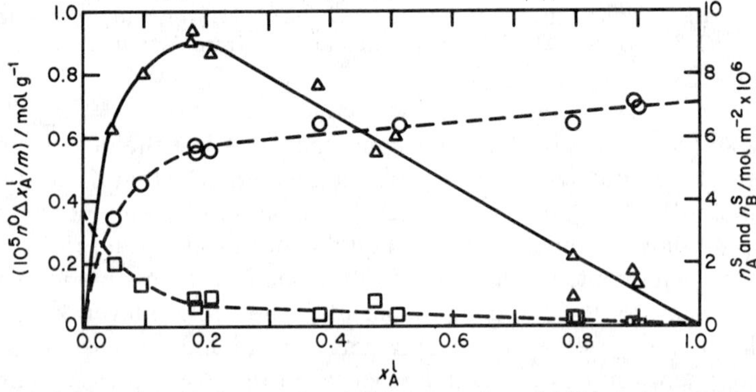

Figure 4 Adsorption isotherms for the system ethanol (A)/p-xylene (B)/rutile at 298.15 K; △ composite isotherm; ○ individual isotherm for ethanol; □ individual isotherm for p-xylene. (Experimental data of Parfitt and Wiltshire, 1964)

V. Lattice and Related Models of Liquid Mixtures and of Adsorbed Layers

Most of the available models of the solid/solution interface were developed at a time when the properties of liquids at the molecular level were poorly understood. Consequently, the best that can be expected of them is that they predict the qualitative behaviour of the interface, and with suitable adjustment of the parameters a reasonable fit with experimental data can be obtained. The problem is that the adjusted parameters lose their identity.

These models are generally based on either a lattice model of the surface and of the solution phase, or a homogeneous "surface phase" without any structural detail and in equilibrium with a structureless liquid. The former are discussed in this section, the latter in the following section.

Before proceeding with the lattice models, a digression on Monte Carlo (MC) simulation is in order. It is fair to say that MC simulations have made a major contribution to our understanding of liquids (Barker and Henderson, 1976). The MC procedure has been described many times but the review by Wood (1968) is possibly the most useful.

The Monte Carlo recipe requires only that for any given arrangement of molecules (configuration) of the system the total internal energy be known or be calculable. A modification of the configuration is then made (strict rules apply to this procedure) and the difference in energy between the new and old configuration is ascertained. This energy difference is then used in a test to decide whether the new configuration will be accepted or the old retained. Further trial configurations are then made and the string of accepted configurations is said to form a Markov chain. Concurrent with the process of forming the Markov chain, the required mechanical properties of each accepted configuration are evaluated. The chain is terminated when the averages (over all accepted configurations) of the mechanical properties are considered to be independent of the arbitrary, initial configuration.

The important feature of the MC procedure is that the resultant average properties are exact (in a thermodynamic sense) for the prescription used to evaluate the configurational energy. In principle there is no limit to the complexity of the system to be studied but in practice computer time and costs fix a limit. At the time of writing, computing costs have severely restricted the number of MC simulations of the solid/liquid interface, and to the author's knowledge there is none in which the liquid contains more than one component. However, these MC simulations of the solid/liquid interface are relevant to models of the solid/solution interface.

In order to realize an MC simulation it is necessary to be able to evaluate the internal energy for a given configuration of the system. If the system is stationary the internal energy may be partitioned in the following manner:

(i) the potential energy of interaction with a gravitational or other external field; (ii) the kinetic energy of thermal motion of the molecules; (iii) the potential energy of interaction between the molecules. The average energy of the first two contributions can be obtained directly. The third is a combination of a repulsive term at small intermolecular separations, the ubiquitous dispersion energy (always attractive), and the electrostatic contribution for the permanent electric moments (charges, dipoles, etc.). If the molecules have any permanent electric moment, the electrostatic interaction energy is readily evaluated (Bötcher, 1973). The *a priori* evaluation of the combined repulsive and dispersion energies is practical only for a system consisting of simple molecules. Fortunately, a good approximation to this combined energy term is obtained as a sum of pairwise interactions between all pairs of molecules within the system. Representations of the pairwise intermolecular potential functions for rare-gas atom interactions are now known with considerable accuracy. It is also known that the simple Lennard-Jones (LJ) potential is a reasonable representation of this combined interaction. The LJ potential is defined by

$$u(r) = 4\varepsilon[(\sigma/r)^{12} - (\sigma/r)^6] \tag{69}$$

where $u(r)$ is the energy of interaction of two particles separated by a distance r, and ε and σ are adjustable parameters. Most MC simulations use an LJ potential to evaluate the combined repulsive and dispersion force contribution to the intermolecular energy.

In addition to the macroscopic properties such as pressure, density, compressibility, and energy, certain microscopic (molecular) information can be extracted from an MC simulation. The particle distribution functions are examples. The singlet distribution function $\rho_i^{(1)}(\mathbf{r}_1)$ is defined as the average number density of component i at the point in space described by the vector $\mathbf{r}_1$ ($\mathbf{r}_1$ consists of the elements x_1, y_1, z_1, the coordinates of point 1 with respect to the rectangular cartesian axes $OXYZ$). The pair distribution function $\rho_{ij}^{(2)}(\mathbf{r}_1, \mathbf{r}_2)$ is the average number density for pairs of particles in which species i is at the point $\mathbf{r}_1$ and species j at $\mathbf{r}_2$. The radial distribution function $g_{ij}^{(2)}(\mathbf{r}_1, \mathbf{r}_2)$ is defined by

$$\rho_{ij}^{(2)}(\mathbf{r}_1, \mathbf{r}_2) = \rho_i^{(1)}(\mathbf{r}_1)\, \rho_j^{(1)}(\mathbf{r}_2)\, g_{ij}^{(2)}(\mathbf{r}_1, \mathbf{r}_2) \tag{70}$$

The radial distribution function (RDF) has a simple physical interpretation; if we consider a particle of component i at $\mathbf{r}_1$, then $g_{ij}^{(2)}(\mathbf{r}_1, \mathbf{r}_2)$ is the ratio of the actual probability of finding a particle of component j at $\mathbf{r}_2$ to the probability of finding a particle of component j at $\mathbf{r}_2$ regardless of the occupancy of $\mathbf{r}_1$. Thus the RDF is a measure of the correlation between pairs of particles. Particle distribution functions of higher order can be defined but the additional complexity is seldom matched by an increased understanding of molecular structure. Indeed, if the singlet and radial distribution functions are known, the intermolecular interactions are restricted to pairs of molecules, and the form of this interaction is known, then all the mechanical properties of the system can be evaluated directly (Rice, 1967).

In a bulk phase the particle distribution functions are independent of spatial position, but not of the thermodynamic state. Figure 5 shows the radial distribution functions from MC simulations for three bulk phases all at the same temperature (131.74 K), all interacting via the same LJ potential (appropriate to argon, $\varepsilon/k_B = 119.76T$, where k_B is the Boltzmann constant,

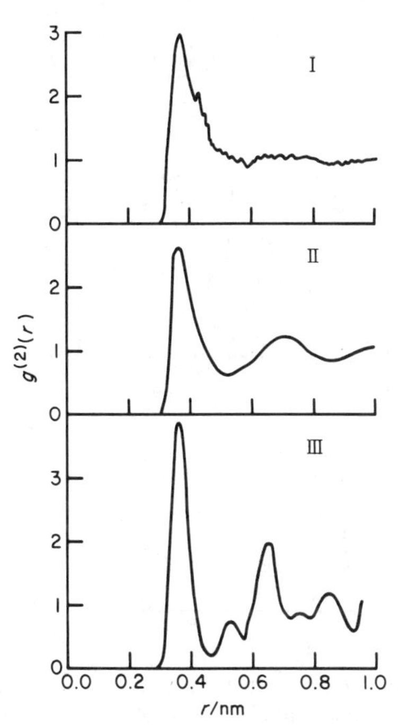

Figure 5 Monte Carlo simulation of the radial distribution function $g^{(2)}(\mathbf{r})$ for bulk argon at 131.74 K: (I) gas phase at 11.8 atm (1 atm = 101 325 Pa); (II) liquid phase at 769 atm; (III) crystal phase at 3140 atm.

$\sigma = 0.3405$ nm), but at three different pressures (11.8, 769.0, and 3140 atm). The RDF I is appropriate to a gas phase and the single peak suggests strong clustering in pairs. The RDF II shows a series of peaks and is typical of a liquid phase where there is some short-range order. The RDF III describes the very strong ordering found in a face-centred cubic structure (the order extends far beyond the cut-off distance of the figure). These figures clearly show the type and extent of molecular correlations in different bulk phases and are important guides in the description of structure at a solid/fluid interface.

Figure 6(a) shows the singlet distribution function for a MC simulation of liquid argon in contact with solid carbon dioxide at 131.74 K and 769 atm. The coordinate z shows the distance from the bounding surface of the

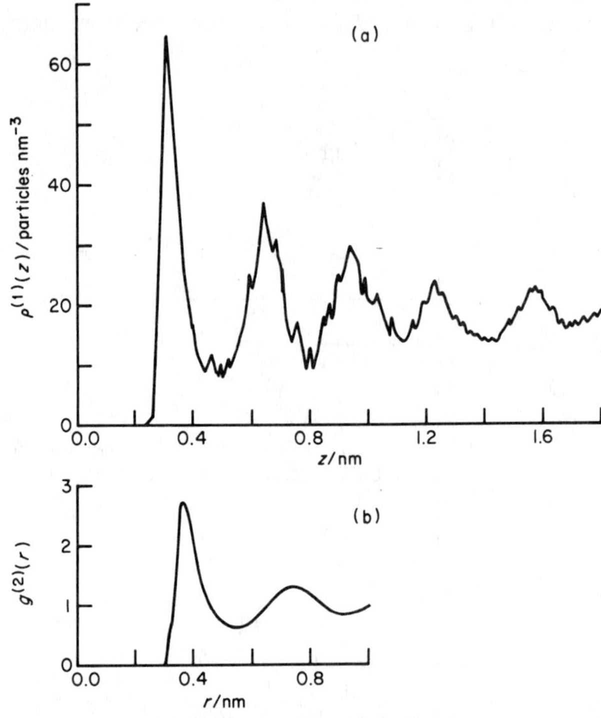

Figure 6 (a) Singlet distribution function $\rho^{(1)}(\mathbf{z})$ for argon, from a Monte Carlo simulation of a solid carbon dioxide/liquid argon interface at 131.74 K and 769 atm. The boundary of the solid is at $z = 0$. (b) Radial distribution function $g^{(2)}(\mathbf{r})$ for argon in the layer adjacent to the solid [first peak in Fig. 6(a)].

solid. At least five peaks can be seen (ignore the MC "noise") which indicate the presence of five layers of argon in the interface. Note that the layers are not well defined, since the density does not drop to zero between layers. Figure 6(b) shows the RDF for pairs of particles whose centres are both in the first layer adjacent to the solid. Comparison of this RDF with RDF II in Fig. 5 shows that the liquid particles in the first layer are arranged in a liquid-like structure.

Figure 7(a) and (b) have the same information as Fig. 6(a) and (b) except that the solid phase is now graphite. The interaction between argon and graphite is much stronger than the interaction of argon and carbon dioxide. This is evident by the much sharper definition of the layers; the density is zero between the first and second. The RDF for particles in the first layer [Fig. 7(a)] indicates a crystalline structure. (The RDF in Fig. 7(b) differs in detail from the RDF III in Fig. 5 because the packing in an essentially two-dimensional layer differs from that in a three-dimensional crystal.) The RDF for particles in the second layer (not shown) is midway between that appropriate for a liquid and a close-packed layer (Lane and Spurling, 1981). In support of the reliability of these observations we note

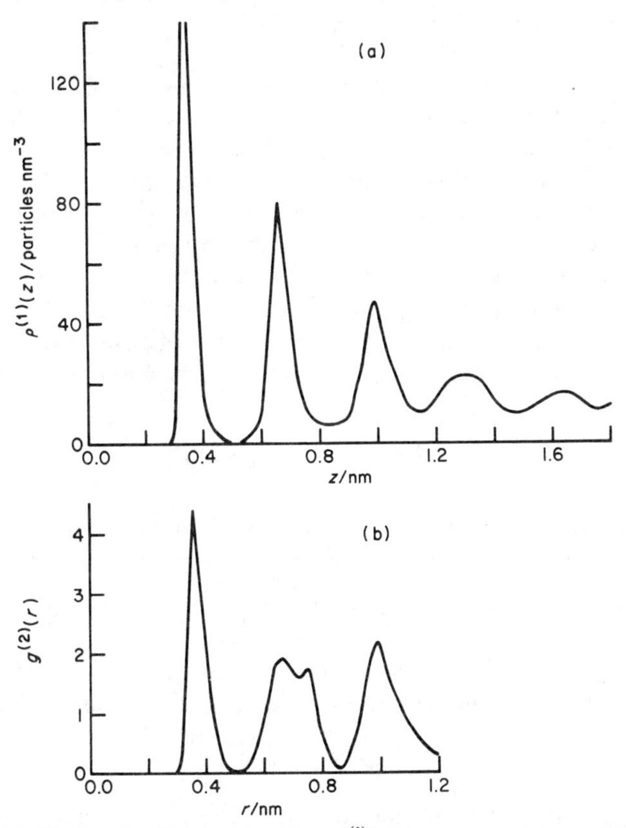

Figure 7 (a) Singlet distribution function $\rho^{(1)}(\mathbf{z})$ for argon, from a Monte Carlo simulation of a graphite/liquid argon interface at 131.74 K and 769 atm. The centres of the carbon atoms in the outermost solid layer are in a plane at $z = 0$. (b) Radial distribution function $g^{(2)}(\mathbf{r})$ for argon in the layer adjacent to the solid [first peak in Fig. 7(a)].

that earlier MC simulations (Lane and Spurling, 1976, 1978) of the graphite/krypton vapour interface produced adsorption isotherms in close agreement with the experimental isotherms of Thomy and Duval (1970), without any need to fit parameters to the experimental adsorption data.

Results of MC simulations, similar to those shown in Fig. 6, have been reported by Snook and Henderson (1978) and Snook and van Megen (1979).

The MC simulations have given some insight into the structure of a solid/liquid interface in which the constituent molecules interact via simple but realistic potentials. It is evident that a lattice model of an interface receives some support from the MC simulations but will err in being too orderly. It is unfortunate that the lattice model of an interface generally requires the simultaneous use of a lattice model of the liquid phase. Almost by definition, the lattice model of a liquid must be a contradiction. Nevertheless, a lattice model of the solid/solution interface is the simplest model available that includes intermolecular interactions specifically.

In a lattice model of a bulk liquid, each lattice site can be either vacant, or occupied by a single entity which may be an atom, molecule, or segment of a polymer. For simplicity, this discussion is restricted to two entities, molecules of type A or B, each molecule occupying a single site with vacancies being excluded. In addition, the intermolecular interaction is restricted to nearest neighbour molecules. The description is brief; a more complete account has been given by Guggenheim (1952, 1966).

If a macroscopic system can exist in any one of r states (microscopic), each with a distinctive energy E_i, the partition function Q is defined by

$$Q = \sum_{i=1}^{r} \exp(-E_i/k_B T) \tag{71}$$

where k_B is the Boltzmann constant. The fundamental postulate of statistical thermodynamics is that the Helmholtz free energy is given by

$$A = -k_B T \ln Q \tag{72}$$

Following Guggenheim (1952) the partition function is now split into terms appropriate to the independent degrees of freedom, giving

$$Q = Q_{int} Q_{tr} \tag{73}$$

where Q_{int} includes all the internal degrees of freedom and molecular rotations, and Q_{tr} the combined contribution of the kinetic energy of vibration and the intermolecular interactions. In a mixture, Q_{int} is a function only of the numbers of molecules of each species, whereas Q_{tr} is also a function of the relative proportions of each species.

It is the essence of the simple lattice model that the translational partition function is separable into contributions from each pair of nearest neighbour molecules on the lattice. Each molecule has z_1 nearest neighbours; $z_1 = 6$ for a simple cubic lattice, $z_1 = 8$ for a body-centred cubic lattice, and $z_1 = 12$ for a face-centred or hexagonal close-packed lattice. Thus a lattice of pure component containing N_A molecules will have $N_{AA} = \frac{1}{2}z_1 N_A$ nearest neighbour pairs, and the partition function may be written as

$$Q(N_A) = Q_{int}(N_A)\,(q_{AA}^{2/z_1})^{N_A} \tag{74}$$

where $Q_{int}(N_A)$ is the internal partition function for N_A molecules and q_{AA}^{2/z_1} is the contribution of each AA nearest neighbour pair. Similarly, for N_B molecules of pure component B,

$$Q(N_B) = Q_{int}(N_B)\,(q_{BB}^{2/z_1})^{N_B} \tag{75}$$

If the components are now mixed, some AB nearest neighbour pairs will form at the expense of AA and BB pairs, and a new term q_{AB}^{2/z_1} must be introduced. If, for one specified arrangement of molecules on the lattice, the number of AB pairs is given by

$$N_{AB} = z_1 X \tag{76}$$

then a simple count requires that

$$N_{AA} = \tfrac{1}{2}z_1(N_A - X) \tag{77}$$

$$N_{BB} = \tfrac{1}{2}z_1(N_B - X) \tag{78}$$

The contribution Q_p of this particular configuration to the partition function is then given by

$$Q_p = Q_{int}(N_A)\,Q_{int}(N_B)\,(q_{AA}^{2/z_1})^{N_{AA}}\,(q_{BB}^{2/z_1})^{N_{BB}}\,(q_{AB}^{2/z_1})^{N_{AB}} \tag{79}$$

On substitution of equations (74)–(78) and rearrangement, from (79) we obtain

$$Q_p = Q(N_A)\,Q(N_B)\exp(-Xw/k_BT) \tag{80}$$

w being the interchange free energy defined by

$$w = -k_BT\ln[q_{AB}/(q_{AA}q_{BB})^{1/2}] \tag{81}$$

The first two terms of the product on the right-hand side of equation (80) may be recognized as the respective partition functions for the unmixed components. If there are $g(N_A, N_B, X)$ distinguishable configurations having the same value of X, combination of equations (80) and (71) gives the partition function for the mixture as

$$Q(N_A, N_B) = \sum_X g(N_A, N_B, X)\,Q(N_A)\,Q(N_B)\exp(-Xw/k_BT) \tag{82}$$

If N_A and N_B are large, it has been shown (Guggenheim, 1952, 1957) that very little error is introduced if the sum in equation (81) is replaced by the largest term. By using this information, the Helmholtz free energy of mixing ΔA is given by (81), (74), and (71) as

$$\Delta A = -k_B T \ln g(N_A, N_B, X^*) + X^* w \tag{83}$$

where X^* is the value of X in the largest term.

Unfortunately, it is necessary to approximate for $g(N_A, N_B, X^*)$ and X^*. The Bragg–Williams approximation is the simplest and assumes completely random mixing, so that

$$g(N_A, N_B, X^*) = N!/(N_A! + N_B!) \tag{84}$$

$$X^* = N_A N_B / N = N x_A x_B \tag{85}$$

and

$$N = N_A + N_B \tag{86}$$

Since the lattice dimensions are fixed, there is no volume change on mixing, so by the definition of the Gibbs function in equation (16),

$$\Delta G = \Delta A \tag{87}$$

Combining equations (83) and (87) with the Bragg–Williams approximations, and using Stirling's approximation ($\ln N! = N \ln N - N$, for large N), ΔG is given by

$$\Delta G = N k_B T (x_A \ln x_A + x_B \ln x_B) + N x_A x_B w \tag{88}$$

Taking $N = L$ (the Avogadro constant) and noting that $L k_B = R$ (the gas constant), the molar Gibbs function of mixing is obtained from (88) as

$$\Delta_M G = RT(x_A \ln x_A + x_B \ln x_B) + L x_A x_B w \tag{89}$$

By combining (89) with (25) and (26), the excess molar Gibbs function is given by

$$G_m^E = L x_A x_B w \tag{90}$$

The molar excess entropy and enthalpy are obtained from (90) through equations (27) and (28). Equation (90) is used to obtain w by fitting to experimental values of G_m^E as a function of composition. For an ideal solution $G_m^E = 0$, so equation (90) requires that $w = 0$ be a necessary condition for ideality.

The total Gibbs function of the solution is given by definition as

$$G = N_A \mu_A^0 + N_A \mu_B^0 + \Delta G \tag{91}$$

Combining equations (88) and (91) and differentiating with respect to N_A gives the chemical potential of A as

$$\mu_A = \mu_A^0 + k_B T \ln x_A + (1 - x_A)^2 w \tag{92}$$

Similarly, the chemical potential for B is

$$\mu_B = \mu_B^0 + k_B T \ln x_B + (1 - x_B)^2 w \tag{93}$$

Comparison of equations (92) and (93) with (22) gives the activity coefficient as

$$f_i = \exp[(1 - x_i)^2 w / k_B T] \tag{94}$$

and indicates clearly that positive values of w lead to positive deviations from Raoult's law ($f_i > 1$) and negative w to negative deviations ($f_i < 1$). Usually, positive values of w correspond to endothermic and negative w to exothermic mixing of A with B. There can be exceptions to this rule for heats of mixing.

The lattice model of the solid/solution interface consists of N molecules of solution in r layers, the rth layer being in contact with a solution reservoir of mole fraction x_A. Any molecule not in the first layer has $\boldsymbol{l} z_1$ nearest neighbours in the same layer and $\boldsymbol{m} z_1$ nearest neighbours in each of the adjacent layers, and hence

$$\boldsymbol{l} + 2\boldsymbol{m} = 1 \tag{95}$$

In the first layer each molecule has $\boldsymbol{l}$ near neighbours in the first layer, $\boldsymbol{m}$ in the second layer, and a single contact with the solid. The partition functions q_{AS} and q_{BS} respectively describe the interactions of an A or B particle with the solid.

There are no restrictions on the compositions $x_{A,1}, x_{A,2}, \ldots, x_{A,r}$ in the r layers other than the condition that equilibrium exist between the layers and with the bulk solution reservoir where the composition x_A fixes the chemical potentials μ_A and μ_B. Because the layer concentrations are not fixed, it is necessary to introduce the grand partition function Ξ, defined for this system by

$$\Xi = \sum_{x_{A,1}} \ldots \sum_{x_{A,r}} Q(N/r, x_{A,i=1,r}) \times \exp\left(N/r \sum_i x_{A,i} \mu_A \right) \exp\left(N/r \sum_i x_{B,i} \mu_B \right) \tag{96}$$

The equivalent expression to (72) relates Ξ to the thermodynamic properties (Guggenheim, 1952) by

$$\gamma A_s - pV = -k_B T \ln \Xi \tag{97}$$

The partition function is given (Lane, 1968) by

$$Q(N/r, x_{A,i=1,r}) = \sum_X g(N/r, x_{A,i=1,r}, X) Q_p \tag{98}$$

with

$$Q_p = Q(N_A)\, Q(N_B) \exp(-wX/k_BT) q_{BS}^{N/r} \times \exp[-(N/r) x_{A,1} v/k_BT] \tag{99}$$

and

$$v = -k_BT \ln[(q_{AS}/q_{BS})(q_{BB}/q_{AA})^m] \tag{100}$$

Again, the largest term dominates the multiple sum in (96) and the other terms can be ignored (Guggenheim, 1957). This term can be evaluated by maximizing Ξ with respect to the $x_{A,i}$, i.e. by simultaneous solution of the r partial differential equations

$$\partial \Xi / \partial x_{A,i} = 0 \tag{101}$$

In the Bragg–Williams approximation of completely random mixing within layers,

$$g(N/r, x_{A,i=1,r}, X^*) = \prod_{i=1}^{r} \frac{(N/r)!}{[x_{A,i}(N/r)]![x_{B,i}(N/r)]!} \tag{102}$$

where Π indicates the product over all terms. The maximum value of X is given by

$$X^* = (N/r) \sum_{i=1}^{r} \sum_{j=i-1}^{i+1} a_j x_{A,i} x_{B,j} \tag{103}$$

where

$$\left. \begin{aligned} a_0 &= 0 \\ a_i &= \boldsymbol{l} \\ a_{i-1} = a_{i+1} &= \boldsymbol{m} \end{aligned} \right\} \tag{104}$$

Note that X defined here differs by a factor $\frac{1}{2}z_1$ from the X defined by Lane (1968).

By substituting (102)–(104) into (99) and (98) the partition function $Q(N/r, x_{A,i=1,r})$ is obtained. This is then substituted into (96), together with the expressions (92) and (93) for the chemical potentials, and the grand partition function is determined according to the condition (101).

The result of this process is the second-order difference equation

$$\frac{w}{k_{\mathrm{B}}T}\left[\sum_{j=i-1}^{i+1} a_j(1-2x_{\mathrm{A},j})-(1-2x_{\mathrm{A}})\right]$$
$$+\ln\frac{x_{\mathrm{A},i}}{1-x_{\mathrm{A},i}}-\ln\frac{x_{\mathrm{A}}}{1-x_{\mathrm{A}}}+b_i\frac{v}{k_{\mathrm{B}}T}=0 \quad (1\leqslant i\leqslant r) \tag{105}$$

where

$$\left.\begin{aligned} b_1 &= 1 \\ b_{i\neq 1} &= 0 \end{aligned}\right\} \tag{106}$$

These equations may be solved by an iterative process (Lane and Johnson, 1967); the initial approximation is generally made by setting all $x_{\mathrm{A},i}=x_{\mathrm{A}}$.

The surface excess n_{A}^{σ}, with respect to a dividing surface placed between the solid and the first layer, is obtained from equation (41) as

$$n_{\mathrm{A}}^{\sigma}=N/(rL)\sum_{i=1}^{r}(x_{\mathrm{A},i}-x_{\mathrm{A}}) \tag{107}$$

and n_{B}^{σ} by

$$n_{\mathrm{B}}^{\sigma}=N/(rL)\sum_{i=1}^{r}(x_{\mathrm{B},i}-x_{\mathrm{B}})=-n_{\mathrm{A}}^{\sigma} \tag{108}$$

According to (44) this dividing surface gives the reduced adsorption directly, and by (58) can be identified with the experimental quantity $n^0\Delta x/m$.

Using this same dividing surface with the definition (40) and the relation (72), the surface excess Helmholtz free-energy is given by

$$A^{\sigma}=-k_{\mathrm{B}}T\ln[Q(N/r,x_{\mathrm{A},i=1,r})/Q(Nx_{\mathrm{A}},Nx_{\mathrm{B}})] \tag{109}$$

where the partition functions are evaluated using equations (98) and (82) and the appropriate Bragg–Williams approximations. The surface tension is then evaluated from (109), (52), (92), (93), (107), and (108) as

$$\gamma A_{\mathrm{s}}=\gamma_{\mathrm{B}}^{0}A_{\mathrm{s}}+(N/r)\left(x_{\mathrm{A},1}v+k_{\mathrm{B}}T\left[\sum_{i=1}^{r}x_{\mathrm{A},i}\ln(x_{\mathrm{A},i}/x_{\mathrm{A}})\right.\right.$$
$$+(1-x_{\mathrm{A},i})\ln\{(1-x_{\mathrm{A},i})/(1-x_{\mathrm{A}})\}$$
$$\left.\left.+(w/k_{\mathrm{B}}T)\left\{\sum_{j=i-1}^{i+1}a_jx_{\mathrm{A},i}(1-x_{\mathrm{A},j})-x_{\mathrm{A},i}+2x_{\mathrm{A},i}x_{\mathrm{A}}-x_{\mathrm{A}}^{2}\right\}\right]\right) \tag{110}$$

Equation (110) is applicable to any set of $x_{\mathrm{A},i=1,r}$ but can be simplified if the layer concentrations satisfy (105).

When the solution reservoir contains only component A, then (110) immediately gives

$$v = (\gamma_A^0 - \gamma_B^0)a_s \tag{111}$$

a_s being the area per lattice site in layer 1. Although neither γ_A^0 nor γ_B^0 is directly measurable, their difference can be obtained from the composite isotherm. This is achieved by substituting (11), (58), and (59) into (50) and integrating over the whole composition range, giving

$$(\gamma_A^0 - \gamma_B^0) = -\frac{1}{A_s}\int_0^1 \frac{n^0\Delta x}{m}\frac{1}{1-x_A}RT\left(\frac{1}{x_A} + \frac{1}{f_A}\frac{df_A}{dx_A}\right) dx_A \tag{112}$$

where A_s is now the area of m grams of solid.

In the Bragg–Williams approximation $w/k_BT > 2$ leads to partial miscibility, while $w/k_BT < -2$ produces order–disorder phenomena (Guggenheim, 1952). It is therefore necessary to use values of w/k_BT between these two limits in order to ensure complete miscibility. There is no limit on the value of v, but for convenience we define A to be the component that makes v negative.

In all the following calculations of the model properties, a hexagonal-close-packed or face-centred cubic structure is assumed for the liquid and interfacial layers. This structure has $z_l = 12$, $\boldsymbol{l} = 0.5$, and $\boldsymbol{m} = 0.25$.

Figure 8 shows the adsorption isotherms for systems with $N/r = L$, $w/k_BT = 0.0$, and $v/k_BT = -0.5, -1.0, -2.0$, and -5.0 respectively. In all

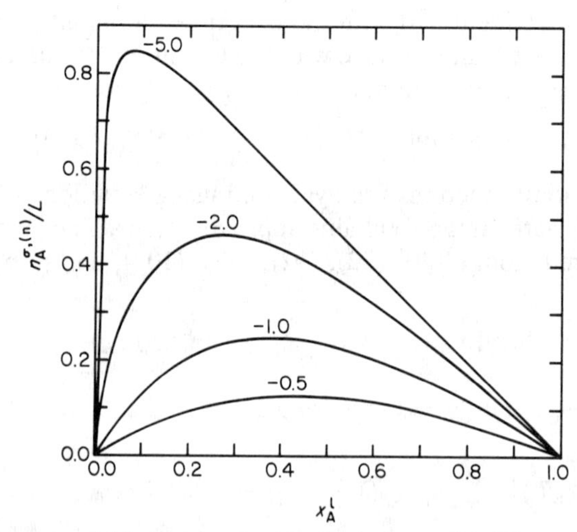

Figure 8 Lattice model isotherms for ideal solutions with $N/r = L$, $w/k_BT = 0$, and $v/k_BT = -0.5, -1.0, -2.0$, and -5.0. The isotherms are labelled with values of v/k_BT.

these examples the solution is ideal, and the adsorption is accounted for completely by the difference between the interaction of the two liquid components with the solid. The choice of A as the component with the lower solid/pure liquid surface tension leads to preferential adsorption of component A over the whole solution composition. Increasing the absolute value of v leads to an increase in preferential adsorption of A at all bulk phase concentrations.

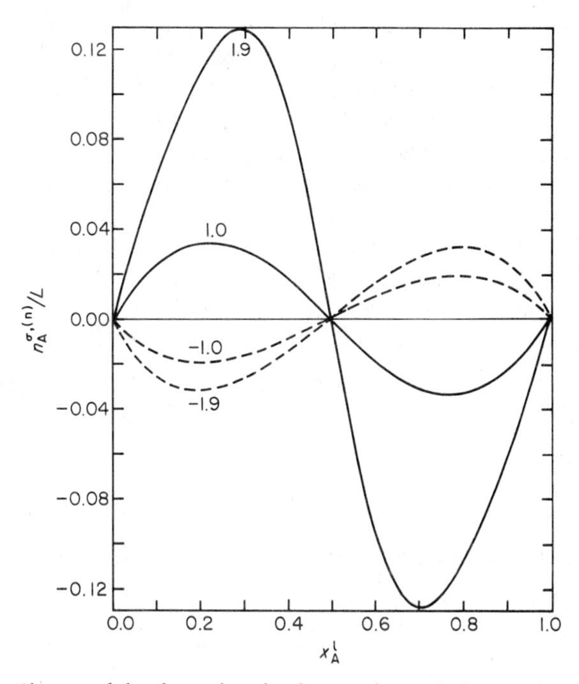

Figure 9 Lattice model adsorption isotherms for strictly regular solutions with $N/r = L$, $v/k_BT = 0$, and $w/k_BT = -1.9$, -1.0, 1.0, and 1.9. The isotherms are labelled with the values of w/K_BT.

Figure 9 shows adsorption isotherms for systems with $N/r = L$, $v/k_BT = 0$, and $w/k_BT = -1.9$, -1.0, 1.0, and 1.9 respectively. In all cases the two pure components produce the same surface tension with the solid, and preferential adsorption is determined completely by the solution non-ideality. Several important features are apparent in Fig. 9:

(1) For a given value of w/k_BT, $\Gamma_A^{(n)}(x_A) = -\Gamma_A^{(n)}(1 - x_A)$.
(2) At any given solution composition the sign for the adsorption of A is reversed with a change in sign of w.

(3) At any given solution composition the absolute value of the adsorption of A is less for negative values of w than it is for the equivalent positive value of w.

The first feature is a direct result of the antisymmetry, about the composition $x_A = 0.5$, of the difference equation (105) with $v/k_BT = 0$.

A qualitative explanation of the second feature can be given in terms of the behaviour of the bulk solution when it is placed in contact with the solid. With positive values of w, the surface is enriched with the minority component, moving the bulk solution further from the equimolar composition $x_A = 0.5$, which has the highest excess free-energy. This behaviour is a precursor of phase separation (which occurs when $w/k_BT > 2.0$), where the smaller of the two phases is richest in the minority component. Conversely, with negative values of w the surface is enriched with the majority component, the bulk solution moving closer to the free-energy favoured composition $x_A = 0.5$.

The third feature embodies an important principle that is relevant to all interfaces. Consider a system consisting of an interface between two phases α and β, both phases being in contact with a large reservoir containing all system components at fixed chemical potential. A function P is defined by

$$P = U - T(S^\alpha + S^\beta + S^\sigma) + p^\alpha V^\alpha + p^\beta V^\beta - \sum_i (n_i^\alpha + n_i^\beta + n_i^\sigma)\mu_i \quad (113)$$

and, by comparison with (10) and (49),

$$P = \gamma A_s \quad (114)$$

Differentiating (113) and comparing with (5) and (48), we obtain

$$\mathrm{d}P = -(S^\alpha + S^\beta + S^\sigma)\,\mathrm{d}T + V^\alpha\,\mathrm{d}p^\alpha + V^\beta\,\mathrm{d}p^\beta + \gamma\,\mathrm{d}A_s - \sum_i (n_i^\alpha + n_i^\beta + n_i^\sigma)\,\mathrm{d}\mu_i \quad (115)$$

The function P is a Legendre transform of U and has a minimum value at equilibrium at fixed $T, p^\alpha, p^\beta, A_s$, and the μ_i. Consider a situation in which the bulk phases α and β have established equilibrium and the only possible variations occur in the interface. The Gibbs–Duhem equation (11) applies to each bulk phase, so

$$\mathrm{d}P = -S^\sigma\,\mathrm{d}T + \gamma\,\mathrm{d}A_s - \Sigma n_i^\sigma\,\mathrm{d}\mu_i \quad (116)$$

When equilibrium is established, P will adopt its minimum value, and since T, A_s, and the μ_i have been fixed, by (114) γ must have adopted its minimum value. This result applies to any equilibrium system at constant temperature, where the position of the interface is fixed. In particular, it applies to all systems relevant to this chapter. Although, in measuring adsorption from solution, a change in solution composition is observed, the final solid/

solution interface is indistinguishable from that of a system containing the same amount of solid but a very much larger volume of solution of composition identical with that of the final state of the smaller system.

Returning to the third feature of the isotherms in Fig. 9, we consider the specific case with $N/(rL) = 1$, $x_A = 0.8$, and $w/k_B T = \pm 1.9$. For a freshly formed surface, with each layer having the liquid composition, the surface tension is calculated by (110) as $\gamma A_s/RT = \pm 0.076$, the sign being opposite to the sign of w. In both cases the relaxation to the equilibrium concentration profile must lower the surface tension. Thus the antisymmetry with w, observed for the freshly formed surface, has been overridden by the thermodynamic requirement detailed above. At equilibrium, $\gamma A_s/RT = -0.0876$ when $w/k_B T = 1.9$ and 0.0712 when $w/k_B T = -1.9$. The layer relaxation is dependent upon the sign of w, and the two equilibrium concentration profiles are shown in Fig. 10. For positive w, the concentration profile is monotonic on passing from the first to the rth layer, each

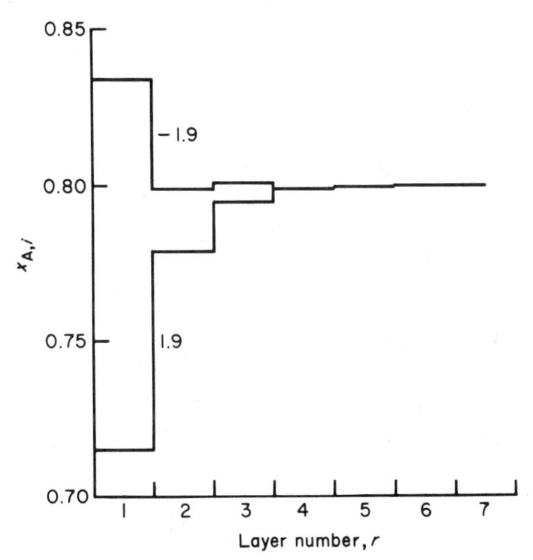

Figure 10 Concentration profile of component A at the solid/solution interface for a lattice model with $N/r = L$, solution composition $x_A = 0.8$, $v/k_B T = 0$, and $w/k_B T = 1.9$ or -1.9. The profiles are labelled with the values of $w/k_B T$.

layer making a negative contribution to the absolute magnitude of the adsorption. For negative values of w, the layer concentrations oscillate about the bulk concentration with layers making opposite contributions to the absolute magnitude of the adsorption. This behaviour for negative w increases the number of energetically favoured AB nearest neighbour pairs, but reduces the contribution of non-ideality to the absolute magnitude of the adsorption.

The combination of non-zero values of w and v leads to complexity in detail but some simple generalizations can be made. It is not necessary for v to be zero in order to get a change in sign of the adsorption at some liquid concentration $x_{A,az}$. At this liquid composition, all layers have the concentration $x_{A,az}$, hence the term surface azeotrope. From equation (105), the azeotropic concentration is given by

$$x_{A,az} = 0.5(1 - v/mw) \tag{117}$$

and, since this concentration must lie between 0 and 1, the condition for surface azeotrope formation is

$$-1 \leqslant v/mw \leqslant 1 \tag{118}$$

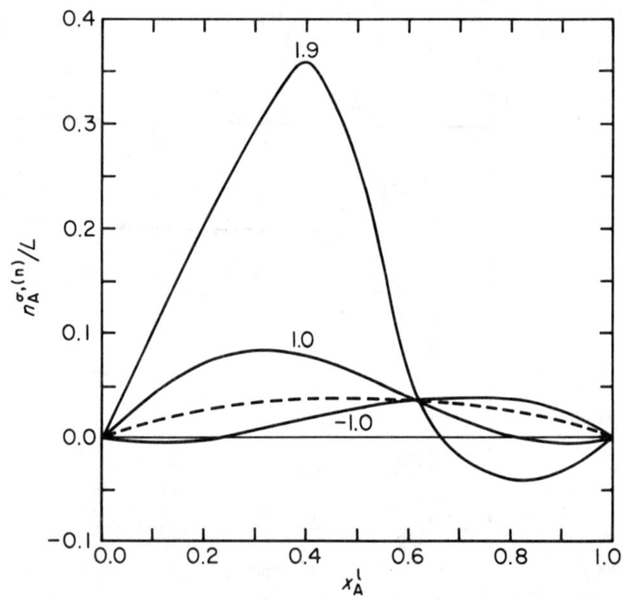

Figure 11 Lattice model adsorption isotherms with $N/r = L$, $v/k_BT = -0.15$, and $w/k_BT = -1.0$, 1.0, and 1.9. The full curves are labelled with the appropriate values of w/k_BT. The reference curve (broken line) corresponds to an ideal solution with $w = 0$

The full curves in Fig. 11 show isotherms with azeotrope points. The parameters are $N/r = L$, $v/k_BT = -0.15$, and $w/k_BT = -1.0$, 1.0, and 1.9. The reference curve (broken) has the same area and value of v, but is appropriate to an ideal solution with $w = 0$. It is evident from Figs 9 and

11 that, with values of v small enough for azeotrope formation, the isotherms for a given w' and v' are given approximately by a linear combination of the isotherm for w'/k_BT and $v = 0$ with that for $w = 0$ and v'/k_BT.

Figure 12 shows the adsorption isotherms (full curves) for $N/r = L$, $v/k_BT = -1$, and $w/k_BT = -1$, 1, and 1.9 respectively. The broken curve shows the isotherm for an ideal solution. The value of v is too large to satisfy the condition (118) for azeotrope formation, and all isotherms show positive adsorption of component A at all solution concentrations.

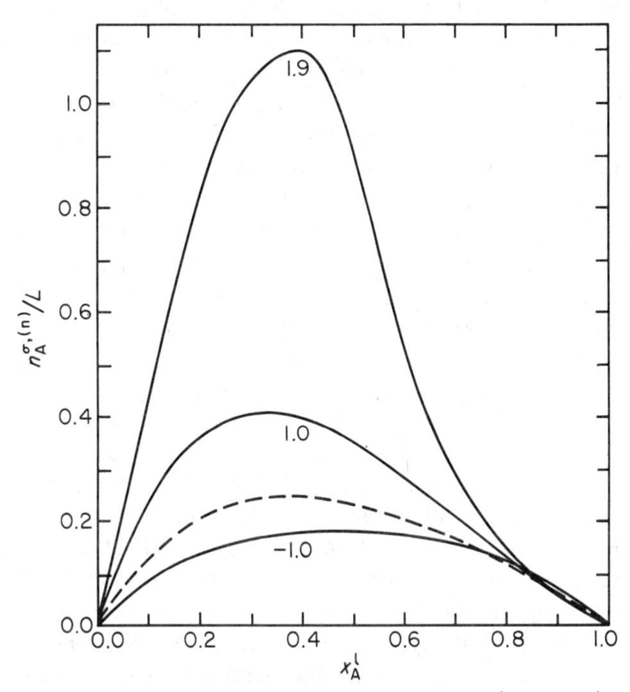

Figure 12 Lattice model adsorption isotherms with $N/r = L$, $v/k_BT = 1$, and $w/k_BT = -1.0$, 1.0, and 1.9. The full curves are labelled with the appropriate values of w/k_BT. The reference curve (broken line) corresponds to an ideal solution with $w = 0$.

Although the shape of the isotherm is still given by the linear combination approximation used for the surface azeotropes of Fig. 11, this approximation is of no value in any quantitative assessment of the isotherm. An important new feature of the isotherms in Fig. 12 is the similarity of all of them for $0.8 < x_A < 1.0$. This is a consequence of the form of the difference equation (105). Any pair of isotherms described by (105), and having the same value of v, will intersect at some value of $x_A \neq 0$, or 1, provided only that both values of w satisfy $-2.0 \leqslant w/k_BT \leqslant 2.0$. With $v = 0$, the intersection occurs at $x_A = 0.5$ (see Fig. 9), and as v becomes more negative the crossover concentration increases ($x_A \to 1$ as $v \to -\infty$) and is dependent upon the values of w. Therefore it is only at concentrations $x_A < 0.5$ that the non-ideality of the solute is important, and, as expected, negative values of w depress and positive values of w raise the isotherm with respect to the isotherm for an ideal solution with the same value of v. The deviation of the isotherm from the reference ideal isotherm increases with the absolute magnitude of w.

Comparison of Figs 8, 9, 11, and 12 with the Schay and Nagy classification of Fig. 2, and noting the comments above, indicates that the lattice model can mimic the shape of the experimental isotherms.

It is difficult to find any published experimental isotherms suitable for making a critical, quantitative assessment of the model. In addition to the requirement of precise experimental adsorption data, it is necessary for the solid surface to be smooth and homogeneous, and its surface area per unit mass must be known; the solution must consist of molecules that are approximately spherical, and of equal size, and their solution thermodynamic properties must be known. The adsorption isotherm, and associated experimental data, reported by Ash *et al.* (1973, 1975) for the system graphitized carbon black/benzene/cyclohexane fulfil these requirements. The experimental adsorption data at 313.15 K are shown in Fig. 13 by the square symbols, the component A being benzene. The surface area of adsorbent is reported as 86 $m^2\,g^{-1}$, and the area of a layer containing 1 mol of liquid as $2.41 \times 10^5\,m^2$, giving $N/rL = 3.57 \times 10^{-4}\,mol\,g^{-1}$ of adsorbent. The other model parameters are $w/k_BT = 0.474$ and $v/k_BT = -0.718$, derived through equations (111) and (112). The agreement between the model and experimental isotherms is reasonable.

There is often a temptation to adjust the parameters to improve the agreement between model and experiment. The broken line of Fig. 13 is an example of this; the parameter w has been left unchanged but N/rL has been decreased to $1.46 \times 10^{-4}\,mol\,g^{-1}$ adsorbent and, by (111), $v/k_BT = -1.75$. Yielding to this temptation has improved agreement between model and experiment, but has not improved our understanding of adsorption from solution. It is better to accept the poorness of fit and improve our understanding of the phenomenon.

Using a derivation similar to that for (112), and combining with (58), the surface tension at any concentration $x_A = \alpha$, where $0 \leqslant \alpha \leqslant 1$, is given by

$$\gamma(x_A = \alpha) = \gamma_B^0 - \int_0^\alpha \Gamma_A^{(n)} u \, dx_A \tag{119}$$

with

$$u = [1/(1 - x_A)](\partial \mu_A / \partial x_A) \geqslant 0 \tag{120}$$

The inequality in (120) arises because $1/(1 - x_A)$ is positive for all permitted values of x_A, and according to the condition for stability with respect to diffusion (Prigogine and Defay, 1954) $\partial\mu_A/\partial x_A \geqslant 0$, this derivative becoming zero only at the consolute point. It is advantageous to introduce, from the calculus, Rolle's theorem (Courant, 1951). If a function $\phi(x)$ is continuous in the closed interval $a \leqslant x \leqslant b$, and differentiable in the open

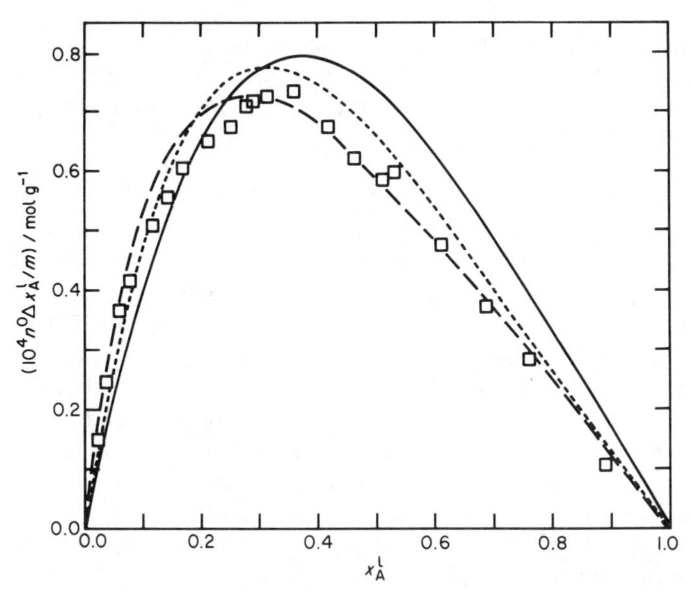

Figure 13 Comparison of the lattice model adsorption isotherms with the experimental data (Ash *et al.*, 1973) for the system benzene (A)/cyclohexane (B)/Graphon at 313.15 K. The curves represent the model isotherms. (——) $N/rL = 3.57 \times 10^{-4}$ mol g^{-1} adsorbent; $w/k_BT = 0.474$; $v/k_BT = -0.718$. (– – –) $N/rL = 1.46 \times 10^{-4}$ mol g^{-1}; $w/k_BT = 0.474$; $v/k_BT = -1.75$. (- - - - -) $N/rL = 3.57 \times 10^{-4}$ mol g^{-1}; $w_{11} = 0.0$, all other $w_{ij} = 0.474$; $v/k_BT = -0.718$.

interval $a < x < b$, and in addition $\phi(a) = \phi(b) = 0$, there exists at least one point ξ in the interior of the interval at which $\phi'(\xi) = 0$ ($\phi'(x)$ is the derivative of $\phi(x)$ with respect to x). It follows from this theorem that, if there is only one point ξ in this interval, there is no point η in the interior of the interval for which $\phi(\eta) = 0$. (If there were a point η, Rolle's theorem could be applied separately to the intervals $a \leqslant x \leqslant \eta$ and $\eta \leqslant x \leqslant b$, giving at least two distinct values for ξ.)

A function $\phi(x_A)$ is defined by

$$\phi(x_A) = \gamma^m - \gamma^e \tag{121}$$

where the superscripts m and e indicate model and experiment respectively. Equation (119) is applicable to both model and experiment, and by differentiation with respect to x_A gives

$$\phi'(x_A) = -(\Gamma_A^{(n),m} - \Gamma_A^{(n),e})u \tag{122}$$

From equations (110), (111), and (119) it is evident that

$$\phi(x_A = 0) = \phi(x_A = 1) = 0 \tag{123}$$

Applying Rolle's theorem to the function $\phi(x_A)$ requires that there be at least one concentration $x_A = \xi$ for which $\phi'(x_A = \xi)$ vanishes. Excluding the special case of the consolute point, this condition applied to equations (122) and (123) requires that

$$\Gamma_A^{(n),m}(x_A = \xi) = \Gamma_A^{(n),e}(x_A = \xi) \tag{124}$$

Examination of Fig. 13 indicates that the model and experimental adsorption isotherms do satisfy (124) by crossing at a single point, with $\xi \approx 0.2$. Using the extension of Rolle's theorem, the single value of ξ, together with the definition of $\phi(x_A)$ in (121), indicates that the model and experimental surface tensions are only in agreement at $x_A = 0$ or 1. Furthermore, the condition that $\phi'(x_A = \xi) = 0$ is also the condition that $\phi(x_A = \xi)$ represents either a maximum or minimum for $\phi(x_A)$, and the absolute value $|\phi(x_A = \xi)|$ represents the largest deviation between the model and experimental surface tensions. Again from the calculus, the function $\phi(x_A)$ is a maximum or a minimum according to whether the second derivative $\phi''(x_A)$ is negative or positive at $x_A = \xi$. Differentiation of (122) and substitution with the result (124) gives

$$\phi''(x_A = \xi) = -u\, \mathrm{d}(\Gamma_A^{(n),m} - \Gamma_A^{(n),e})/\mathrm{d}x_A \tag{125}$$

From Fig. 13 it is obvious that, at $x_A = \xi$, $(\Gamma_A^{(n),m} - \Gamma_A^{(n),e})$ is increasing with x_A, and since u is always positive, $\phi''(x_A = \xi) < 0$, and the function $\phi(x_A)$ has a maximum at $x_A = \xi$. Using this result, with the properties (123) it is evident that $\phi(x_A) \geqslant 0$ for all permissible values of x_A. According to (121) this result demonstrates that the model surface tension is greater than the experimental value at all compositions $0 < x_A < 1$.

It was noted earlier that the surface tension must be a minimum consistent with a given solution composition. In the present form of the multilayer model, the only relaxation mechanism contributing to a reduction in surface tension is the change in concentration of the layers near the solid. In a real system there may be other factors contributing to a reduction in surface tension; these factors include a change in molecular packing near the solid, or preferential orientation of molecules whose interaction with other molecules is not spherically symmetric. Thus it is not surprising that this comparison of model and experimental adsorption isotherms leads to the conclusion that the model overestimates the magnitude of the surface tension. This same conclusion was reached in the only other rigorous comparison of the model with suitable experimental data (Lane, 1975).

It is possible to modify the multilayer model to account for the change in intermolecular interaction with molecular orientation near the solid. This is achieved by introducing interchange parameters w_{ij} for the interaction of particles in layer i with those in layer j. Provided that these w_{ij} are independent of concentration, the integrity of the derivation of the model properties is not affected, and it is sufficient to replace w by the appropriate w_{ij}, for example in equation (105). It has been argued by Ash *et al.* (1975) that benzene molecules are adsorbed in a "flat" orientation by graphite. They suggest that $w_{11} \approx 0$ while all the other w_{ij} have the bulk value of 0.474. The short-dash line in Fig. 13 shows the adsorption for the multilayer model with this choice of parameters for w_{ij}, and the correct experimental value for a_s. The agreement with the experimental adsorption isotherm is better than for the multilayer model with a single w (full line). The analysis based on Rolle's theorem can be applied to the adsorption isotherms for the models with variable w and fixed w ($x_A = \xi \approx 0.3$) and separately for the model with variable w and the experimental curve ($x_A = \xi \approx 0.1$). Combining the two analyses gives the result that the surface tension for the model with variable w is closer to the experimental values for all mixture compositions. In the following section it will be shown that, for a given value of w and x_A in the bulk solution, the surface tension at the solid/solution interface decreases with w_{11}.

In spite of the obvious shortcomings of this model, it is unlikely to be grossly inaccurate in predicting the surface properties of solutions of approximately spherical molecules of similar size. It does have the advantage that it includes intermolecular interactions specifically, is consistent with all thermodynamic requirements, and gives some insight into the interplay of surface and bulk thermodynamic properties.

It is of relevance to the following section to note that it is only for ideal solutions ($w = 0$), and surface azeotropes, that the lattice model degenerates to a monolayer model with $x_{A,2}, x_{A,3}, \ldots, x_{A,r} = x_A$. This was first noted by Defay and Prigogine (1950). The first systematic development of a multilayer model is generally attributed to Ono and Kondo (1960) while Ash *et al.* (1968) extended the multilayer model to include very simple polymer molecules.

VI. Other Models of the Solid/Solution Interface

The special restrictions of the multilayer model prevent it from being applied to most experimental adsorption systems. In the absence of a more general, molecular model of the solid/solution interface it has been common practice to adopt a simple, macroscopic model. There are many variants of this model but the one developed in this section is similar to that described by Everett (1965a). This particular choice is arbitrary.

The basic concept is that the interface may be approximated by a thin, homogeneous layer of liquid. The layer is bounded on one side by the solid and on the opposite side by the homogeneous bulk solution. The Gibbs dividing surface is placed between the surface layer and the bulk liquid, so that the surface excess Γ_i^s or n_i^s of the ith liquid component is the total of that component in the layer (the surface excess of solid is ignored). The superscript s for the surface quantities indicates that the model is based on a surface phase (Everett, 1972).

A partial molar area a_i^s, of species i in the layer, is defined by

$$a_i^s = \partial A_s / \partial n_i^s \tag{126}$$

the positions of the layer dividing surfaces remaining fixed during any possible variation of A_s or n_i^s. According to equations (20) and (21), the a_i^s satisfy

$$\sum_i n_i^s a_i^s = A_s \tag{127}$$

or

$$\sum_i \Gamma_i^s a_i^s = 1 \tag{128}$$

The surface mole fraction x_i^s of component i is defined by

$$x_i^s = \Gamma_i^s / \sum_i \Gamma_i^s \tag{129}$$

The surface activity coefficient f_i^s is defined by

$$\mu_i^s = \mu_i^{0,s} + RT \ln f_i^s x_i^s + (\gamma_i^0 - \gamma) a_i^s \tag{130}$$

$$f_i^s (x_i^s = 1) = 1 \tag{131}$$

and, by (37) and (22),

$$\mu_i^{0,s} = \mu_i^{0,1} \tag{132}$$

where $\mu_i^{0,1}$ is the chemical potential of pure liquid i.

Some comment on the inclusion of the term $(\gamma^0 - \gamma_i) a_i^s$ in the definition of the surface chemical potential is appropriate since another form has been used (Butler, 1932). Defining a surface excess energy variable B^s by

$$B^s = A^s - \gamma A_s \tag{133}$$

(note that A^s is the Helmholtz free-energy of the surface layer) and by differentiating B^s and combining with (53), we obtain

$$dB^s = -S^s\,dT - A_s\,d\gamma + \sum_i \mu_i\,dn_i \tag{134}$$

A Maxwell type relation can be obtained from the second derivative of B^s which, on combination with (134) and (126), gives

$$\frac{\partial^2 B^s}{\partial \gamma \partial n_i^s} = -a_i^s = \frac{\partial \mu_i}{\partial \gamma} \tag{135}$$

Assuming a_i to be constant, the chemical potential difference $\Delta\mu_{i,\gamma}$ due to the mechanical variable, surface tension, changing from its value for pure component i to its current value is given by integrating (135) as

$$\Delta\mu_{i,\gamma} = \int_{\gamma^0}^{\gamma} \left(\frac{\partial \mu_i}{\partial \gamma}\right)_{d\gamma} = -a_i^s(\gamma - \gamma_i^0) \tag{136}$$

It is readily recognized that (135) is the surface analogue of the bulk relation

$$(\partial \mu_i/\partial p) = V_i \tag{137}$$

and that (130) is the surface analogue of the relation

$$\mu_i = \mu_i^{0,1}(p^0) + RT\ln f_i^1 x_i^1 + V_i(p - p^0) \tag{138}$$

for a bulk solution whose pressure p differs from that for the standard state p^0.

Returning to the development of the surface layer model, equations (130), (37), and (22) are combined and on rearrangement give the surface tension as

$$\gamma = \gamma_i^0 + (RT/a_i^s)\ln(f_i^s x_i^s / f_i^1 x_i^1) \tag{139}$$

the superscript (l) identifying the variables appropriate to the liquid phase. If there are two components A and B, the relation (139) for each component may be combined to eliminate the unknown solid/solution surface tension. Rearranging the resulting expression gives

$$\left(\frac{f_A^s x_A^s}{f_A^1 x_A^1}\right)^{a^{\ominus,s}/a_A^s} \left(\frac{f_B^1 x_B^1}{f_B^g x_B^g}\right)^{a^{\ominus,s}/a_B^s} = K \tag{140}$$

where $a^{\ominus,s}$ is standard partial molar area and

$$K = \exp[-(\gamma_A^0 - \gamma_B^0)a^{\ominus,s}/RT] \tag{141}$$

Equation (140) is sometimes expressed in the form

$$\left(\frac{f_A^s x_A^s}{f_A^1 x_A^1}\right)^{1/r} \left(\frac{f_B^1 x_B^1}{f_B^s x_B^s}\right) = K' \tag{142}$$

where

$$r = a_A^s / a_B^s \tag{143}$$

and

$$\ln K' = (a_B^s / a_A^{\ominus,s}) \ln K \tag{144}$$

The form of equation (140) has led to the description "law of mass action" being applied to this class of model (Kipling, 1965). The constant K can be evaluated from the adsorption isotherm by integrating according to (112) and substituting the result for $(\gamma_A^0 - \gamma_B^0)$ in (141). It is interesting that for a monolayer model, and equal sized molecules so that $a_A = a_B = a^{\ominus}$, then (141) and (111) may be combined to give

$$K = \exp(-v/k_B T) \tag{145}$$

which is the link between the layer and lattice models. (Note that, in obtaining (141), a_s in the lattice model is an area per molecule, whereas $a^{\ominus}$ is an area per mole.)

Besides satisfying (130) the surface activity coefficients are subject to a further self-consistency restraint. Differentiation of (130) and substitution in the Gibbs adsorption isotherm (50) gives at constant temperature, after using the relations (128) and (129),

$$\sum_i x_i^s \, d \ln f_i^s = 0 \tag{146}$$

This result is the surface analogue of the relation

$$\sum_i x_i^l \, d \ln f_i^l = 0 \tag{147}$$

obtained at constant temperature by combining the Gibbs–Duhem equation (11) with the derivatives of equation (22).

Surface activity coefficients may be obtained from experimental adsorption isotherms. Equation (139) is rearranged as

$$\ln f_A^s = a_A(\gamma - \gamma_A^0)/RT + \ln(f_A^l x_A^l / x_A^s) \tag{148}$$

By reversing the direction of integration in (119) the surface tension difference in (148) is given by

$$\gamma - \gamma_A^0 = -RT \int_1^{x_A^l} [\Gamma_A^{(n)}/(1 - x_A^s)] [\partial \ln(f_A^l x_A^l)/\partial x_A^l] \, dx_A^l \tag{149}$$

Combining (148) and (149) gives the surface activity coefficient

$$\ln f_A^s = \ln(f_A^l x_A^l / x_A^s) - a_A \int_1^{x_A^l} [\Gamma_A^{(n)}/(1 - x_A^l)] \times [\partial \ln(f_A^l x_A^l)/\partial x_A^l] \, dx_A^l \tag{150}$$

It is important to note that in (150) all quantities on the right-hand side are realizable experimentally, with the exception of the surface mole fraction x_A^s. Although this term is defined in (129), the actual magnitude is not determined until the position of the layer dividing surface, between the surface layer and bulk solution, is fixed. If τ is the distance between this dividing surface and the dividing surface defined by (44) (which gives the reduced adsorption $\Gamma_A^{(n)}$ directly), then (41) and (42) give the surface layer adsorption as

$$\Gamma_i = \Gamma_i^{(n)} + \tau c_i^l \tag{151}$$

where c_i^l is the molar density of component i in the solution. (The thickness τ is related to the a_i^s by $\tau = \Sigma x_i^s V_i^s / \Sigma x_i^s a_i^s$.) Combining (151), (129), and (44), the surface layer mole fraction is given by

$$x_i^s = x_i^l + \Gamma_i^{(n)}/\tau c^l \tag{152}$$

where $c^l = \sum_i c_i^l$, and $\tau c^l = 1/\Sigma x_i^s a_i^s$.

From (152) it is clear that $x_i^s \to x_i^l$ as $\tau \to \infty$, but that for small values of τ the adsorption term $\Gamma_i^{(n)}$ dominates. It is also clear that, for a two component solution, non-zero values of $\Gamma_i^{(n)}$ require one component (say B) to have $\Gamma_B^{(n)} < 0$. Thus, if τ is made small enough, by (152) $x_B^s < 0$, which is unacceptable. It has been common practice to assign τ a value consistent with a monomolecular surface layer, but this may not be sufficient to ensure positive x_i^s for all species. Indeed, for the systems Graphon/benzene/ethanol and Graphon/n-heptane/ethanol, Brown *et al.* (1975) found it necessary to set τ at values corresponding to surface layers with a thickness of three and four molecular diameters respectively.

A further requirement must be imposed on the thickness τ. Rusanov (1971) has shown that for a two-component solution it is also necessary that $\partial x_A^s / \partial x_A^l$ be positive. In his paper, Rusanov summarized some earlier (Rusanov, 1967), very elegant thermodynamic reasons for this requirement. In simple terms it may be argued that, since the condition of thermodynamic stability for the bulk solution (Prigogine and Defay, 1954) requires that $\partial \mu_A / \partial x_A^l > 0$, this also requires that $\partial \mu_A^s / \partial x_A^s > 0$ for the surface layer, and since by (37) $\mu_A^s = \mu_A^l$, then $\partial x_A^s / \partial x_A^l > 0$.

The model has provided the relationship (150) for the formal evaluation of surface activity coefficients from experimental data. In addition, the model may be used to analyse, and interpret, theoretical predictions of the surface activity coefficients. Any theoretical function $f_i^s(x_i^s)$ relating surface activity coefficients with surface concentration must satisfy (146). This is not a severe restriction since there is an unlimited set of functions fulfilling this requirement. If the variables associated with any two of these functions are identified by the superscripts p and q, a function $\phi(x^l_A)$ may be defined by

$$\phi(x^l_A) = \gamma^p - \gamma^q \tag{153}$$

According to (149) and (152) the derivative function is

$$\phi'(x^l_A) = -(x^{s,p}_A - x^{s,q}_A)u\tau c^l \tag{154}$$

where u is defined by (120) as being equal to $(\partial\mu_A/\partial x^l_A)/(1-x^l_A)$. By applying the definition (130) to both theoretical functions p and q, the function $\phi(x^l_A) = 0$ at $x^l_A = 0, 1$. By Rolle's theorem (see preceding section) there is at least one solution composition $x^l_A = \xi$ for which $\phi'(x^l_A) = 0$, so by (153), and the stability requirement that $u > 0$ (τc^l is always positive),

$$x^{s,p}_A = x^{s,q}_A = \eta\,;\, x^l_A = \xi \tag{155}$$

Since both functions p and q have the same value of K or K', by equation (142)

$$(f^{s,p}_A)^{1/r}/f^{s,p}_B = (f^{s,q}_A)^{1/r}/f^{s,q}_B = [(f^l_A)^{1/r}/f^l_B] \times [K'(\xi)^{1/r}]/(1-\xi) \tag{156}$$

Thus Rolle's theorem shows that there must be always one solution concentration $0 < x^l < 1$ for which the two adsorption isotherms cross, and that at this point the ratio $(f^s_A)^{1/r}/f^s_B$ is the same for both theoretical functions. If there is only one intersection, the converse of Rolle's theorem (see preceding section) requires that the surface tension, predicted by one surface activity function, is lower than that predicted by the other function at all solution concentrations. Combining the relations (139) and (156) gives the surface tension difference at the crossover as

$$(\gamma^p - \gamma^q) = RT\ln(f^{s,p}_A - f^{s,q}_A)/a^s_A\,;\, x^s_A = \eta \tag{157}$$

Thus the theoretical function predicting the lower surface activity coefficient f^s_A at $x^s_A = \eta$ predicts the lower surface tension over the whole solution concentration range.

It is possible for the two theoretical adsorption isotherms to cross at more than one point. The relations (155) and (156) must be satisfied at each point. In addition, from successive intersections the surface tension difference must be zero at some intermediate concentration. This more complex situation will not be discussed further.

The condition $\phi(x^l_A) = 0$ at $x^l_A = \zeta$ indicates that the function $\phi(x^l_A)$ has either a maximum or a minimum at that point. The second derivative $\phi''(x^l_A)$ at this point is obtained by differentiating (154) with respect to x^l_A, and combining with equations (152) and (154), giving

$$\phi''(x^l_A) = -u\,\mathrm{d}(\Gamma^{(n),p}_A - \Gamma^{(n),q}_A)/\mathrm{d}x^l_A\,;\, x^l_A = \xi \tag{158}$$

If $\gamma^{p} < \gamma^{q}$, the function $\phi''(x_A^l)$ has a minimum, and according to the calculus $\phi''(x_A^l) > 0$ at $x_A^l = \xi$. Since u is always positive, the derivative $d(\Gamma_A^{(n),p} - \Gamma_A^{(n),q})/dx_A^l < 0$. If there is only one value of ξ, this result requires that, if $\gamma^{p} < \gamma^{q}$, $\Gamma^{(n),p} > \Gamma^{(n),q}$ for $0 < x_A^l < \xi$ and $\Gamma^{(n),p} < \Gamma^{(n),q}$ for $\xi < x_A^l < 1$. Although it may seem paradoxical, this analysis shows that the one solution concentration $x_A^l = \xi$, at which the two surface activity coefficients $f_i^{s,p}$ and $f_i^{s,q}$ predict the same adsorption, is the concentration at which they predict the maximum difference in the surface tension.

The right-hand side of equation (150), through which the experimental surface activity coefficients are evaluated, may be chosen as one of the two "theoretical" functions. On comparing a predicted adsorption isotherm with that obtained by experiment, we expect the two curves to intersect at least once at some $x_A^l = \xi$. By observing whether the predicted isotherm has lower or higher values of $\Gamma_A^{(n)}$ for $0 < x_A^l < \xi$, it is possible to state whether the predicted surface tension is higher or lower than the experimental value. This result is extremely useful in assessing the merit of a particular function for the surface activity coefficients.

It is reasonable to assume that the function, relating the surface activity coefficients to surface concentration, will bear some relation to the equivalent function in the solution. The simplest choice is to equate these two functions. However, this choice may not properly account for any modification of the intermolecular interaction energies in the surface layer. This possibility may be accounted for in the following fashion.

The activity coefficients are separated into two parts,

$$f_i = f_i^* t_i \tag{159}$$

where f_i^* is the activity coefficient in an athermal reference system (no heat of mixing) with components of size and shape similar to the real solution, and t_i is the thermal perturbation arising from a non-zero heat of mixing in the real system. Because the activity coefficients in the reference system must satisfy all the thermodynamic requirements, the restrictions (146) or (147) must apply to the f_i^*, and hence

$$\sum_i x_i \, d \ln f_i^* = 0 \tag{160}$$

Furthermore, differentiating (159) and combining with (160) and either (146) or (147) gives

$$\sum_i x_i \, d \ln t_i = 0 \tag{161}$$

Finally the definitions of the activity coefficients in (22), (130), (131), and (159) require that, for both the solution and surface phases,

$$f_i = f_i^* = t_i = 1 \, ; x_i = 1 \tag{162}$$

The relations (159)–(162) are general and apply to any homogeneous mixture. The following assumptions are peculiar to this model. The first is that variations in f_i^* with composition are the same for both the solution and surface phase, so that for $0 \leqslant \alpha \leqslant 1$,

$$\left(\frac{\partial \ln f_i^{*,\mathrm{s}}}{\partial x_i^{\mathrm{s}}}\right)_{(x_i^{\mathrm{s}}=\alpha)} = \left(\frac{\partial \ln f_i^{*,\mathrm{l}}}{\partial x_i^{\mathrm{l}}}\right)_{(x_i^{l}=\alpha)} \tag{163}$$

The second is that the thermal parts of the activity coefficients are related by

$$\left(\frac{\partial \ln t_i^{\mathrm{s}}}{\partial x_i^{\mathrm{s}}}\right)_{(x_i^{\mathrm{s}}=\alpha)} = k_{\mathrm{t}}\left(\frac{\partial \ln t_i^{\mathrm{l}}}{\partial x_i^{\mathrm{l}}}\right)_{(x_i^{l}=\alpha)} \tag{164}$$

where k_{t} is a constant. For (164) to satisfy (146) it is necessary that k_{t} be the same for both components of the mixture. Noting (162), each of the equations (163) and (164) may be integrated from $x_i = 1$ to $x_i = \alpha$, giving

$$f_i^{*,\mathrm{s}}(x_i^{\mathrm{s}} = \alpha) = f_i^{*,\mathrm{l}}(x_i^{\mathrm{l}} = \alpha) \tag{165}$$

$$t_i^{\mathrm{s}}(x_i^{\mathrm{s}} = \alpha) = [t_i^{\mathrm{l}}(x_i^{\mathrm{l}} = \alpha)]^{k_{\mathrm{t}}} \tag{166}$$

The simple choice of setting the surface activity coefficients equal to the solution activity coefficient at the same solution concentration is equivalent to setting $k_{\mathrm{t}} = 1$.

If two values $k_{\mathrm{t}}^{\mathrm{p}}$ and $k_{\mathrm{t}}^{\mathrm{q}}$ are chosen, according to (156) the predicted adsorption isotherms will intersect when

$$\left(\frac{(t_{\mathrm{A}}^{\mathrm{l}})^{1/r}}{t_{\mathrm{B}}^{\mathrm{l}}}\right)^{k_{\mathrm{t}}^{\mathrm{p}}}_{(x_{\mathrm{A}}^{\mathrm{l}}=\eta)} = \left(\frac{(t_{\mathrm{A}}^{\mathrm{l}})^{1/r}}{t_{\mathrm{B}}^{\mathrm{l}}}\right)^{k_{\mathrm{t}}^{\mathrm{q}}}_{(x_{\mathrm{A}}^{\mathrm{l}}=\eta)} \tag{167}$$

the terms in $f_{\mathrm{A}}^{*,\mathrm{s}}$ and $f_{\mathrm{B}}^{*,\mathrm{s}}$ being independent of the choice of k_{t}. Since $k_{\mathrm{t}}^{\mathrm{p}} \neq k_{\mathrm{t}}^{\mathrm{q}}$, equation (167) can only be satisfied if the quantity within the larger brackets is unity, or

$$(t_{\mathrm{A}}^{\mathrm{l}})^{1/r}(x_{\mathrm{A}}^{\mathrm{l}} = \eta) = (t_{\mathrm{B}}^{\mathrm{l}})(x_{\mathrm{A}}^{\mathrm{l}} = \eta) \tag{168}$$

Therefore, the assumptions (163) and (164) ensure that, for this model, all predicted isotherms for a given system intersect at a single point $x_{\mathrm{A}}^{\mathrm{s}} = \eta$, the value of η being given by the solution of equation (168). (This point may, or may not, lie on the experimental adsorption isotherm.) According to (157) the surface tensions of the predicted isotherms p and q satisfy

$$(\gamma^{\mathrm{p}} - \gamma^{\mathrm{q}}) = (k_{\mathrm{t}}^{\mathrm{p}} - k_{\mathrm{t}}^{\mathrm{q}})\, RT \ln(t_{\mathrm{A}}^{\mathrm{l}})(x_{\mathrm{A}}^{\mathrm{l}} = \eta) \tag{169}$$

thus if $k_{\mathrm{t}}^{\mathrm{p}} < k_{\mathrm{t}}^{\mathrm{q}}$, the adsorption isotherm p will predict the lower surface tension at all solution phase compositions if $(t_{\mathrm{A}}^{\mathrm{l}})(x_{\mathrm{A}}^{\mathrm{l}} = \eta) > 1$ and vice versa if $(t_{\mathrm{A}}^{\mathrm{l}})(x_{\mathrm{A}}^{\mathrm{l}} = \eta) < 1$.

Finally, the predicted isotherms can exhibit surface azeotropic formation ($x_A^s = x_A^l = \theta$) provided that

$$\left(\frac{(t_A^l)^{1/r}}{t_B^l}\right)^{(k_t - 1)}_{(x_A^l = \theta)} = K' \tag{170}$$

The terms in f_i^*, x_i^l, and x_i^s all cancel in the full expression (142). Thus the choice of setting $k_t = 1$ can only produce a surface azeotrope if $K' = 1$, when there is no preferential adsorption at any solution composition. For non-trivial values of K', a surface azeotrope can only be predicted if $k_t \neq 1$. Note that k_t is limited in the values that may be placed on it. If $k_t = 1$, the surface phase activity coefficients have the same composition properties as the solution phase, and since this is completely miscible the surface phase will be too. As the magnitude of $(k_t - 1)$ increases, the possibility of phase separation arises, or $\partial\mu_A^s/\partial x_A^s < 0$, for some x_A^s.

The partition of the activity coefficients in (159) is not unique, and it is necessary to define an operational method to remove this arbitrariness. (This problem arises with the definition of f_i itself; two common procedures are popular and involve setting $f_i = 1$ at either $x_i = 0$ or $x_i = 1$. The latter choice has been used in this chapter.) According to equation (26), the activity coefficients defined by setting $f_i = 1$ at $x_i = 1$ satisfy

$$G_m^E/RT = x_A \ln f_A + x_B \ln f_B \tag{171}$$

Differentiating with respect to x_A and combining with (147) gives

$$\partial(G_m^E/RT)/\partial x_A = \ln f_A - \ln f_B \tag{172}$$

Equations (171) and (172) may be solved simultaneously giving

$$\begin{aligned} \ln f_A &= G_m^E/RT + (1 - x_A)\,\partial(G_m^E/RT)/\partial x_A \\ \ln f_B &= G_m^E/RT - x_A\,\partial(G_m^E/RT)/\partial x_A \end{aligned} \tag{173}$$

The excess molar Gibbs function G_m^E is related to the molar heat of mixing H_m^E and the excess molar entropy of mixing S_m^E (see equations (26)–(28)) by

$$G_m^E = H_m^E - TS_m^E \tag{174}$$

In an athermal mixture $H_m^E = 0$ and $G_m^E = -TS_m^E$, so the athermal part of the activity coefficient f_i^*, by analogy with (171), is defined by

$$-S_m^E/R = x_A \ln f_A^* + x_B \ln f_B^* \tag{175}$$

The thermal contribution to the activity coefficient is equated with the thermal contribution H_m^E to G_m^E, thus

$$H_m^E/RT = x_A \ln t_A + x_B \ln t_B \tag{176}$$

By following the procedure used to separate the complete activity coefficients f_A and f_B, the terms f_i^* and t_i are obtained as

$$\begin{aligned}\ln f_A^* &= -S_m^E/R + (1 - x_A)\,\partial(-S_m^E/R)/\partial x_A \\ \ln f_B^* &= -S_m^E/R - x_A\,\partial(-S_m^E/R)/\partial x_B\end{aligned} \tag{177}$$

and

$$\begin{aligned}\ln t_A &= H_m^E/RT + (1 - x_A)\,\partial(H_m^E/R)/\partial x_A \\ \ln t_B &= H_m/RT - x_A\,\partial(H_m^E/R)/\partial x_A\end{aligned} \tag{178}$$

It must be emphasized that the definitions (175) and (176) do not constitute the only legitimate way to partition the activity coefficient. This choice has the advantage of simplicity but the disadvantage that for some solutions the f_i^* may be temperature dependent, suggesting a small residual thermal component.

Many solutions have molar excess thermodynamic properties that are conveniently fitted by the expressions

$$-S_m^E/R = \sum_i x_i \ln(\phi_i/x_i) \tag{179}$$

where ϕ_i is the volume fraction of component i, defined by

$$\phi_i = x_i V_i / \sum_j x_j V_j \tag{180}$$

and

$$H_m^E/RT = \phi_A \phi_B (w/k_B T)(x_A V_A + x_B V_B)/V_B \tag{181}$$

The equations (179) and (181) are known as Flory's approximation (Flory, 1941, 1942) and the term w is an interchange energy, when viewed in the context of the lattice model of mixtures of molecules of different sizes (Guggenheim, 1952).

The expression (179) leads to an athermal contribution to the activity coefficient given by

$$f_i^* = (\phi_i/x_i)\exp(1 - \phi_i/x_i) \tag{182}$$

and (181) to a thermal contribution

$$\begin{aligned}t_A &= \exp[(V_A/V_B)(w/k_B T)\,\phi_B^2] \\ t_B &= \exp[(w/k_B T)\,\phi_A^2]\end{aligned} \tag{183}$$

Assuming that the ratio V_A/V_B is the same for both the liquid and the surface phase, the constant k_t is given by (183) and (166) as

$$k_t = w^s/w^l \tag{184}$$

the superscripts (s and l) identifying the surface and liquid phases respectively. If both types of molecule lie flat in the surface phase, and the thickness τ is of monomolecular thickness, the surface molar volume $V_i^s = \tau a_i$ and hence by (143),

$$V_A^s / V_B^s = r \tag{185}$$

Combining equations (182)–(185) with the relation (142) gives on rearrangement

$$\begin{aligned}(1/r)\ln(\phi_A^s/\phi_A^l) - \ln[(1-\phi_A^s)/(1-\phi_A^l)] + w^l[k_t(1-2\phi_A^s)\\ -(1-2\phi_A^l)] - \ln K' = 0\end{aligned} \tag{186}$$

Equation (186) may be solved numerically. The common point ($x_A^s = \eta$) on all the isotherms described by (185), having fixed values for w^l and K' but differing values of k_t, is given by (183) and (168) as

$$\phi_A^s = \phi_B^s = 0.5 \tag{187}$$

which on combining with (180) and (185) gives the surface layer concentration η at the crossover as

$$\eta = V_B^s/(V_A^s + V_B) = 1/(r+1) \tag{188}$$

If $w^{s,p}$ and $w^{s,q}$ respectively correspond to the two choices k_t^p and k_t^q, combination of equations (169), (183), (184), and (188) gives at the adsorption isotherm crossover

$$(\gamma^p - \gamma^q) = Lr[r/(r+1)]^2(w^{s,p} - w^{s,q}) \tag{189}$$

where L is the Avogadro constant. The right-hand side of (189) is positive or negative depending upon whether $w^{s,p}$ is greater or less than $w^{s,q}$. If $w^{s,p} < w^{s,q}$, then $\gamma^p < \gamma^q$ at the crossover point. Furthermore, since (188) allows only one solution for η, the corollary of Rolle's theorem requires that $\gamma^p < \gamma^q$ for all solution compositions $0 < x_A^l < 1$. Finally, if $w^{s,p} < w^{s,q}$, then

$$\Gamma_A^{(n),p} > \Gamma_A^{(n),q} \text{ for } 0 < x_A^l < \xi$$

$$\Gamma_A^{(n),p} < \Gamma_A^{(n),q} \text{ for } \xi < x_A^l < 1$$

The conditions for the formation of a surface azeotrope are obtained by combining equations (183) and (170), giving the surface azeotrope composition θ as

$$\phi_A^s = \phi_A^l = \theta = 0.5 - 0.5k_B T \ln K'/[w^l(k_t - 1)] \tag{190}$$

Since θ can only have values between 0 and 1, azeotrope formation is observed if k_t satisfies

$$1 - |\ln K' k_B T/w^l| < k_t < 1 + |\ln K' k_B T/w^l| \qquad (191)$$

the straight brackets || indicating the modulus of the enclosed term.

All the features of the surface phase model are illustrated by Figs 14–17, in which the model adsorption isotherms are compared with the experimental adsorption data reported by Ash *et al.* (1975). Note that the component A has been chosen to have the larger molecules in the mixture so that K', defined by (142), may be the reciprocal of the value reported by Ash *et al.* (1975).

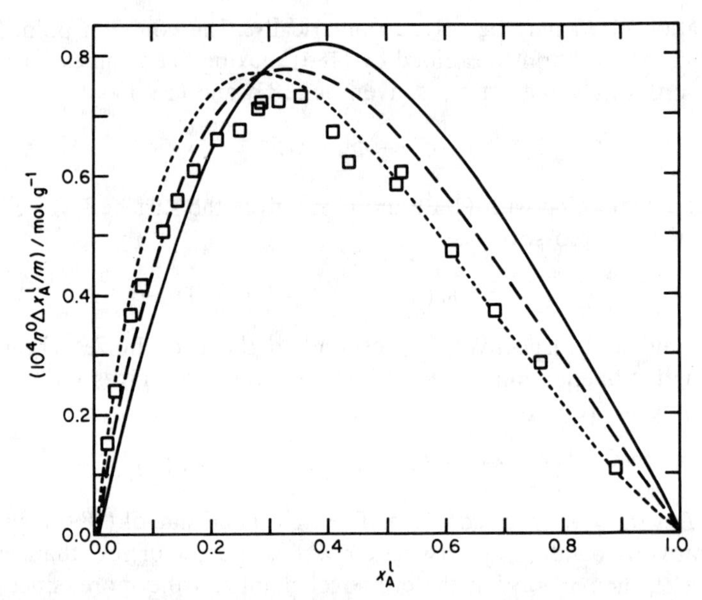

Figure 14 Comparison of the surface phase model adsorption isotherms with experimental data (Ash *et al.*, 1973) for the system benzene (A)/cyclohexane (B)/ Graphon at 313.15 K. All model curves have $a_A = a_B = 2.41 \times 10^5\ m^2\ mol^{-1}$, $K' = 2.05$, $r = 1$, and $w^l/k_B T = 0.474$. (——) $k_t = 1$; (– – –) $k_t = 0.5$; (- - - - -) $k_t = 0$.

For every system, the model isotherms are shown by a full line for $k_t = 1$, by a long-dash line for $k_t = 0.5$, and by a short-dash line for $k_t = 0$. All the formal requirements of the model are satisfied. In each case the isotherms for the three values of k_t intersect at a single point at $x_A^l = \xi$, $x_A^s = \eta$. By chance, all the solutions have a thermal contribution to the activity coefficient $t_A^l > 1$ at a solution concentration $x_A^l = \eta$. Thus for solution concentrations $0 < x_A^l < \xi$, the adsorption of component A increases with decrease in k_t. As a consequence, the model surface tension, at any solution concentration, decreases with k_t.

Figure 14 shows experimental data for the solution benzene (A)/cyclohexane (B) adsorbed on Graphon at 313.15 K. The two solution components have similar molar volumes so $r = 1$. The molar areas, defined by (126), are $a_A = a_B = 2.41 \times 10^5\ m^2\ mol^{-1}$, the solid surface area is $86\ m^2\ g^{-1}$, and the equilibrium constant $K' = 2.05$. The expressions (179) and (181) with $w^l/k_B T = 0.474$ give a reasonable approximation for S_m^E and H_m^E. The model curves intersect at $x_A^l = \xi = 0.285$ and $x_A^s = \eta = 0.5$. It is evident from Fig. 14 that the best agreement between experimental and model isotherms would be obtained with a value of k_t between 0.25 and 0.5. However, the agreement can never be good over the middle range of solution composition since the model crossover point does not fit the experimental data.

Experimental adsorption data for the system n-heptane (A)/benzene (B)/Graphon at 343.15 K are shown in Fig. 15. As before, the

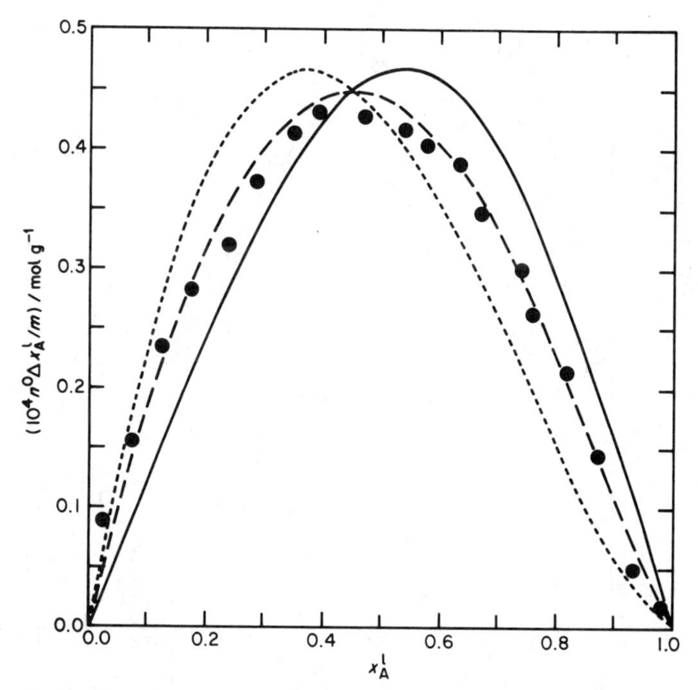

Figure 15 Comparison of the surface phase model adsorption isotherms with experimental data (Ash *et al.*, 1973) for the system n-heptane (A)/benzene (B)/Graphon at 343.15 K. All model curves have $a_B = 2.41 \times 10^5\ m^2\ mol^{-1}$, $r = 1.5$, $K' = 0.662$, and $w^l/k_B T = 0.382$. (——) $k_t = 1$; (– – –) $k_t = 0.5$; (- - - - -) $k_t = 0$.

molar surface area of benzene is $2.41 \times 10^5\ m^2\ mol^{-1}$, the solid surface area $86\ m^2\ g^{-1}$, but $r = 1.5$. The adsorption constant $K' = 0.662$, so the smaller molecular species, benzene, is preferentially adsorbed. The solution thermodynamic data are approximated by (179) and (181) with $w^l/k_B T = 0.382$. The model crossover point occurs at $x^l_A = \xi = 0.449$ and $x^s_A = \eta = 0.286$. In this example the model crossover point is close to the experimental data, so good agreement between model and experiment may be obtained by adjusting k_t. Indeed, the model curve for $k_t = 0.5$ does fit the experiment data very well.

Experimental adsorption data for a system with the same components, n-heptane/benzene/Graphon, but at the temperature 283.15 K, are shown in Fig. 16. The solution thermodynamic properties are reasonably fitted by setting $w^l/k_B T = 0.666$. Although the constant $K' = 0.820$ is less than unity, the smaller component (benzene) is only adsorbed preferentially

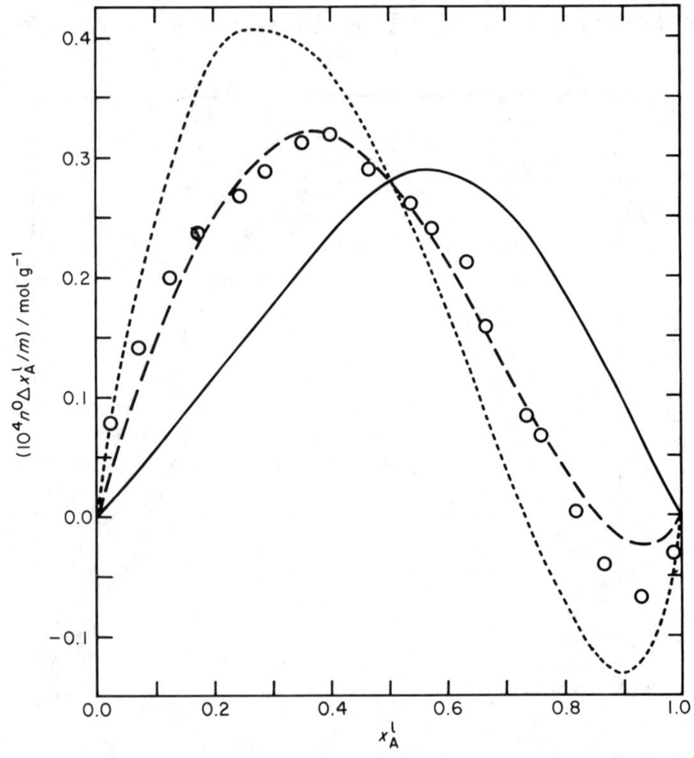

Figure 16 Comparison of surface phase model adsorption isotherms with experimental data (Ash *et al.*, 1973) for the system n-heptane (A)/benzene (B)/Graphon at 283.15 K. All curves have $a_B = 2.41 \times 10^5\ m^2\ mol^{-1}$, $r = 1.5$, $K' = 0.820$, and $w^l/k_B T = 0.666$. (——) $k_t = 1$; (– – –) $k_t = 0.5$; (- - - - -) $k_t = 0$.

over part of the liquid composition range, the azeotrope occurring at $x^1 = \theta \approx 0.82$. Thus the model curve with $k_t = 1$ cannot give a reasonable fit to the experimental data, since it is precluded from showing a surface azeotrope [see equation (170)]. The model adsorption isotherms with $k_t = 0$, 0.5 both show azeotropic behaviour, the curve with $k_t = 0.5$ providing a better fit to the experimental data.

Figure 17 shows experimental data for the adsorption of n-heptane (A)/cyclohexane (B) by Graphon at 283.15 K (open circles) and 343.15 K

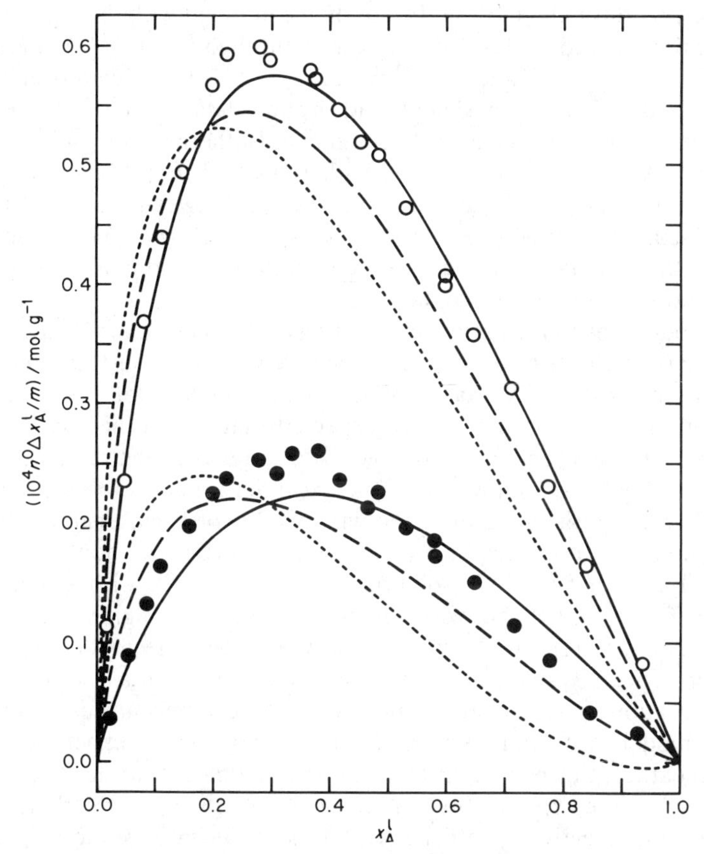

Figure 17 Comparison of surface phase model adsorption isotherms with experimental data (Ash *et al.*, 1975) for the system n-heptane (A)/cyclohexane (B)/Graphon. At 283.15 K, the experimental data are shown by the open circles and the model curves have $a_B = 2.41 \times 10^5$ m^2 mol^{-1}, $r = 1.5$, $K' = 1.97$. At 343.15 K, the experimental data are shown by filled circles, and the model curves have the same values for a_B and r but $K' = 1.30$. The solution thermodynamic data are calculated according to equation (192). (——) $k_t = 1$; (– – –) $k_t = 0.5$; (- - - - -) $k_t = 0$.

(full circles). The solution thermodynamic data are poorly represented by the approximations (179) and (181). Consequently, the expressions given by Crutzen *et al*. (1950) to fit their own experimental thermodynamic data have been used. These expressions are

$$
\begin{aligned}
S^E_m/R &= x_A(1 - x_A)[-0.448 + 0.141(2x_A - 1) - 0.091(2x_A - 1)^2] \\
H^E_m/R &= x_A(1 - x_A)[130.8 - 40.8(2x_A - 1) + 26.2(2x_A - 1)^2]
\end{aligned}
\tag{192}
$$

and both S^E_m and H^E_m are independent of temperature over the range of interest. These expressions were used to calculate the solution activity coefficients through (173), and were then combined with the experimental adsorption data to evaluate the K' values through (144), (141), and (112). The size ratio $r = 1.5$, and at 283.15 K, $K' = 1.97$. It is evident from Fig. 17 that at 283.15 K the experimental adsorption data are best fitted by $k_t = 1$, rather than by $k_t = 0.5$ or 0. The model isotherms cross at $x^l_A = 0.187$ and $x^s_A = 0.363$. At 343.15 K, and with $K' = 1.30$, Fig. 17 shows that the best fitting curve would have k_t with a value between 0.5 and 1.0. The model isotherms at 343.15 K cross at $x^l_A = 0.291$, $x^s_A = 0.363$, and by contrast with the situation at the lower temperature, the common point is well removed from the experimental data.

It is pertinent to recall the earlier observation that, if $k_t = 1$, the surface activity coefficients have the same dependence on the surface phase concentrations as the bulk phase activity coefficients have on the bulk phase concentration. It may have been expected that the surface phase model with $k_t = 1$ would give a reasonable fit to the experimental data. Examination of Figs 14–17 shows, with one exception, that the best fit between model and experiment is obtained with a value of k_t closer to 0.5 than to 1.0. The exception, n-heptane (A)/cyclohexane (B)/Graphon at 283.15 K, is notable in that the bulk solution is "pseudo-ideal", with activity coefficients $f^l_A \approx f^l_B \approx 1.0$, but with a non-zero heat of mixing. Examination of the remaining systems shows that, in every case, the model adsorption isotherms with $k_t = 1$ (full curves) cut the experimental isotherms at a concentration $x^l_A = \xi$ in such a manner that, for concentrations $0 < x^l_A < \xi$, the model adsorption of component A is less than the experimental. As shown earlier, this behaviour indicates that the model surface tensions are greater than the experimental values. In all these systems the thermal part of the solution activity coefficients $t^l_i > 1$, so reducing k_t reduces the model surface tension, thereby bringing the model isotherms into better agreement with experiment. The combination of $t^l_i > 1$ and $k_t < 1$, when substituted into equations (159), (165), and (166), leads to the inequality $1 < f^s_i(x^s_i = \alpha) < f^l_i(x^l_i = \alpha)$, where $0 < \alpha < 1$. This result implies that the surface phase is "more ideal" than the solution phase, and has been reported on many occasions (Everett, 1973; Kiselev and Khopina, 1969; Nagy and Schay,

1963; Sircar and Myers, 1970). It is not clear whether this behaviour would extend to solutions for which $t_i^{\mathrm{l}} < 1$.

In the pure liquids A and B, the thermal parts of the activity coefficients are defined to be $t_{\mathrm{A}}^{\mathrm{l}} = t_{\mathrm{B}}^{\mathrm{l}} = 1$. If, on mixing, there is a change in the total enthalpy, for the mixture $t_{\mathrm{A}}^{\mathrm{l}} \neq 1$ and $t_{\mathrm{B}}^{\mathrm{l}} \neq 1$. The change in enthalpy on mixing is approximately equal to the change in the total energy of intermolecular interactions, as A–B replace some A–A and B–B intermolecular interactions. On mixing, the A and B molecules in a bulk phase are subject to a change of average molecular environment which is equal in every direction, whereas A and B molecules in a surface phase have an essentially unchanged molecular environment in the direction of the solid. This simple argument provides a qualitative molecular explanation of the observation that the thermal parts of the activity coefficients are "more ideal" in the surface than in the bulk phase. It is not obvious how the argument can be quantified since it requires that $t_{\mathrm{A}}^{\mathrm{s}}$ and $t_{\mathrm{B}}^{\mathrm{s}}$ be functions of both $x_{\mathrm{A}}^{\mathrm{s}}$ and $x_{\mathrm{A}}^{\mathrm{l}}$, a requirement that generally contravenes the thermodynamic consistency relation (161).

There is evidence that the intermolecular interaction between an adsorbed molecule and a solid body is modified by the presence of other adsorbed molecules (Everett, 1965b; McLachlan, 1964; Richmond and Sarkies, 1973; Sinanoglu and Pitzer, 1960). A simple summary of this phenomenon has been given by Pierotti and Thomas (1971), who indicate that the adsorbed molecule/solid interaction has an additional repulsive term if the adsorbed molecules are equidistant from the solid, and an additional attractive term if one is directly above the other. At a solid/solution interface, both terms would be needed with a consequent reduction in overall magnitude of the effect. The modifying behaviour is operative for both pure liquids and their mixtures, so the net result is determined by the difference in the modification when a neighbouring adsorbed molecule is exchanged for one of a different species. At the present time, no precise estimate is possible of the magnitude, or sign, of the corrections to the surface activity coefficients introduced by this phenomenon.

In the preceding section it was shown that, at equilibrium, the solid/solution surface tension adopts the minimum value consistent with the given temperature and solution composition. Any relaxation mechanism, operating in the surface phase but not in the solution, is such as to reduce the surface tension below the value expected in the absence of the mechanism. If any intermolecular interactions, involving either or both A and B type molecules, are spherically asymmetric, the presence of the solid can favour particular orientations in the surface phase thereby providing a relaxation mechanism and thereby lowering the surface tension.

The surface phase model has been used to estimate the area of a solid/solution interface. Everett (1965a) has shown that for equal-sized molecules, forming both an ideal solution and surface phase, equation (140) may be rearranged as

$$x_A^l x_B^l/(x_A^s - x_A^l) = x_A^l + 1/(K-1) \tag{193}$$

According to (152),

$$\Gamma_A^{(n)} = (x_A^s - x_A^l)/a_m \tag{194}$$

where a_m is the area of a surface phase having one mole of mixture. If the number of moles of component in the surface phase between 1 gram of solid and solution is n^s/m, equations (193), (194), and (68) may be combined to give

$$x_A^l x_B^l/(n^0 \Delta x_A/m) = (m/n^s)[x_A^l + 1/(K-1)] \tag{195}$$

All quantities on the left-hand side of (195) are measurable. If $x_A^l x_B^l/(n^0 \Delta x_A/m)$ is plotted against x_A^l, the resulting curve should be linear, with slope m/n^s. By assigning a value to the cross-sectional area of the molecules, the area per gram of solid is evaluated from n^s/m. The adsorption coefficient K is given as the ratio (intercept at $x_A^l = 1$)/(intercept at $x_A^l = 0$).

Schay (1970) and Foti *et al.* (1974) have extended this procedure to include mixtures with molecules of equal size. The separation factor S is defined by

$$S = x_A^s x_B^l / x_B^s x_A^l \tag{196}$$

By repeating the procedure used for ideal mixtures, the following relation is obtained,

$$x_A^l x_B^l/(n^0 \Delta x_A/m) = (m/n_A^{s(0)})[(S - 1/r)x_A^l + 1/r]/(S-1) \tag{197}$$

with $n_i^{s(0)}/m$ being the number of moles of pure component i that can fit into the surface phase associated with 1 gram of solid. Comparison of (196) and (142) indicates that S is generally a function of concentration, so plotting $x_A^l x_B^l/(n^0 \Delta x_A^l/m)$ against x_A^l will produce a curve with slope $M(x_A^l)$ at x_A^l given by

$$M(x_A^l) = \frac{m}{n_A^{s(0)}}\left[\frac{S - 1/r}{S-1} + \left(\frac{(1/r - 1)x_A^l - 1/r}{(S-1)^2}\right)\frac{dS}{dx_A^l}\right] \tag{198}$$

If the resulting curve has a nearly linear section, it is reasonable to expect that in this section S is nearly constant. Since A has been chosen as the larger species $0 < 1/r < 1$ and, if $S \gg 1$ (component A is strongly adsorbed), then

$$M(x_A^l) \approx m/n_A^{s(0)} \qquad (S \gg 1) \tag{199}$$

If $S \approx 0$ (component B is strongly adsorbed), then

$$M(x_A^l) \approx m/n_B^{s(0)} \qquad (S \approx 0) \tag{200}$$

Both (199) and (200) are restricted to near-linear sections of the curve.

The intercept $C(x_A^l = 1)$ at x_A^l is given by (197) as

$$C(x_A^l = 1) = (m/n_A^{s(0)})S/(S-1) \tag{201}$$

regardless of any variation of S with x_A^l. If S is large, then

$$C(x_A^l = 1) \approx m/n_A^{s(0)} \qquad (S \gg 1) \tag{202}$$

Similarly, if $S \approx 0$ and $S \ll 1/r$,

$$C(x_A^l = 0) \approx -m/n_B^{s(0)} \qquad (S \approx 0, S \ll 1/r) \tag{203}$$

where $C(x_A^l = 0)$ is the intercept of the curve at $x_A^l = 0$. It is clear from this analysis that plotting $x_A^l x_B^l/(n^0 \Delta x_A/m)$ against x_A^l provides a useful estimate of the solid surface area (or $n_i^{s(0)}/m$), for mixtures of unequal sized molecules, provided that one of the components is strongly adsorbed. Foti *et al*. (1974) suggest some alternative methods for estimating solid/solution areas from experimental adsorption data, where the adsorption strengths of the two components are more evenly balanced.

The Polanyi (1914) potential theory was originally proposed to model the adsorption of gases by a solid. It can be modified to describe adsorption from completely miscible solutions (Hansen and Fackler, 1953). The fundamental premise is that each component i is subject to an adsorption potential $\varepsilon_i(z)$ that is a function of distance z from the solid ($z = 0$). The adsorption potentials are defined by reference to the solid/pure component i, and therefore include both the adsorbate–adsorbent and the adsorbate–adsorbate interactions of the pure liquid. The modification of the adsorption potentials by the introduction of another component is accounted for by the introduction of the distance-dependent mole fractions $x_i(z)$ and activity coefficients $f_i(z)$. This model is obviously better than the surface phase model, but unhappily it is usually cast in a form that leads to a chemical potential $\mu_i(z)$ that, at equilibrium, is not uniform throughout the solution and surface region [see equation (37)]. This problem can be overcome by acknowledging that the real surface tension does not act in a single plane (Melrose, 1968), and defining the chemical potential, by analogy with (130), as

$$\mu_i^s(z) = \mu_i^{0,l} + RT \ln f_i(z)x_i(z) + [\sigma_i^0(z) - \sigma(z)]a_i \tag{204}$$

where the $\sigma_i^0(z)$ are related to the surface tension by

$$\gamma_i^0 = \int_0^\infty \sigma_i^0(z)\,dz \tag{205}$$

and a similar expression relates $\sigma(z)$ to γ.

It is evident from (204) that for the pure liquid $\mu_i^0(z) = \mu_i^{0,1}$. This alternative approach will produce activity coefficients different from those defined by chemical potentials that are dependent upon z, even for the pure liquid. Any further development of the Polanyi potential theory is left to the interested reader.

In the preceding section, the concept of molecular distribution functions was introduced in the discussion of Monte Carlo simulations. It has been shown by Buff (1955, 1956) that, if the molecular distribution functions and the form of the intermolecular interactions are known, all the properties of a surface may be evaluated. If the total intermolecular energy of interaction is satisfactorily accounted for by adding the interactions of pairs of molecules, it is only necessary to know the relevant singlet and pairwise distribution functions (or the radial distribution functions). If there are two approximately spherical solution components A and B, both singlet distribution functions $\rho_A^{(1)}(z)$ and $\rho_B^{(1)}$ are required. The pair density functions $\rho_{AA}^{(2)}(\mathbf{r}_1, \mathbf{r}_2)$, $\rho_{BB}^{(2)}(\mathbf{r}_1, \mathbf{r}_2)$, and $\rho_{AB}^{(2)}(\mathbf{r}_1, \mathbf{r}_2)$ are required, $\mathbf{r}_1$ and $\mathbf{r}_2$ being the coordinates of the two molecules undergoing pairwise interaction. Some progress has been made in the development of models that will provide the necessary distribution functions for a given set of intermolecular potential functions. Two of the most advanced are those of Perram and Smith (1977) and Chan *et al.* (1979). For mathematical convenience, they choose "sticky hard sphere" intermolecular potential functions. Although physically unrealistic, these functions combine both repulsive and attractive forces, and can be made to reproduce realistic macroscopic properties such as second virial coefficients. The particle distribution functions are obtained via the Percus–Yevick approximation for liquids. The results of this model can only be checked by reference to data generated with a Monte Carlo simulation using the "sticky hard sphere" potentials. Although this approach is promising, it is a long way from providing information about real solid/solution interfaces.

VII. Porous Solids

It is evident that porous solids introduce a complexity which mitigates against a detailed molecular description of adsorption from solution. Nevertheless, porous solids are of considerable technological importance. There are at least two types of surface involved, internal within the pores and an external surface similar to that of non-porous solids. Provided that the pores are accessible to both components, the surface phase model may be used. The only necessary modification is to replace the surface areas by pore volumes, and in order to maintain consistency the surface tensions are replaced by pore pressures. Thus (130) now becomes

$$\mu_i^{\mathrm{p}} = \mu_i^{0,1} + RT \ln f_i^{\mathrm{p}} x_i^{\mathrm{p}} + (p_i^{0,\mathrm{p}} - p_i^{\mathrm{p}}) V_i^{\mathrm{p}} \tag{206}$$

the superscript p indicating the pore phase. Equation (206) refers only to solution in the pores. When equilibrium is established, the pore concentrations satisfy

$$\left(\frac{f_{\mathrm{A}}^{\mathrm{p}} x_{\mathrm{A}}^{\mathrm{p}}}{f_{\mathrm{A}}^{1} x_{\mathrm{A}}^{1}}\right)^{V_{\mathrm{B}}/V_{\mathrm{A}}} \left(\frac{f_{\mathrm{B}}^{1} x_{\mathrm{B}}^{1}}{f_{\mathrm{B}}^{\mathrm{p}} x_{\mathrm{B}}^{\mathrm{p}}}\right) = K'^{,\mathrm{p}} \tag{207}$$

If the volume of the external surface phase is significant, when compared with the pore volume, it is necessary to consider the additional surface phase equilibrium described by (142). In general $K'^{,\mathrm{p}} = K'$, where K' is the surface phase equilibrium constant. It is noted that Kipling (1965) was less than enthusiastic with this approach to porous solids.

References

Adams, L. H. (1915). *J. Am. Chem. Soc.* **37**, 1181.
Ash, S. G., Everett, D. H. and Findenegg, G. H. (1968). *Trans. Faraday Soc.* **64**, 2639.
Ash, S. G., Bown, R. and Everett, D. H. (1973). *J. Chem. Thermodyn.* **5**, 239.
Ash, S. G., Bown, R. and Everett, D. H. (1975). *J. Chem. Soc. Faraday Trans. 1* **71**, 123.
Barker, J. A. and Henderson, D. (1976). *Rev. Mod. Phys.* **48**, 587.
Böttcher, C. J. F. (1973). *Theory of Electric Polarization*, 2nd edn, Vol. 1, p. 18ff, p. 92ff, Elsevier, Amsterdam.
Brown, C. E., Everett, D. H. and Morgan, C. J. (1975). *J. Chem. Soc. Faraday Trans. 1* **71**, 883.
Buff, F. P. (1955). *J. Chem. Phys.* **23**, 419.
Buff, F. P. (1956). *J. Chem. Phys.* **25**, 146.
Butler, J. A. V. (1932). *Proc. Roy. Soc.* A**135**, 348.
Callen, H. B. (1960). *Thermodynamics*, John Wiley, New York and London.
Candler, C. (1951). *Modern Interferometers*, p. 480, Hilger and Watts, Glasgow.
Chan D. Y. C., Pailthorpe, B. A., McCaskill, J. S., Mitchell, D. J. and Ninham, B. W. (1979). *J. Colloid Interface Sci.* **72**, 27.
Courant, R. (1951). *Differential and Integral Calculus* (E. J. McShane, trans.), 2nd edn, Vol. 1, p. 104, Blackie and Son, London and Glasgow.
Crutzen, J. L., Haase, R. and Sieg, L. (1950), *Z. Naturforsch.* **59**, 600.
Defay, R. and Prigogine, I. (1950). *Trans. Faraday Soc.* **46**, 199.
Defay, R., Prigogine, I., Bellemans, A. and Everett, D. H. (1966). *Surface Tension and Adsorption*, Longmans, London.
Everett, D. H. (1950). *Trans. Faraday Soc.* **46**, 453.
Everett, D. H. (1964). *Trans. Faraday Soc.* **60**, 1803.
Everett, D. H. (1965a). *Trans. Faraday Soc.* **61**, 2478.
Everett, D. H. (1965b). *Discuss. Faraday Soc.* **40**, 177.
Everett, D. H. (1972). *Pure Appl. Chem.* **31**, 579.
Everett, D. H. (1973). In *Colloid Science*, Specialist Periodical Reports, Vol. 1, p. 49, Chemical Society, London.

Flory, P. J. (1941). *J. Chem. Phys.* **9**, 660.
Flory, P. J. (1942). *J. Chem. Phys.* **10**, 51.
Foti, G., Nagy, L. G. and Schay, G. (1974). *Acta Chim. Acad. Sci. Hung.* **80**, 25.
Gibbs, J. W. (1961). *The Scientific Papers of J. Willard Gibbs*, Vol. 1, pp. 55–371, Dover, New York.
Guggenheim, E. A. (1952). *Mixtures*, Oxford University Press, London.
Guggenheim, E. A. (1957). *Thermodynamics*, North Holland, Amsterdam.
Guggenheim, E. A. (1966). *Applications of Statistical Mechanics*, Ch. 6, Oxford University Press, London.
Hansen, R. S. and Fackler, W. V. (1953). *J. Phys. Chem.* **57**, 634.
Kipling, J. J. (1965). *Adsorption from Solutions of Non-Electrolytes*, Ch. 4, Academic Press, London and New York.
Kiselev, A. V. and Khopina, V. V. (1969). *Trans. Faraday Soc.* **65**, 1936.
Kurbanbekov, E., Larionov, O. G., Chinutov, K. V. and Yudilevich, M. D. (1969). *Russ. J. Phys. Chem.* **43**, 916.
Lane, J. E. (1968). *Aust. J. Chem.* **21**, 827.
Lane, J. E. (1975). *Faraday Discuss. Chem. Soc.* **59**, 55.
Lane, J. E. and Johnson, C. H. J. (1967). *Aust. J. Chem.* **20**, 611.
Lane, J. E. and Spurling, T. H. (1976). *Aust. J. Chem.* **29**, 2103.
Lane, J. E. and Spurling, T. H. (1978). *Aust. J. Chem.* **31**, 933.
Lane, J. E. and Spurling, T. H. (1981). *Aust. J. Chem.* **34**, 1.
McLachlan, A. D. (1964). *Mol. Phys.* **7**, 381.
Melrose, J. C. (1968). *Ind. Eng. Chem.* **60**, 53.
Nagy, L. G. and Schay, G. (1963). *Acta Chim. Acad. Sci. Hung.* **39**, 365.
Ono, S. and Kondo, S. (1960). *Handbuch der Physik* (S. Flügge, ed.), Vol. 10, p. 134, Springer, Berlin.
Parfitt, G. D. and Wiltshire, I. J. (1964). *J. Phys. Chem.* **68**, 3545.
Perram, J. W. and Smith, E. R. (1977). *Proc. Roy. Soc.* (London) *A***353**, 193.
Pierotti, R. A. and Thomas, H. E. (1971). In *Surface and Colloid Science* (E. Matijević, ed.), Vol. 4, p. 93, Wiley-Interscience, New York.
Polanyi, M. (1914). *Verh. Deut. Physik. Ges.* **16**, 1012.
Prigogine, I. and Defay, R. (1954). *Chemical Thermodynamics* (D. H. Everett, trans.), p. 220, pp. 311–326, pp. 381–392, Longmans, London, New York, Toronto.
Rice, O. K. (1967). *Statistical Mechanics, Thermodynamics and Kinetics*, p. 329ff, p.359, Freeman, San Francisco.
Richmond, P. and Sarkies, K. W. (1973). *J. Phys.* (*C*) **6**, 401.
Rusanov, A. I. (1967). *Phase Equilibria and Surface Phenomena*, Ch. 6 (in Russian), Chimia, Leningrad.
Rusanov, A. I. (1971). *Progr. Surf. Membr. Sci.* **4**, 57.
Schay, G. (1970). In *Proceedings of the International Symposium on Surface Area Determination* (D. H. Everett and R. H. Ottewill, eds), p. 273, Butterworths, London.
Schay, G. and Nagy, L. G. (1966). *Acta Chim. Acad. Sci. Hung.* **50**, 207.
Sinanoglu, O. and Pitzer, K. S. (1960). *J. Chem. Phys.* **32**, 1279.
Sircar, S. and Myers, A. L. (1970). *J. Phys. Chem.* **74**, 2828.
Snook, I. K. and Henderson, D. (1978). *J. Chem. Phys.* **68**, 2134.
Snook, I. K. and van Megen, W. (1979). *J. Chem. Phys.* **70**, 3099.
Thomy, A. and Duval, X. (1970). *J. Chim. Phys. Phys. Chim. Biol.* **67**, 1101.
Wood, W. W. (1968). *The Physics of Simple Liquids* (H. N. V. Temperley, J. S. Rowlinson and G. S. Rushbrooke, eds), p. 115, North Holland, Amsterdam.

3. Adsorption of Nonionic Surfactants

J. S. CLUNIE and B. T. INGRAM

I. Introduction

The term "surfactant" is generally understood to refer to long-chain molecules containing both hydrophilic and lipophilic moieties, i.e. amphiphilic or soap-like structures. Compared with most other solutes, surfactants behave in a distinctive manner in water. Firstly, by adsorption, they significantly modify interfacial properties at unusually low bulk solution concentrations compared with species that are not surface-active. Secondly, by cooperative self-association, surfactants aggregate in dilute aqueous solution over a very narrow concentration range to form micelles. Thus, at low surfactant concentrations only individual amphiphilic molecules exist in aqueous solution whereas above the critical micelle concentration (c.m.c.), where aggregation first occurs, both individual molecules and micellar clusters coexist in dynamic equilibrium (Goodman and Walker, 1979). At still higher solution concentrations, micelles of varying size and shape pack together with characteristic symmetries to form lyotropic mesomorphic phases (Corkill and Goodman, 1969). Recently it has been shown (Laughlin, 1978) how general phase criteria can be used to determine whether polar functional groups will be "operative" as hydrophilic moieties in paraffin chain molecules, i.e. whether or not surfactant behaviour will be exhibited in aqueous solution by such structures.

Surfactants are usually broadly classified according to the nature of their hydrophilic groups, viz. ionic or nonionic. The ionic class can be further subdivided into anionic, cationic, and zwitterionic types (Ottewill, 1979), although the latter may alternatively be included in the nonionic category (Laughlin, 1978). In this chapter we shall mainly consider the adsorption of nonionic surfactants from dilute aqueous solution, because surfactants exhibit their particularly distinctive properties in water, and most published work has been concerned with dilute aqueous systems. We shall, however, refer briefly to other solvent systems.

II. Adsorbates

On a purely practical basis, over 250 types of nonionic surfactant are known (Mukerjee and Mysels, 1970; McCutcheon, 1980), but we shall restrict our discussion mainly to well characterized materials rather than deal at length with ill-defined heterogeneous commercial mixtures. By far the commonest type of nonionic surfactant favoured in adsorption studies is that comprising a polyoxyethylene glycol group, $-(OCH_2CH_2)_yOH$, linked to either an alkyl, C_xH_{2x+1}, or an alkylphenyl, $C_xH_{2x+1}\cdot C_6H_4$, hydrocarbon chain. For brevity in this chapter, the alkyl polyoxyethylene glycol monoethers will be designated as C_xE_y, where x is the number of carbon atoms in the alkyl chain and y is the number of oxyethylene units in the polyoxyethylene chain (E). The corresponding average will be designated $\bar{y}$ for a commercial surfactant with a distribution of polyoxyethylene chains. The alkylphenyl polyoxyethylene glycol monoether series contains a phenyl ring between the alkyl group and the polyoxyethylene chain, and is here abbreviated to $C_x\phi E_y$. With C_xE_y and $C_x\phi E_y$ types of molecule it is important to remember that the E_y grouping is in reality a small polymer, and that in aqueous solution (and possibly also at surfaces) various chain configurations can be adopted, viz. fully extended, helical, meander or random coil (Tanford *et al.*, 1977).

Relatively few adsorption studies have been reported for nonionic surfactants which possess a small compact hydrophilic group such as is usually encountered with ionic surfactants. Dimethyldodecyl phosphine oxide, $DC_{12}PO$, is perhaps the one notable exception (Mast and Benjamin, 1969), although one other homologous series based on the difunctional sulphinylalkanol hydrophilic group, $-SO(CH_2)_nOH$, has also found favour in adsorption studies (Corkill *et al.*, 1967). Long-chain alkanols have featured in a number of studies, but we shall not discuss them in any detail because they lack some of the properties (e.g. micellization) that would allow them to be classified strictly as surfactants (Laughlin, 1978).

The distinctive solution behaviour of nonionic surfactants can be conveniently summarized in the phase diagram (Clunie *et al*, 1969) for $C_{12}E_6$ in water (Fig. 1). The formation of micelles occurs in very dilute solution—the c.m.c. for $C_{12}E_6$ is 87 μmol dm^{-3} at 25°C (Balmbra *et al.*, 1962)—and the factors responsible for the formation of micelles have been reviewed in depth by Goodman and Walker (1979). With increase in temperature, changes in hydration and chain configuration lead to changes in micelle size, shape, and polydispersity, with eventual phase separation into two coexisting isotropic solutions at the lower consolute temperature or "cloud point". Similar phase behaviour is encountered in the $DC_{12}PO$–water system (Herrmann *et al.*, 1966).

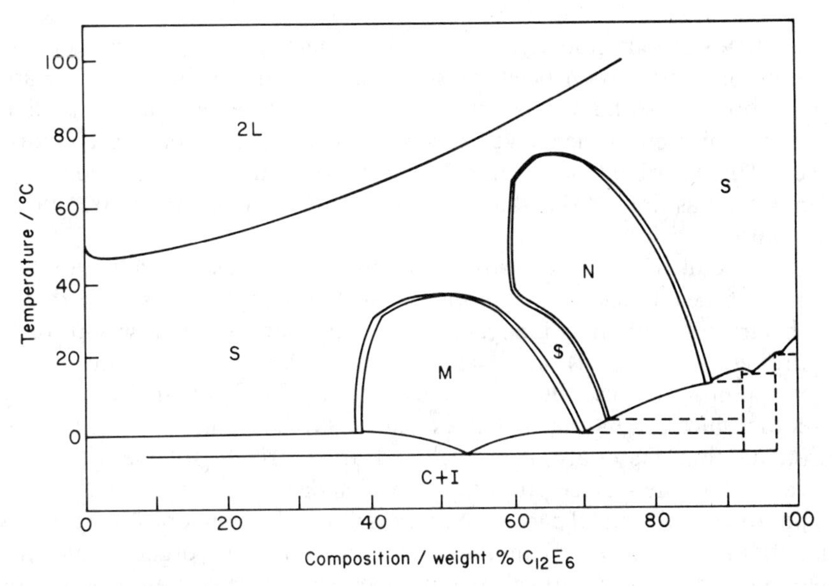

Figure 1 Phase diagram for $C_{12}E_6$ + H_2O: 2L, coexisting isotropic liquids; S, isotropic solution; M, middle phase; N, neat phase; I, ice; C, crystals. (Clunie *et al.*, 1969)

In comparing various nonionic surfactants, and in particular those based on the polyoxyethylene glycol head group, it has proved useful to adopt the "hydrophilic–lipophilic balance" (HLB) classification scheme of Griffen (1949), whereby an empirical HLB number can be simply calculated from the relative weighting of the hydrophilic and lipophilic moieties in the surfactant molecule. For polyoxyethylene surfactants, the HLB = weight % of polyoxyethylene glycol ÷ 5. Phase behaviour (and correlated properties) vary with HLB number, and for water-soluble C_xE_y compounds the HLB may be conveniently derived from lower consolute data. It should also be noted that the latter can be affected by the presence of a third component in the system, that is by the type and amount of organic solubilizate, e.g. urea, and simple inorganic electrolyte, e.g. sodium bromide (Shinoda, 1967). Thus, in considering the adsorption of nonionic surfactants at solid surfaces, it is important not to ignore bulk phase behaviour, and in particular the possibility of phase separation ("nucleation") phenomena as phase boundaries are approached (Corkill *et al.*, 1966a).

III. Adsorbents

Adsorbents used in studies with nonionic surfactants range from materials like graphitized carbon black or silver iodide, utilized because they are good model adsorbents, to much less well characterized materials, like cotton or pigment, which have been used because of their practical importance. The adsorbents are generally fine powders although other physical forms such as fibres (Mast and Benjamin, 1969; Kissa and Dettre, 1975) have also been used.

It is useful to divide the adsorbents into two types depending on the nature of the surface and its interaction with adsorbate. Those adsorbents with surfaces containing ionogenic sites or dipolar molecular groups (e.g. hydroxyl or carbonyl) are classified as "polar" or, alternatively, as "hydrophilic" because they have a high affinity for water. In this category are the silicates, inorganic oxides and hydroxides, natural fibres, and proteinaceous materials. Also included are sparingly soluble salts, like silver iodide, the surface properties of which depend on the particular ions exposed on the crystal faces. "Non-polar" or "hydrophobic" adsorbents are usually carbonaceous materials such as carbon blacks, organic pigments, and polymers. These are also sometimes classified as "low energy" or "non-specific" adsorbents because they interact with adsorbate through van der Waals dispersion forces rather than the more specific, and generally stronger, dipolar or electrostatic forces. Metals are usually assigned to the high energy category.

In classifying adsorbents it is important to realize that a solid may have several quite different types of surface depending on past treatment. These different types can sometimes coexist on the same adsorbent giving a non-uniform or heterogeneous surface. Such differences can result from (i) surface oxidation, which can give polar sites on the surface of an otherwise non-polar solid like carbon black, or reactive oxides on metal surfaces, (ii) heating, which can remove polar groups, e.g. hydroxyls, from a substance such as silica rendering it hydrophobic, (iii) the presence of potential-determining ions in the adsorbate solution which can alter the surface charge on hetero-ionic solids, e.g. silver iodide crystals, and amphiprotic adsorbents, e.g. metal oxides, (iv) the exposure of different crystal planes and edges with different adsorption characteristics, as in polymorphic materials such as titanium dioxide, and (v) the presence of adsorbed or deposited materials which may be on the surface as accidental contaminants or deliberately introduced to alter the surface properties as in the cases of silanated silicas and smectite clay adsorbents pretreated with trialkyl ammonium salts. An adsorbent can also have different surface geometries—a low-area form where the surface is mostly external, and a high-area form where the surface is mainly internal, as in porous solids, swelling clays, and zeolites.

Despite these considerations many adsorption studies with nonionic surfactants have been carried out with poorly characterized adsorbents. As yet, little use has been made of modern spectroscopic methods of surface analysis (Parfitt and Rochester, 1976), and adsorbent characterization has mostly been confined to determination of specific surface area (Sing, 1976).

IV. Adsorption Process

A. EXPERIMENTAL METHODS

1. *Adsorption Isotherm*

For the very dilute solutions used in most work with nonionic surfactants the number of moles of surfactant, n_2^s, adsorbed on unit mass of solid is given (Aveyard and Haydon, 1973) to a good approximation by

$$n_2^s = n_0 \Delta x_2 / m$$

where n_0 is the total number of moles of solution before adsorption, m is the mass of insoluble adsorbent, and Δx_2 is the change in mole fraction of surfactant in solution resulting from adsorption. Δx_2 has been obtained by measuring changes in refractive index, either by interferometry or differential refractometry (Mathai and Ottewill, 1966; Corkill *et al.*, 1966, 1967), UV absorbance (Kuno and Abe, 1961; Akers and Riley, 1974; Schwuger and Smolka, 1977), and surface tension (Ottewill and Walker, 1968). Gravimetric, IR absorbance (Fukushima and Kumagai, 1973), colorimetric (Wååg, 1968; Klimenko and Koganovskii, 1973), and radiotracer techniques, particularly useful for very low concentrations (Corkill *et al.*, 1967; Mast and Benjamin, 1969), have also been used. The less accurate method, measurement of the amount of surfactant on the solid after drying, has also been employed occasionally; when it has been used in conjunction with solution depletion techniques, involving gravimetric (Schott, 1964), radiotracer (Gordon and Shebs, 1968; Mast and Benjamin, 1969), and carbon analysis methods (Schott, 1964), satisfactory mass balance has been found.

If the solute contains more than one surface-active component its composition in solution may change as a result of preferential adsorption of the different surface-active species, and the concentration dependence of the measured property may alter. The apparent adsorption value could therefore depend on the particular method of determining the concentration. This is an important consideration in assessing adsorption data for nonionic surfactants because so much work has been carried out using impure and multicomponent surfactants. The excess surface concentration of surfactant, Γ_2^s, is related to n_2^s by $\Gamma_2^s = n_2^s/A_s$, where A_s is the specific surface area of the adsorbent. The specific surface area is generally determined by BET vapour adsorption analysis (Brunauer *et al.*, 1938) using nitrogen or krypton, although occasionally other vapours such as benzene or water have been used. However, the surface of solid that is accessible to the gas used in the BET determination is not necessarily the same as that available to the surfactant in solution. For example, in a porous solid the area accessible to a large surfactant molecule is likely to be much less than that available to a small gas molecule. On the other hand, in solids like smectite clay or cotton, which can swell in water, or in particulate agglomerates, which can break into smaller fragments, especially in surfactant solutions, the area available to adsorbate may be much larger than that given by the BET analysis. Specific surface areas can be determined using the adsorption of dyes or simple amphiphilic molecules, such as fatty acids or alcohols, and it might be supposed that these surface areas would correspond more closely to those accessible to the nonionic surfactant. However, doubt concerning which value to use for the effective cross-sectional area of the "standard" adsorbate still makes the calculation of Γ_2^s rather uncertain.

2. *Other Techniques*

In only a few adsorption studies has determination of the isotherm been supplemented by other techniques. These have included measurement of enthalpies of immersion (Skewis and Zettlemoyer, 1960; Corkill *et al.*, 1966, 1967; Klimenko *et al.*, 1979), X-ray diffraction determination of the interlayer spacing in smectite clays (Schott, 1964; Platikanov *et al.*, 1977), and examination of the IR spectra of nonionic surfactants on oxide surfaces (Fukushima and Kumagai, 1973; Kumagai and Fukushima, 1976a). However, complementary studies, particularly those concerned with sol stabilization by nonionic surfactants, have usually included measurements of hydrodynamic and electrokinetic properties, and a recent study of the orientation of adsorbed surfactant (Bisio *et al.*, 1980) combined hydrodynamic, electrokinetic, and contact angle measurements.

B. GENERAL FEATURES

1. *Experimental Observations*

The adsorption isotherms of nonionic surfactants are generally Langmuirian, like those of most other highly surface-active solutes adsorbing from dilute solution. They are reversible with little hysteresis. However, the isotherms are often the stepped L4 type of Langmuir isotherm (Giles *et al.*, 1960) rather than the simple L2 type (see Chapter 1, Fig. 2). Figure 2 shows an idealized L4 type equilibrium adsorption isotherm for nonionic

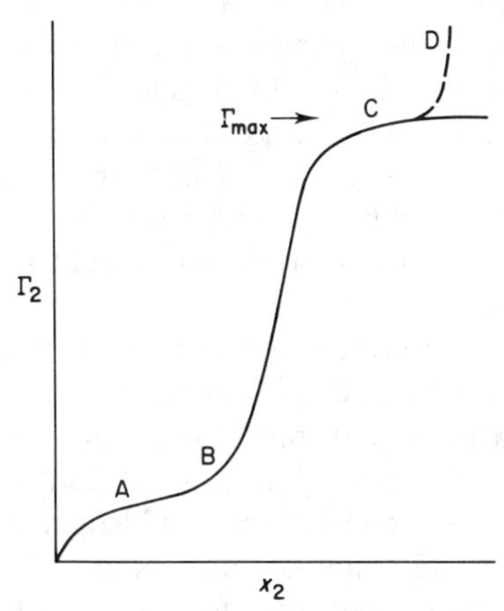

Figure 2 Idealized L4 type isotherm for adsorption of nonionic surfactants at the solid/solution interface.

surfactants. The first plateau, A, has been found in only a few systems, although these include both polar (Schott, 1964; Mathai and Ottewill, 1966; Klimenko and Koganovskii, 1973) and non-polar (Corkill *et al.*, 1966b, 1967; Klimenko and Koganovskii, 1973) adsorbents with alkyl sulphinyl ethanol or alkyl polyoxyethylene glycol monoether adsorbates. However, experimental insensitivity in this very low concentration region may have obscured the presence of a plateau in some studies. The inflexion and sharp increase in adsorption at B generally occur at bulk solution concentrations that are close to the c.m.c. and are followed by a second plateau region, C. A further, usually very marked, increase in adsorption, D, is sometimes observed at higher concentrations.

Occasionally plateau C is absent from such systems, and the isotherm then has a sigmoidal shape—L3 type (Giles *et al.*, 1960)—which is more usually characteristic of capillary condensation phenomena. The adsorption isotherms of fatty alcohols are often of this type (Hansen and Craig, 1954).

The molecular structure of the surfactant influences the shape of the isotherm in various ways. Within an homologous series it is found that increasing the length of the hydrocarbon chain generally increases the magnitude of adsorption, Γ_{max}, at the plateau C. Increasing the size of the polar head group, e.g. by adding ethylene oxide groups, tends to decrease Γ_{max}. These trends parallel those found in adsorption at the solution/vapour interface (Crook *et al.*, 1963; Lange, 1965; Barry and El Eini, 1976; Donbrow, 1975). Data from the few experiments using alkyl polyoxyethylene glycol monoethers in which both the alkyl chain and the number of ethoxy groups have been varied (Corkill *et al.*, 1966a; Mathai and Ottewill, 1966) suggest that for a non-polar adsorbent Γ_{max} may decrease with increasing HLB value of the surfactant, but that the relationship may not hold strictly for polar adsorbents. Increasing the alkyl chain length in an homologous series decreases the c.m.c. and also the concentration at which the adsorption plateau, C, is reached. Isotherms for members of the same homologous series are roughly superimposable when the "reduced adsorption", Γ_2^s/Γ_{max}, is plotted against "reduced concentration", x_2/c.m.c. (Hansen and Craig, 1954).

The effect of adsorbate structure on the early stages of adsorption has not been very well investigated. However, it seems that adsorption at A decreases with increasing size of the head group (Klimenko *et al.*, 1974b) but that, in contrast to the behaviour at C, it also decreases as the length of the alkyl chain is increased (Klimenko and Koganovskii, 1973; Klimenko *et al.*, 1974b). The initial slope of the isotherm tends to increase with increasing alkyl chain length but the effect of the hydrophilic moiety seems to depend on the type of head group. For example, increasing the number of ethoxy groups increases the initial gradient in adsorption on acetylene black (Klimenko *et al.*, 1974b), but increasing the number of methylene groups between the polar groups of sulphinyl ethanol surfactants adsorbed on Graphon has no effect on the slope (Corkill *et al.*, 1966b, 1967). There has been no systematic study of the effect of surfactant structure on those isotherms that show the steep rise in adsorption, D, at higher concentrations.

Although there have not been many experiments directly comparing the adsorption of the same surfactants on different adsorbents (Klimenko and Koganovskii, 1973) it is possible to identify certain trends in the influence of adsorbent by examining data from different workers. As the adsorbent becomes more polar the first plateau, A, occurs at lower adsorption values (Klimenko and Koganovskii, 1973), and indeed in some studies adsorption appears to occur only at concentrations above the c.m.c. (Kuno and Abe, 1961). Γ_{max} is generally two or three times as large for polar adsorbents as it is for the same surfactant on a non-polar adsorbent (Corkill *et al.*, 1966a; Mathai and Ottewill, 1966). Adsorption does not seem to be very sensitive to surface charge (Mathai and Ottewill, 1966; Watanabe *et al.*, 1963).

Investigations of temperature effects have featured in only a few studies but it seems that increasing temperature increases adsorption above the c.m.c., and the isotherm is more likely to exhibit the adsorption increase, D, as temperature is raised (Corkill *et al.*, 1966a; Kumagai and Fukushima, 1976b). Below the c.m.c., the effect of temperature is more obscure because of lack of experimental data. An experiment with octyl sulphinyl ethanol adsorbed on Graphon (Corkill *et al.*, 1966b, 1967) indicated that, below the c.m.c., adsorption decreased with increasing temperature.

It is important to consider the influence that other solute species, particularly surface-active ones, may have on the adsorption of nonionic surfactants, because most commercially available nonionic surfactants are mixtures. Commercial polyethoxylated surfactants are usually mixtures of homologues with an almost Poisson distribution of ethoxy groups (Shachat and Greenwald, 1967). Although some adsorption studies (e.g. Corkill *et al.*, 1966, 1967; Mathai and Ottewill, 1966; Ottewill and Walker, 1968) have utilized single-component polyethoxylated surfactants, most have used mixtures, sometimes with a narrowed homologue distribution (e.g. Schott, 1964). However, there has been no direct comparison of the adsorption isotherms of single- and multi-component nonionic surfactants on solids. McCracken and Datyner (1977) concluded from theoretical considerations that, even if there were to be selective adsorption of homologues from a Poisson distribution polyethoxylated surfactant mixture, the apparent maximum adsorption would not differ significantly from that of a single-component surfactant that had the same number of ethoxy groups as the number average for the mixture. A similar conclusion was reached concerning adsorption at the solution/air interface, although the apparent adsorption calculated from surface tension measurements has been found to be significantly smaller for single-component surfactants than for the normal polydisperse mixture (Crook *et al.*, 1963; Donbrow, 1975).

A study of the adsorption of a nonionic/anionic surfactant mixture on porous charcoal (Schwuger and Smolka, 1975, 1977) showed that the adsorption isotherm of the nonionic was only altered slightly by the anionic, although the adsorption of the latter was markedly affected by the nonionic surfactant and exhibited a maximum in its isotherm. Indeed, the most likely cause of a maximum or, more rarely, a minimum in the adsorption isotherm of a surfactant is the presence of other surface-active species. It has been suggested that the reason for this is associated with changes in the relative monomer concentrations of surfactants above the c.m.c. caused by mixed micelle formation (Trogus *et al.*, 1979).

Little is known about the effect of non-surface-active solutes on the adsorption of nonionic surfactants on solids. Electrolytes can alter the solubility, surface activity, and aggregation properties of nonionic surfactants (Shinoda *et al.*, 1963), and it is probable that they will have an effect on adsorption at the solid/solution interface. Thus an electrolyte that "salts out" a surfactant would probably increase its adsorption (Doren *et al.*, 1975). However, on polar surfaces some inorganic ions may be strongly adsorbed and actually displace surfactant. An example is the hydrolysed lanthanum ion (Mathai, 1963; Ottewill, 1967). Non-electrolytes which at high concentrations can alter solvent structure might also be expected to have an effect on nonionic surfactant adsorption (Schwuger, 1971).

pH has been shown to have some influence on the adsorption of nonionic surfactants on surfaces with carboxyl groups (Krings *et al.*, 1974). At high pH (>5.2) adsorption is low but it is increased tenfold at lower pH. This effect was ascribed to hydrogen-bonding between carboxyl and surfactant ether groups.

Surfactants can solubilize water-insoluble organic materials by incorporating them in micelles, and the effect of solubilizate on the adsorption of nonionic surfactants has been investigated. It was found that the presence of either a polar (Koganovskii *et al.*, 1974) or a non-polar (Koganovskii *et al.*, 1976) solubilizate had no significant effect on the maximum adsorption of surfactant on acetylene black but increased the adsorption on silica gel (Koganovskii *et al.*, 1975). In both cases the solubilizate was adsorbed with the surfactant, but on acetylene black the ratio of solubilizate to surfactant was greater than in the bulk solution whereas on silica gel it was less.

Despite the fact that most nonionic surfactants, unlike ionic ones, are soluble in a wide range of non-aqueous liquids, very little work has been done to investigate their adsorption from non-aqueous media. A direct comparison between adsorption from aqueous and non-aqueous solutions was carried out using polyethoxylated nonyl phenols in cyclohexane (Kuno *et al.*, 1964). Langmuirian isotherms were found on calcium carbonate and sigmoidal ones on carbon black. This contrasted with the behaviour in aqueous systems where Langmuirian isotherms were found on carbon and sigmoidal ones on calcium carbonate, although other workers (Akers and Riley, 1974) later found Langmuirian isotherms with the aqueous calcium carbonate systems. A study using contact-potential change as a measure of adsorption indicated Langmuirian isotherms for nonionic surfactants adsorbing from toluene on to metals (Lunina *et al.*, 1976). However, interpretation of work with non-aqueous solvents is complicated by the likelihood of there always being some water present because of the great difficulty of completely dehydrating the surfactant.

Calorimetric studies have been confined to systems with a non-polar adsorbent such as graphitized carbon black (Skewis and Zettlemoyer, 1960; Corkill *et al.*, 1966, 1967; Klimenko *et al.*, 1979). This is an ideal adsorbent for such studies because, having a homogeneous low-energy surface, its heat of immersion in water is low (about 35 mJ m^{-2}) and the heat changes due to surfactant adsorption can be measured fairly easily. The total heat change on immersing the solid in the surfactant solution is equal to the heat of wetting plus the heat change due to dilution resulting from adsorption. (In the case of micellar solutions this latter heat will also include a contribution from the heat of de-micellization.) In general, the dilution effect is small and the heat of wetting is easily determined.

Corkill *et al.* (1966, 1967) found that the absolute value of the heat of wetting of Graphon in solutions of n-alkyl hexaoxyethylene glycol monoethers and alkyl sulphinyl alkanols, at concentrations sufficient to give surface saturation, increased with increasing surfactant alkyl chain length. The absolute value of the heat also increased linearly with increasing adsorption of surfactant and then became constant at surface concentrations near those at which inflexion points were seen in some of the adsorption isotherms but well below the surface saturation values.

Measurement of the interlayer spacing in smectite clays has been another useful source of information about the adsorption process. The basal spacing increased in two steps of 0.42 nm as dodecyl polyoxyethylene glycol monoethers were progressively adsorbed on montmorillonite clay (Schott, 1964). This distance is similar to the thickness of a surfactant molecule lying flat on the surface. The surface concentrations at which the two spacings were observed coincided with the plateau regions in a two-step Langmuir (L4) isotherm where the first plateau was reached just below the c.m.c. However, a later study (Platikanov *et al.*, 1977) found changes in basal spacing that were much larger and more consistent with a vertical orientation of the surfactant. This work also revealed a series of stepwise decreases in interlayer spacing as temperature increased.

Most nonionic surfactants lack groups that are strongly spectroscopically absorbing, and there is very little spectroscopic information on their surface properties. However, some infrared measurements carried out on polyethoxy monolaurates adsorbed on titanium dioxide (Fukushima and Kumagai, 1973) showed that the peak due to the stretching vibration of the surfactant ester group had been shifted from 1738 cm^{-1} to 1722 cm^{-1}. Moreover, the spectrum of octadecanol adsorbed on silica shows that there is an interaction between hydroxyl and silanol groups (Low and Lee, 1972).

Experimental data on the kinetics of adsorption of nonionic surfactants are very sparse. However, it has been reported (Klimenko *et al.*, 1975b) that the concentration of nonionic surfactant in contact with a silica gel decreases at a rate that is approximately proportional to the average surfactant concentration when the surfactant is below its c.m.c. but that the rate is almost constant above the c.m.c. Schwuger and Smolka (1975, 1977) observed that in a mixed anionic/nonionic surfactant system the anionic surfactant was able to diffuse to the surface faster than the nonionic but was subsequently displaced by the latter. Mast and Benjamin (1969), studying the time dependence of adsorption of radioactive dimethyldodecyl phosphine oxide on cotton, rayon, and nylon fibres, concluded that the surfactant was diffusing into the fibres. A similar conclusion was reached in earlier studies with ethoxylated surfactants (Gordon and Shebs, 1968).

2. *Theory*

Nonionic surfactants are physically adsorbed, rather than chemisorbed. However, they differ from many other surface-active solutes in that quite small changes in concentration, temperature, or molecular structure of the adsorbate can have a large effect on the adsorption. This is due to adsorbate–adsorbate and adsorbate–solvent interactions which cause solute

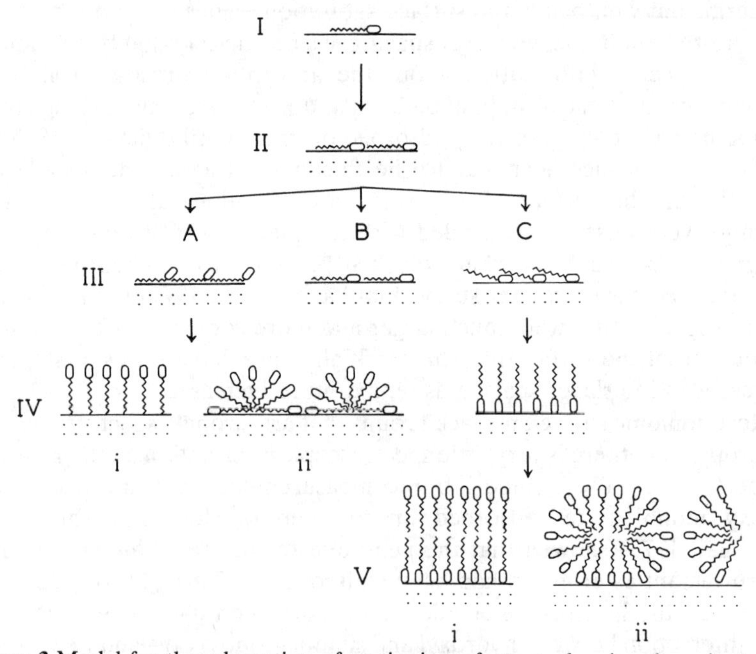

Figure 3 Model for the adsorption of nonionic surfactant, showing the orientation of surfactant molecules at the surface. I–V are the successive stages of adsorption, and sequences A–C correspond to situations where there are respectively weak, intermediate, and strong interactions between the adsorbent and the hydrophilic moiety of the surfactant

aggregation in bulk solution and which lead to changes in orientation and packing of surfactant at the surface. In Fig. 3 we show a general scheme of the most likely orientation changes undergone by nonionic surfactants adsorbed from aqueous solution. The three adsorption isotherms corresponding to the different adsorption sequences shown in Fig. 3 are illustrated in Fig. 4.

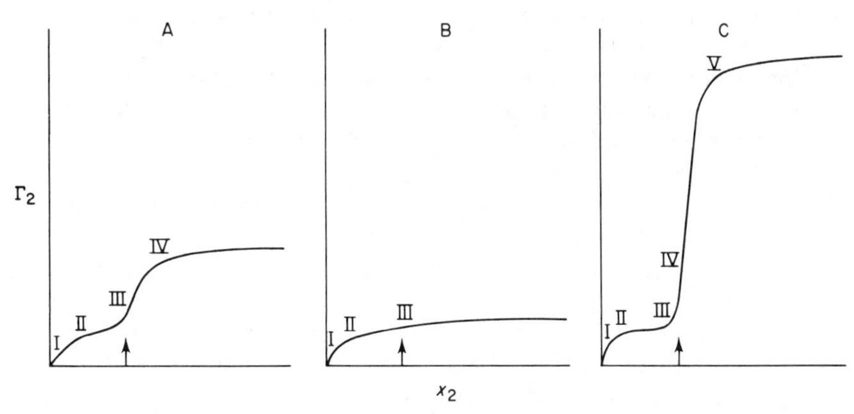

Figure 4 Adsorption isotherms corresponding to the three adsorption sequences shown in Fig. 3 (I–V), indicating the different orientations; the c.m.c. is indicated by an arrow.

In the first stage of adsorption [Fig. 3(I)] the surfactant is adsorbing on a surface where there are very few other adsorbate molecules and where consequently adsorbate–adsorbate interactions are negligible. Adsorption occurs because of van der Waals interaction (principally dispersion forces) and therefore, in the case of an aqueous solvent, it is determined mainly by the hydrophobic (i.e. lipophilic) moiety of the surfactant. Nevertheless, the polar groups of the surfactant may have some interaction with the surface, and hydrophilic ethylene oxide groups can have a slight positive adsorption even on a non-polar adsorbent (Corkill *et al.*, 1966a; Klimenko *et al.*, 1974b). When the interaction is due to dispersion forces the heats of adsorption are relatively small and correspond to the heat liberated by replacing surface solvent molecules with surfactant (Corkill *et al.*, 1966, 1967). At this stage the adsorbate tends to lie flat on the surface because its hydrophobic portion is positively adsorbed, as are also most types of nonionic hydrophilic head groups, especially large polyethylene oxide ones. With the molecule lying parallel to the surface, the adsorption energy will increase in almost equal increments for each additional carbon atom in its alkyl chain, and the initial slope of the isotherm will increase correspondingly (Traube's rule). The same can also happen with each additional ethylene oxide group (Klimenko *et al.*, 1974b).

The approach to monolayer saturation with the molecules lying flat [Fig. 3(II)] is accompanied by a gradual decrease in the slope of the adsorption isotherm. Although most of the "free" solvent molecules will have been displaced from the surface by the time the monolayer is complete, the surfactant itself will probably still be hydrated at this stage, so there will continue to be solvent molecules in the interfacial layer (Corkill *et al.*, 1967). Increase in the size of the surfactant molecule by, for example, lengthening an alkyl or polyethylene oxide chain will decrease the adsorption. On the other hand, increasing temperature should increase the adsorption because de-solvation reduces the effective size of the adsorbate (Volkov, 1974, 1976). In general, increasing temperature reduces the solubility of nonionic surfactants but increases their surface activity. This is also true of fatty alcohols (Bartell and Donahue, 1952; Bartell *et al.*, 1951).

The subsequent stages of adsorption are increasingly dominated by adsorbate–adsorbate interactions, although it is the adsorbate–adsorbent interaction that initially determines how the adsorption progresses when stage II is complete. The adsorbate–adsorbent interaction depends on the nature of the adsorbent and on the hydrophilic–lipophilic balance in the surfactant. When the hydrophilic group is only weakly adsorbed it will be displaced from the surface by the alkyl chains of adjacent molecules [Fig. 3(III)A]. This can occur more readily when the adsorbent is non-polar or when the surfactant has a short polyethylene oxide chain rather than a long one (i.e. low HLB). However, if there is a strong attraction between the hydrophilic group and the surface, such as may occur with polar adsorbents like silicates and oxides, the alkyl chain is displaced [Fig. 3(III)C]. The intermediate situation [Fig. 3(III)B] occurs when neither type of displacement is favoured and the adsorbate then remains flat on the surface. In this case adsorption might not be expected to change with subsequent increase in surfactant concentration, except perhaps from a slight increase due to surfactant de-solvation. However, it has been suggested (Kuno *et al.*, 1964; Barclay and Ottewill, 1970) that multilayers of horizontally oriented adsorbate can be formed. Nevertheless, apart from some adsorption data for smectite clay (Schott, 1964, 1967; Barclay and Ottewill, 1970), there does not seem to be very much evidence for the existence of layers of nonionic surfactant molecules lying flat on the surface at high bulk solution concentrations.

There is some evidence that transitions from one adsorption sequence to another can be brought about by changing the adsorbate rather than the adsorbent. Experiments on carbon black (Klimenko *et al.*, 1974b) suggest that a transition from type A to type B adsorption might occur with alkyl polyoxyethylene glycol monoethers when the ratio of ethoxy groups to carbon atoms in the alkyl chain exceeds 6, and an example of adsorption changing from type C to type A when the surfactant alkyl chain is lengthened may have been provided by the experiments of Mathai and Ottewill (1966) on silver iodide. (However, see Section IV.C for an alternative explanation of the silver iodide observations.) It could be conjectured that other factors which might change the adsorption sequence are the presence of other types of surfactant and changes to the solvent. For example, a strongly adsorbed co-surfactant might convert a hydrophilic solid to a hydrophobic one. Hence the adsorption of a nonionic surfactant in the presence of, say, a cationic surfactant, which can adsorb strongly on some polar substrates through an ion-exchange mechanism, would follow sequence A whereas without the cationic surfactant it would follow sequence C. Decreasing the polarity of the solvent would favour displacement of the alkyl chains from a non-polar adsorbent, so whereas adsorption might progress by sequence A in water it would change to B or C in a less polar solvent.

The change in amount adsorbed in the third stage of adsorption is unlikely to be large, but as the concentration of surfactant in the bulk solution approaches the c.m.c. there will be a tendency for the alkyl chains of the adsorbed molecules to aggregate. This will cause the molecules to become vertically oriented and there will be a large increase in adsorption (stage IV). This increase is probably not entirely caused by the change in orientation. The lateral forces due to alkyl chain interactions in the adsorbed layer will compress the head group, and for an ethylene oxide chain this will result in a less coiled, more extended conformation. There may also be some de-solvation. However, even in the close-packed monolayer the ethoxy chain will not be fully extended (van Voorst Vader, 1960; Rösch, 1967; Tanford *et al.*, 1977). The longer the surfactant alkyl chain the greater will be the cohesion force and hence the smaller the surfactant cross-sectional area. This may explain why saturation adsorption increases with increasing alkyl chain length as well as with decreasing polyethylene oxide chain length. With non-polar adsorbents the adsorption energy of a methylene group is almost the same as its micellization energy (Klimenko *et al.*, 1974b), so the surface aggregation process can occur quite easily even at concentrations below the bulk solution c.m.c. With polar adsorbents, however, the head group may be quite strongly bound to the surface, and partial detachment of a large ethoxy chain from the surface, needed for close-packing of the alkyl chains [Fig. 3(IV)C], may not be achieved until the surfactant concentration is above the c.m.c. When the adsorption layer is like that shown in Fig. 3(IV)C the surface will be hydrophobic.

The interactions occurring in the adsorption layer during the fourth and subsequent stages of adsorption are similar to the interactions in the bulk solution where enthalpy changes caused by increased alkyl–alkyl interactions balance those due to head group interactions and de-solvation processes. For this reason the heat of adsorption becomes constant, although adsorption increases with increasing temperature because the head group becomes less solvated and more compact (Corkill *et al.*, 1966, 1967; Volkov, 1974, 1976).

The parallel between bulk solution micellization and the aggregation processes occurring at the surface has been most strongly emphasized by Klimenko and Koganovskii (Klimenko *et al.*, 1974a; Koganovskii *et al.*, 1979; Klimenko, 1979, 1980). They have suggested that above the c.m.c. adsorbed surfactant forms micellar aggregates on the surface. To support this hypothesis they cite the fairly close agreement between measured saturation adsorption and adsorption calculated on the assumption that the surface is covered with close-packed half-micelles, i.e. aggregates with the same size and shape as a bulk solution micelle divided in half as shown in Fig. 3(IV)A(ii). However, the saturation adsorption values can be explained equally well by the planar monolayer model [Fig. 3(IV)A(i)], because measured adsorption values are generally not very different from those derived from the tensions at the air/solution surface where the existence of hemi-spherical micelles is unlikely. The transport of solubilizate to the surface has also been explained with a surface micelle model (Koganovskii *et al.*, 1974, 1975, 1976), although again a planar monolayer model could also account for the observations. Unfortunately there are at present no experimental techniques that could differentiate unambiguously between the two structures and establish the correct model. Data on the kinetics of adsorption (Klimenko *et al.*, 1975b) indicate that, even if there are micelles on the surface, they are not deposited from bulk solution but result from the association of surfactant molecules already adsorbed as monomer.

In the case of ionic surfactants it has been suggested (see Chapter 6) that true adsorption may not occur at all on certain polar mineral substrates, and that the apparent adsorption at low concentrations is simply an ion-exchange process which is followed at concentrations near the c.m.c. by hemi-micelle formation. In these systems "adsorption" is solely a surface nucleation phenomenon resulting from surfactant–surfactant or surfactant–solvent interactions rather than surfactant–adsorbent interactions. The absence of adsorption below the c.m.c in certain nonionic surfactant–polar adsorbent systems may indicate a similar process in which van der Waals attraction between surfactant and surface is opposed by a repulsion originating in a thick surface solvation layer. Adsorption is thus prevented until surfactant aggregation effects dominate. However, most nonionic surfactants contain polar groups which are likely to be adsorbed quite strongly on polar surfaces by, for example, hydrogen-bonding to surface hydroxyls, so the apparent absence of adsorption below the c.m.c.

is more probably due to lack of experimental precision at very low concentrations.

For non-polar adsorbents, Fig. 3(IV)A(i or ii) represents the final adsorption state, but for polar adsorbents Fig. 3(IV)C is only an intermediate stage. Increasing surfactant concentration causes the adsorption of a further layer,oriented as shown in Fig. 3(V)C(i), which is promoted by the elimination of alkyl chain–solvent interactions and by alkyl–alkyl attraction. The surface of the solid thus regains its hydrophilic character and there is an increase in the stability of dispersions (Ottewill, 1967). Increasing the length of the alkyl chain increases lateral compression so adsorption increases with increasing alkyl chain length and decreasing ethoxy chain length. However, the saturation adsorption is usually more than twice as large as in the monolayer (Mathai and Ottewill, 1966), which may mean that the compression effect of alkyl chain interactions is greater in a bilayer than in a monolayer. Several temperature dependent structures for the bilayer, corresponding to alternative packing arrangements of different chain conformations, also result from alkyl chain interactions (Weiss, 1980). The surface micelle model for adsorption on polar solids above the c.m.c. is shown in Fig. 3(V)C(ii). The micelles are assumed to be slightly distorted versions of those in bulk solution but to have similar aggregation numbers (e.g. Iler, 1955; Kuno *et al.*, 1964).

It is important to note that adsorption sequence C may be followed by nonionic surfactants even on adsorbents that might be regarded as hydrophobic. For example, on a polycarbonate membrane, which although bearing a slight negative charge (1 ionogenic group per 80 nm^2) had a non-zero air/water contact angle, the adsorption of $C_8\phi E_{\bar{9}}$ was shown to follow sequence C rather than sequence A (Bisio *et al.*, 1980).

The large adsorption values found for some systems on raising the temperature are probably caused by the hetero-flocculation or surface nucleation of large surfactant aggregates. In general, very steep isotherms and high adsorption on polar and non-polar substrates seem to be associated with the formation of two phases in the bulk solution (Corkill *et al.*, 1966a; Kumagai and Fukushima, 1976b). The surfactant "phase" on the surface probably resembles that in the solution. Therefore, above the first cloud point, or lower consolute temperature, an adsorbed monolayer or bilayer is probably covered with large, essentially hydrophilic, surfactant aggregates. At higher temperature these could change to a lamellar mesophase structure (Kumagai and Fukushima, 1976b) which, at even higher temperatures, near the second cloud point, might convert to an amorphous surfactant phase containing very little water. Such transitions are likely to occur on the surface at slightly lower temperatures than in bulk solution, and these surface nucleation effects may be accentuated on porous solids by capillary condensation. The sigmoidal shape of fatty alcohol isotherms

(Bartell and Donahue, 1952; Hansen and Craig, 1954) is another manifestation of this phenomenon. The surface nucleates the phase separation of the sparingly soluble alcohol at concentrations slightly below its solubility limit in bulk solution.

Although consideration of the theoretical aspects of nonionic surfactant adsorption has mostly been confined to qualitative descriptions, a quantitative theory, based on a three-stage adsorption model, has been developed by Klimenko (1978b, 1979) for the adsorption of alkyl polyoxyethylene glycol monoethers on non-polar adsorbents. Klimenko assumes that in the first stage surfactant displaces solvent and lies flat on the surface, although he postulates that there is some interaction between adsorbate molecules due to alkyl chain attraction. A modified Langmuir adsorption equation is used for this stage (Klimenko, 1978b):

$$c_2 K_a = \left[\frac{\Gamma_2}{\Gamma_2^* - \Gamma_2(1 - a_1/a_2')}\frac{a_1}{a_2'}\right] f_a' \tag{1}$$

where c_2 is the equilibrium concentration of surfactant in the bulk solution, Γ_2 is the surface excess concentration at c_2, Γ_2^* is the surface excess at the c.m.c., K_a is a constant, a_1 and a_2' are the effective cross-sectional areas of the solvent and adsorbate molecules in the surface, and f_a' is an adsorbate surface activity coefficient. The term inside the square brackets is a type of surface "concentration" which is defined as the ratio of numbers of adsorbed surfactant molecules to number of solvent molecules in the equilibrium interfacial layer (Koganovskii and Levchenko, 1966). The constant K_a allows for the adsorbate–adsorbent interactions and is therefore related to the energy of adsorption at infinite dilution. The coefficient f_a', which accounts for adsorbate–adsorbate interactions, is assumed to have the following dependence on Γ_2 in this first stage of adsorption (Klimenko, 1978a, b):

$$f_a' = \exp\left[\frac{\Gamma_2}{\Gamma_2^* - \Gamma_2} - \frac{K_2\Gamma_2}{\Gamma_2^*}\right] \tag{2}$$

where K_2 is an adsorbate–adsorbate interaction constant. Equation (1) is an adaptation of an isotherm used by Koganovskii (Koganovskii and Levchenko, 1966; Brown and Everett, 1975). Combination of equations (1) and (2) gives an equation that is essentially the same as one used by Hill (1946) for gas adsorption and corresponds to a two-dimensional van der Waals equation of state (Koganovskii *et al.*, 1977).

When all the solvent molecules have been displaced and the surface is covered with a close-packed monolayer of horizontal adsorbate molecules, the second stage begins and the ethoxy chains are progressively displaced by the alkyl chains of adjacent molecules. This allows the surface "concentration" to increase by an amount (Klimenko *et al.*, 1975a):

$$\frac{\Gamma_2 - \Gamma_2'}{\Gamma_2'(a_E'/a_1) - (\Gamma_2 - \Gamma_2')(a_2/a_1)}$$

where Γ_2' is the surface excess at the beginning of the second stage, a_E' is the cross-sectional area of the ethoxy chain lying flat on the surface, and a_2 is the area of the surface covered by each surfactant molecule. A simple model for the displacement of the ethoxy chain gives a_2 equal to $[a_2' - a_E(1 - \cos\alpha)]$ where α is the smallest angle that the tilted ethoxy chain can make with the surface (Klimenko *et al.*, 1974b). The adsorption isotherm for the second stage is thus (Klimenko, 1978b):

$$c_2 K_a = \left[\frac{\Gamma_2'}{\Gamma_2^* - \Gamma_2'(1 - a_1/a_2')}\frac{a_1}{a_2'} + \frac{\Gamma_2 - \Gamma_2'}{\Gamma_2' a_E' - (\Gamma_2 - \Gamma_2')a_2} a_1\right] f_a'' \qquad (3)$$

The activity coefficient, f_a'', is assumed to include contributions from ethoxy chain interactions and therefore to differ from f_a' in its dependence on Γ_2. The logarithm of f_a'' is arbitrarily assumed to have a linear dependence on Γ_2:

$$\ln f_a'' = \ln(f_a')_{\Gamma_2'} + [\ln(f_a'')_{\Gamma_2^*} - \ln(f_a')_{\Gamma_2'}]\Gamma_2/(\Gamma_2^* - \Gamma_2') \qquad (4)$$

In this equation f_a' with subscript Γ_2', the maximum value of f_a' which is reached when $\Gamma_2 = \Gamma_2'$, is obtained by substituting the value for Γ_2' into equation (2); f_a'' with subscript Γ_2^*, the maximum value of f_a'' achieved when $c_2 = c^*$, the c.m.c., and $\Gamma_2 = \Gamma_2^*$, can be obtained by substitution into equation (3).

In the final adsorption stage, which starts at the c.m.c., Klimenko (1979) assumes that adsorbed surfactant associates into hemi-micelles on the surface. By considering the equilibrium between molecules in the bulk solution and "free" positions in these surface micelles, he derives a simple Langmuir isotherm:

$$c_2 K_a^* = \Gamma_2/(\Gamma_2^\infty - \Gamma_2) \qquad (5)$$

where Γ_2^∞ is the maximum surface excess, i.e. the surface excess when the surface is covered with close-packed hemi-micelles. The adsorption constant K_a^* is taken to be inversely proportional to the c.m.c. because it is assumed that the surface micelles are so similar to the bulk solution micelles that the adsorption energy will be almost equal to the energy of micellization. There is no activity coefficient because it is assumed that above the c.m.c. deviations from ideality in the surface and in the bulk solution derive from a similar micellar association effect with the result that the two activity coefficients will cancel each other in the adsorption equation.

However, although this assumption may be correct at concentrations far above the c.m.c., it seems less likely to hold near the c.m.c. where the major proportion of surfactant in the bulk solution will still be in the monomer state. Nevertheless, despite this reservation and some doubts (Brown and Everett, 1975) concerning the theoretical basis for the area correction term in equation (1), the three isotherms have given good agreement with experiment for several nonionic surfactants over large concentration ranges (Klimenko, 1978b, 1979), although this may owe something to the semi-empirical form of the activity coefficients.

Kinetic aspects of nonionic surfactant adsorption have received little theoretical attention. The time taken to reach equilibrium when a surfactant is adsorbed from dilute solution on a solid is usually several hours and generally decreases with increasing concentration of surfactant. Several transport mechanisms occur in adsorption. In most experiments there is some agitation or stirring, and this ensures that adsorbate is carried near the external surfaces of the solid by convection rather than diffusion. However, surfactant then diffuses through an unstirred layer to reach the surface. In the early stages of adsorption this step is probably rate-determining because the unstirred region can be quite extensive, particularly if the solid is porous. It is usual (Kipling, 1965) to distinguish between diffusion within pores ("internal" diffusion) and diffusion through the thin layer adjacent to the external surface ("external" diffusion), although the effective thickness of the external layer also depends on surface geometry as well as on hydrodynamic factors such as solvent viscosity and rate of stirring. The adsorption step is probably much faster than the diffusion to the surface although, as coverage increases, desorption and surface orientation effects will slow down the net rate of adsorption (Ward and Tordai, 1946; Hansen. 1960). The Langmuir model (Davies and Rideal, 1963) gives a simple form for the rate of adsorption:

$$d\theta/dt = k_a(1 - \theta)c - k_d\theta \qquad (6)$$

where θ is the surface coverage at time t, c is the solution concentration, and k_a and k_d are the adsorption and desorption rate constants respectively. However, interactions between adsorbed molecules and the dependence of adsorbate orientation on θ make it unlikely that in practice k_a and k_d would be independent of θ.

After the surfactant has been adsorbed two further transport processes may occur. The first is surface diffusion (the spreading of adsorbate along the surface) which can play an important role in the attainment of adsorption equilibrium in porous solids (Schwuger, 1971). The second is diffusion of the surfactant into the solid, which has been found to happen with nonionic surfactants and some polymeric adsorbents (Mast and Benjamin, 1969).

Probably the most important theoretical deduction made from the few kinetic studies carried out with nonionic surfactants concerns the nature of the adsorption process above c.m.c. The observation (Klimenko *et al.*, 1975b) that the rate of adsorption is dependent on surfactant monomer concentration, not total concentration, suggests that adsorption of micelles does not occur to any great extent.

C. SPECIFIC SYSTEMS

1. *Non-polar Adsorbents*

(a) *Carbon black*. This has been the adsorbent most frequently used in studies with nonionic surfactants. An early investigation of n-alkanol adsorption from aqueous solution on to Spheron 6 and Graphon, a graphitized carbon black with reproducible, uniform, graphite basal plane surfaces (Hansen and Craig, 1954), produced isotherms of the L3 type. These were superimposable for the higher alcohols when adsorptions were plotted against a "reduced concentration", i.e. actual concentration divided by solubility. Adsorption values at the "knee" of the Graphon isotherms corresponded approximately to the values that would be given by a close-packed monolayer of alcohol molecules oriented with their long axes horizontal. As the solution approached saturation the adsorptions on both adsorbents rose rapidly above the values that could be given by any kind of monolayer packing, even one vertically oriented. The adsorption of fatty alcohols on non-polar substrates seems to parallel the bulk solution behaviour. Simple alcohols do not form micelles. On a non-polar adsorbate, therefore, they would not be expected to be able to form a stable vertically oriented monolayer of the type shown in Fig. 3(IV)A(i).

In an early study with nonionic surfactants, Abe and Kuno (1962) measured the adsorption of a series of commercial polyethoxylated nonyl phenols ($C_9\phi E_{\bar{y}}$, $\bar{y} = 5$ to 30). Their carbon black adsorbent had an average particle size of 20 nm and an argon BET surface area of 241 $m^2\,g^{-1}$. All the surfactants had simple Langmuirian isotherms reaching surface saturation below the c.m.c. Adsorption increased with decreasing number of ethoxy units, and areas per molecule calculated from the maximum adsorption were very much larger than the cross-sectional areas obtained for adsorption at the solution/air interface (Hsiao *et al.*, 1956). For example, $C_9\phi E_{\overline{9}}$ and $C_9\phi E_{\overline{10}}$ had areas of 2.44 and 2.75 nm^2 respectively on carbon but 0.53 and 0.60 nm^2 at the solution/air interface. Nevertheless, in all cases the dispersions of carbon particles were stable when surface saturation had been achieved. It was concluded that the surfactant was adsorbed by attachment of the alkyl chain, although Ottewill (1967) has suggested that the ethoxy chain might also have been lying extended or loosely coiled on the surface. However, it is difficult to reconcile the large areas with the observed stabilization of the dispersions, particularly in view of the much smaller area obtained in later work with similar surfactants. This may indicate that the area of solid accessible to the surfactant was less than the BET specific surface area, perhaps because of porosity.

Kuno *et al.* (1964) later examined the adsorption of polyethoxylated nonyl phenols from cyclohexane on the same carbon black. The isotherms were L3 type and were attributed to multilayer adsorption on polar sites resulting from oxidation and impurities on the carbon surface. The BET equation was used to obtain the surfactant areas in the monolayer, and these were twice as large as the areas found in the aqueous study. Like the latter, they increased with increasing ethoxy chain length.

In a detailed study of adsorption on Graphon, Corkill *et al.* (1966a) used a series of very pure alkyl polyoxyethylene glycol monoethers (C_xE_y). The adsorbent had a nitrogen BET specific surface area of 91 $m^2\,g^{-1}$ and average particle diameter of 25 nm. The isotherms were mostly simple Langmuirian with saturation adsorption reached near or slightly above the c.m.c. The maximum adsorption values increased with increasing alkyl chain length and decreasing ethoxy chain length.

The minimum areas per molecule obtained from the 25°C isotherms are given in Table 1 where they are compared with areas at the solution/air

Table 1 Areas occupied by n-alkyl polyoxyethylene glycol monoethers at adsorption saturation, 25°C

	Area per molecule/nm^2		
	Graphon/solution (Corkill *et al.*, 1966a)	Silver iodide/solution (Mathai and Ottewill, 1966)	Air/solution [equation (7)]
E_6	1.68	—	—
C_6E_6	0.93	—	0.94
C_8E_6	0.81	0.24	0.83
$C_{10}E_6$	0.65	0.18	0.72
$C_{12}E_6$	0.55	0.18	0.61
C_8E_9	1.09	—	1.02
$C_{16}E_9$	—	0.60	0.47

interface. These latter areas have been interpolated from the data of Lange (1965) and Barry and El Eini (1976) using the semi-empirical relationship

$$A_{A/W} = (K_1 - K_2x)\sqrt{y} \tag{7}$$

where $A_{A/W}$ is the minimum area (nm^2) obtained from the surface tension–concentration curve (Gibbs equation), and K_1 and K_2 are empirical constants equal to 0.521 and 0.0228 respectively for C_xE_y surfactants at

25°C. The similarity between the areas at the two interfaces suggests that, at saturation, molecules adsorbed on the solid are vertically oriented as in Fig. 3(IV)A(i). Hexaoxyethylene glycol was also positively adsorbed, indicating that the hydrophilic group had some affinity for the carbon surface. The heat of immersion of the carbon in the C_xE_6 surfactant solutions was measured and found to be constant at concentrations well below those at which adsorption saturations were reached. Table 2 shows the heats of immersion in water, heptane, anhydrous hexaoxyethylene

Table 2 Heats of immersion of Graphon, 25°C. (Corkill *et al.*, 1966a)

Liquid	Heat of immersion/J g^{-1}
Water	−3.1
Heptane	−10.0
Hexaoxyethylene glycol (E_6)	−10.0
Aqueous $C_{12}E_6$ solution	−9.2
Aqueous $C_{10}E_6$ solution	−7.5
Aqueous C_8E_6 solution	−6.3
Aqueous C_6E_6 solution	−5.4

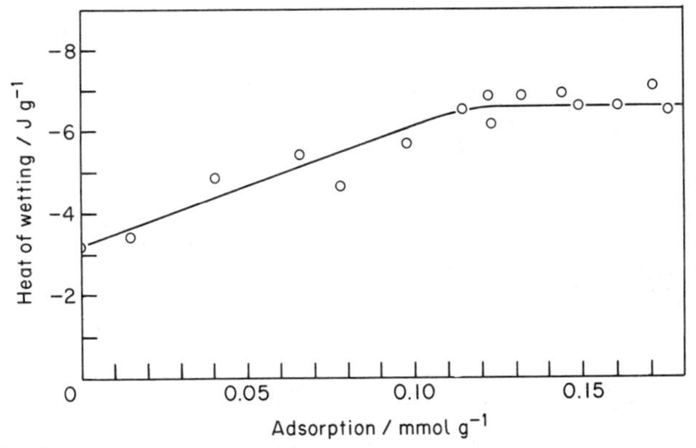

Figure 5 The heat of wetting of Graphon (25°C) by C_8E_6 solutions. (Corkill *et al.*, 1966a)

glycol, and the aqueous surfactant solutions at adsorption saturation. From this comparison Corkill *et al.* (1966a) concluded that some of the hydration water associated with the surfactant head group in the bulk aqueous phase was also present in the adsorbed layer, and that the proportion of water of hydration increased as the alkyl chain length decreased. The heat of wetting for C_8E_6 is shown in Fig. 5 as a function of surface excess. It was suggested that the initial linear change in heat with coverage corresponds to the replacement of water at the interface by the hydrated surfactant molecule until the surface is saturated with horizontal adsorbate molecules. The plateau value for the heat is reached at a molecular area of 1.32 nm^2 which is close to the cross-sectional area of a C_8E_6 molecule lying flat. The constancy of the heat at higher concentrations was attributed to the balance between decreasing enthalpy, due to elimination of alkyl/solution interface as the molecules become vertically oriented, and increasing enthalpy associated with desolvation of the adsorbing molecules. The adsorption thus occurs with a net increase in entropy. Such a process would be analogous to micellization (Corkill *et al.*, 1964; Corkill and Goodman, 1969; Goodman and Walker, 1979). However, adsorption of the surfactants increased with increasing temperature (Fig. 6). This could not have been predicted from the calorimetric measurements and is not seen in physical adsorption from

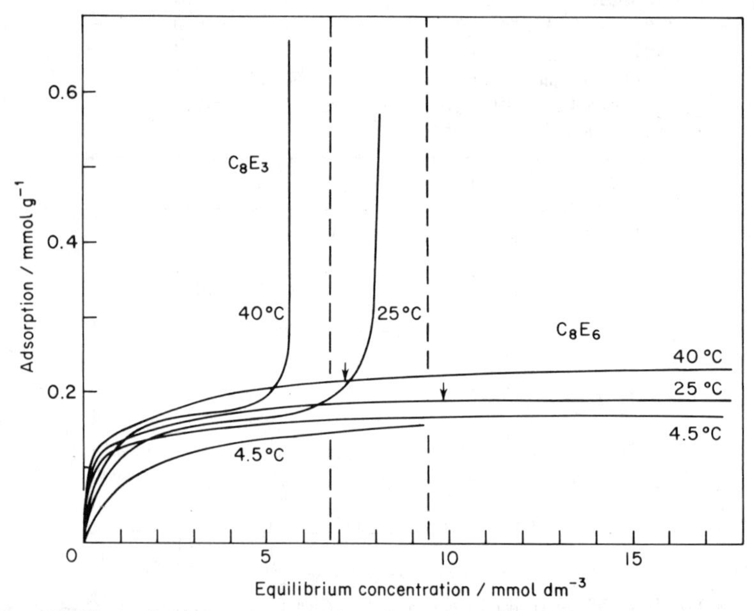

Figure 6 Effect of temperature on the adsorption of C_8E_3 and C_8E_6 on Graphon; arrows indicate c.m.c., and broken lines show phase separation boundaries. Experimental points are omitted for clarity. (Corkill *et al.*, 1966a)

single-component phases, i.e. gas on solid. Corkill *et al.* suggested that the adsorbing species is actually the solvated surfactant molecule, which is essentially different at each temperature because the surfactant–solvent interaction in polyethoxylated surfactants is very temperature sensitive (see Section II). Increasing temperature gradually desolvates the head group, making it less hydrophilic and more compact, and this increases surface activity and saturation adsorption values. The importance of the surfactant–solvent interaction was also apparent in the effect of temperature on the adsorption of C_8E_3. At 4.5°C the adsorption isotherm was simple Langmuirian but at 25°C and 40°C there was a very steep rise in adsorption at high concentrations (Fig. 6). This is characteristic of a surface condensation effect, and it was pointed out that with C_8E_3 the steep rise occurs at concentrations slightly below those at which surfactant phase separation occurs in the bulk solution. These concentrations are shown by broken lines in Fig. 6.

In a later study with the same adsorbent, Corkill *et al.* (1966b, 1967) used alkylsulphinyl alkanol surfactants, $C_xH_{2x+1}SO(CH_2)_nOH$. They found that the saturation adsorption, like that of the ethoxylated surfactants, increased with increasing alkyl chain length and decreasing head group size. From the magnitude of the areas occupied by the surfactant at surface saturation (Table 3), and the observation that the carbon black

Table 3 Adsorption data for n-alkylsulphinyl alkanols on Graphon, 25°C. (Corkill *et al.*, 1967)

	Area per molecule/nm^2				
Adsorbate	at isotherm inflexion	from horizontal model	at saturation	from vertical model	$-\Delta\bar{H}_2/$ kJ mol^{-1}
$C_4H_9SO(CH_2)_2OH$	—	0.58	~0.83	0.25	—
$C_6H_{13}SO(CH_2)_2OH$	0.84	0.70	0.44	0.25	42
$C_8H_{17}SO(CH_2)_2OH$	1.08	0.83	0.25	0.25	59
$C_{10}H_{21}SO(CH_2)_2OH$	~1.68	0.98	—	0.25	109
$C_8H_{17}SO(CH_2)_3OH$	1.31	0.90	0.32	0.25	84
$C_8H_{17}SO(CH_2)_4OH$	1.68	0.98	0.36	0.25	117

dispersions were well stabilized, it was deduced that at surface saturation the surfactant was vertically oriented. The minimum areas were slightly larger than the smallest cross-sectional area of the fully extended surfactant molecules (0.25 nm^2). This difference was attributed to head group solvation, and it was suggested that the extent of this would depend on the number of methylene groups between the two polar groups. Most of the adsorption isotherms were L4 type with an inflexion near the c.m.c. This is clearly illustrated in the isotherm for octylsulphinylethanol (Fig. 7) where the use of radioactive adsorbate allowed accurate adsorption measurements to be made at very low surface coverage. Table 3 lists the surfactant areas

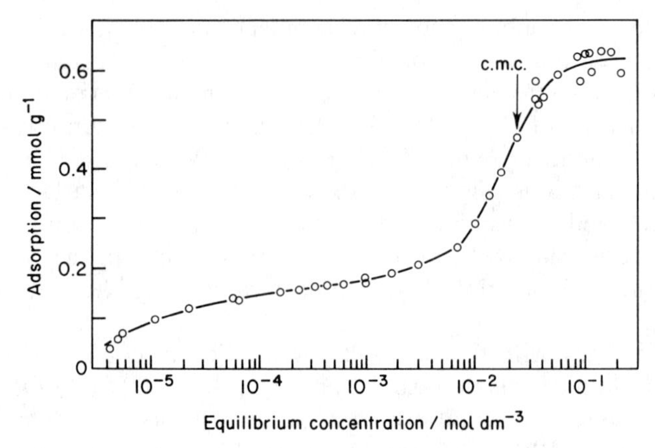

Figure 7 Adsorption of octylsulphinylethanol on Graphon, 25°C. (Corkill *et al.*, 1967)

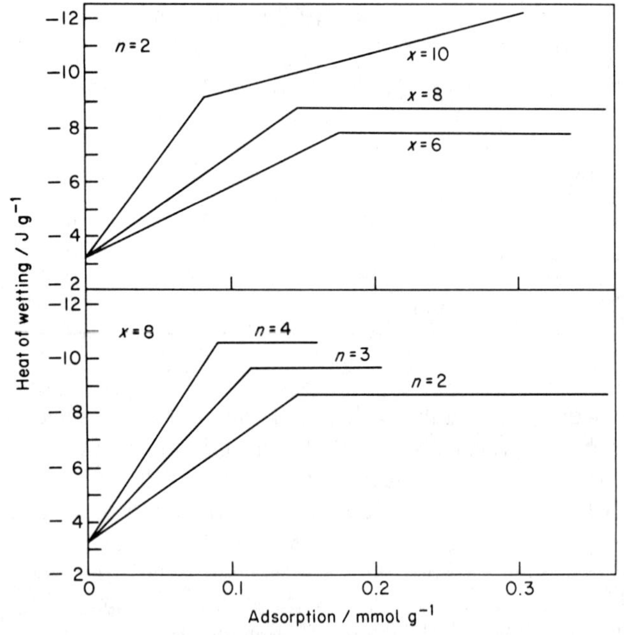

Figure 8 The heat of wetting of Graphon by n-alkyl sulphinyl alkanol, $C_xH_{2x+1}SO(CH_2)_nOH$, solutions, 25°C. Experimental points are omitted for clarity. (Corkill *et al.*, 1967)

at the inflexion points of the isotherms and compares them with areas obtained from models of the horizontal surfactant molecules. It can be seen that the latter areas are somewhat smaller than the experimental areas, a difference that may be attributable to surfactant solvation. The heat of wetting for the alkylsulphinyl alkanols (Fig. 8) changes linearly with increasing surface coverage until adsorption reaches a value that corresponds fairly closely to the inflexion point of the isotherm. Thereafter the heat is constant except for that of decylsulphinylethanol which continues to change but less steeply. It was suggested that this may have been due to the surfactant being quite close to its solubility limit. The change in heat with coverage was greater for the longer alkyl chains and larger number of head group methylenes. The partial molar heat of adsorption of the surfactant ($\Delta\bar{H}_2$) was calculated by assuming that the surface concentration of "free" solvent became zero when the heat of wetting became constant. $\Delta\bar{H}_2$ had large negative values which increased as x and n increased (Table 3). Corkill *et al.* concluded that the alkylsulphinyl alkanol system was very similar to the ethoxylated alcohol systems they had studied previously. The surfactant is initially adsorbed as a highly hydrated molecule lying parallel to the surface, and the initial enthalpy changes are associated with the displacement of solvent molecules from the surface. When complete monolayer coverage has been achieved, adsorbate–adsorbate interactions, akin to those giving rise to micellization in the bulk solution, cause orientation changes and a large increase in adsorption near the c.m.c. The heat of adsorption becomes constant because the negative enthalpy due to elimination of alkyl/solution interface is offset by the positive enthalpy due to de-solvation.

The idea that surfactant adsorption occurs in several stages was developed further by Klimenko and Koganovskii (Klimenko and Koganovskii, 1973; Klimenko *et al.*, 1974a, b). They studied the adsorption of various ethoxylated alcohols on a non-porous acetylene carbon black which had a benzene vapour specific surface area of 130 $m^2\,g^{-1}$ (Klimenko and Koganovskii, 1973; Klimenko *et al.*, 1974a,b, 1975a). All their adsorption isotherms had points of inflexion close to the c.m.c. The molecular areas at the inflexion were smaller than the areas obtained from models of the surfactant lying horizontal on the surface. This difference was attributed to the partial displacement of the ethoxy chains by the alkyl chains of adjacent molecules. It was suggested (Klimenko *et al.*, 1974b) that the angle between the tilted hydrophilic groups and the surface was proportional to the difference in adsorption energies of the alkyl and ethoxy chains. These energies were taken to be equal to the number of methylene or ethoxy units multiplied by the incremental adsorption energies, $\Delta F(CH_2)$ and $\Delta F(EO)$, which were determined by comparing the adsorption energies of various C_xE_y surfactants at infinite dilution (Koganovskii and Levchenko, 1966). The values found for $\Delta F(CH_2)$ and $\Delta F(EO)$ were 2.45 and 0.43 kJ mol^{-1} respectively. The value for $\Delta F(CH_2)$ is similar to that obtained in earlier work with anionic surfactants and alkanols and is a little smaller than the corresponding incremental energy of micellization of 3.0 kJ mol^{-1} (Corkill *et al.*, 1964). Klimenko (1978b, 1979) subsequently developed theoretical isotherms using the concept of multi-stage adsorption [see Section IV.B.2] and tested them quite successfully against the experimental data for the C_xE_y surfactants. It was found (Klimenko, 1978b) that the adsorbate–adsorbent interaction constant K_a in equation (1) increased as x and y increased, which is of course in accord with both alkyl and ethoxy portions of the surfactant being positively adsorbed at low coverage. On the other hand, the adsorbate–adsorbate interaction constant K_2 in equation (2) decreased as y increased, perhaps because attraction between alkyl chains is offset by ethoxy chain repulsion.

Adsorption was also studied at concentrations above the c.m.c. where it was postulated that the processes causing aggregation in the bulk of the solution would lead to a similar type of association on the surface. Consequently the surface would eventually be covered with hemi-spherical or semi-ellipsoidal "hemi-micelles" in a square close-packed arrangement in which each surface micelle contains half the number of molecules found in a micelle in the bulk solution. Using the experimental values of the maximum adsorption and micellar aggregation numbers derived from sedimentation equilibrium and light-scattering measurements, Klimenko *et al.* (1974a) calculated the cross-sectional areas of the surface micelles and found that they agreed well with the areas of the bulk solution micelles (Klimenko, 1979). Adsorption studies were carried out with aqueous solutions in which naphthalene or α-naphtholphthalein had been solubilized by nonionic surfactant (Koganovskii *et al.*, 1974, 1976). The ratio of solubilizate to surfactant on the surface was found to be greater than in the bulk solution. From this it was deduced that solubilizate is carried to the surface by micelles and then partitions between surface micelles and vacant sites on the surface. An alternative explanation is that the solubilization capacity of the adsorption layer is greater than that of bulk solution micelles because surfactant on the surface has a different orientation, packing density, and perhaps degree of solvation, compared with the micelle.

Kumągai and Fukushima (1976b) studied the adsorption of pure $C_{12}E_4$, $C_{12}E_5$, and $C_{12}E_6$ on carbon black with a specific surface area of 19 m^2 g^{-1}. They were interested in the effect of temperature on adsorption and dispersion stability. They found that at temperatures below the first cloud point (lower consolute temperature) the isotherms were simple Langmuirian reaching maximum adsorption above the c.m.c. The minimum area for $C_{12}E_5$ was 0.53 nm^2 which is close to the area of 0.55 nm^2 for these surfactant molecules at the air/water interface obtained from equation (7). Above the cloud point there was an enormous increase in adsorption although the particles were still well stabilized. However, when the temperature was increased to the point at which surfactant formed a mesomorphic phase in solution, the particles then coagulated.

Mast and Benjamin (1969) carried out one of the few studies not based on ethylene oxide adducts. They examined the adsorption of radioactive dimethyl dodecyl phosphine oxide on various hydrophobic and hydrophilic solids. They found that, on carbon black with a nitrogen specific surface area of $7\,m^2\,g^{-1}$, the phosphine oxide surfactant had a simple Langmuirian isotherm (Fig. 9) which reached a plateau just above the c.m.c. The minimum area occupied by the surfactant molecule was $0.35\,nm^2$ which is close to the air/water value of $0.41\,nm^2$ (Ingram and Luckhurst, 1979). No significant change in adsorption was observed on increasing the temperature from 1.5°C to 40°C despite the latter being near the lower consolute temperature for the system.

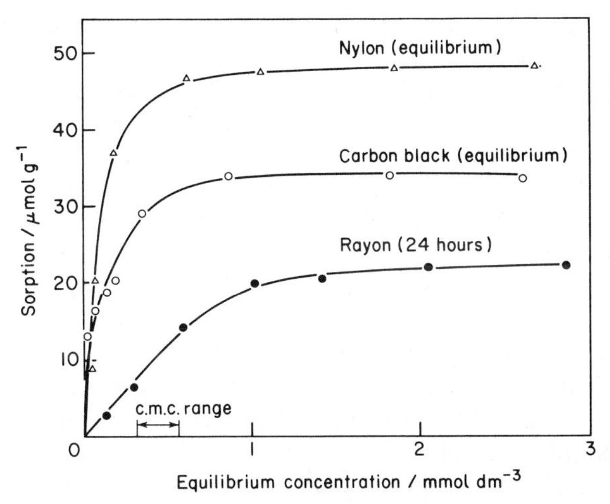

Figure 9 Sorption of dimethyl dodecyl phosphine oxide on nylon, carbon black, and rayon, 26.7°C. (Mast and Benjamin, 1969)

The work of Schwuger and Smolka (1975, 1977) on the adsorption of ethoxylated *p*-n-octyl phenols, $C_8\phi E_y$, is distinguished by including a study of mixed surfactant adsorption. They examined the adsorption of the single nonionic surfactants, with y = 8, 10, and 12, and a mixture of $C_8\phi E_8$ and sodium dodecyl sulphate on porous charcoal with a BET specific surface area of $1400\,m^2\,g^{-1}$. In the single-surfactant systems they found simple

Langmuirian isotherms in which maximum adsorption increased as ethoxy chain length decreased and was unaffected by the presence of 0.2 mol m^{-3} sodium chloride. However, the minimum areas obtained from these isotherms were very much larger than the biggest cross-sectional areas of the molecules. By assuming that the surfactant was actually occupying only the same area as it would at the air/water interface, it was deduced that only about 20% of the BET surface area was accessible to surfactant. It was noted that the adsorption energies at infinite dilution, obtained from the isotherms, decreased with increasing number of ethoxy groups. This is contrary to the findings of Klimenko and Koganovskii (Klimenko *et al.*, 1974b, Klimenko, 1978b) and is probably caused by the absence of measurements at sufficiently low concentrations to include the region where surfactant adsorbs in a horizontal orientation. The experiments with mixtures of nonionic and anionic surfactant showed that, on the whole, the presence of anionic surfactant has little effect on the adsorption of nonionic surfactant, whereas the nonionic surfactant significantly reduces the adsorption of the anionic at high concentrations but increases it at low concentrations. The enhanced anionic adsorption at low coverages was attributed to the nonionic surfactant reducing repulsion between adsorbed anions and thereby increasing their surface activity. At higher concentrations the nonionic surfactant displaces the anionic one so that the adsorption of the latter displays a maximum. The kinetics of adsorption in the mixed surfactant system were also studied. It was found that, although the initial rate of adsorption of the ionic surfactant was faster than that of the nonionic ($C_8\phi E_{12}$), probably because the latter was much bigger and diffused more slowly to the surface, eventually the nonionic surfactant displaced the ionic surfactant. This gave rise to a maximum in the adsorption versus time curve for the sodium dodecyl sulphate.

(b) *Polymers.* Polystyrene latex has frequently been used in studies concerned with dispersion stability, and a few of these have included adsorption experiments with nonionic surfactants. Ottewill and Walker (1968) examined the adsorption of pure $C_{12}E_6$ on a polystyrene latex with a mean particle size of 44 nm and specific surface area, from electron microscopy, of 105 $m^2\,g^{-1}$. They found a Langmuirian isotherm which reached a plateau near the c.m.c. and gave a minimum area of 0.4 nm^2 for the surfactant. This is smaller than the 0.55 nm^2 found on Graphon (Corkill *et al.*, 1966a). Nevertheless it was thought to be compatible with a vertically oriented monolayer [Fig. 3(IV)A(i)]. Sedimentation experiments enabled the hydrodynamic radius of the particles to be determined, and from this a value of 5 ± 0.5 nm was deduced for the thickness of the adsorbed layer. The fully extended surfactant molecule is about 4 nm long, so this thickness is in agreement with the vertically oriented monolayer model. The sedimentation coefficient decreased markedly in the concentration region where adsorption was increasing rapidly. The monolayer density calculated from the measured surface concentrations and thickness was only 0.26 g cm^{-3} which suggests that there was a high water content in the surface. This was probably water of hydration.

In a more recent study, Harris (1980) used a small-angle neutron scattering method to determine the adsorption of $C_{10}E_5$ on polystyrene latex. He found limiting areas of $0.71 \pm 0.07\ nm^2$ and $0.63 \pm 0.06\ nm^2$ per adsorbed molecule at 25°C and 40°C respectively. These values are similar to the areas at the air/water interface [0.66 nm^2 from equation (7)] and therefore more consistent with monolayer adsorption and with the Graphon work of Corkill *et al.* Furthermore, the maximum adsorption was reached at concentrations some way above the c.m.c., which is also more in accord with carbon black measurements.

Mast and Benjamin (1969) studied the adsorption of dimethyl dodecyl phosphine oxide on various polymer powders and fibres. The adsorption isotherms were all of the simple Langmuir type. On polyethylene powder and nylon (Fig. 9) maximum adsorption was reached near the c.m.c., but on microporous polystyrene bead the adsorption plateau was only reached at concentrations well above the c.m.c. The maximum adsorption value on the polystyrene was about half that found for carbon black. However, it is probable that not all of the 615 $m^2\ g^{-1}$ determined by nitrogen adsorption was accessible to the surfactant. The maximum adsorption on polyethylene powder and on polyester fibre was similar to the carbon black value, but the maximum adsorption on nylon fibres was about 40 times as large and adsorption equilibrium took several days to be achieved. It was therefore concluded that the surfactant was soluble in the nylon.

Gordon and Shebs (1968) also studied the adsorption of nonionic surfactants on nylon and polyester fibres. They used radioactive ethoxylated dodecanol with an average of 9.5 ethoxy groups, doubly labelled with tritium and carbon-14 in the alkyl and ethoxy chains respectively. The adsorption isotherm for polyester was simple Langmuirian, and analysis of the supernatant solution and adsorbent indicated that the surfactant components with the shorter ethoxy chains were being preferentially adsorbed. However, the surfactant appeared to be soluble in nylon and consequently gave high sorption values.

A recent investigation of surfactant adsorption on track-etched polycarbonate membrane (Bisio *et al.*, 1980) ingeniously combined determination of adsorption layer thickness by filter viscometry with streaming potential and air/solution contact angle measurements to provide information about the orientation of adsorbed anionic, cationic, and nonionic surfactant molecules. It was deduced that the nonionic surfactant $C_8\phi E_{\bar{9}}$ was adsorbed horizontal to the surface below the c.m.c., became vertical near the c.m.c. [as in Fig. 3(IV)C], and formed a bilayer above the c.m.c. [as in

Fig. 3(V)C(i)]. It is noteworthy that there was sufficient interaction between the head group of the nonionic surfactant and what might be regarded as a fairly low energy or hydrophobic surface to ensure that adsorption followed sequence C of Fig. 3 instead of sequence A. The cationic surfactant also followed sequence C but the anionic surfactant appeared to follow sequence B on the weakly anionic surface.

(c) *Organic pigments.* Although nonionic surfactants are used to improve pigment dispersibility there have been very few published studies of nonionic surfactant adsorption on organic pigments. Aristov *et al.* (1978) used the pigment β copper phthalocyanine, which had a nitrogen BET specific surface area of 75 $m^2\,g^{-1}$, as the adsorbent for a series of ethoxylated alcohols of the type $C_xE_{\overline{12}}$ where x was 1, 5, 8, 12, or 17. Apart from the first member of the series, which cannot really be classed as a surfactant and which showed a maximum in its adsorption isotherm, all the adsorbates gave simple Langmuir isotherms with surface saturation being reached at concentrations well above the c.m.c. The maximum adsorption increased linearly with increasing alkyl chain length. It was suggested that at saturation the surface of the hydrophobic adsorbent was covered with hemi-spherical micelles. Good agreement between the radius calculated for the surface micelle using the maximum experimental adsorption value and the radius of the micelle in bulk solution obtained by viscometry was taken as confirming this model of surface aggregation. However, as in the other cases discussed earlier, the experimental areas at saturation were little different from the minimum areas for the surfactant molecule at the air/solution interface calculated from equation (7). Therefore, the monolayer model [Fig. 3(IV)A(i)] would be as likely as the surface micelle model [Fig. 3(IV)A(ii)].

In a later study (Aristov *et al.*, 1979) ethoxylated octyl phenols were adsorbed on both the α and β forms of copper phthalocyanine. Surface saturation was achieved well above the c.m.c. Maximum adsorption increased with decreasing ethoxylation, and there was little difference in the maximum adsorption for the two adsorbates. However, the isotherms were S2 type (Giles *et al.*, 1960) and not the L2 type of the earlier study. It was deduced from the initial slopes of the isotherms that surfactant interacts more strongly with the β form. Once again fairly good agreement between experimental and theoretical maximum adsorption was obtained using a close-packed hemi-spherical surface micelle model.

2. *Polar and Metal Adsorbents*

(a) *Silica/silicates.* One of the earliest sets of adsorption measurements reported is that of Hsiao and Dunning (1955) on sand using ill-defined commercial nonionic surfactant mixtures based on nonylphenyl ethoxylates.

McCracken and Datyner (1977) have subsequently re-analysed Hsiao and Dunning's data assuming a Poisson distribution of ethoxylate chains in the adsorbate mixtures. They found that the calculated close-packed area per adsorbed molecule at the solid/solution interface for a heterogeneous nonionic surfactant is close to that expected for the equivalent homogeneous nonionic surfactant. Furthermore, this result was relatively insensitive to reasonable variations in assumptions concerning selective adsorption processes in the heterogeneous mixture case.

Although other studies on the adsorption of nonionic surfactants have been carried out on better defined substances, e.g. quartz (Dunning, 1957; Doren *et al.*, 1975), silica gel (Koganovskii *et al.*, 1975; Wolf and Wurster, 1970), pyrogenic and chemically modified silicas (Seng and Sell, 1977), nevertheless the adsorbates used have still tended to be ill-defined technical grade materials, viz. alkylphenyl ethoxylates $C_x \phi E_{\bar{y}}$ with x = 9, 10, 12, and $\bar{y}$ varying from 7 to 40. Generally, Langmuir adsorption isotherms of the L2 type have been found at ordinary temperatures in this work, with plateau adsorption occurring at the c.m.c. The adsorption tends to be greater the higher the pH, which has been ascribed (Doren *et al.*, 1975) to changes in the surface charge characteristics of the substrate. The average area per adsorbed molecule on hydrophilic substrates at saturation adsorption was found (Wolf and Wurster, 1970) to increase with increasing $E_{\bar{y}}$ at fixed C_x from 0.55 nm^2 for $C_9 \phi E_{\bar{7}}$ to 2.90 nm^2 for $C_9 \phi E_{\overline{40}}$ irrespective of the specific surface area of the silica. On the basis of molecular model dimensions, Seng and Sell (1977) have postulated that an E_y chain adopts a meander rather than an extended zig-zag configuration when adsorbed on hydrophilic silica. In an interesting comparison using various chemically modified silicas (through reaction of the surface silanol groups with appropriate organic reagents to confer increasing hydrophobicity) it has been established (Seng and Sell, 1977) that with decreasing surface polarity (i.e. increasing hydrophobicity) adsorption of nonionic surfactant increases with a decrease in the average surface area per adsorbate molecule at saturation (e.g. 0.30 nm^2 for $C_9 \phi E_{\bar{7}}$ on trimethyl silica). The conclusion drawn was that the hydrophobic chain adsorbs on a hydrophobic surface via van der Waals attraction whereas the hydrophilic chain adsorbs on a hydrophilic surface principally via hydrogen bonding—a view also supported by the similar studies of Rupprecht (1978).

The adsorption of $C_8 \phi E_{\overline{9.5}}$ on quartz (specific surface area 2.5 m^2 g^{-1}) decreases as the pH increases, and this has been ascribed (Doren *et al.*, 1975) to changes in the quartz surface. With increase in pH above the zero point of charge (in this case, approximately pH = 2) the silanol sites on the quartz surface are progressively changed into silicate anions. On the assumption that the nonionic surfactant adsorbs through H-bonding between the polyoxyethylene group and the silanols on the quartz surface, this can then explain the reduced adsorption with increased pH—the surface coverage at the plateau at pH 9 is roughly one-third of that at pH 3.5. It has also been shown that the presence of thiocyanate ion (a "salting-in" electrolyte) decreases adsorption whereas the presence of sulphate ion (a "salting-out" electrolyte) increases adsorption of nonionic surfactant on quartz. These effects have been attributed to changes in solution phase and not at the quartz surface—a view that can be supported by the work of Corkill *et al.* (1968) on the thermodynamics of micellization in the presence of a third component.

The reported data for the adsorption of nonionic surfactants on silicates, especially on the internal surfaces of swelling clays like montmorillonite, appear to be more reliable in that better defined systems have been studied. Adsorption isotherms at 25°C for $C_{12}E_{\overline{14}}$ and for $C_{12}E_{\overline{30}}$ from aqueous solutions on sodium and calcium montmorillonite have been determined by Schott (1964) using gravimetric and surface tension measurements. He also used X-ray diffraction methods to determine the interlayer spacing between the montmorillonite lamellae, the surfaces of which carry sorbed surfactant. The adsorption is Langmuirian to above the c.m.c., after which it develops a sigmoid shape, the inflexion occurring at the composition corresponding to the formation of a bilayer complex. The X-ray diffraction data, presented as a plot of interlamellar spacing against concentration of surfactant adsorbed, show a stepped profile. The first step of 0.44 nm corresponds to a single monolayer of adsorbate lying flat and fully extended between the lamellae, while the second step of 0.84 nm corresponds to a bilayer of adsorbate lying parallel to the silicate lamellae. Under these conditions there seems to be no ambiguity about the orientation of the adsorbate. Schott (1964) has suggested that either secondary valence forces or molecular packing considerations determine the orientation.

Barclay and Ottewill (1970) have determined the adsorption isotherm for *pure* $C_{12}E_6$ on sodium montmorillonite at 25°C (Fig. 10). Saturation adsorption corresponding to an area of 0.53 nm^2 per adsorbed molecule was reached at concentrations well above the c.m.c. The area at the c.m.c. was 1.44 nm^2 which corresponds to a horizontally extended orientation for the adsorbate. It was suggested that, in view of Schott's work, those values might indicate that the saturated surface was covered with horizontally oriented dimers.

Further X-ray studies of nonionic surfactant adsorption on n-hexadecyl-ammonium treated montmorillonites, and the effect of temperature on this interaction, have been reported by Weiss and coworkers (Platikanov *et al.*, 1977). Using $C_xE_{\bar{y}}$ with x = 16, 18, and $\bar{y}$ = 2, 10, 20, they confirmed that nonionic C_xE_y surfactants adsorb as bilayers on the internal surfaces with the alkyl chains perpendicular to the treated silicate surface, and they have compared the results with the adsorption from the nonionic melt, and from benzene solution. The interlamellar spacing and maximum amount of surfactant adsorbed were the same irrespective of sorption method used. With $\bar{y} = 2$, the alkyl chain and one ethylene oxide group are vertically oriented to the treated silicate surface while the terminal ethylene oxide monoglycol group is attached to the mineral surface. Even more interesting were the temperature dependent phase transitions found for the adsorbate

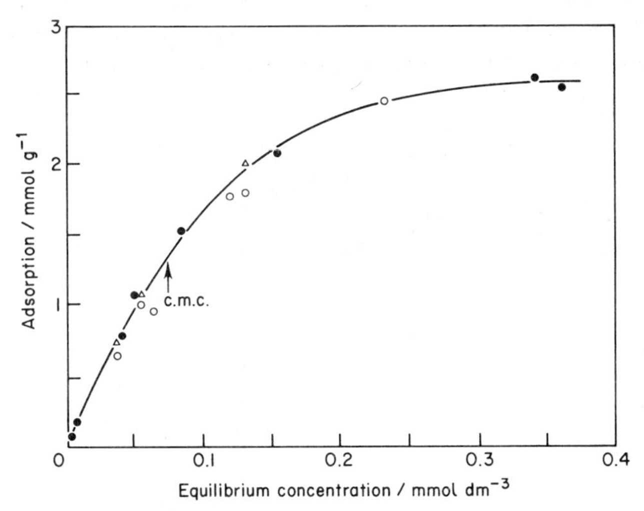

Figure 10 Adsorption of $C_{12}E_6$ on sodium montmorillonite from 0.1 mol dm^{-3} sodium chloride solution, 25°C: ○, centrifuged at 370 km s^{-2}; ●, centrifuged at 120 km s^{-2}; △, not centrifuged. (Barclay and Ottewill, 1970)

by X-ray interlamellar spacing and line-width measurements. These observed β–α phase transitions with temperature are associated with a partial desorption of surfactant, a decrease in packing density of adsorbate, and molecular configuration changes to a more liquid-like alkyl chain still vertically oriented but now with all the polar head groups attached to the substrate. In the opinion of Platikanov *et al.*, previous data indicating mono- and bi-layers of flat-lying surfactants refer to metastable phases which derive from steric hindrances associated with lattice expansion. As with cationic surfactants, the complexes of montmorillonite with nonionic surfactants can form mixed intercalation complexes with other long-chain compounds of appropriate size and shape, e.g. n-hexadecanol.

Weiss (1980) has reviewed data from his laboratory on the intercalation complexes of $C_xE_{\bar{y}}$ with x = 12–18, $\bar{y}$ = 2–75 and various host lattices. This work included adsorption studies from both aqueous and non-aqueous solutions, as well as from the nonionic melt, but with the emphasis on the nature and structure of the sorbed surfactant layers as determined by X-ray (001) basal spacing measurements. Evidence was advanced for the identity of sorbed amphiphile structures on internal and external silicate surfaces in terms of phase transitions, packing densities, and chain configurations with the proviso that differences in temperature range of existence and stability for internal, *vis-à-vis* external, surface structures do occur (Fig. 11).

This work of Weiss (1980) on nonionic intercalation complexes with layer silicates clearly represents a major advance in the study of adsorbed layers. In particular, it has been demonstrated that the contribution of both

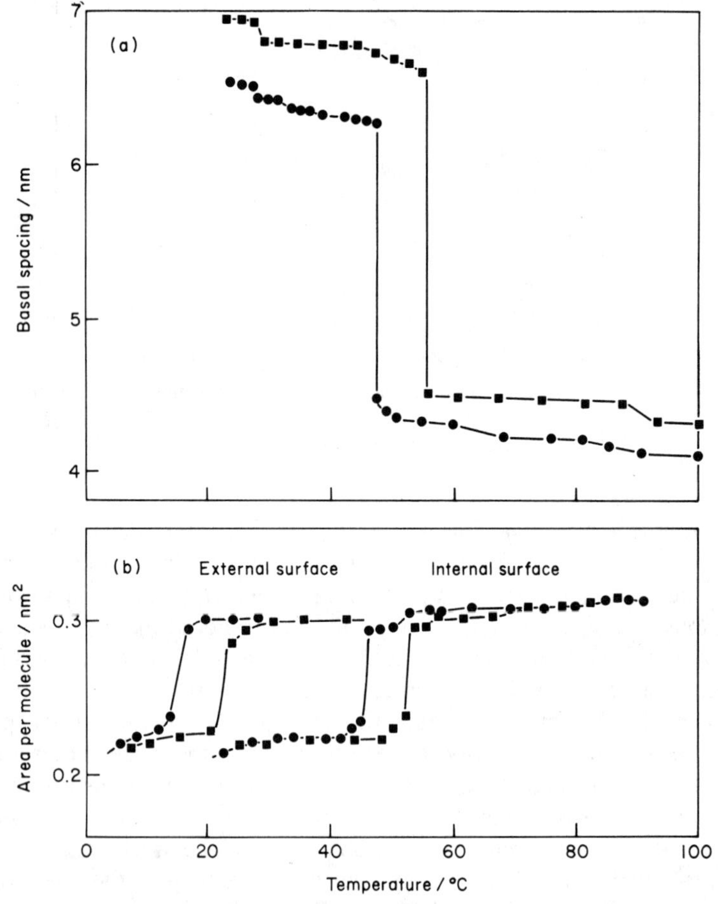

Figure 11 (a) Basal spacing of $C_{16}E_2$ (●) and $C_{18}E_2$ (■) intercalates in n-$C_{16}H_{33}NH_3^+$ montmorillonite. (b) Area per adsorbed surfactant molecule on internal and external surfaces. (Weiss, 1980)

the alkyl and polyoxyethylene chain, in terms of configurational possibilities and associated energetics, to the structure of adsorbed nonionic surfactant layers is much more complicated than thought hitherto. Depending on the molecular structure of the adsorbate, the temperature, and degree of surface coverage, different phases and chain conformations have been distinguished while observed ageing phenomena over a period of several weeks have been shown to derive from disorder–order rearrangements in the adsorbed layer.

(b) *Inorganic pigments*. Kuno and Abe (1961) were the first to determine nonionic surfactant adsorption onto calcium carbonate. They used a series of commercial $C_9\phi E_{\bar{y}}$ compounds and, as Ottewill (1967) has pointed out, this together with limitations in experimental technique has probably obscured several interesting features. No adsorption was found for $C_9\phi E_{\bar{4}}$ and with compounds having a polyoxyethylene chain greater than $E_{\bar{8}}$. With $C_9\phi E_{\bar{5}}$ and $C_9\phi E_{\bar{6}}$ the curves were essentially simple Langmuirian (type L2) above the c.m.c. with areas per molecule at the plateau of 0.155 nm^2 for $C_9\phi E_{\bar{5}}$ and 0.21 nm^2 for $C_9\phi E_{\bar{6}}$. Since these areas are below the close-packed monolayer area of about 0.6 nm^2 for these molecules, it has been suggested that multilayer adsorption is occurring. Ottewill (1967) has suggested that there may be sufficient attraction between a polar surface and the dipoles of the polyethyleneoxide chain that the first layer of nonionic surfactant is attached to the polar surface largely by the E chain, thus permitting the adsorption of a second layer to screen the hydrocarbon chains from the aqueous solution. With $C_9\phi E_{\bar{7}}$, sigmoid adsorption curves (type L3) were obtained, and the explanation for this is probably the same as for the adsorption of pure C_8E_3 on Graphon at elevated temperatures, viz. surface nucleation followed by condensation as the lower consolute temperature is approached.

Some doubts, however, have been cast on Kuno and Abe's work by Akers and Riley (1974) who also used a series of commercial surfactants, $C_9\phi E_{\bar{y}}$ with $\bar{y} = 8.5$ to 50, and purified ground chalk (crystallographically calcite). All the adsorption isotherms were Langmuirian (type L2) with the plateau region occurring above the c.m.c. (Fig. 12). The marked increase in calculated surface area per average adsorbate molecule with increasing E, from 0.28 nm^2 for $C_9\phi E_{\overline{8.5}}$ to 9.22 nm^2 for $C_9\phi E_{\overline{50}}$, suggests that the polyoxyethylene chain is adsorbed in a flat configuration corresponding to an extended chain. While Akers and Riley's data yield a more consistent pattern, clearly much can be gained from further work based on pure, well characterized adsorbates with proper attention given to solution phase properties.

The adsorption of various commercial nonionic ethoxylated surfactants on the rutile and anatase forms of titanium dioxide has been reported (Fukushima and Kumagai, 1973; Kumagai and Fukushima, 1976a). For polyethoxylated monolaurate esters $C_{12}COE_{\bar{y}}$ ($\bar{y}$ = 3–10) adsorbing on rutile surface-treated with Al_2O_3, SiO_2, or ZnO, and on "small particle size" rutile (specific surface area 53.9 $m^2\,g^{-1}$), Langmuir isotherms (type L2) were found with plateau adsorption increasing as $E_{\bar{y}}$ decreases. Surprisingly, no observable adsorption of $C_{12}E_{\bar{y}}$ on treated or untreated rutile was found, and there was no measurable adsorption of either the polyethoxylated lauryl ethers or lauric esters on untreated anatase. It was concluded that adsorption was dependent on the functional groups present in the adsorbate, and that carbonyl groups interact more strongly with the surface than ether oxygens (as indicated by IR spectral data).

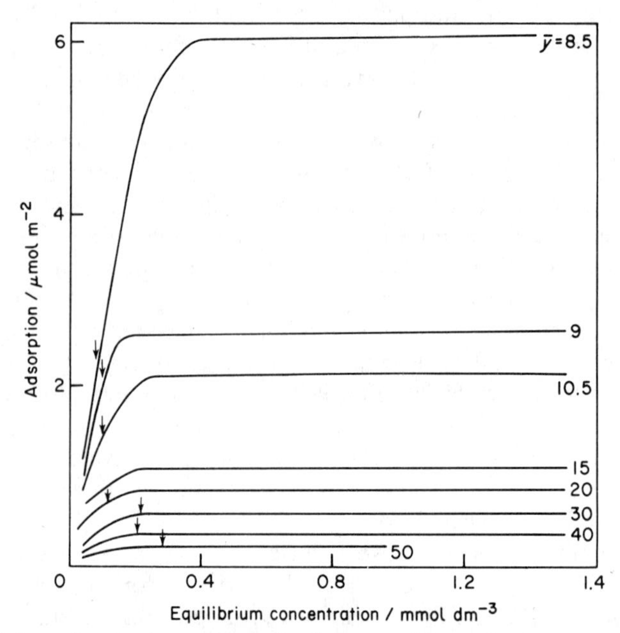

Figure 12 Adsorption isotherms for polyethoxynonyl phenols, $C_9\phi E_{\bar{y}}$, onto calcium carbonate from water, 27.5°C; the arrows indicate c.m.c. Experimental points are omitted for clarity. (Akers and Riley, 1974)

These adsorption studies have been extended to other homologous series of commercial nonionic surfactants containing an amino or an amido group between the C_{12} alkyl chain and the polyoxyethylene chain $E_{\bar{y}}$ with $\bar{y}$ = 2–10. Significantly, the laurylamine ethoxylates were found to adsorb on untreated anatase whereas none of the other series gave any detectable adsorption. For $C_{12}NE_{\bar{y}}$ the adsorption on anatase (specific surface area 9.4 $m^2\,g^{-1}$) was Langmuirian (type L2), saturation adsorption decreasing with increasing $E_{\bar{y}}$ with no detectable adsorption for hydrophilic groupings greater than $E_{\bar{6}}$. Clearly, more work is required with better defined systems in order to clarify the issue of whether ethoxylates interact with titanium dioxide as with other polar substrates.

Only passing mention has been made in the literature to adsorption on other inorganic pigments. Lange (1968) determined the adsorption of $C_{12}E_8$ on black iron oxide (specific surface area $12.5\,m^2\,g^{-1}$). The isotherm indicated fairly weak adsorption—at best an order of magnitude less than that of an anionic surfactant of the same alkyl chain length, sodium dodecyl sulphate, under similar conditions.

(c) *Silver iodide.* This is a hetero-ionic solid which forms well characterized crystalline sols and has often been used as a model colloid. Mathai and Ottewill (1966) studied the adsorption of pure C_8E_6, $C_{10}E_6$, $C_{12}E_6$, and $C_{16}E_9$ on a negatively charged (pI 4) silver iodide powder with a specific surface area of $0.7\,m^2\,g^{-1}$. Apart from $C_{16}E_9$, all the surfactants had L4 type isotherms with inflexion points and steep increases in adsorption close to the c.m.c. (Fig. 13). The maximum adsorption corresponded to areas of 0.24, 0.18, and 0.18 nm^2 for the C_8E_6, $C_{10}E_6$, and $C_{12}E_6$ molecules

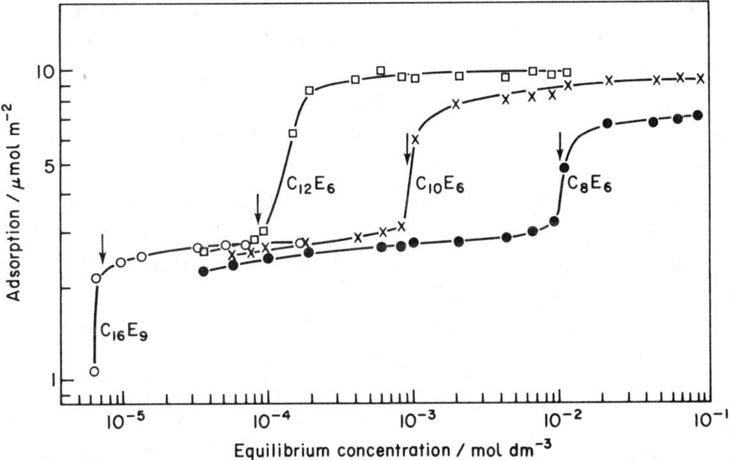

Figure 13 Adsorption of n-alkyl polyoxyethylene glycol monoethers on silver iodide (25°C) from $0.1\,mmol\,dm^{-3}$ potassium iodide solution; arrows indicate c.m.c. (Mathai and Ottewill, 1966)

respectively (Table 1). The minimum area for $C_{16}E_9$, which had a simple Langmuir isotherm, was 0.60 nm^2 per molecule. The minimum air/solution interfacial areas for these molecules are 0.83, 0.72, 0.61, and 0.47 nm^2 respectively. It was suggested that the E_6 surfactants probably form multilayers on the surface but that $C_{16}E_9$ forms a monolayer. All the sols were well stabilized at saturation adsorption, indicating that the ethoxy chains of the outermost surface layer were directed towards the solvent phase. The maximum adsorption values for the E_6 surfactants were somewhat larger than would be expected if the molecules formed a bilayer with the same packing area as molecules at the air/solution interface, although cooperative effects between the alkyl chains of the two layers might operate to produce closer packing than in a monolayer. The areas of the E_6 surfactants at the first adsorption plateau, just below the c.m.c., were about 0.6 nm^2 per molecule. This corresponds to the area occupied by a vertically oriented monolayer. The adsorption behaviour of the $C_{16}E_9$ seems at first to be anomalous because the adsorption appears to be proceeding via sequence A in Fig. 3. However, it does not seem very likely that a molecule with an HLB about the same as that of the $C_{10}E_6$, and a large ethoxy chain to bind it to the polar surface, should not adsorb via sequence C. An explanation for the low adsorption might therefore be that the ethoxy chains of the first adsorption layer cannot detach sufficiently from the surface to give a close-packed vertical monolayer. The bilayer thus forms with the alkyl chains of the second layer lying alongside those of the first, i.e. interdigitated.

Later work with positively charged (pAg 3) silver iodide (Hough 1973), however, suggested that, on this type of surface, adsorption of nonionic surfactants takes place via sequence A. The isotherm of $C_{12}E_6$ at 25°C was simple Langmuirian (L2 type) and the plateau, which occurred near the c.m.c., corresponded to an area per molecule of 0.43 nm^2. Dimethyl dodecyl phosphine oxide also gave an L2 type isotherm, and the fact that its limiting area of 0.29 nm^2 was somewhat smaller than the air/solution monolayer value was ascribed to the influence of the solid surface on the ordering of adsorbed surfactant molecules in a monolayer rather than to bilayer formation.

(d) *Metals.* Although there have been several studies concerned with the adsorption of alcohols and nonionic surfactants from organic solvents onto metals (Bowden and Moore, 1951; Cook and Hackerman, 1951; Daniel, 1951; Lunina *et al.*, 1976, 1980; Romina *et al.*, 1979) information on adsorption from aqueous solution is sparse. Only mercury appears to have received any attention, differential capacity measurements (dropping mercury electrode) being the favoured method of study. In general, the main features of adsorption of alkylpolyoxyethylene glycol monoethers at the mercury/water interface resemble those at the silver iodide/water interface (Watanabe *et al.*, 1963). There is some evidence to suggest that multilayer adsorption occurs above the c.m.c., and that the adsorption is not very dependent on the potential of the interface on either side of the zero point of charge. The adsorption of inorganic ions at the mercury/aqueous solution interface was found to be inhibited by the adsorption of $C_{12}E_9$ (Watanabe *et al.*, 1963).

More recent studies on the adsorption of commercial $C_{12}E_{\bar{y}}$ ($\bar{y} = 4, 5, 8$) via the dropping mercury electrode (Mueller and Doerfler, 1977) have yielded a surface excess, calculated from the electrocapillary curves, of 13–22 $\mu mol\ m^{-2}$, i.e. about six-fold greater than that at the air/water interface, viz. 3 $\mu mol\ m^{-2}$. This has been taken to indicate multilayer adsorption but there is no clear indication of whether alkyl or polyoxyethylene groups are adsorbed at the mercury surface or what the orientation might be. It has been hypothesized that the alkyl chain could possibly be adsorbed in the positive range while the polyoxyethylene chain could be adsorbed in the negative range of the electrocapillary curves.

(e) *Textile fibres.* Much of the early work on the adsorption of commercial alkyl and alkylphenyl polyoxyethylene glycols on various fibrous substrates (e.g. cotton, wool, silk, rayon, nylon, etc.) has been reviewed by Schönfeldt (1969). This early work showed, in general, that nonionic surfactants of the C_xE_y and $C_x\phi E_y$ types were adsorbed less than anionic surfactants, with nonionic surfactant adsorption decreasing as the E_y chain length was increased. Using doubly radioactively labelled surfactants, Gordon and Shebs (1968) showed that shorter E_y chain molecules were selectively sorbed by cotton, a result confirmed by Schott (1967) who further deduced from specific adsorption data that $C_{12}E_{14}$ formed a close-packed monolayer with the surfactant molecules lying flat on the cotton surface, as in the analogous case for montmorillonite as substrate.

Much more definitive data, on the other hand, derive from the study of the sorption of a highly pure nonionic surfactant, dimethyl dodecyl phosphine oxide ($DC_{12}PO$) by cotton, nylon, rayon, and polyester fibres (Mast and Benjamin, 1969). The phase properties of $DC_{12}PO$ in water are known, and sorption measurements were carried out as a function of concentration, temperature, and also time, using radioactively labelled adsorbate. With hydrophilic cotton and viscose rayon, the adsorption of $DC_{12}PO$ was found to increase with temperature with most of the sorption occurring above the c.m.c. (Fig. 14). This was explained as possibly due to a "surface micellization" process. The presence of inorganic electrolyte (NaCl) did not affect this sorption process. In terms of kinetics, cotton took about 8 hours to reach equilibrium sorption in an aqueous solution containing 0.05% sorbate, and the plot of uptake against square-root of time was linear in the early stages and resembled that for dyes. This was taken to indicate that a diffusion process is occurring with an average diffusion coefficient of $6 \times 10^{-16}\ m^2\ s^{-1}$ at 26.7°C, which is similar to values published for the sorption of dyes.

On nylon, as on carbon black (Fig. 9), sorption of $DC_{12}PO$ was found to be essentially constant above the c.m.c. although 4 days were required in order to reach equilibrium with nylon. On polyester, sorption was similar to that on nylon but was more than an order of magnitude less for similar fibre geometry (Table 4). The data obtained for polyester, like those for carbon black, were interpreted in terms of monolayer coverage with a surface area per molecule of 0.35 nm^2. The data for nylon, however, were treated in terms of a diffusion process as for cotton. Interestingly it was pointed out that similar rate processes have been reported (Gordon and Shebs, 1968) for C_xE_y compounds, and furthermore that the quantities of C_xE_y sorbed (in decreasing order) by nylon, cotton, and polyester fabric are very similar to the amounts of $DC_{12}PO$ sorbed by the same fabrics.

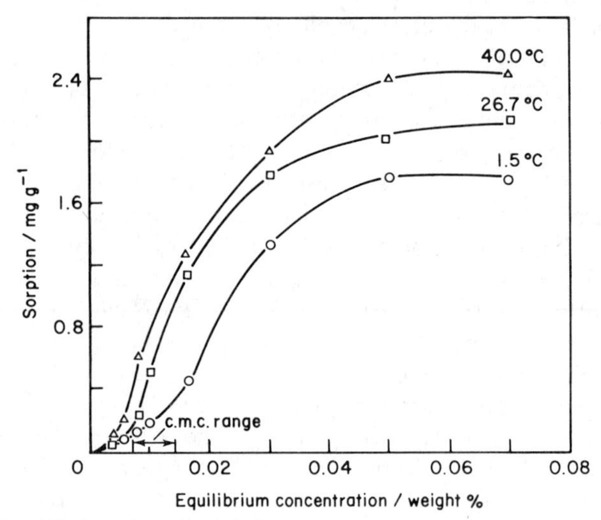

Figure 14 Equilibrium sorption of dimethyl dodecyl phosphine oxide on cotton at various temperatures. (Mast and Benjamin, 1969)

Table 4 Sorption data for dimethyl dodecyl phosphine oxide on fibrous substrates, 26.7°C. (Mast and Benjamin, 1969)

Adsorbent	Specific surface area/m^2 g^{-1}	Equilibrium sorption/mg g^{-1}
Cotton	0.72–30	2.1
Rayon	—	7.6
Nylon	0.25	11.8
Polyester	0.23	0.3

V. Consequences of Adsorption

The principal effects of adsorption of nonionic surfactants at the solid/solution interface are to be found in modified surface energetics (e.g.

wetting, spreading, etc.) and in the stabilization of particulate dispersions. Despite the importance of this adsorption process in many technological applications, relatively little fundamental work has been published on the structure and properties of these adsorbed layers (Weiss, 1980) and on how adsorbed layers modify surface properties (Tadros, 1980). On the other hand, much more attention has been directed to understanding the stabilizing effects of adsorbed nonionic surfactants on particulate dispersions in aqueous media (Ottewill, 1967). This is the phenomenon of steric stabilization or "protection" of lyophobic colloidal dispersions against coagulation by electrolytes due to the presence of adsorbed nonionic surfactant at the surface of the disperse phase (Ottewill, 1973). Theoretically it has been shown (Napper, 1968) that the free energy of interaction between two particles arising from the interaction of nonionic adsorbates depends on the thickness of the adsorbed layer, an enthalpy term which characterizes the interaction of the stabilizer with the solvent, and an entropy of mixing parameter for the stabilizing layers. This theory has been successfully used to explain the stability of several lyophobic aqueous dispersions containing nonionic surfactants as stabilizer (Ottewill, 1973). The effects of adsorbed nonionic surfactants on electrokinetic (Ottewill, 1967) and rheological (Goodwin, 1975) properties of lyophobic aqueous dispersions have also been reviewed, while recently a correlation has been shown to exist between the viscosity and electrophoretic mobility for aqueous colloidal silica stabilized by $C_9\phi E_{\overline{12.5}}$ (Rupprecht and Hofer, 1979). Interestingly, viscosity and electrophoretic mobility maxima occurred at about 50% surface coverage. A method has also been proposed recently (Kallay *et al.*, 1979) for distinguishing between the steric and electrical effects of adsorbed nonionic surfactant on colloid stability.

VI. Applications

It would be a formidable task beyond the scope of this chapter to try and review the many applications of nonionic surfactants that derive from their adsorption at solid surfaces. Like ionic surfactants, nonionic surfactants (principally the alkylphenol ethoxylates and alcohol ethoxylates) find widespread use in a large number of diverse fields, and it is not always easy to differentiate utility based on adsorption, wetting, and dispersion stabilization properties from utility based on other properties such as emulsification, solubilization, controlled foaming, and system compatibility (due to absence of electrical charge). Nevertheless, it is possible to recognize numerous applications based on adsorption of nonionic surfactants (especially alkoxylates) in industries as diverse as textiles, resins, plastics, paints, paper, detergents, pharmaceuticals, cosmetics, agricultural chemicals, food, oil recovery, mineral technology, and many others. Details of such applications (including patent literature) can be found in the excellent book by Schönfeldt (1969). The various uses and fields of application of commercially available nonionic surfactants are conveniently listed in McCutcheon's annual publications.

VII. Summary and Conclusions

Despite the considerable practical importance of many processes involving the adsorption of nonionic surfactants on solids, there have been relatively few studies to elucidate the adsorption mechanisms. The picture of nonionic surfactant adsorption that has emerged so far shows that high adsorption occurs at very low surfactant concentrations on a wide variety of substrates, and that the adsorption process in aqueous solution can generally be divided into two parts depending on the aggregation state of the surfactant in bulk solution. At low concentrations (below the c.m.c.) surfactant displaces water from the adsorbent surface, giving rise eventually (near the c.m.c.) to a monolayer of surfactant molecules oriented with their long axes parallel to the surface. At still higher concentrations interactions between adsorbed molecules, similar to those promoting micellization in the bulk solution, lead to closer packing and a further steep rise in adsorption. The orientation adopted by the adsorbed surfactant during this process depends on whether or not there is a specific attraction between the surfactant and the solid. On a completely non-polar surface the surfactant molecule will be vertically oriented with its hydrophilic portion directed towards the aqueous phase. However, on a surprisingly large number of solids the hydrophilic moiety appears to bind quite strongly to polar sites—probably through hydrogen bonding—and in such cases the surfactant orients with its hydrophobic portion outwards towards the aqueous phase. Hydrophobic interactions then promote the build-up of a further adsorption layer oriented this time with the polar head groups towards the aqueous phase. Thus on both types of solid the final adsorption state is one in which the surface has acquired a hydrophilic character. Consequently the adsorption of nonionic surfactant can have a profound effect on the interactions between surfaces such as occur in wetting, detergency, and particulate dispersion.

Several questions concerning the adsorption mechanism still remain to be answered. A contentious issue concerns the configuration of adsorbed surfactant above the c.m.c. Does the surfactant orient in one or more layers which essentially replicate the geometry of the solid surface or does it form a close-packed assembly of semi-micelles or complete micelles similar in structure to those in the bulk solution? Although the present writers believe the layer model to be more probable, the question cannot be fully resolved without the development of new techniques. However, small-angle neutron scattering, used with selective deuteration of different parts of the surfactant molecule, may give further information about the structure of the adsorption layer. There is also a need for more information on the nature of the molecular interactions within the adsorption layer itself. Recent work (Weiss, 1980) has indicated that several temperature dependent "phase" changes may occur in nonionic surfactant layers adsorbed between clay lamellae. Is this generally characteristic of nonionic surfactant adsorbed on other surfaces? Calorimetric studies might help to provide an answer. There is also scope for greater use of spectroscopic techniques to provide information about the specific forces acting between nonionic surfactants and polar surfaces, and to ascertain what part, if any, solvation molecules play in this interaction. More precise adsorption measurements are required to determine whether on some polar surfaces there really is no surfactant adsorption below the c.m.c. Finally, there is a general need for more studies that combine several investigative techniques using several types of adsorbent and which seek to exploit the fact that *both* molecular moieties of a nonionic surfactant may be members of separate homologous series, both of which can be changed incrementally and systematically to alter surface properties.

References

Abe, R. and Kuno, H. (1962). *Kolloid-Z.* **181**, 70.

Akers, R. J. and Riley, P. W. (1974) *J. Colloid Interface Sci.* **48**, 162.

Aristov, B. G., Vasina, A. F. and Frolov, Y. G. (1978). *Kolloidn. Zh.* **40**, 965.

Aristov, B. G., Vasina, A. F. and Frolov, Y. G. (1979). *Kolloidn. Zh.* **41**, 965.

Aveyard, R. and Haydon, D. A. (1973). *An Introduction to the Principles of Surface Chemistry,* p. 211, Cambridge University Press.

Balmbra, R. R., Clunie, J. S., Corkill, J. M. and Goodman, J. F. (1962). *Trans. Faraday Soc.* **58**, 1661.

Barclay, L. M. and Ottewill, R. H. (1970). *Spec. Discuss. Faraday Soc.* **1**, 138, 164.

Barry, B. W. and El Eini, D. I. D. (1976) *J. Colloid Interface Sci.* **54**, 339.

Bartell, F. E. and Donahue, D. J. (1952). *J. Phys. Chem.* **56**, 665.

Bartell, F. E., Thomas, T. L. and Fu, Y. (1951). *J. Phys. Colloid Chem.* **55**, 1456.

Bisio, P. D., Cartledge, J. G., Keesom, W. H. and Radke, C. J. (1980). *J. Colloid Interface Sci.* **78**, 225.

Bowden, F. P. and Moore, A. C. (1951). *Trans. Faraday Soc.* **47**, 900.

Brown, C. E. and Everett, D. H. (1975). In *Colloid Science* (D. H. Everett, ed.), Specialist Periodical Reports, Vol. 2, p. 52, Chemical Society, London.

Brunauer, S., Emmett, P. H. and Teller, E. (1938). *J. Am. Chem. Soc.* **60**, 309.

Clunie, J. S., Goodman, J. F. and Symons, P. C. (1969). *Trans. Faraday Soc.* **65**, 287.

Cook, E. L. and Hackerman, N. (1951). *J. Phys. Chem.* **55**, 549.

Corkill, J. M. and Goodman, J. F. (1969) *Adv. Colloid Interface Sci.* **2**, 297.

Corkill, J. M., Goodman, J. F. and Harrold, S. P. (1964). *Trans. Faraday Soc.* **60**, 202.
Corkill, J. M., Goodman, J. F. and Tate J. R. (1966a). *Trans. Faraday Soc.* **62**, 979.
Corkill, J. M., Goodman, J. F. and Tate, J. R. (1966b). In *Wetting*, S.C.I. Monograph No. 2, p. 363, Society of Chemical Industry, London.
Corkill, J. M., Goodman, J. F. and Tate, J. R. (1967). *Trans. Faraday Soc.* **63**, 2264.
Corkill, J. M., Goodman, J. F. and Tate, J. R. (1968). In *Hydrogen-Bonded Solvent Systems* (A. K. Covington and P. Jones, eds), p. 181, Taylor and Francis, London.
Crook, E. H., Fordyce, D. B. and Trebbi, G. F. (1963). *J. Phys. Chem.* **67**, 1987.
Daniel, S. G. (1951). *Trans. Faraday Soc.* **47**, 1345.
Davies, J. T. and Rideal, E. K. (1963). In *Interfacial Phenomena*, Ch. 4, p. 154, Academic Press, London.
Donbrow, M. (1975). *J. Colloid Interface Sci.* **53**, 145.
Doren, A., Vargas, D. and Goldfarb, J. (1975). *Inst. Min. Metall. Trans.* (*C*) **84**, C34.
Dunning, H. N. (1957). *Chem. Eng. Data* **2**, 88.
Fukushima, S. and Kumagai, S. (1973). *J. Colloid Interface Sci.* **42**, 539.
Giles, C. H., MacEwan, T. H., Nakhwa, S. N. and Smith, D. (1960). *J. Chem. Soc.* 3973.
Goodman, J. F. and Walker, T. (1979). In *Colloid Science* (D. H. Everett, ed.), Specialist Periodical Reports, Vol. 3, p. 230, Chemical Society, London.
Goodwin, J. W. (1975). In *Colloid Science* (D. H. Everett, ed.), Specialist Periodical Reports, Vol. 2, p. 246, Chemical Society, London.
Gordon, B. E. and Shebs, W. T. (1968). 5th International Congress of Surface Activity, Barcelona, **3**, 155.
Griffen, W. C. (1949). *J. Soc. Cosmet. Chem.* **1**, 311.
Hansen, R. S. (1960). *J. Phys. Chem.* **64**, 637.
Hansen, R. S. and Craig, R. P. (1954). *J. Phys. Chem.* **58**, 211.
Harris, N. M. (1980). D. Phil. Thesis, University of Oxford.
Hermann, K. W., Brushmiller, J. G. and Courchene, W. L. (1966). *J. Phys. Chem.* **70**, 2909.
Hill, T. L. (1946). *J. Chem. Phys.* **14**, 441.
Hough, D. B. (1973). Ph.D. Thesis, University of Bristol.
Hsiao, L. and Dunning, H. N. (1955). *J. Phys. Chem.* **59**, 538.
Hsiao, L., Dunning, H. N. and Lorenz, P. B. (1956). *J. Phys. Chem.* **60**, 657.
Iler, R. K. (1955). In *The Colloid Chemistry of Silica and Silicates,* p. 251, Cornell University Press, New York.
Ingram, B. T. and Luckhurst, A. H. W. (1979). In *Surface Active Agents,* p. 89, Society of Chemical Industry, London.
Kallay, N., Krznarić, I. and Čegelj, Ž, (1979). *Colloid Polym. Sci.* **257**, 75.
Kipling, J. J. (1965). *Adsorption from Solutions of Non-Electrolytes,* Ch. 13, p. 230, Academic Press, London.
Kissa, E. and Dettre, R. H. (1975). *Text. Res. J.* **45**, 773.
Klimenko, N. A. (1978a). *Kolloidn. Zh.* **40**, 994.
Klimenko, N. A. (1978b). *Kolloidn. Zh.* **40**, 1105.
Klimenko, N. A. (1979). *Kolloidn. Zh.* **41**, 781.
Klimenko, N. A. (1980). *Kolloidn. Zh.* **42**, 561.
Klimenko, N. A. and Koganovskii, A. M. (1973). *Kolloidn. Zh.* **35**, 772.

Klimenko, N. A., Tryasorukova, A. A. and Permilovskaya, A. A. (1974a). *Kolloidn. Zh.* **36**, 678.
Klimenko, N. A., Permilovskaya, A. A. and Koganovskii, A. M. (1974b). *Kolloidn. Zh.* **36**, 788.
Klimenko, N. A., Permilovskaya, A. A. and Koganovskii, A. M. (1975a). *Kolloidn. Zh.* **37**, 969.
Klimenko, N. A., Permilovskaya, A. A., Tryasorukova, A. A. and Koganovskii, A. M. (1975b). *Kolloidn. Zh.* **37**, 972.
Klimenko, N. A., Polyakov, V. Y. and Permilovskaya, A. A. (1979). *Kolloidn. Zh.* **41**, 1081.
Koganovskii, A. M. and Levchenko, T. M. (1966). *Kolloidn. Zh.* **28**, 225.
Koganovskii, A. M., Klimenko, N. A. and Tryasorukova, A. A. (1974). *Kolloidn. Zh.* **36**, 861.
Koganovskii, A. M., Klimenko, N. A. and Tryasorukova, A. A. (1975). *Kolloidn. Zh.* **37**, 560.
Koganovskii, A. M., Klimenko, N. A. and Tryasorukova, A. A. (1976). *Kolloidn. Zh.* **38**, 165.
Koganovskii, A. M., Klimenko, N. A. and Chobanu, M. M. (1977). *Kolloidn. Zh.* **39**, 358.
Koganovskii, A. M., Klimenko, N. A. and Chobanu, M. M. (1979). *Kolloidn. Zh.* **41**, 1003.
Krings, P., Schwuger, M. J. and Krauch, C. H. (1974). *Naturwiss.* **61**, 75.
Kumagai, S. and Fukushima, S. (1976a). *J. Colloid Interface Sci.* **56**, 227.
Kumagai, S. and Fukushima, S. (1976b). In *Colloid and Interface Science* (M. Kerker, ed.), Vol. 4, p. 91, Academic Press, New York.
Kuno, H. and Abe, R. (1961). *Kolloid-Z.* **177**, 40.
Kuno, H., Abe, R. and Tahara, S. (1964). *Kolloid-Z.* **198**, 77.
Lange, H. (1965). *Kolloid-Z.* **201**, 131.
Lange, H. (1968). 5th International Congress of Surface Activity, Barcelona, **2**, 35.
Laughlin, R. G. (1978). In *Advances in Liquid Crystals.* (G. H. Brown, ed.), Vol. 3, pp. 42, 99, Academic Press, New York.
Low, M. J. D. and Lee, P. H. (1972). U.S. Nat. Tech. Inform. Service AD Reports, No. 751585.
Lunina, M. A. Romina, N. N. and Korenev, A. D. (1976). *Kolloidn. Zh.* **38**, 586.
Lunina, M. A., Khachaturyan, M. A. and Kolesov, I. K. (1980). *Kolloidn. Zh.* **42**, 140.
McCracken, J. R. and Datyner, A. (1977). *J. Colloid Interface Sci.* **60**, 201.
McCutcheon's Detergents and Emulsifiers Annual (1980), International Edition, M.C. Publishing Co, Glen Rock, N.J.
Mast, R. C. and Benjamin, L. (1969). *J. Colloid Interface Sci.* **31**, 31.
Mathai, K. G. (1963), Ph.D. thesis, University of Cambridge.
Mathai, K. G. and Ottewill, R. H. (1966). *Trans. Faraday Soc.* **62**, 750, 759.
Mueller, E. and Doerfler, H. D. (1977). *Tenside* **14**, 75.
Mukerjee, P. and Mysels, K. J. (1970). *Critical Micelle Concentrations of Aqueous Surfactant Systems,* NSRDS-NBS36, U.S. Dept. Commerce.
Napper, D. H. (1968). *Trans. Faraday Soc.* **64**, 1701.
Ottewill, R. H. (1967). In *Nonionic Surfactants* (M. J. Schick, ed.), Ch. 19, p. 627, Dekker, New York.
Ottewill, R. H. (1973). In *Colloid Science* (D. H. Everett, ed.), Specialist Periodical Reports, Vol. 1, p. 173, Chemical Society, London.

Ottewill, R. H. (1979). In *Surface Active Agents,* p. 1, Society of Chemical Industry, London.
Ottewill, R. H. and Walker, T. (1968). *Kolloid-Z.* **227**, 108.
Parfitt, G. D. and Rochester, C. H. (1976). In *Characterization of Powder Surfaces* (G. D. Parfitt and K. S. W. Sing, eds), p. 57, Academic Press, London.
Platikanov, D., Weiss, A. and Lagaly, G. (1977). *Colloid Polym. Sci.* **255**, 907.
Romina, N. N., Lunina, M. A. and Korenev, A. D. (1979). *Kolloidn. Zh.* **41**, 1197.
Rösch, M. (1967). In *Nonionic Surfactants* (M. J. Schick, ed.), Ch. 22, p. 753, Dekker, New York.
Rupprecht, H. (1978). *Progr. Colloid Polym. Sci.* **65**, 29.
Rupprecht, H. and Hofer, J. (1979). In *Surface Active Agents,* p. 37, Society of Chemical Industry, London.
Schönfeldt, N. (1969). *Surface Active Ethylene Oxide Adducts,* Ch. 3, p. 130, Ch. 4, p. 386, Pergamon Press, Oxford.
Schott, H. (1964). *Kolloid-Z.* **199**, 158.
Schott, H. (1967). *J. Colloid Interface Sci.* **23**, 46.
Schwuger, M. J. (1971). *Ber. Bunsenges. Phys. Chem.* **75**, 167.
Schwuger, M. J. and Smolka, H. G. (1975). Proc. IUPAC Internat. Conf. on Colloid and Surf. Sci., Budapest, Vol. 1, p. 247.
Schwuger, M. J. and Smolka, H. G. (1977). *Colloid Polym. Sci.* **255**, 589.
Seng, H.-P. and Sell, P.-J. (1977). *Tenside* **14**, 4.
Shachat, N. and Greenwald, H. L. (1967). In *Nonionic Surfactants* (M. J. Schick, ed.), p. 8, Dekker, New York.
Shinoda, K. (1967). In *Solvent Properties of Surfactant Solutions* (Surfactant Science Series, Vol. 2), Ch. 2, Dekker, New York.
Shinoda, K., Nakagawa, T., Tamamushi, B. and Isemura, T. (1963). *Colloidal Surfactants*, Academic Press, New York.
Sing, K. S. W. (1976). In *Characterization of Powder Surfaces* (G. D. Parfitt and K. S. W. Sing, eds), p. 1, Academic Press, London.
Skewis, J. D. and Zettlemoyer, A. C. (1960). 3rd International Congress of Surface Activity, Cologne, **2**, 401.
Tadros, Th.F. (1980). *Adv. Colloid Interface Sci.* **12**, 141.
Tanford, C., Nozaki, Y. and Rohde, M. F. (1977). *J. Phys. Chem.* **81**, 1555.
Trogus, F. J., Schechter, R. S. and Wade, W. H. (1979). *J. Colloid Interface Sci.* **70**, 293.
Van Voorst Vader, F. (1960). *Trans. Faraday Soc.* **56**, 1078.
Volkov, V. A. (1974). *Kolloidn. Zh.* **36**, 941.
Volkov, V. A. (1976). *Kolloidn. Zh.* **38**, 135.
Wååg, Å. (1968) 5th International Congress of Surface Activity, Barcelona, **3**, 143.
Ward, A. F. H. and Tordai, L. (1946). *J. Chem. Phys.* **14**, 453.
Watanabe, A., Tsuji, F. and Ueda, S. (1963). *Kolloid-Z.* **193**, 39.
Weiss, A. (1980). *Chem. and Ind.* (London) 382.
Wolf, F. and Wurster, S. (1970). *Tenside* **7**, 140.

4. Adsorption of Polymers

G. J. FLEER and J. LYKLEMA

I. General Features

A. INTRODUCTION

In many respects, polymer adsorption and the adsorption of small molecules differ drastically. This difference is due to the large number of configurations that a macromolecule can assume, both in the bulk of a solution and at an interface. For flexible polymers, the entropy loss per molecule upon adsorption is greater than for small molecules or for stiff polymer chains; on the other hand, the decrease in energy is also much higher owing to the many possible attachments per chain. Since it is impossible (and unnecessary) to describe each configuration at any instant in detail, statistical methods must be used to approach polymer adsorption theoretically. Also from the experimental viewpoint there are substantial differences with the corresponding low molecular weight case, since the mere establishment of adsorbed amounts is not quite enough to describe the (average) state of the macromolecular adsorbate; in addition, information is required on such matters as bound fraction, layer thickness, and segment density distribution.

In this chapter the current state of the art on polymer adsorption is discussed, restricting the discussion to uncharged macromolecules. After exposing some general trends, the modern theoretical developments will be critically compared. It is not feasible to treat this part without prior knowledge of polymer solution theory. Although some relevant features of this theory will be discussed here, it is beyond the scope of the present chapter to discuss it all in detail; the reader is referred to textbooks (Flory, 1953; Tanford, 1961; Morawetz, 1965; Yamakawa, 1971). The theoretical sections will be followed by a discussion of some typical techniques, characteristic examples, and applications.

B. PRESENT STATE

The domain of polymer adsorption on solids can briefly be characterized as a state of confusing development. The developments include increasingly sophisticated theories and experiments, but the confusion is due to the fact that many investigators are unfamiliar with relevant underlying principles, so in choosing experimental conditions often insufficient attention has been paid to crucial variables. Consequently, such studies have only limited meaning.

It is all too easy to measure an adsorption isotherm, since the availability of an analytical technique suffices to complete the experiment. However, important issues, such as adsorption reversibility (against exchange both with solvent and with other solutes), time effects, (homo)dispersity of the polymer, and characterization of the adsorbent, are often not considered. As a consequence, many relevant features cannot be ascertained with confidence, so only the qualitative trends may be taken as established (see Section I.C). On the theoretical side, the confusion stems in part from the unfamiliarity of many investigators with the methods of statistical thermodynamics, or with the condensed mathematical tools that are used to explain (or hide!) physical properties. In their book, Lipatov and Sergeeva (1972) treat the Silberberg and Hoeve theories as completely different, whereas, as we hope to show below, these theories are very similar and differ only in details of the assumptions and in the presentation.

The subject of polymer adsorption on solids has not been extensively reviewed in recent years. The book by Lipatov and Sergeeva (1972) covers all aspects of this chapter in some detail, but it is somewhat uncritical. Ash (1973) reviewed the literature published during 1970–1. Miller and Bach (1973), in their review of biopolymer adsorption, also pay attention to polymer adsorption theory, emphasizing Silberberg's theory. Recently, Vincent and Whittington (1981) reviewed the field, covering both theoretical and experimental aspects.

C. QUALITATIVE TRENDS

Notwithstanding the caveat on the quality of many experiments, a number of qualitative, or sometimes semi-quantitative, trends can be formulated that appear to be obeyed by most polymer adsorption systems. The most important of them are:

(1) Many polymers adsorb from solution onto a variety of surfaces.
(2) Polymer adsorption isotherms have a high-affinity character, i.e. in the initial part of the isotherm, at undetectably low concentrations, the adsorbed amount rises steeply, while at higher concentrations it reaches a (pseudo)plateau.
(3) The plateau level is of the order of a few mg m^{-2}, corresponding to 2–5 equivalent monolayers. Sometimes a "sharp" isotherm is obtained, in which after a very steep initial rise an essentially horizontal platform is reached, but more often a continuing slight increase with concentration is observed. (Whether or not a horizontal platform is found depends obviously on the scale of the concentration axis; over the usual range of up to a few hundred p.p.m. a platform may seem horizontal, whereas it could show a non-zero slope over several decades of the concentration.)
(4) Adsorption increases with decreasing solvent quality.
(5) Adsorption increases with molecular weight in a poor solvent, and is rather insensitive to molecular weight in a good solvent.
(6) The influence of the temperature is small or absent.
(7) It is difficult or impossible to desorb polymers by dilution, but they can be exchanged against other polymers or against low molecular weight solutes.
(8) The influence of the adsorbent area/solution volume ratio has been given little attention, but the few available treatments show that isotherms become sharper as the ratio decreases, i.e. the vertical part becomes longer and the plateau becomes more horizontal.
(9) Polymer adsorption is a slower process than adsorption of low molecular weight substances, especially if the molecular weight distribution is wide.

It may be repeated that the above are no more than trends and are not without exceptions. Still, they constitute a general framework reflecting underlying principles that ultimately must also be covered by theory. Qualitatively, concerning these nine points the following comments can be made:

(1) Since, on adsorption, a polymer molecule loses part of its conformational entropy, a certain minimum (or critical) adsorption free energy per segment is required to secure attachment. Because of the many segments contacting the adsorbent, the adsorbed amount increases strongly if the adsorption free energy exceeds its critical value.
(2) and (3) These are generally predicted by modern theories (see Section IV). According to these theories, monodisperse polymers display a sharp adsorption isotherm. The rounded shape of some experimental isotherms is related to heterodispersity effects (see Section V).
(4)–(6) These follow also from theory, but the effect of solvent quality and of molecular weight is quantitatively somewhat different in the various theoretical treatments.
(7) This is partly due to the high affinity character; for heterodisperse polymers the preferential adsorption of long chains must be taken into account as we shall show in Section V.
(8) This may be a heterodispersity effect, but partial flocculation of the adsorbent dispersion may also play a role.
(9) That polymer adsorption is a slower process than adsorption of low molecular weight substances can intuitively be understood because of their lower coefficient of diffusion and because time may be needed to undergo conformational changes. Heterodispersity also plays a role owing to the exchange of molecules of different molecular weight, which may be the rate-determining step. In our opinion, the kinetics and dynamics of polymer adsorption deserve more attention.

II. Definitions and Parameters in the Description of Polymer Adsorption

A. SOME DEFINITIONS AND SYMBOLS

In an adsorbed linear macromolecule three types of segment sequence are distinguished: trains, loops, and tails (see Fig. 1). As far as we are aware, this distinction was first proposed by Jenkel and Rumbach (1951). A train is a series of consecutive segments, all in contact with the surface. A loop consists of segments only in contact with the solvent; it is bound by a train on each side. A tail is terminally bound to a train; the other end dangles in the solution. Depending on such parameters as solvent quality, molecular weight, free energy of adsorption per segment, etc., a certain distribution of lengths of loops, trains, and tails will develop; theories should give these distributions.

For different types of macromolecule these terms remain valid in principle, although the composition of loops, trains, and tails may depend on the molecular architecture. For instance, in an ABAB block copolymer in which A consists of very surface-active segments, but where B is built from segments with a high affinity to the solvent, as a trend the trains will be enriched in sequences A and the loops and tails in B. For a comb-type grafted polymer with an A type backbone, this backbone may be entirely in the interface with the teeth being the tails.

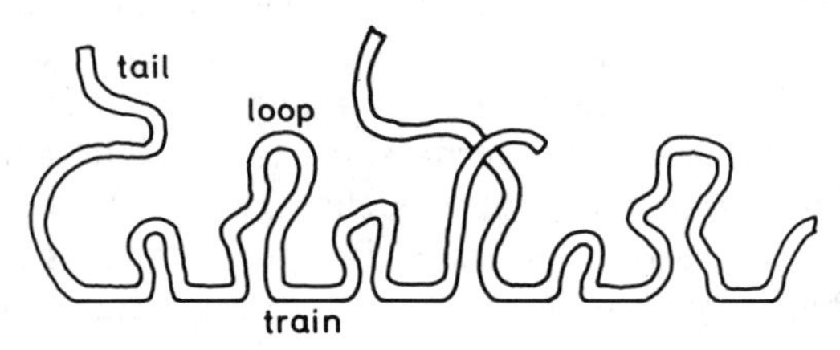

Figure 1 Adsorbed polymer molecules with loops, trains, and tails.

Extending the relevant I.U.P.A.C. symbols (1972) we shall express the adsorbed amount as a dimensionless quantity θ, the total surface coverage; it is the number of segments belonging to adsorbed chains, divided by the maximum possible number of segments in a monolayer. The direct surface coverage, θ_1, is the fractional coverage of the first layer, i.e. the number of train segments expressed as a fraction of a monolayer saturated with trains. Obviously, $\theta_1 \leqslant 1$. With these definitions, the bound fraction, p, the fraction of segments in direct contact with the surface, follows as

$$p = \theta_1/\theta \tag{1}$$

For volume fractions the symbol ϕ will be used. The reference state is the bulk solution, where $\phi = \phi_*$. In some cases a symbol for the volume fraction in the various adsorbate layers is useful. For instance, ϕ_1 represents the occupied volume fraction in the first layer. In lattice theories ϕ_1 is identical to θ_1. Generally, the volume fraction in the ith layer from the surface is ϕ_i. The number of segments per polymer chain is represented by r; clearly, r is proportional to M, the molecular weight.

B. THE PARAMETERS χ AND χ_s

The quality of the solvent is measured through the Flory–Huggins interaction parameter χ (Flory, 1953). Theoretically, its value ranges from below zero for a very good solvent, through 0 in an athermal solvent, to 0.5 in an ideally poor or θ-solvent. Experimentally, in most cases a value between 0.3 and 0.5 is found, but for aqueous solutions the range is much more restricted and usually covers values between 0.45 and 0.5; hence the experimental value is less discriminative than would be expected on the basis of the wide structural and compositional variety of the polymers in the various solvents.

The physical meaning of χ can be visualized from an exchange process in which a segment in pure bulk polymer is exchanged with a solvent molecule in pure solvent, thereby counting all non-covalent interaction energies. We indicate the coordination number by z and take the solvent to be component 1, the polymer being component 2. Then, in the exchange process mentioned, z contacts (1–1) and z contacts (2–2) are broken, and $2z$ contacts (1–2) are formed, corresponding to an exchange enthalpy $z(2h_{12} - h_{11} - h_{22})$, where the h terms stand for the interaction enthalpies per contact. The parameter χ is now defined as the net enthalpy change (in units of kT) per solvent molecule, i.e. per z contacts (1–2). Since in the above exchange process $2z$ contacts (1–2) are formed, we have

$$\chi = z(h_{12} - \tfrac{1}{2}h_{11} - \tfrac{1}{2}h_{22})/kT \tag{2}$$

It follows that χ is an exchange parameter. If $\chi > 0$, the solvent is poor and "like" contacts are preferred over "unlike" contacts. Despite this unfavourable enthalpy contribution, polymer solutions can be thermodynamically stable provided that the entropy of mixing overcompensates for this higher enthalpy. It can be shown that for $\chi > 0.5$ a solution of infinitely long chains becomes unstable; for short chains the critical value of χ is greater than 0.5 (Flory, 1953), approaching 2 in solutions of monomers.

Equations for polymer solutions are invariably derived under some restricting simplifying assumptions. One of them is that, as in the example given above, an invariant coordination number (z) is used, so differences in the specific partial volume (and other specific properties) between solvent and polymer are ignored. In this approximation, interaction energies and enthalpies are identical. Another simplification is that series expansions are usually cut off after the second virial term, which implies neglect of higher-order interactions. If such theoretical equations are applied to real systems, using χ as an adjustable parameter, all these theoretical imperfections accure in χ so that in such cases χ has a more composite nature than in equation (2); it may even contain entropy contributions, especially if solvent orientation effects play a role.

As discussed above, χ measures the segment–solvent interaction. In polymer adsorption theories, a similar parameter is used for the interaction of a segment with the surface. Adsorption is also an exchange process; upon adsorption of a segment, a solvent molecule is displaced from the surface. Silberberg (1968) introduced the adsorption enthalpy parameter, χ_s. Its physical meaning can be visualized as follows. If a solvent molecule at a surface site is exchanged with a segment at a "solution" site, the segment–solvent interaction (for which χ is a measure) changes. However, by choosing a suitable exchange process this latter interaction may be eliminated in such a way that χ_s can be defined independent of χ. To that end, we consider the special situation in which both sites have an equal number of solvent and solute contacts; the local volume fraction around both sites must then be 0.5. Furthermore, we make a distinction within the z contacts that a segment or a solvent molecule on the surface has with its neighbours; z' of them are with the surface (denoted by s) and $(z - z')$ with "solution" sites.

The parameter χ_s can now be defined as the net enthalpy change (in units of kT) of an exchange process, in which a segment on the surface, having $\frac{1}{2}(z - z')$ contacts with solvent molecules and $\frac{1}{2}(z - z')$ contacts with other segments, is exchanged with a solvent molecule in the solution, having $\frac{1}{2}z$ contacts with other solvent molecules and $\frac{1}{2}z$ with segments. As a result of this process, the segment replaces z' contacts (s–2) by $\frac{1}{2}z'$ contacts (2–2) and $\frac{1}{2}z'$ contacts (1–2), whereas the solvent molecule loses $\frac{1}{2}z'$ contacts (1–2) and $\frac{1}{2}z'$ contacts (1–1), but gains z' contacts (s–1); the other $(z - z')$ contacts remain unaltered before and after the exchange, both for the solvent molecule and for the segment, and give no contribution to the overall enthalpy. Then the net enthalpy of this exchange is $z'(h_{s1} - h_{s2} + \frac{1}{2}h_{22} - \frac{1}{2}h_{11})$ or

$$\chi_s = z'(h_{s1} - h_{s2} + \tfrac{1}{2}h_{22} - \tfrac{1}{2}h_{11})/kT \tag{3}$$

Defined in this way, χ_s is positive if a segment is preferred over a solvent molecule by the adsorbent. The statement in Section I.C, that a minimum adsorption enthalpy will be needed if polymers are to adsorb, can be rephrased more appropriately as the existence of a critical χ_s value, χ_{sc}. In Section IV it will be shown that the critical adsorption energy is of the order of a few tenths of kT per segment.

In practical cases, χ_s is again a composite quantity containing for instance also the additive parts of the segment adsorption entropy, such as those due to solvent orientation (leading to hydrophobic bonding), but of course not the loop–train–tail conformational entropy.

III. Theoretical Models

A. INTRODUCTION

The first theories (Simha *et al.*, 1953; Frisch and Simha, 1956, 1957; Silberberg, 1962, 1967; DiMarzio, 1965; DiMarzio and McCrackin, 1965;

Hoeve *et al.*, 1965; Rubin, 1965, 1966; Roe, 1965, 1966; Motomura and Matuura, 1969; Motomura *et al.*, 1971a, 1971b) on polymer adsorption treat the case of an isolated chain on a surface. In these models the segment–surface interaction and the conformational statistics of an adsorbed molecule, consisting of trains, loops, and (sometimes) tails, is considered in detail, but the interaction between the segments and between different molecules is neglected. Since the chains do not "feel" each other, there is no limit to the number of segments that can be accommodated on the surface, and the adsorbed amount increases linearly with increasing solution concentration, without bounds (Hoeve *et al.*, 1965).

Although these theories provide a suitable starting point for more realistic models, they have little relevance in practice. Even in very dilute solutions the segment concentration near the surface is usually so high that the direct surface coverage approaches saturation. Therefore, it is imperative to include the interaction of segments with each other and with the solvent.

We consider only theories that account explicitly for these segment–segment and segment–solvent interactions, so this chapter differs from other recent reviews (Dickinson and Lal, 1980; Vincent and Whittington, 1981) where many details about the adsorption of isolated molecules can be found. We restrict ourselves to four theories: Hoeve (1966, 1970, 1971); Silberberg (1968); Roe (1974); Scheutjens and Fleer (1979, 1980, 1982). The theories of Hoeve and Silberberg start from the statistics of an isolated chain, whereby end effects (tails) are neglected. In the partition function for the system of many adsorbed and free molecules they account for polymer–solvent interaction using the ideas of the Flory–Huggins theory (Flory, 1953; Tanford, 1961; Morawetz, 1965; Yamakawa, 1971) for polymer solutions. The outline of both theories is therefore the same; differences occur in details of the model. Roe and Scheutjens and Fleer do not start from a model for the mode of adsorption of individual molecules, but derive the partition function for the mixture of free and adsorbed polymer chains and solvent molecules by calculating the number of ways in which the chains and solvent molecules may be arranged in an arbitrary but (fixed) preassigned concentration gradient near the surface; after maximization of the partition function the equilibrium concentration profile is obtained. Also in these theories the Flory–Huggins model is used to account for the segment–segment and segment–solvent interaction. In the derivation Roe makes more restrictive assumptions than Scheutjens and Fleer. The latter theory is able to calculate the complete distribution of polymer conformations near the surface (including those conformations in which tails are present), whereas Roe's model gives only the overall segment concentration profile, without information about the distribution of trains, loops, and tails.

B. THE THEORIES OF HOEVE AND OF SILBERBERG

The theories of Silberberg and Hoeve were developed over the period 1962–1971. Silberberg (1962) was the first to introduce a statistical theory for an isolated adsorbed chain consisting of loops, trains, and tails. In the derivation of the partition function he made the assumption of a narrow loop size distribution. Hoeve *et al.* (1965) pointed out that this assumption is incorrect; these authors presented an improved theory for the adsorption of isolated chains. Hoeve (1966) extended this theory to include the interaction between the segments in θ-solvents. Silberberg (1967) corrected his partition function and presented an elaborate account of the self-exclusion of isolated polymer molecules. In a following contribution (Silberberg, 1968) he extended the treatment to interacting chains, neglecting end effects, and obtained numerical results for θ-solvents ($\chi = 0.5$) and athermal solvents ($\chi = 0$). In later years, Hoeve (1970, 1971) derived approximate analytical expressions for polymer adsorption at any value of the polymer–solvent interaction parameter χ.

In this section we review the most important aspects of both theories in such a way that the common features and the differences between the two models show up clearly. To this end, it appears advantageous to follow roughly Hoeve's approach. In the second part of this section we show what modifications were introduced by Silberberg. It will be shown that the two theories are very similar, with respect to the derivation and to the results.

An adsorbed molecule (r segments) is considered as consisting of alternate sequences of m loops and $m + 1$ adsorbed trains. End effects are neglected so tails are assumed to be absent; for long chains, $m \approx m + 1$. Let us have n_l loops of l segments each, and m_t trains of t segments each, then

$$\sum_l n_l = \sum_t m_t = m \tag{4}$$

$$\sum_l l n_l + \sum_t t m_t = r \tag{5}$$

According to elementary statistics, there are $m!/\Pi_l n_l!$ ways of arranging the m loops and $m!/\Pi_t m_t!$ ways of arranging the m trains over an adsorbed chain. We define ω_l as the ratio of the partition function of a loop of length l to the partition function of a sequence of l segments in the solution. The quantity ω_l is smaller than unity since the number of possible random walks starting from and returning to the surface is much less than that of unrestricted walks in the solution. Similarly, we define ν_t as the partition function of a train of length t with respect to the solution. This quantity contains a factor that is smaller than unity since a segment in the two-dimensional surface layer has less conformational freedom than a segment in a three-dimensional solution, and a Boltzmann factor expressing the preference of a segment for the surface layer due to the adsorption energy. We may write for the ratio between the partition function of one isolated adsorbed molecule and that of a chain in solution

$$q = \sum_{n_l, m_t} (m!)^2 \prod_l \frac{\omega_l^{n_l}}{n_l!} \prod_t \frac{\nu_t^{m_t}}{m_t!} \tag{6}$$

where the summation extends over all possible combinations of n_l and m_t consistent with the boundary constraints in equations (4) and (5).

Let the system contain n polymer molecules, of which n_σ are adsorbed and n_* in the solution†

$$n_\sigma + n_* = n \tag{7}$$

The partition function for the whole system can now be written as

$$Q = n! \sum_{n_\sigma, n_*} q^{n_\sigma} \frac{V_\sigma^{n_\sigma}}{n_\sigma!} \frac{V^{n_*}}{n_*!} \mathrm{e}^{-\Delta G_\mathrm{m}/kT} \tag{8}$$

The factor q^{n_σ}, where q is given by equation (6), accounts for the conformational entropy and the adsorption energy of the n_σ adsorbed molecules. The factors $V_\sigma^{n_\sigma}$ and V^{n_*} are related to the configurational entropy of the centres of gravity of the chains; V_σ is the volume of one monolayer on the surface, and V is the solution volume. The factors $n!/n_\sigma!n_*!$ are the numbers of ways in which the chains may be distributed over the adsorbed and solution state. Finally, the Boltzmann factor $\exp(-\Delta G_\mathrm{m}/kT)$ accounts for the polymer–solvent interaction. ΔG_m is the free energy of mixing; it contains an entropy term originating from the random mixing of segments and solvent molecules in the surface region (accounting for the mutual exclusion of segments), and an energy term describing the segment–solvent interaction as expressed by the χ parameter. For the evaluation of ΔG_m the Flory–Huggins theory can be used, provided that an assumption is made about the overall distribution of segments in the surface region. Hence, if expressions are found for the quantities ω_l, ν_t, and ΔG_m, equations (6) and (8) can be maximized, taking into account the boundary constraints in equations (4), (5), and (7), in order to find equilibrium properties of the system, among which are θ, θ_1, and p. In passing, it is noted that the configurational part of equation (8) does not contain factorials pertaining to the solvent. This implies that the configurational entropy of the solvent enters only through the mixing term ΔG_m.

† Although it is customary to use subscripts to indicate components and superscripts for phases, in this chapter we prefer the use of subscripts for phases, including the layer index. The reason is that, in polymer adsorption systems encompassing many layers but only two components, it is rather likely that a superscript denoting a layer might be confused with an exponent.

1. *Hoeve's Model*

Hoeve treats the loops as random walks restricted by the presence of the surface. He assumes that the loops have a Gaussian distribution. His expression for ω_l reads

$$\omega_l = cl^{-3/2} \tag{9}$$

where the constant c is a composite quantity, accounting for chain flexibility and the probability of the transition from a loop to a train and conversely. For a model chain in a lattice the value of c can be calculated from the lattice parameters (Hoeve, 1965, 1977). With increasing chain stiffness c becomes smaller.

For the evaluation of ν_t Hoeve assumes that any train segment contributes the same factor σ so that

$$\nu_t = \sigma^t \tag{10}$$

The value of σ should be constant for a given polymer/solvent/surface system. As mentioned above, σ contains the adsorption energy and an entropy factor due to the reduction in conformational freedom when a free segment is adsorbed.

Hoeve evaluates ΔG_m by assuming an exponential concentration profile in the adsorbed layer. The exponential concentration profile was shown to apply for isolated chains; this case can be approximated by setting ΔG_m in equation (8) equal to zero. Hence this assumption implies that, although the adsorbed layer is more extended for interacting chains than for isolated molecules, the exponential dependence is retained. In Section IV we shall see that this is a reasonable assumption if the adsorbed layer consists mainly of loops, but that it is not valid if the tail fraction becomes important.

The final equations of Hoeve can be summarized as follows. The results are expressed in terms of a parameter λ where λr is the difference in conformational and adsorption free energy (in units of kT) between an adsorbed chain and a chain in the bulk solution. Adsorption occurs only for negative values of λ. The parameter λ occurs in the equations through the Truesdell (1945) functions $f_{-1/2}(\lambda)$ and $f_{-3/2}(\lambda)$, defined as

$$f_{-1/2}(\lambda) = \sum_{i=1}^{\infty} i^{-1/2}\, e^{i\lambda} \qquad f_{-3/2}(\lambda) = \sum_{i=1}^{\infty} i^{-3/2}\, e^{i\lambda} \tag{11}$$

For long chains $\lambda \rightarrow 0$ and the Truesdell functions approach the limits $f_{-1/2}(\lambda) = (\pi/-\lambda)^{1/2}$ and $f_{-3/2}(\lambda) = 2.612$, respectively. For a given value of λ, the bound fraction p can be calculated as

$$p^{-1} = 1 + cf_{-1/2}(1 + cf_{-3/2})^{-1} \tag{12}$$

Then θ_1 follows from the implicit equation

$$[2\chi - \tfrac{1}{2}(\tfrac{1}{2} - \chi)K/p]\theta_1 + \ln(1 - \theta_1) + [\lambda + \ln\sigma(1 + cf_{-3/2})] = 0 \quad (13)$$

where K is given by

$$K = 2(6\pi)^{1/2}\frac{\delta}{l_s}\left(\frac{c}{1 + cf_{-3/2}}\right) \quad (14)$$

Here δ/l_s is the ratio between the thickness δ of a segment and the length l_s of a (statistical) segment. From θ_1 and p, θ follows through equation (1). Then the equilibrium bulk volume fraction may be calculated from the adsorption isotherm equation

$$\theta = \phi_* \, e^{-r[\lambda + \frac{1}{2}(\frac{1}{2} - \chi)K\theta_1]} \quad (15)$$

The concentration profile in the adsorbed layer is given by

$$\begin{aligned} &\text{trains } (x < \delta): \phi_1 = \theta_1 \\ &\text{loops } (x > \delta): \phi = K\theta_1 \, e^{-Kpx/\delta} \end{aligned} \quad (16)$$

where x is the distance from the surface. From this equation the root-mean-square thickness $t_{rms} = \langle x^2 \rangle^{1/2}$ of the adsorbed layer can for long chains (i.e. $\lambda \to 0$) be calculated as

$$t_{rms} = l_s(-6\lambda)^{-1/2} \quad (17)$$

In this way all the equilibrium properties of the system may be found for any ϕ_*, in terms of the parameters c, σ, and χ. Hoeve gives also the average loop length $\langle l \rangle$, the average train length $\langle t \rangle$, and the loop and train distribution. His expressions are

$$\langle l \rangle = f_{-1/2}/f_{-3/2} \quad (18)$$

$$\langle t \rangle = 1 + 1/cf_{-3/2} \quad (19)$$

$$n_l/m = f_{-3/2}^{-1} l^{-3/2} \, e^{\lambda l} \quad (20)$$

$$m_t/m = cf_{-3/2}(1 + cf_{-3/2})^{-t} \quad (21)$$

Obviously, the numbers of loops and trains per adsorbed molecule are given by $m\langle t \rangle = pr$, or $m\langle l \rangle = (1 - p)r$.

2. *Silberberg's Model*

Silberberg's approach is similar to that of Hoeve, but his model is slightly more sophisticated. He writes the partition function for an isolated molecule in a form that is somewhat different from equation (6). However, his equations may be transformed so as to show more clearly the analogies and differences with the Hoeve model.

Silberberg assigns an adsorption energy $-kT\chi_s$ to each train segment. This gives rise to an energy factor $\exp(\chi_s \Sigma_t t n_t)$ in the partition function for an isolated molecule. For the sake of comparison we use the equivalent approach of introducing a Boltzmann factor $\exp(t\chi_s)$ in the partition function ν_t of a train of length t.

In the Silberberg theory, the number of possible arrangements for a loop, a train, and a free chain can be written generally as

$$\Omega_s = \gamma_x z_x^s s^{-a_x} \tag{22}$$

where x is a subscript to denote the type of the sequence; for a loop x is replaced by L, for a train by T, and for a chain in solution by $*$. In this equation, s is the length of the loop, train, or free chain, γ_x is a constant allowing for end effects, z_x is the effective coordination number, and a_x is a numerical constant. For example, under θ-conditions Silberberg takes for a loop $z_L = z_*$, $a_L = 3/2$, $\Omega_l = \gamma_L z_*^l l^{-3/2}$, and for a train $z_T = z_1$, $a_T = 0$, $\Omega_t = \gamma_T z_1^t$, where z_* *and* z_1 are the effective coordination numbers in the bulk of the solution and in the two-dimensional layer adjacent to the surface, respectively. The partition function q for an isolated adsorbed molecule with respect to the same in solution is now found by replacing in equation (6) ω_l by Ω_l, ν_t by $\Omega_t e^{t\chi_s}$, and dividing by $\Omega_r = \gamma_* z_*^r r^{-a_*}$ for a free molecule. Realizing that $\Pi_l \gamma_L^{n_l} = \gamma_L^m$ and $\Pi_t \gamma_T^{m_t} = \gamma_T^m$, and that $\Pi_l z_*^{l n_l} \Pi_t z_1^{t m_t} = z_*^r \Pi_t (z_1/z_*)^{t m_t}$, we obtain, apart from the factor $\gamma_* r^{-a_*}$, the same partition function q as Hoeve [see equation (6)] if we substitute in this equation

$$\omega_l = \gamma_L \gamma_T l^{-3/2} \tag{23}$$

(θ-solvent)

$$\nu_t = [(z_1/z_*) \exp \chi_s]^t \tag{24}$$

Comparing these expressions with (9) and (10) we see that for a θ-solvent both theories use essentially the same partition functions for an isolated molecule, with

$$c = \gamma_L \gamma_T \tag{25}$$

$$\sigma = (z_1/z_*) \exp \chi_s \tag{26}$$

From these equations it is again seen that the flexibility parameter c is related to the probability of a loop–train and train–loop transition, and that σ, as stated before, contains an entropy term z_1/z_* and a Boltzmann factor e^{χ_s} accounting for the adsorption energy.

For athermal solvents ($\chi = 0$), Silberberg uses values for the exponents a_L and a_T that are different from those under θ-conditions, since self-exclusion effects in the random walk then play a role. In this case $a_L = 4/3$ and $a_T = 1/3$. Following similar reasoning as above, we now find the following expressions for ω_l and ν_t

$$\omega_l = \gamma_L \gamma_T l^{-4/3} \tag{27}$$

(athermal solvent)

$$\nu_t = t^{1/3}[(z_l/z_*)\, e^{\chi_s}]^t \tag{28}$$

so for an athermal solvent the analogy between both theories is less than for a θ-solvent.

An important further difference between Hoeve's and Silberberg's models is the concentration profile on the basis of which the free energy of mixing is evaluated. In contrast to the exponential segment density distribution used by Hoeve, Silberberg assumes a step function: the segment density is taken to be constant over a distance corresponding to half the average loop length, beyond which it drops discontinuously to its value in the bulk solution. In this surface phase only segments of adsorbed molecules are assumed to be present.

The results of the Silberberg theory cannot be expressed analytically, but can only be obtained numerically. Hitherto, only the two cases $\chi = 0$ and $\chi = 0.5$ have been elaborated. As will be shown in Section IV, the general trends are in most cases the same as those predicted by the Hoeve theory. This is, of course, not surprising in view of the common basis of both approaches.

As indicated above, both theories use various approximations. The conformational properties of adsorbing chains are probably more accurately represented in Silberberg's treatment. On the other hand, Silberberg's assumption of a step function for the segment density distribution seems to be more approximate than Hoeve's exponential distribution. In view of these uncertainties, both theories should only be cautiously applied to experimental systems. For comparison with experiment, Hoeve's model has the distinct advantage that the theoretical results can be given as a set of relatively simple analytical equations, especially if the parameter λ is small enough to allow the use of the limiting values for the Truesdell functions $f_{-1/2}$ and $f_{-3/2}$.

C. THE THEORIES OF ROE AND OF SCHEUTJENS AND FLEER

These theories have the common feature that, as in Silberberg's approach, a lattice model is employed. They derive the partition function of the system by calculating the number of ways in which n polymer chains and n^0 solvent molecules can be arranged in a given concentration gradient near the surface; after maximization of the partition function the equilibrium properties of the system are obtained.

First we define a few lattice parameters. The lattice is divided into M layers parallel to the surface, each containing L sites. The layers are numbered $i = 1, 2, \ldots, M$, where $i = 1$ is the layer adjoining the surface, and layer M is situated in the bulk solution. Each lattice site is occupied by either a polymer segment or a solvent molecule. If n_i and n_i^0 are the numbers of segments and solvent molecules in layer i, respectively, we have

$$n_i + n_i^0 = L \quad \text{and} \quad rn + n^0 = ML \tag{29}$$

where $n^0 = \Sigma_{i=1}^M n_i^0$ and $rn = \Sigma_{i=1}^M n_i$. The volume fraction ϕ_i for segments and ϕ_i^0 for solvent molecules in layer i are given by

$$\phi_i = n_i/L \quad \text{and} \quad \phi_i^0 = n_i^0/L \tag{30}$$

respectively. We shall use the notation $\{\phi_i\}$ to indicate the complete set of M volume fractions ϕ_i.

If z is the coordination number of the lattice, each site has z neighbours, a fraction λ_0 of which are in the same layer and a fraction λ_1 in each of the two neighbouring layers. Obviously, $\lambda_0 + 2\lambda_1 = 1$. In a simple cubic lattice $z = 6$, $\lambda_0 = 2/3$, $\lambda_1 = 1/6$; in a hexagonal lattice $z = 12$, $\lambda_0 = 1/2$, $\lambda_1 = 1/4$.

In calculating the free energy of the mixture of n polymer chains and n^0 solvent molecules near the surface it is useful to take as the reference state n chains in pure disoriented bulk polymer and n^0 solvent molecules in pure solvent, of which L are in contact with the surface. For the ratio between the partition function for the mixture and that for the reference state (using only the two maximum terms) we write

$$Q = (\Omega/\Omega^+)\, e^{-\Delta U/kT} \tag{31}$$

where Ω is the number of ways of arranging the polymer and the solvent near the surface in accordance with the assigned concentration gradient, and Ω^+ is the number of possible arrangements of n chains over rn lattice sites in amorphous bulk polymer. Since there is only one way of placing n^0 indistinguishable solvent molecules in n^0 lattice sites in the reference state, a term Ω^+(solvent) in equation (31) can be omitted. The quantity Ω^+ has been derived by Flory (1953) and is given by

$$\Omega^+ = \frac{(rn)!}{n!}\left(\frac{z}{rn}\right)^{(r-1)n} \tag{32}$$

In equation (31), ΔU is the energy difference between the mixture and the reference state. It contains a term $-n_1\chi_s kT$ for the n_1 polymer segments in contact with the surface [see equation (3)], and a mixing term due to the n_{12} segment–solvent contacts which, according to equation (2), can be written as $n_{12}\chi kT/z$. In a homogeneous system of n^0 solvent molecules and a segment volume fraction ϕ we can simply write $n_{12} = zn^0\phi$. In a concentration gradient, ϕ_i depends on the layer number i. If we assume random mixing in each layer (Bragg–Williams or mean field approximation), each solvent molecule in layer i has $z\lambda_0\phi_i$ contacts with segments in layer i, $z\lambda_1\phi_{i+1}$ with layer $i+1$, and $z\lambda_1\phi_{i-1}$ with layer $i-1$. Then the average volume fraction ϕ has to be replaced by the weighted average $\langle\phi_i\rangle$ over the three layers $i-1$, i, and $i+1$

$$\langle\phi_i\rangle = \lambda_1\phi_{i-1} + \lambda_0\phi_i + \lambda_1\phi_{i+1} \tag{33}$$

where $\langle\phi_i\rangle$ may be called the site volume fraction. For the energy difference ΔU we can write

$$\Delta U/kT = -n_1\chi_s + \chi\sum_{i=1}^{M} n_i^0\langle\phi_i\rangle \tag{34}$$

The key for the further evaluation is the derivation of the combinatory factor Ω. It is in this derivation that the models of Roe and of Scheutjens and Fleer differ. Once an expression for Ω is obtained, the partition function ratio Q may be maximized to find the equilibrium properties of the system. As will be shown below, Roe expresses Ω in the overall concentration profile $\{\phi_i\}$; in his approximation the information about polymer conformations is lost. On the other hand, Scheutjens and Fleer account fully for all possible chain conformations. Their expressions are therefore slightly more complicated, but as a result the equilibrium distribution of chain conformations can also be obtained.

1. *Roe's Model*

Roe (1974) calculates Ω using the assumption that the spatial distribution of any chain segment does not depend on its ranking number in the chain. More specifically, the contribution to ϕ_i due to each segment, whether it is in the middle part of the chain or near one of the chain ends, is assumed to be equal to ϕ_i/r, for any layer i. This equal distribution of all segments of a chain is correct for a homogeneous system (e.g. in the bulk solution) but not near a surface, since the surface imposes restrictions that are not the same for end and middle segments. In effect, as shown by Scheutjens and Fleer (1980), this assumption is more or less equivalent to the neglect of end effects.

On the basis of this assumption, Roe calculates the number of possibilities of placing n chains and n^0 solvent molecules with a given overall segment concentration gradient $\{\phi_i\}$. The outcome is

$$\Omega = \prod_{i=1}^{M} \frac{L!}{n_i! n_i^0!} \langle\phi_i\rangle^{n_i(1-1/r)} \tag{35}$$

The M factors $\langle\phi_i\rangle^{n_i(1-1/r)}$ account for the fact that a fraction $(1 - 1/r)$ of the polymer segments can be placed only on sites immediately adjacent to a previously placed segment, so that the average segment volume fraction $\langle\phi_i\rangle$, which according to equation (33) is an average over the layers $i - 1$, i, and $i + 1$, occurs in the equation. Because of this, Roe's partition function is given in terms of the overall segment concentration profile $\{\phi_i\}$ only, without specifying the various chain conformations contributing to this profile.

Equation (31) can be maximized with respect to the volume fractions ϕ_i in each layer, taking into account the boundary constraints given in equation (29). From the resulting M implicit equations the equilibrium concentration profile $\{\phi_i\}$ can be calculated. These final equations are

$$\begin{aligned} &2\chi(\langle\phi_i\rangle - \phi_*) \\ &\quad + (\chi_s + \lambda_1\chi)\delta_{1,i} + \ln[(\phi_*/\phi_*^0)(\phi_i^0/\phi_i)] \\ &\quad + (1 - 1/r)[\ln(\langle\phi_i\rangle/\phi_i) + \langle\phi_i/\langle\phi_i\rangle\rangle - 1] = 0 \qquad (i = 1, 2, \ldots, M) \end{aligned} \tag{36}$$

where $\delta_{1,i}$ is the Kronecker delta, equal to unity if $i = 1$ and zero if $i \neq 1$, and

$$\langle\phi_i/\langle\phi_i\rangle\rangle = \lambda_1(\phi_{i-1}/\langle\phi_{i-1}\rangle) + \lambda_0(\phi_i/\langle\phi_i\rangle) + \lambda_i(\phi_{i+1}/\langle\phi_{i+1}\rangle) \tag{37}$$

By virtue of (33), $\langle\phi_i/\langle\phi_i\rangle\rangle = 1$ in a homogeneous environment (or in a linear concentration profile). Equation (36) may be solved for $\{\phi_i\}$ by a numerical iteration procedure. From this, the direct surface coverage ϕ_1 $(= \theta_1)$ follows immediately. The total surface coverage θ cannot be found directly, since the contribution of unadsorbed chain segments to ϕ_i is not known. However, an excess surface coverage θ^σ may be defined†

$$\theta^\sigma = \sum_{i=1}^{M} (\phi_i - \phi_*) \tag{38}$$

For strong adsorption from dilute solution θ^σ will be only slightly smaller than θ (see Fig. 2). In that case the bound fraction may be approximated by $p = \theta_1/\theta^\sigma$.

† In distinction with the subindex σ to indicate the adsorbed state, we use the superindex σ to indicate Gibbs surface excesses.

2. *Scheutjens and Fleer's Model*

Scheutjens and Fleer (1979) use a different derivation for Q, accounting for all possible chain conformations; therefore, end effects are not neglected. A "conformation" is characterized by specifying the layer numbers where each of the successive chain segments finds itself. Thus, in the conformation

$$(1, i)\,(2, j)\,(3, k) \ldots (r-1, l)\,(r, m) \tag{39}$$

the first chain segment is in layer i, the second in j, the third in k, etc. Since a segment is situated next to its preceding segment, j is one of the layers $i-1$, i, or $i+1$; similarly, $k = j-1$, j, or $j+1$, etc. Different conformations are labelled by subscripts c, d, ... Each conformation contains many different "arrangements" (specified lattice sites for each of the segments). If we place a chain of the conformation of equation (39) in an empty lattice, we have for the first segment L possibilities, for the second segment there are $\lambda_0 z$ (if $j = i$) or $\lambda_1 z$ (if $j = i \pm 1$) possibilities, for the third segment we have $\lambda_0 z$ (if $k = j$) or $\lambda_1 z$ (if $k = j \pm 1$) possibilities, etc. Here the approximation is made that immediate step reversals are allowed. Generally, the number of different arrangements in a specified conformation c is given by $Lz^{r-1}\omega_c$, where

$$\omega_c = \lambda_0^{q_c} \lambda_1^{r-1-q_c} \tag{40}$$

Here, q_c is the number of bonds in conformation c that lie within one layer, and $(r - 1 - q_c)$ is the number of bonds perpendicular to the surface. The quantity q_c is known for each conformation c.

In contrast to Roe, Scheutjens and Fleer specify not only the overall concentration profile $\{\phi_i\}$, but also the complete set of conformations $\{n_c\}$. Thus, it is calculated in how many ways $n_c, n_d, n_e, \ldots$ chains (with $\Sigma_c n_c = n$) and n^0 solvent molecules may be arranged in the lattice. Their expression for Ω is

$$\Omega = \left(\frac{z}{L}\right)^{(r-1)n} \prod_{i=1}^{M} \frac{L!}{n_i^0!} \prod_c \frac{\omega_c^{n_c}}{n_c!} \tag{41}$$

where the second multiple product extends over all the possible conformations. The difference with Roe's model [compare equation (35)] is that the volume fraction ϕ_i does not occur explicitly in Ω (although, of course, $\{\phi_i\}$ is fixed if all the n_c are given).

The maximum value of Q is found by differentiating equation (31) with respect to n_c, taking into account the boundary constraints $\Sigma_c n_c = n$ and $rn + n^0 = ML$ [equation (29)]. This is a typical distinction with Roe's analysis where Q is maximized with respect to ϕ_i. The result is

$$\frac{n_c}{L} = \frac{\omega_c \phi_*}{r} \prod_{i=1}^{M} p_i^{r_{i,c}} \tag{42}$$

where $r_{i,c}$ is the number of segments that conformation c has in layer i, and the "free segment probability" p_i is given by

$$p_i = \frac{\phi_i^0}{\phi_*^0} \, \mathrm{e}^{2\chi(\langle\phi_i\rangle - \phi_*)} \, \mathrm{e}^{(\chi_s + \lambda_1 \chi)\delta_{1,i}} \tag{43}$$

Note that the quantity p_i occurs implicitly also in Roe's equation; the first three terms of equation (36) can be replaced by $\ln(p_i \phi_* / \phi_i)$.

The free segment probability expresses the preference of a free (detached) segment for a site in layer i over a site in the bulk solution. It is a measure for the probability of finding a segment in layer i if there were no covalent bonds that restrict its position in the chain, i.e. if the segment were free. For monomers ($r = 1$), equation (42) reduces to $n_i/\mathrm{L} = \phi_* \mathrm{p}_i$ (in that case $n_c = n_i$), so $p_i = \phi_i / \phi_*$. Thus p_i gives the distribution for free segments.

The physical meaning of p_i can also be visualized from an exchange process in which a segment in the bulk solution is exchanged with a solvent molecule in layer i. The equilibrium constant of this process is $\phi_i \phi_*^0 / \phi_* \phi_i^0 = \exp(-\Delta u_i / kT)$ where Δu_i is the energy difference for the exchange. (Recall that in lattice theories energy and enthalpy are identical, and that in the present model only nearest neighbour interactions contribute to Δu_i. In real systems, Δu_i might contain an entropy contribution due to solvent orientation effects.) Then p_i can be written as $(\phi_i^0 / \phi_*^0) \exp(-\Delta u_i / kT)$; the factor ϕ_i^0 / ϕ_*^0 accounts for the entropy of mixing and Δu_i represents the energy of mixing and, for the first layer, the energy of interaction with the surface [see equation (43)].

For polymer chains, each segment of the chain can be assigned a quantity p_i according to the layer number i where the segment finds itself. From equation (42) we see that the probability that conformation c of an $r$mer occurs is proportional to the multiple product of r weighting factors $p_i, p_j, p_k, \ldots,$ one for each segment; the frequency of the factor p_i in this product equals $r_{i,c}$, i.e. the number of segments that conformation c has in i. The quantity p_i is used as a weighting factor for each of the chain segments in i. According to equation (43), this weighting factor contains three contributions. The first, ϕ_i^0 / ϕ_*^0, accounts for the entropy of mixing, as indicated above. The second is a Boltzmann factor representing the tendency of segments to accumulate in (or avoid, depending on the sign of χ) layers of high segment concentration. The last contribution, which differs from unity only for the layer adjacent to the surface, expresses the preferential adsorption of a segment over a solvent molecule. Since p_i is defined with respect to the bulk solution, $p_* = 1$.

The next problem is to find the concentration profile $\{\phi_i\}$. As stated above, for monomers equation (42) reduces to $\phi_i = p_i\phi_*$ ($i = 1, 2, \ldots, M$). From this set of M implicit equations in ϕ_i, the concentration profile can be calculated, and previous results for the adsorption of monomers from regular solutions are recovered. For $\chi = 0$, $\phi_i = \phi_*$ for $i \geqslant 2$ and a Langmuir type equation is obtained (Everett, 1964); for $\chi \neq 0$, the equations are identical to those of Ono and Kondo (1960) and Lane (1968) for monomer adsorption from a regular solution.

For polymers the situation is more complex since the volume fraction ϕ_i is the sum of the contributions $\phi_i(s)$ due to each of the chain segments with ranking number s ($s = 1, 2, \ldots, r$)

$$\phi_i = \sum_{s=1}^{r} \phi_i(s) \tag{44}$$

The contribution $\phi_i(s)$ of the sth chain segments is proportional to the probability that the sth segment of any chain is found in layer i. The latter probability can be found from the joint probability that the walk of the first s steps of a chain (starting from the one chain end) *and* the walk of the last $r - s + 1$ steps (starting from the other chain end) both end in i. Scheutjens and Fleer derive the following relation

$$\phi_i(s) = \frac{\phi_*}{rp_i} p(i, s)\, p(i, r - s + 1) \tag{45}$$

where the end segment probability $p(i, s)$ gives the probability that any chain of s segments ends in layer i. If the end segment of an $s$mer is in i, the penultimate segment (ranking number $s - 1$) can only be in one of the layers $i - 1$, i, or $i + 1$. The probability that the sth segment is in i *and* the ($s - 1$)th in $i - 1$ is $p_i\lambda_1 p(i - 1, s - 1)$, that of both segments s *and* $s - 1$ in i equals $p_i\lambda_0 p(i, s - 1)$, and the probability that s is in i *and* $s - 1$ in $i + 1$ is $p_i\lambda_1 p(i + 1, s - 1)$. In this way, $p(i, s)$ may be expressed in end segment probabilities for a chain with one segment less

$$p(i, s) = p_i[\lambda_1 p(i - 1, s - 1) + \lambda_0 p(i, s - 1) + \lambda_1 p(i + 1, s - 1)] \tag{46}$$

This is a recurrent relation. For instance, $p(i - 1, s - 1)$ is, in turn, related to $p(j, s - 2)$ ($j = i - 2, i - 1, i$). Proceeding this way, $p(i, s)$ may eventually be expressed as a function of $p(j, 1) = p_j$ ($j = 1, 2, \ldots, M$) so that all the end segment probabilities occurring in equations (44) and (45) combined can be found from all the p_i. This is conveniently expressed in a matrix formalism first introduced by Rubin and DiMarzio (Rubin, 1965; DiMarzio and Rubin, 1971). In this formalism, $p(i, s)$ is the ith component of a column vector $\mathbf{p}(s)$ which follows from the matrix multiplication

$$\mathbf{p}(s) = \mathbf{w}^{s-1}\mathbf{p}(1) \tag{47}$$

where the matrix $\mathbf{w}$ has the elements

$$w_{ij} = \lambda_{j-i} p_i \tag{48}$$

with $\lambda_{j-i} = \lambda_0$ if $j = i$ and $\lambda_{j-i} = \lambda_1$ if $j = i \pm 1$; otherwise $\lambda_{j-i} = 0$. The column vector $\mathbf{p}(1)$ has the components $p(i, 1) = p_i$.

From equations (44) and (45) the equation for the concentration profile $\{\phi_i\}$ follows

$$\phi_i = \frac{\phi_*}{r p_i} \sum_{s=1}^{r} p(i, s)\, p(i, r - s + 1) \qquad (i = 1, 2, \ldots, M) \tag{49}$$

Through the matrix procedure the end segment probabilities $p(i, s)$ can be expressed in terms of p_i, where, in turn, p_i is a function of ϕ_i [equation (43)]. Thus we have again M implicit equations in M values of ϕ_i from which the equilibrium concentration profile $\{\phi_i\}$ may be found by numerical iteration. This has to be done by computer. Since for an $r$mer $r - 1$ matrix multiplications per iteration are required, there is in practice an upper limit to the chain length for which calculations can be made. Up to now, results for chains up to $r = 5000$ have been obtained.

From the results of Scheutjens and Fleer not only the excess surface coverage θ^σ as defined in equation (38) can be computed, but also the total surface coverage θ. This may be defined as

$$\theta = \sum_{i=1}^{M} (\phi_i - \phi_i^{\mathrm{f}}) \tag{50}$$

where ϕ_i^{f} is the segment volume fraction in layer i due to free, unadsorbed chains. For the layers close to the surface, ϕ_i^{f} is smaller than ϕ_* because most (or all, for $i = 1$) of the segments in these layers belong to adsorbed chains. The difference between ϕ_i, ϕ_i^{f}, and ϕ_* (and hence the difference between θ and θ^σ) is illustrated in Fig. 2.

The volume fraction ϕ_i^{f} can be found from the equilibrium values of p_i using the same matrix procedure [equations (45)–(47)] by setting the first component of the vector $\mathbf{p}(1)$ equal to zero. Then also the elements of the matrix $\mathbf{w}$ for $i = 1$ become zero. The physical background of this is that segments of free chains cannot be in the first layer.

Since the concentrations of all the conformations contributing to the equilibrium set $\{n_c\}$ are in principle known from the M values of p_i [equation (42)], all kinds of averages may be calculated. Scheutjens and Fleer (1980) have shown how this principle may be used to obtain the average train, loop, and tail lengths, and the train, loop, and tail size distributions. The separate segment concentration profiles due to loops and tails can also be obtained, as well as the root-mean-square extension due to loops and trains and the overall root-mean-square layer thickness. Some typical results will be presented in the following section.

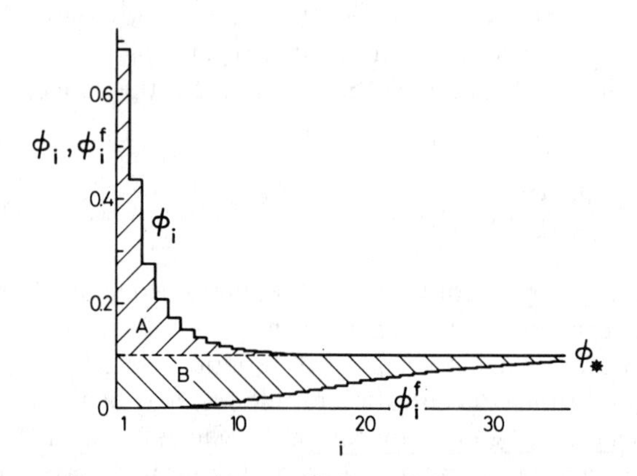

Figure 2 The overall concentration profile ϕ_i and that due to non-adsorbed chains ϕ_i^f near an adsorbing surface. Area A is equal to θ^σ, whereas the sum of A and B equals θ. In order to show the difference between θ and θ^σ more clearly, a rather high bulk volume fraction ($\phi_* = 0.1$) was chosen in this example. The profiles have been calculated from the theory of Scheutjens and Fleer for $r = 1000$, $\chi_s = 1$, $\chi = 0.5$, and $\lambda_0 = 0.5$ (hexagonal lattice).

IV. Results of Theory

In this section we compare the outcome of the four theories described before as closely as possible. We present results for the total surface coverage θ, the direct surface coverage θ_1, the bound fraction $p = \theta_1/\theta$, the root-mean-square layer thickness t_{rms}, and the distribution of segments near the surface as a function of the chain length r, the equilibrium bulk volume fraction ϕ_*, and the interaction parameters χ and χ_s.

First we have to make a choice for the chain flexibility parameter c ($= \gamma_L \gamma_T$), in such a way that a comparison with the lattice theories of Roe and of Scheutjens and Fleer can be made. Hoeve computed the parameter c for a flexible model chain in a cubic lattice, obtaining 0.120 (1965) or 0.102 (1977) depending on details of the model. Since Silberberg gives most of his results for $\gamma_L \gamma_T = 0.1$, we use also in Hoeve's equations the value $c = 0.1$. Moreover, the ratio δ/l_s in equation (14) is taken to be unity. In order to express Hoeve's σ-parameter in Silberberg's χ_s we have to specify the ratio z_1/z_* [equations (24) and (26)]. For this ratio we choose the values used by Silberberg, i.e. 5/11 for a θ-solvent and 0.4 for an athermal solvent.

In the lattice theories of Roe and of Scheutjens and Fleer, a value has to be assigned to the lattice parameter λ_0 ($=1 - 2\lambda_1$), which is essentially identical to the parameter z_1/z_*. For both theories it has been shown that the results are rather insensitive to the lattice type. Since most of the published results apply to a hexagonal lattice, we present here only data for this lattice ($\lambda_0 = 0.5$).

The results of the Silberberg theory have been taken from the figures of the original paper (1968). Results for the other three theories have been calculated in the authors' laboratory.

Figure 3 gives the total surface coverage θ and the direct surface coverage θ_1 as a function of χ_s, for $r = 100$, $\phi_* = 10^{-4}$ in an athermal ($\chi = 0$) and a θ-solvent ($\chi = 0.5$). All curves show the same general trend: adsorption occurs only if χ_s exceeds a critical value χ_{sc}; at low $\chi_s - \chi_{sc}$ the surface coverage increases steeply with increasing χ_s, but at high χ_s values the surface becomes saturated ($\theta_1 \rightarrow 1$), with the total surface coverage approaching a limiting value.

As stated in Section II.B, the critical value χ_{sc} corresponds to the minimum adsorption energy necessary to balance the unfavourable entropy per segment in the surface layer as compared to the same in the bulk of the solution. In Hoeve's model this critical value for χ_s can be derived from equation (13). For adsorbing chains $\theta_1 > 0$, so $\lambda + \ln \sigma(1 + cf_{-3/2}) > 0$. For very weak adsorption $-\lambda \rightarrow 0$ and $f_{-3/2} \rightarrow 2.612c$. Hence the minimum value for σ is $(1 + 2.612c)^{-1}$. Using equation (26) we find that in Hoeve's model $\chi_{sc} = -\ln[(z_1/z_*)(1 + 2.612c)]$. The difference in χ_{sc} between an athermal and a θ-solvent (see Fig. 3) thus reduces to a different choice for the ratio z_1/z_* in both cases. For $c = 0.1$, $\chi_{sc} = 0.556$ if $z_1/z_* = 5/11$ ($\chi = 0.5$) and $\chi_{sc} = 0.648$ if $z_1/z_* = 0.4$ ($\chi = 0$). In lattice theories, the critical adsorption energy for infinitely long chains can be derived (DiMarzio and Rubin, 1971; Roe, 1974; Scheutjens and Fleer, 1982) as $\chi_{sc} = -\ln(1 - \lambda_1)$, corresponding to $\chi_{sc} = 0.288$ for a hexagonal lattice. (For short chains, χ_{sc} is smaller; if $r = 1$, $\chi_{sc} = 0$.)

For high χ_s and $r = 100$, the limiting value for θ at high χ_s ranges from 1.3 to 1.7 monolayers for $\chi = 0.5$ and from 1.2 to 1.5 monolayers for $\chi = 0$ [see Fig. 3(a)]. As expected, the adsorbed amount is greater in poorer solvents. The limit of θ at high χ_s refutes the intuitive notion that at very high attractive energies all segments would be in contact with the surface. Such a flat conformation does occur for isolated molecules, but not for non-zero equilibrium concentrations where the chains are competing for sites on the surface. For entropic reasons, a considerable fraction of the segments is then present in loops (and tails).

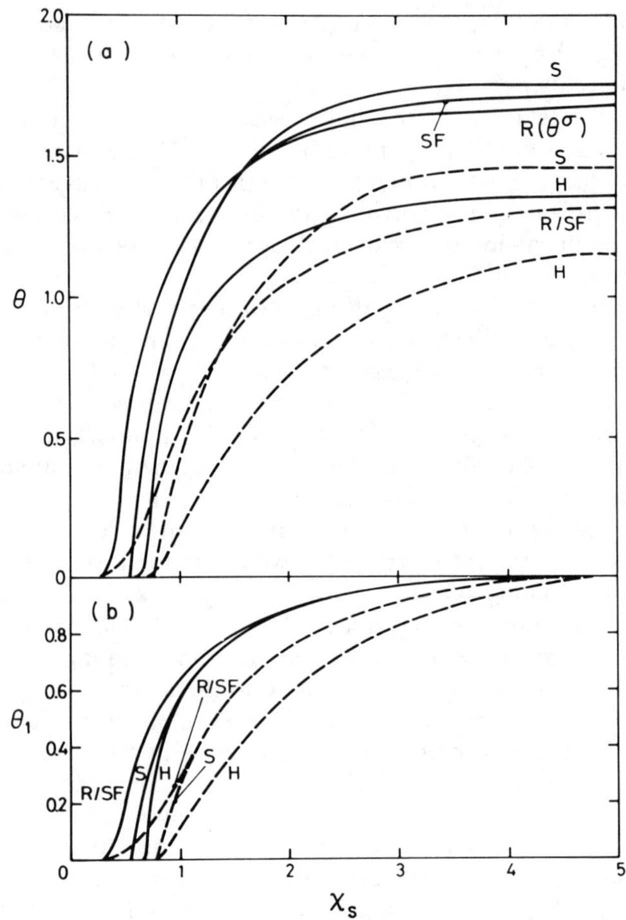

Figure 3 Total surface coverage θ (a), and direct surface coverage θ_1 (b), as a function of the adsorption energy parameter χ_s according to various theories: R = Roe, S = Silberberg, H = Hoeve, SF = Scheutjens and Fleer; $r = 100$, $\phi_* = 10^{-4}$, $c = 0.1$ (H, S), $\lambda_0 = 0.5$ (R, SF). Full curves, $\chi = 0.5$; dashed curves, $\chi = 0$.

The balancing of entropy and energy is somewhat different in the four theories. Hoeve's model predicts the lowest θ, and Silberberg's theory the highest. The results according to Roe and Scheutjens and Fleer are intermediate and differ only little between each other for this rather low chain length. From Roe's theory, only results for the excess surface coverage θ^σ are obtainable. For solutions as dilute as $\phi_* = 10^{-4}$ the difference between θ and θ^σ is insignificant (well below 1%).

With respect to θ_1 [see Fig. 3(b)] the different theories agree very well. Differences show up most clearly at low χ_s and are in this region mainly caused by a different value of χ_{sc}. For $\chi = 0.5$ Silberberg's and Hoeve's theories agree quite closely, whereas for $\chi = 0$ the discrepancy is more pronounced. This is easily understood in view of the closer analogy in the underlying equations for θ-solvents as compared to athermal solvents [see equations (23)–(28)].

In the next four figures, we compare several theoretical results for a rather high value of χ_s, such that $\theta_1 \approx 1$. This choice was made because most of Silberberg's results are given for $\chi_s = \infty$. The data of Figs. 4–7 apply to $\chi_s = 5$, which according to Fig. 3 leads to plateau values for θ and θ_1.

Figure 4 gives adsorption isotherms from a θ-solvent for a range of chain lengths according to the Hoeve and Silberberg theories, with the equilibrium volume fraction ϕ_* plotted logarithmically. For not too concentrated solutions, the adsorbed amount depends only weakly on ϕ_*; in all cases the increase in θ is less than 10% per decade in ϕ_*. Thus typical high-affinity adsorption isotherms are obtained. For chains shorter than, say, 3000 segments, the difference in θ between the two theories is not very great, but for long chains the chain length dependence is quite different. This aspect will be dealt with more extensively in the discussion of Fig. 6.

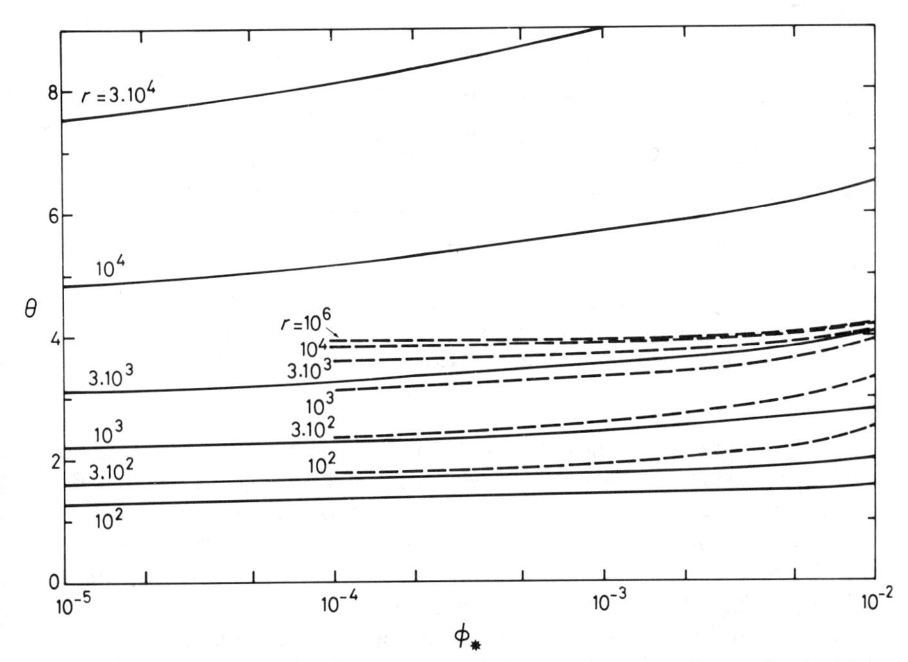

Figure 4 Adsorption isotherms for various chain lengths r, according to the theories of Hoeve (full curves) and Silberberg (dashed curves); $\chi_s = 5$, $\chi = 0.5$, $c = 0.1$.

Adsorption isotherms in the Scheutjens and Fleer theory will be discussed below (Fig. 8).

Figure 5 displays for the same theories the average loop length $\langle l \rangle$ as a function of ϕ_*. The general shape is the same as that of the adsorption isotherms, although the increase of $\langle l \rangle$ with increasing ϕ_* is slightly stronger than that of θ. For chain lengths shorter than 10^4 Hoeve predicts smaller loops than Silberberg; for very long chains the opposite is the case. Also in this respect, the dependence on the chain length is greatly different in the two pictures.

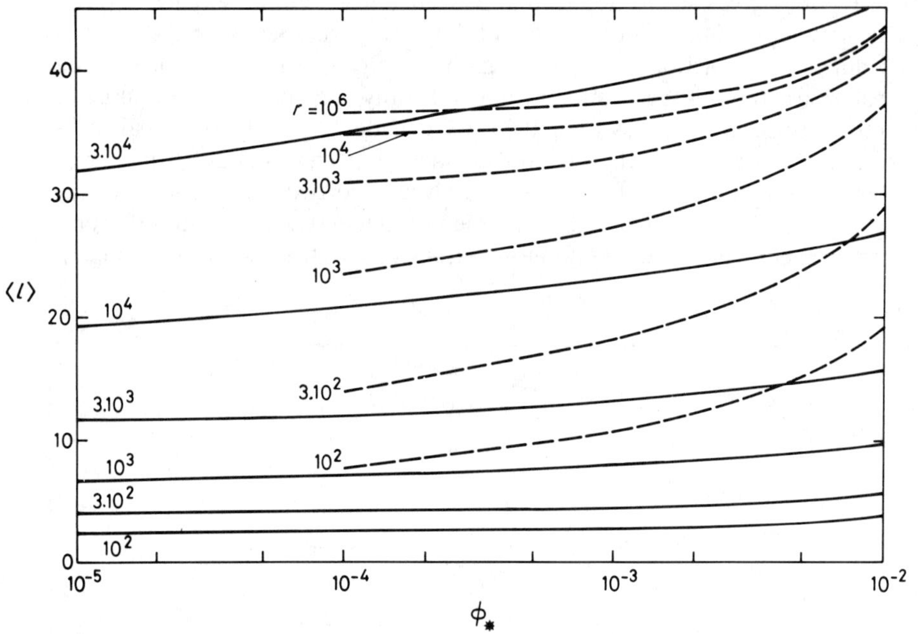

Figure 5 Average loop lengths for various chain lengths r, according to the theories of Hoeve (full curves) and Silberberg (dashed curves); $\chi_s = 5$, $\chi = 0.5$, $c = 0.1$.

Figures 6 and 7 show how, at fixed volume fraction $\phi_* = 10^{-4}$, the total surface coverage θ and the average loop length $\langle l \rangle$ change with chain length. In these figures results are given not only for $\chi = 0.5$ but also for athermal solvents ($\chi = 0$) and, for the Hoeve theory, for two values of χ slightly below 0.5. Moreover, some results from the treatments of Scheutjens and Fleer and of Roe are also indicated. In the theory of Scheutjens and Fleer, only data for chains up to 5000 segments have been computed. As mentioned before, Roe's theory gives only θ^σ, which for $\phi_* = 10^{-4}$ is virtually identical to θ_1 (except for very long chains).

For good solvents, the adsorbed amount is small and only weakly dependent on chain length. This feature shows up clearly in all theoretical treatments, although there are some quantitative differences. For poor solvents the results greatly differ between the four theories. Hoeve predicts the strongest molecular weight dependence; for θ-solvents, θ increases

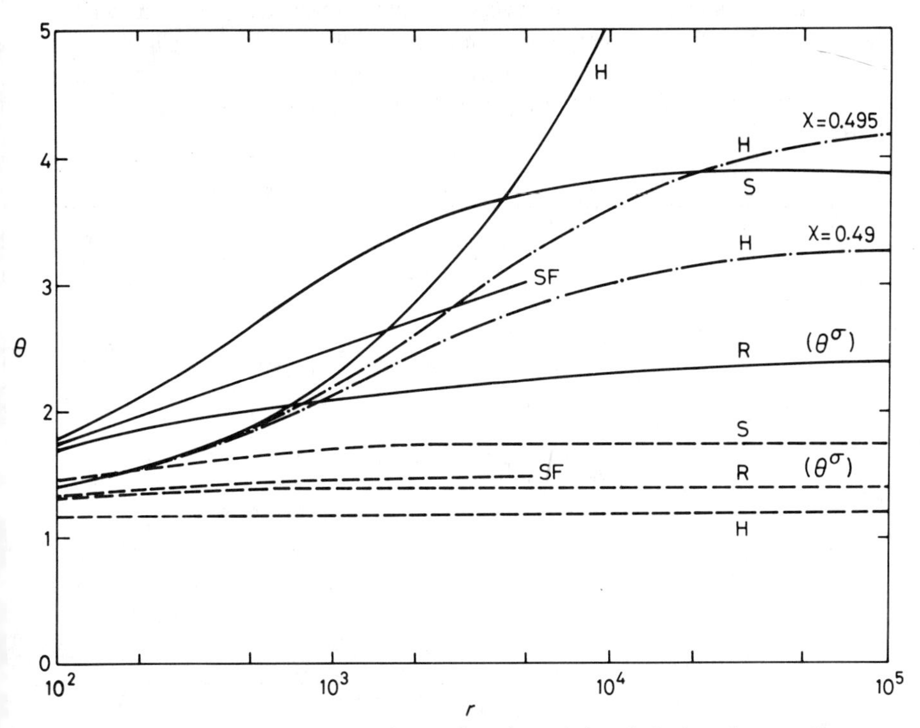

Figure 6 Total surface coverage θ as a function of the chain length according to various theories, at constant bulk volume fraction $\phi_* = 10^{-4}$. R = Roe, S = Silberberg, H = Hoeve, SF = Scheutjens and Fleer. For Roe's theory, the excess surface coverage θ^σ is plotted. Full curves, $\chi = 0.5$; dashed curves, $\chi = 0$; dashed–dotted curves, $\chi = 0.49$ or 0.495 (H). $\chi_s = 5$, $c = 0.1$ (H, S), $\lambda_0 = 0.5$ (R, SF).

proportionally to $\sqrt{r}$, without bounds. This follows from his equations; for long chains $(-\lambda \rightarrow 0,\ f_{-3/2} \rightarrow (\pi/-\lambda)^{1/2})$, the bound fraction p is proportional to $(-\lambda)^{1/2}$ according to equation (12). The direct surface coverage θ_1 reaches a limit for long chains, which depends on χ_s; this limit can be calculated from equation (13). (For high χ_s, $\theta_1 \rightarrow 1$.) Then the total surface coverage $\theta = \theta_1/p$ is proportional to $(-\lambda)^{1/2}$. Equation (15) for $\chi = 0.5$ and constant ϕ_* can now be written as $r\lambda + \frac{1}{2}\ln(-\lambda) =$ constant. Since ln

$(-\lambda)$ changes much less with increasing r (decreasing $-\lambda$) than $r\lambda$, $r\lambda$ is approximately constant or $-\lambda$ is inversely proportional to r. Hence, θ is proportional to $\sqrt{r}$.

Like Hoeve, Silberberg finds a square-root dependence on r for $\chi = 0.5$ and not too long chains ($r < 1000$), but for very great chain lengths θ attains a limit also for $\chi = 0.5$. It is interesting that, for solvents only slightly better than θ-solvents, also Hoeve's treatment predicts a levelling off for large chain lengths. The very high values for θ-solvents (for $r = 10^6$ Hoeve's equations lead to $\theta = 40$ monolayers under the conditions of Fig. 6) are found only if $\chi = 0.5$ exactly; for $\chi = 0.495$ the limiting value for long chains is $\theta = 4.37$.

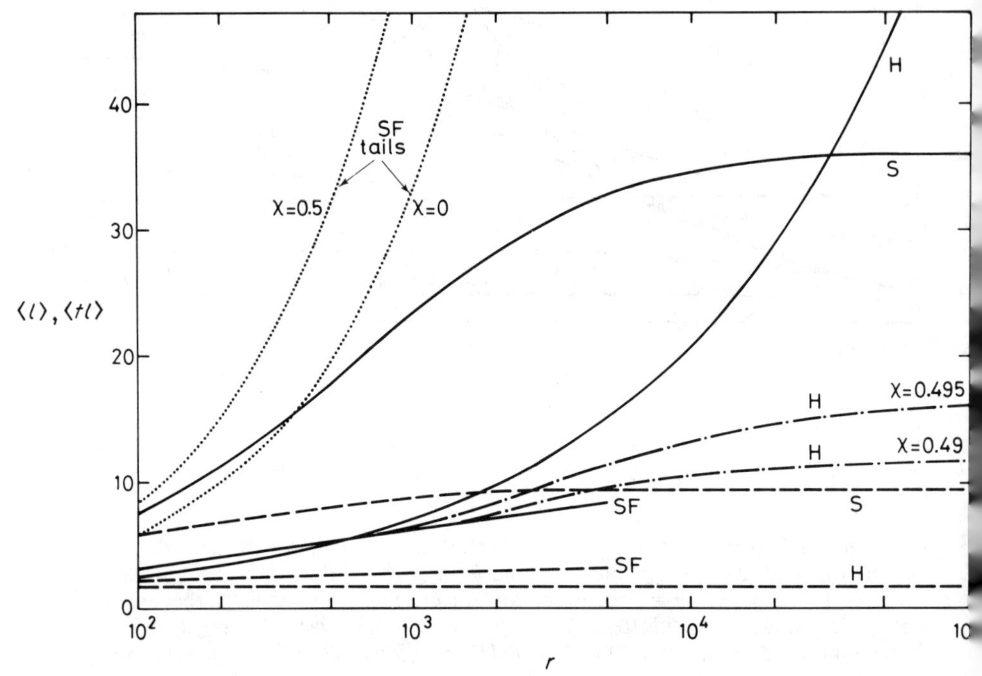

Figure 7 Average loop length as a function of chain length according to various theories. Symbols and conditions as in Fig. 6. The dotted curves are average tail lengths, computed with the theory of Scheutjens and Fleer.

The qualitative differences between the Hoeve and Silberberg theories for (very) long chains adsorbed from θ-solvents are probably related to the different approximations in both models. One might wonder whether there is any physical background from which the limiting behaviour of long chains can be predicted. From polymer solution theory it is known that, for infinitely long chains, θ-conditions lead to phase separation. It seems reasonable to expect that this phase separation is promoted near a surface, because of the extra free energy gained upon adsorption of segments. Then the surface could be considered as a nucleus for phase separation, and θ would have no limit for $r \rightarrow \infty$. If this reasoning is correct, Hoeve predicts the trends better than Silberberg. In a later publication Silberberg (1972) discusses the possibility of multilayer formation around $\chi = 0.5$, which might also be considered as a first step to phase separation. For χ only slightly below 0.5, Hoeve's treatment, like that of Silberberg, leads to a limit in θ, as illustrated above.

According to Roe's model, θ^σ is an increasing function of r also for long chains, but the slope of the θ^σ–$\log r$ curve diminishes with increasing r. This reduction of slope is at least partly caused by the approximations used by Roe. This can be seen from a comparison with the Scheutjens and Fleer theory, which is less approximate since it does not neglect end effects. According to the latter theory, θ is a linear function of $\log r$ for $r > 100$. Unfortunately, no results for chains longer than $r = 5000$ have yet been obtained with this theory, so it is not yet known whether this linear relationship persists for very long chains. Nevertheless it is clear that Roe's model predicts a too strong levelling-off of θ^σ. In view of the above discussion on the limiting behaviour of θ for high r, an upper bound for θ is not to be expected in θ-solvents. In a recent publication, Scheutjens and Fleer (1982) pursue this point in more detail. The total surface coverage θ can be split up into the excess coverage θ^σ and the depletion part θ^d (see Fig. 2, where these quantities correspond to the areas A and B, respectively). It is shown that θ^d is proportional to $\phi_* r^{1/2}$ and thus becomes a relatively important contribution to θ if r is high and ϕ_* not very small. Therefore, in solutions of very long chains at finite concentrations, there is probably no limit for θ. Since θ^d is essentially independent of χ (in contradistinction to θ^σ), this conclusion applies to any solvent quality.

Figure 7 gives the corresponding results for the dependence on r of the average loop length. The general trends are the same as those in Fig. 6. With Roe's model the average loop length cannot be calculated. For not too long chains, Hoeve's model gives values for $\langle l \rangle$ quite close to the more exact Scheutjens and Fleer theory. A new feature in Fig. 7 is the average tail length $\langle tl \rangle$ as obtained from the latter theory. It turns out that $\langle tl \rangle$ considerably exceeds $\langle l \rangle$, and increases more strongly with increasing chain length. In effect, as will be shown in more detail in Fig. 13, $\langle tl \rangle$ increases essentially linearly with increasing r, suggesting that end effects are important also for very long chains.

In Fig. 8 adsorption isotherms and the dependences of the direct surface coverage θ_1 and the bound fraction p on ϕ_* are shown as obtained with the Scheutjens and Fleer theory. Figure 8(a) may be compared with Fig. 4;

the differences are: (i) a more usual value for the adsorption energy parameter χ_s is chosen (to facilitate comparison also four curves are given for $\chi_s = 5$, as in Fig. 4); (ii) curves for short chain lengths ($r = 1$ and 10) are also given; (iii) results are given for volume fractions up to $\phi_* = 1$, corresponding to pure bulk polymer. The last two conditions cannot be analysed with the Hoeve or Silberberg theories since those models are not applicable either to short chains or to concentrated solutions. Roe's theory gives only θ^σ and θ_1, but not θ or p.

For long chains and low volume fractions, the curves of Fig. 8(a) have

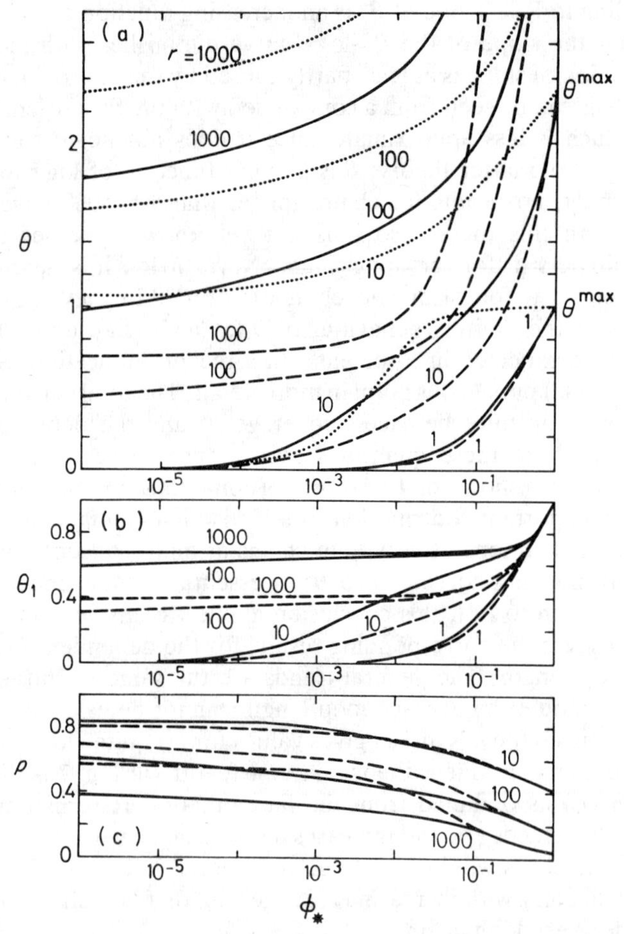

Figure 8 Total surface coverage (a), direct surface coverage (b), and bound fraction (c), as a function of ϕ_* in the Scheutjens and Fleer theory. Full curves, $\chi_s = 1$, $\chi = 0.5$; dashed curves, $\chi_s = 1$, $\chi = 0$; dotted curves, $\chi_s = 5$, $\chi = 0.5$. The value of r is indicated. Hexagonal lattice ($\lambda_0 = 0.5$).

a similar shape to those in Fig. 4. The total surface coverage increases with increasing χ and χ_s, and the adsorption isotherms are of the high affinity type. For short chains (oligomers), the high affinity character is less pronounced, and for monomers ($r = 1$) a Langmuir type adsorption isotherm is obtained. (For $\chi = 0$ and $r = 1$, the equations of Scheutjens and Fleer and also those of Roe reduce to $\phi_1/(1 - \phi_1) = e^{\chi_s} \phi_*/(1 - \phi_*)$ which, for low ϕ_*, is identical to the Langmuir equation; the Langmuirian shape is more easily seen if θ is plotted against ϕ_* on a linear scale.)

One particularly interesting feature of Fig. 8(a) is the limit $\theta^{\max}$ of θ for pure bulk polymer ($\phi_* \rightarrow 1$) which is independent of χ or χ_s. For monomers, clearly $\theta^{\max} = 1$. For $r = 10$, $\theta^{\max} = 2.37$; for $r = 100$, $\theta^{\max} = 6.26$; and for $r = 1000$, $\theta^{\max} = 18.41$. The computations show that, for $r \gtrsim 5$, $\theta^{\max}$ increases linearly with the square-root of the chain length: $\theta^{\max} = 0.64 + 0.562\sqrt{r}$. Under these conditions the segment distribution is purely entropically determined; in fact, the square root dependence is probably related to the undisturbed Gaussian behaviour of polymer chains in pure bulk polymer (Benoit, 1976; DeSantis and Zachmann, 1977). The quantity θ (i.e. the number of monolayers that can be filled with segments of adsorbed chains) should not be confused with θ^σ (i.e. the excess over the bulk density); the latter quantity is equal to zero in pure bulk polymer and $\theta = \theta^d$ in this situation.

Figure 8(b) gives the dependence of the direct surface coverage θ_1 on ϕ_* for $\chi_s = 1$. It is seen that for long chains θ_1 is practically independent of ϕ_* over a wide range of ϕ_*, and only weakly dependent on chain length. This is a common feature of all theoretical models discussed in this chapter.

Figure 8(c) shows the behaviour of the bound fraction p as a function of ϕ_*, r, and χ. The general trends are clear and follow immediately from those in θ and θ_1 through $p = \theta_1/\theta$.

Figure 9 presents a more detailed picture of the chain length dependence of the adsorbed amount according to the theory of Scheutjens and Fleer. The region between $r = 100$ and $r = 1000$ may be compared to the corresponding portion of Fig. 6, applying to a different volume fraction ($\phi_* = 10^{-4}$) and a higher adsorption energy ($\chi_s = 5$). It can be seen that the linear dependence of θ on $\log r$ for $\chi = 0.5$ applies only for relatively long chains; for oligomers the shape of the $\theta(\log r)$ curve is convex with respect to the $\log r$ axis for low χ_s but concave for high χ_s, apparently because with strongly adsorbing surfaces already at low r relatively much material adsorbs.

The difference between θ and θ^σ increases with increasing bulk volume fraction and with increasing chain length. In the inset of Fig. 9 a comparison is made between θ^σ as calculated with the Scheutjens and Fleer and the Roe theories. As has already been discussed in connection with Fig. 6, the flattening of θ^σ (Roe) at high r is too pronounced.

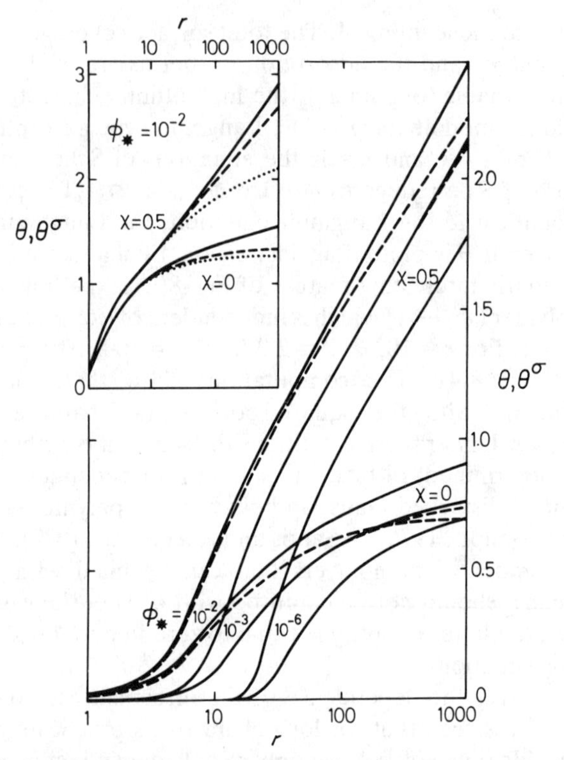

Figure 9 Chain length dependence of the total surface coverage in the theories of Scheutjens and Fleer and of Roe. Full curves, θ (SF); dashed curves, θ^σ (SF); dotted curves, θ^σ(R). The bulk volume fraction ϕ_* is indicated. Hexagonal lattice ($\lambda_0 = 0.5$). Main figure, $\chi_s = 1$; inset, $\chi_s = 3$.

In Fig. 10 it is shown in more detail how the direct surface coverage θ_1 depends on the chain length. For oligomers, θ_1 is small and nearly independent of χ. With increasing r the direct surface coverage increases until a plateau is reached which is independent of chain length and ϕ_*, but which is higher for poorer solvents. These trends have been discussed before.

Figure 11 displays typical concentration profiles in the adsorbed layer, plotted with a logarithmic scale for ϕ_i, according to the theories of Hoeve, Roe, and Scheutjens and Fleer, for $r = 1000$, $\phi_* = 10^{-6}$, and $\chi = 0.5$. The solid curves represent overall profiles. In the Hoeve model the segment concentration (supposed to stem from loops only) decays exponentially, as formulated in equation (16). Of the three models, that of Hoeve predicts the steepest decay. Roe's theory (neglecting also end effects) results in a profile that is not very different from that of Hoeve; also here the profile is more or less exponential. The fact that around $i = 30$ Roe's profile drops slightly below the equilibrium concentration $\phi_* = 10^{-6}$ is related to the (entropic) repulsion of free molecules by segments of adsorbed chains in a steep concentration gradient. A similar trend is found, in principle, in the Scheutjens and Fleer theory, but then the effect is much smaller since the profile is much smoother.

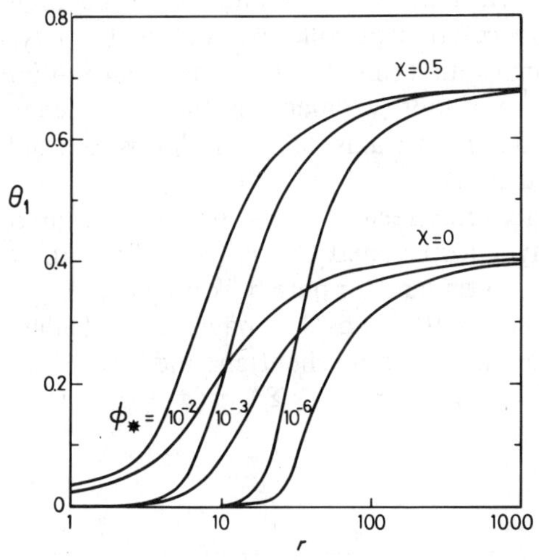

Figure 10 Chain length dependence of the direct surface coverage. Scheutjens and Fleer theory; $\chi_s = 1$.

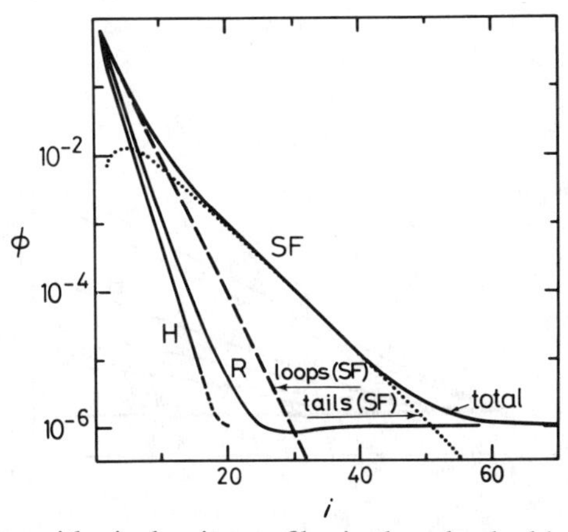

Figure 11 Semi-logarithmic density profiles in the adsorbed layer according to the theories of Hoeve (H), Roe (R), and Scheutjens and Fleer (SF); in the latter case the total density is decomposed into contributions of tails (dotted curve) and loops (dashed curve). $\chi = 0.5$, $\chi_s = 1$, $r = 1000$, $\phi_* = 10^{-6}$; $c = 0.1$ (H), $\lambda_0 = 0.5$ (R, SF).

The theory of Scheutjens and Fleer predicts a much slower decay of the segment concentration. This is mainly due to the contribution of tails, as can be seen from the separate loop and tail segment concentration profiles drawn in Fig. 11. The contribution due to loop segments is indeed exponential, but in the layers beyond $i = 15$ it is very small as compared with the concentration of tail segments. In the outer regions of the adsorbed layer the volume fraction is completely determined by the tails. This is a very important conclusion, which should be borne in mind when the stabilization and flocculation of colloidal systems by polymers are under discussion. In order to underline the dramatic contribution to the profile of a small proportion of tails, we mention that, in the example of Fig. 11, 38% of the segments are in trains, 55.5% in loops, but only 6.5% in tails (see also Figs. 13 and 14).

Figure 12 presents some results on the root-mean-square extension t_{rms} of the adsorbed layer for isolated chains ($\phi_* \rightarrow 0$), a dilute (10^{-6}) and a semidilute (10^{-2}) solution, and for pure bulk polymer ($\phi_* \rightarrow 1$). Two curves (i.e. for $\chi = 0.5$ and $\phi_* = 10^{-6}$ and 10^{-2}) have been calculated with Hoeve's theory; the others are taken from Scheutjens and Fleer. In the latter model, t_{rms} is calculated from $(t_{rms}/l_s)^2 = \theta^{-1} \Sigma_{i=1}^{M} i^2 (\phi_i - \phi_i^f)$.

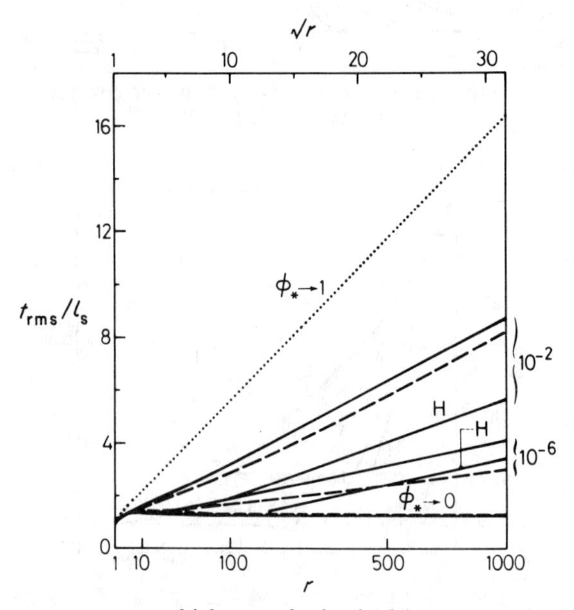

Figure 12 Root-mean-square thickness of adsorbed layers as a function of chain length r, according to Scheutjens and Fleer (SF) and Hoeve (H). Full curves, $\chi = 0.5$; dashed curves, $\chi = 0$; dotted curve, bulk polymer (independent of χ). The bulk solution volume fraction ϕ_* is indicated. $\chi_s = 1$, $c = 0.1$ (H), $\lambda_0 = 0.5$ (SF).

Isolated chains lie very flat. With increasing ϕ_* the layer becomes more extended, and the r.m.s. thickness increases proportionally with $\sqrt{r}$ (compare equation (17) with λ proportional to $1/r$ as discussed above). For $\phi_* = 10^{-6}$ Hoeve's results fall slightly below those of Scheutjens and Fleer; for $\phi_* = 10^{-2}$ the discrepancy is considerable. This fact is mainly caused by Hoeve's neglect of tails; at $\chi = 0.5$ and $r = 1000$ the tail fraction is 6.5% for $\phi_* = 10^{-6}$ and for $\phi_* = 10^{-2}$ it amounts to 21%. Note that the effect of tails is less pronounced in t_{rms} (which is an average over train, tail, and loop segments) than in the concentration profile at large distances (where ϕ_i is solely determined by tails). The very high layer thickness in pure bulk polymer is caused by the very high tail fraction (about 65%) under these conditions.

It may seem somewhat surprising that t_{rms} increases linearly with $\sqrt{r}$, both for $\chi = 0.5$ and $\chi = 0$, whereas the adsorbed amount increases much more weakly with increasing r. Again the tails are responsible for this behaviour; even when θ levels off at high r (which is the case for $\chi = 0$) the layer thickness continues to increase approximately proportionally to $\sqrt{r}$. As will be shown in the next figures, the length of the tails is proportional to the chain length, giving rise to a $\sqrt{r}$ dependence for the mean extension from the surface.

The thicknesses given in Fig. 12 may be compared with the radii of gyration R_g of the coils in solution, which, for a hexagonal lattice, can be approximated by $R_g/l_s = 0.444\sqrt{r}$ (Scheutjens and Fleer, 1980). It turns out that in dilute solutions t_{rms} is lower than R_g by a factor of 2 to 3, a feature that has also been found experimentally.

In Figs. 13 and 14 we consider the composition of the adsorbed layer in terms of trains, loops, and tails more closely. Most of the curves are taken

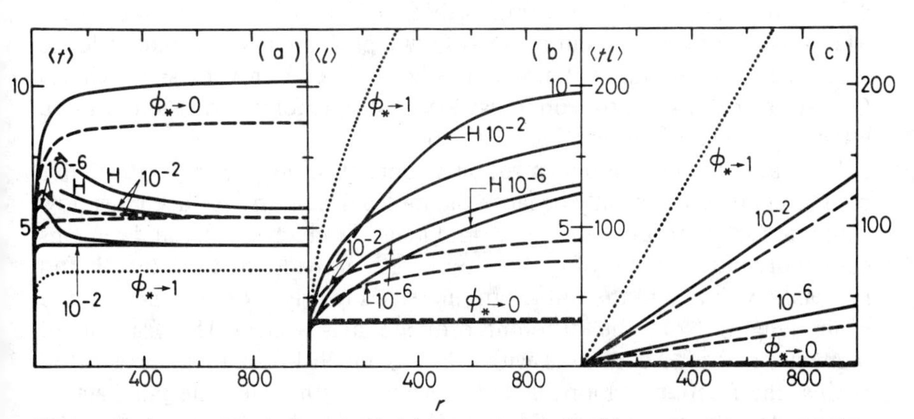

Figure 13 Average train length $\langle t \rangle$ (a), loop length $\langle l \rangle$ (b), and tail length $\langle tl \rangle$ (c), as a function of chain length r according to Scheutjens and Fleer, and Hoeve (H). Symbols and conditions as in Fig. 12.

from Scheutjens and Fleer, but for the sake of comparison a few results of the Hoeve model are also represented. Figure 13 gives the average lengths of trains $\langle t \rangle$, of loops $\langle l \rangle$, and of tails $\langle tl \rangle$, as a function of chain length. Figure 14 shows the corresponding fractions ν_t $(= p)$, ν_l, and ν_{tl} of segments in trains, loops, and tails, respectively. In all cases we consider again $\chi_s = 1$ and $\phi_* \rightarrow 0$ (isolated chains), $\phi_* = 10^{-6}$ and 10^{-2}, and $\phi_* \rightarrow 1$ (bulk polymer).

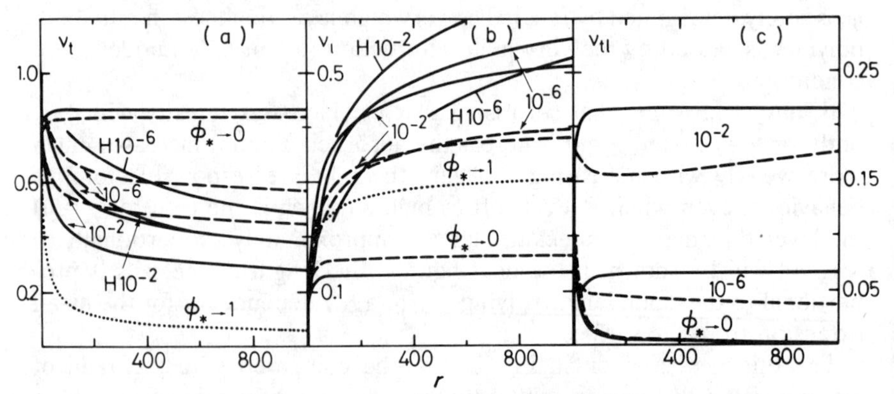

Figure 14 Fraction of segments in trains ν_t (a), loops ν_l (b), and tails ν_{tl} (c), as a function of chain length r according to Scheutjens and Fleer, and Hoeve (H). Symbols and conditions as in Fig. 13.

For isolated chains, the train length is high (8.5 to 10, depending on the solvent quality χ), loops are small (about 1.5) and so are tails (about 2). The fraction of segments in trains is very high (about 85%), that in loops small (15%), and the tail fraction is negligible. For very short chains $(r \lesssim 20)$ some deviations from these trends occur; for a discussion of these effects see Scheutjens and Fleer (1980).

Even at very low, but non-zero, bulk concentration (e.g. $\phi_* = 10^{-6}$) this picture changes drastically because the molecules on the surface must now compete for adsorption sites. The train length is much smaller and remains essentially independent of r, the loop length exceeds the train length and increases with r, and the tail length increases nearly proportionally to the chain length. Note the different ordinate axis scales. The fraction of segments in trains decreases with r and is much lower than for isolated chains, the fraction in loops increases slightly with r, and the tail fraction is considerable and practically constant. All of this underlines that the shapes of adsorbed chains change dramatically if competition occurs, even in solutions as dilute as $\phi_* = 10^{-6}$. For higher ϕ_* (e.g. $\phi_* = 10^{-2}$) these trends are even more pronounced. In pure bulk polymer the contribution of trains (which become as short as 3.5) drops to a very low value (5% for $r = 1000$), the loops become quite long ($\langle l \rangle = 21$ for $r = 1000$) and contain about 30% of the segments, and the tails dominate in the adsorbed layer ($\langle tl \rangle$ is 344 for $r = 1000$) and accommodate 65% of all the segments. Thus a chain adsorbed from bulk polymer may be considered as being divided into three roughly equal sections—two long dangling tails and a middle section in which short trains and long loops alternate.

A comparison with Hoeve's predictions as to the composition of the adsorbed layer is not quite feasible because in his treatment only trains and loops are taken into account. Nevertheless, the qualitative trends as to the dependences of $\langle t \rangle$, $\langle l \rangle$, ν_t, and ν_l on ϕ_* and r are similar to those in the Scheutjens and Fleer theory, as can be seen in Figs. 13(a) and (b) and 14(a) and (b). Quantitatively the agreement is also quite satisfactory, considering the approximations used in Hoeve's model.

Finally, we present a typical example ($r = 1000$, $\phi_* = 10^{-3}$, $\chi = 0.5$, $\chi_s = 1$) of the size distribution of trains, loops, and tails as obtained in the Scheutjens and Fleer theory. Figure 15(a)–(c) give the fractions of trains, loops, and tails of given lengths t, l, and tl, respectively, whereas Figure 15(d)–(f) show the corresponding fraction of segments in trains, loops, and tails of given length. For this example, the average values are $\langle t \rangle = 4.32$, $\langle l \rangle = 7.53$, and $\langle tl \rangle = 89.00$ (indicated by arrows in Fig. 15). For short sequence lengths, the fractions of trains, loops, and tails decrease sharply with increasing lengths t, l, and tl, respectively, but for high t, l, and tl the decay is slower for loops than for trains, whereas that of tails is very slow. These features show up more clearly in the fraction of segments present in trains, loops, and tails; trains exceeding 25 segments have only a small contribution, the number of loop segments in loops as long as 50 segments is still significant, whereas tail sizes as long as 800 segments contribute still substantially. Recalling that in this particular example $r = 1000$, this implies that it is not unlikely to find tails of a length comprising 80% of the entire chain. The tail size distribution is very broad indeed. This again is a feature of great relevance for the quantitative interpretation of steric stabilization of colloidal particles.

We want to compare these results with Hoeve's equations (20) and (21). In order to eliminate the effect of differences in the values for $\langle t \rangle$ and $\langle l \rangle$ between both theories, we express m_t/m and n_l/m in terms of the average train length $\langle t \rangle$ and the average loop length $\langle l \rangle$, respectively. Combining equations (19) and (21) we obtain immediately

$$m_t/m = \langle t \rangle^{-1} (1 - \langle t \rangle^{-1})^{t-1} \tag{51}$$

It turns out that the numerical values obtained by Scheutjens and Fleer for m_t/m, as given in Fig. 15(a), correspond exactly to equation (51). This

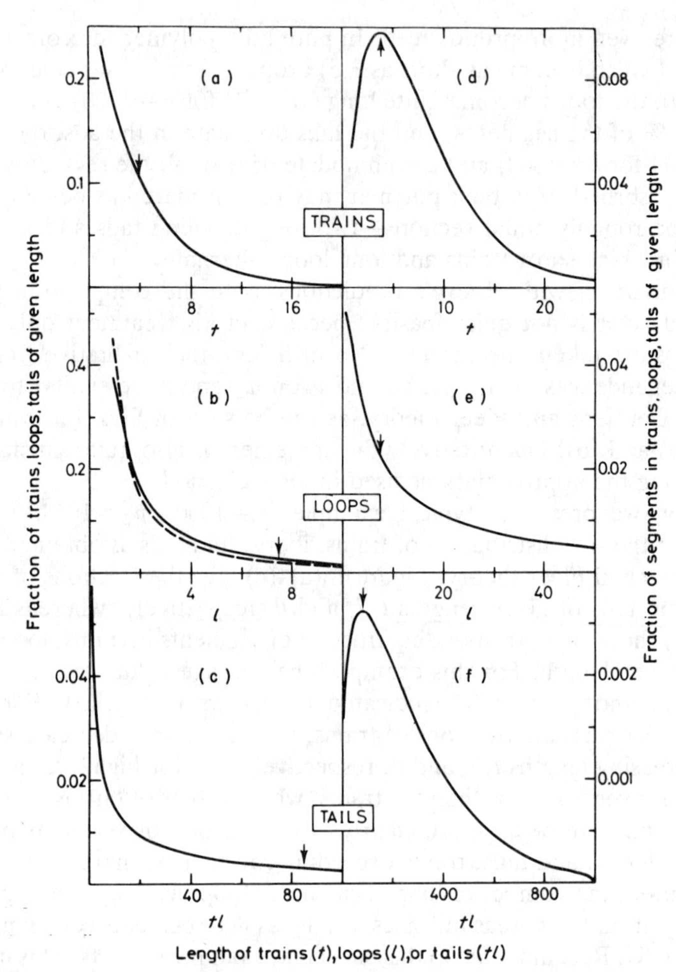

Figure 15 Examples of size distributions of trains, loops, and tails. Scheutjens and Fleer theory, $r = 1000$, $\phi_* = 10^{-3}$, $\chi = 0.5$, $\chi_s = 1$. Left: fractions of tails, trains, and loops of given length. Right: fraction of segments in tails, trains, and loops of given length. The average train, loop, and tail sizes are indicated by the arrows. The dashed curve gives the loop size distribution according to Hoeve (see text).

somewhat surprising result is probably related to the fact that a train may be considered as a two-dimensional walk in which each step has the same weighting factor p_1. For a different combination of ϕ_*, χ_s, and χ a different value for $\langle t \rangle$ is found, but the distribution of train lengths around its average value $\langle t \rangle$ is not affected by these parameters.

For the loop size distribution the comparison is slightly more complicated. In this case the value for $\langle l \rangle$ can be used to find the parameter λ from equation (18). Substituting this value for λ in equation (20), the loop size distribution can be calculated. The result is shown in Fig. 15(b) as the dashed curve. The agreement is not as good as for the train size distribution, but still the Hoeve theory appears to be a good approximation. Full agreement of the two loop size distributions is not to be expected, because in Hoeve's model even the smallest loops are supposed to have a Gaussian distribution.

Summarizing these sections, we conclude that a clear picture of polymers in the adsorbed state is now emerging. The most interesting recent progress is that the effects of finite solution concentration and the concomitant contribution of tails are stressed. This has important consequences for the interpretation of relevant experiments, as will be shown in the following sections.

V. Applicability of Theory to Heterodisperse Polymers

In the preceding sections, the adsorption theory for homodisperse polymers was discussed. Although homodisperse polymers have been used in a few cases, the vast majority of experiments reported in the literature deal with heterodisperse systems in which the distribution of polymer molecular weights is not very narrow. In this section we discuss the effect of polymer heterodispersity on adsorption. We restrict this discussion to the most commonly used type of experiment, i.e. the determination of adsorption isotherms.

If one tries to apply polymer adsorption theories to real systems, one frequently meets difficulties in the interpretation. One of them is that the shape of experimental adsorption isotherms is often much more rounded than that of the sharp high affinity isotherms predicted by theory (Fig. 4). Secondly, one usually observes that, upon dilution of the system with solvent, hardly any desorption, if at all, is detectable. One would expect that, upon decrease of the equilibrium concentration, the adsorbed amount would "follow" the isotherm, leading to lower adsorbed amounts, certainly if the isotherms are rounded. An easy (although not necessarily correct) conclusion is that polymer adsorption is irreversible and that polymer adsorption theories (based on equilibrium thermodynamics) are not applicable to real systems. Thirdly, measured adsorption isotherms depend on the area to volume ratio A/V of the experimental system. This seems to be in conflict with the very essence of an adsorption isotherm, since the latter ought to give a relation between two intensive parameters (viz. the surface concentration and the solution concentration) and should be independent of the extension of the system. In this section, we show that these three points can be explained satisfactorily in terms of equilibrium theories, provided that the polydispersity of the polymer is properly taken into account. We will follow the arguments put forward in a recent paper by Cohen Stuart *et al.* (1980).

Heterodisperse polymers have to be treated as a mixture. In such a mixture, the adsorbability increases with increasing chain length, especially in poor solvents (see Fig. 6). Moreover, there is much theoretical (Cohen Stuart *et al.*, 1980; Roe, 1980; Scheutjens and Fleer, 1982) and experimental (Felter *et al.*, 1969; Felter and Ray, 1970; Howard and Woods, 1972; Sadakne and White, 1973; Vander Linden and van Leemput, 1978b; Cohen Stuart *et al.*, 1980) evidence that long molecules adsorb preferentially over short ones, and that long chains displace short ones from the surface. Therefore, in a polydisperse system where polymer on the surface is *in equilibrium* with polymer in solution, a fractionation occurs; the short molecules enrich in the solution and the long chains on the surface.

These well established facts form a sufficient basis to explain that rounded adsorption isotherms are found for heterodisperse polymers, whereas sharp isotherms are predicted by theory for homodisperse polymers. The principle of the explanation is as follows. If the solution concentration is increased, more big molecules are available to displace the smaller ones from the surface, thereby increasing the adsorbed amount. This increase continues as long as the adsorbed amount increases with molecular weight (of each fraction). In this displacement process the molecular weight distribution of the adsorbate shifts towards higher M and may differ considerably from that in the bulk solution. Since in a given system only a limited surface area is available, at high ϕ_* only the very high molecular weight tail of the weight distribution will be adsorbed. This explains in principle the rounded shape of polymer adsorption isotherms.

The second point mentioned above, i.e. the difference between adsorption and desorption isotherms, is related to the reversibility of the adsorption. Let us first discuss the reversibility for a homodisperse polymer. For such a polymer, the adsorbed amount is, in dilute solutions, practically independent of the solution concentration over several decades of ϕ_* [see Figs. 4 and 8(a)]. This is more easily seen in Fig. 16 which gives, on a double logarithmic scale, a typical adsorption isotherm for a relatively short chain over a very wide range of concentrations (Scheutjens and Fleer, 1982). In this figure, one may distinguish three regions: for $\phi_* \lesssim \phi_*^c$ the adsorbed amount is proportional to ϕ_* (the Henry region where the adsorbed molecules behave as isolated chains); for $\phi_*^c \lesssim \phi_* \lesssim 0.1$ the adsorption depends only very weakly on ϕ_* (the pseudo-plateau region); and for $0.1 \lesssim \phi_* \leqslant 1$ (the concentrated region) θ increases gradually to attain the value for bulk polymer, which is proportional to $\sqrt{r}$. Nearly all polymer adsorption experiments are performed at concentrations corresponding to the pseudo-plateau. In this region the (equilibrium!) adsorbed

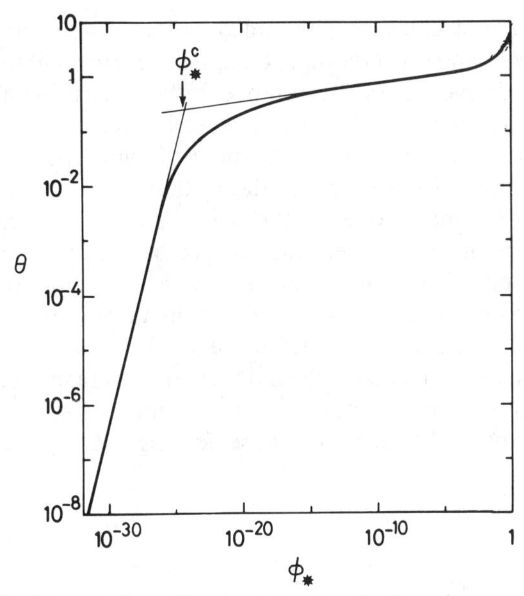

Figure 16 A typical double logarithmic adsorption isotherm over a very wide range of concentrations. The transition concentration ϕ_*^c characterizes the transition from the Henry region to the pseudo-plateau region; $r = 100$, $\chi = 0.5$, $\chi_s = 1$, hexagonal lattice ($\lambda_0 = 0.5$).

amount decreases by only a few per cent if the solution is diluted tenfold. Only at concentrations of the order of ϕ_*^c can appreciable desorption occur. The important point is that ϕ_*^c is *extremely* small, even for relatively short chains; moreover, it decreases exponentially with chain length (Scheutjens and Fleer, 1982). Since such low concentrations cannot normally be obtained by dilution, this means that no detectable desorption will take place. Thus, for monodisperse polymers, the lack of desorption upon dilution with solvent may not be interpreted as an irreversibility, but it is simply a consequence of the shape of the isotherm; the adsorption remains constant down to extremely small equilibrium concentrations.

For heterodisperse polymers adsorbing according to a rounded isotherm, the situation is different; also in this case no desorption is observed. In other words, *ad*sorption isotherms are rounded (as explained above), whereas *de*sorption isotherms are sharp, following thereby the theoretical prediction for monodisperse polymers. In view of the above ideas, the explanation is straightforward. Each component in the polymer mixture adsorbs according to a sharp isotherm, but none of them desorbs unless its concentration in solution is far below the limit of detectability. The only effect of diluting the system with solvent is a lowering of the concentration of the small molecules that were already in the solution; the distribution of the polymer molecules over the adsorbed and the solution state does not change. We can conclude that the *apparent* irreversibility of polymer adsorption (as derived from a difference in the *ad*sorption and *de*sorption isotherms) is not necessarily a true irreversibility; in the above analysis only arguments based upon full attainment of equilibrium were used.

The dependence of the shape of the isotherms on the area to volume ratio A/V can be understood on the basis of the same principles. If, at given total amount of polymer, the surface area in the system is small, only the fraction of polymer with very high molecular weights can be accommodated on the surface, and the amount adsorbed per unit of surface area is high. If, at given volume, more area is available, also smaller molecules may find a place on the surface, the average molecular weight of the adsorbate is lower, and so is the amount adsorbed per unit area. Isotherms at low A/V lie above those for high A/V (see Fig. 17). An

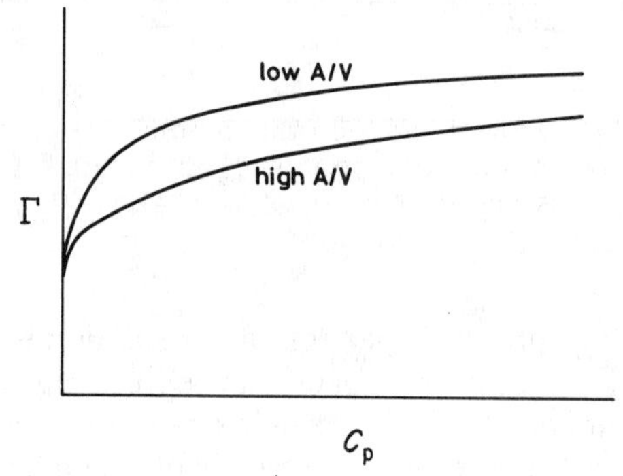

Figure 17 Effect of the surface area/solution volume ratio A/V on the adsorption isotherm of a polydisperse polymer sample.

important consequence is that adsorption results obtained using macroscopic surfaces (as in monolayer films, or in ellipsometry) may not simply be identified with those on colloidally dispersed adsorbents (emulsions, hydrophobic sols); in the first case the adsorbed amount per unit of surface area might be considerably higher than in the latter type of experiment.

On the basis of these ideas we can predict the following general trends. Polymers with a narrow molecular weight distribution adsorb according to relatively sharp isotherms, whereas isotherms for polydisperse samples are rounded. Isotherms become more rounded if (i) the molecular weight distribution is broader, (ii) the dependence of the adsorbed amount on molecular weight is stronger, which implies that isotherms from good solvents are sharper than those from poor solvents, and (iii) the area to volume ratio A/V is higher. Cohen Stuart *et al.* (1980) give a quantitative analysis of these effects. If the molecular weight distribution $w(M)$ and the dependence $\Gamma(M)$ of the adsorbed amount on molecular weight are known, the adsorption isotherm can be calculated. Conversely, from the shape of the isotherm $\Gamma(M)$ can, in principle, be obtained from $w(M)$, or $w(M)$ from $\Gamma(M)$. In this way the foundation is laid for a better understanding of the fractionation effects occurring in polymer adsorption and for a quantitative analysis of heterodisperse polymer adsorption isotherms.

We conclude this section with a very simple example taken from Cohen Stuart *et al.* (1980). We consider a mixture of small (s) and big (b) polymer molecules with molecular weights M_s and M_b, respectively. Let the weight fractions in the mixture be w_s and w_b (with $w_s + w_b = 1$) and the plateau values of the individual sharp isotherms Γ_s and Γ_b (with $\Gamma_b > \Gamma_s$). It can be shown (Cohen Stuart *et al.*, 1980) that for this mixture the adsorption isotherm consists of three linear parts (see Fig. 18). In the initial steep part,

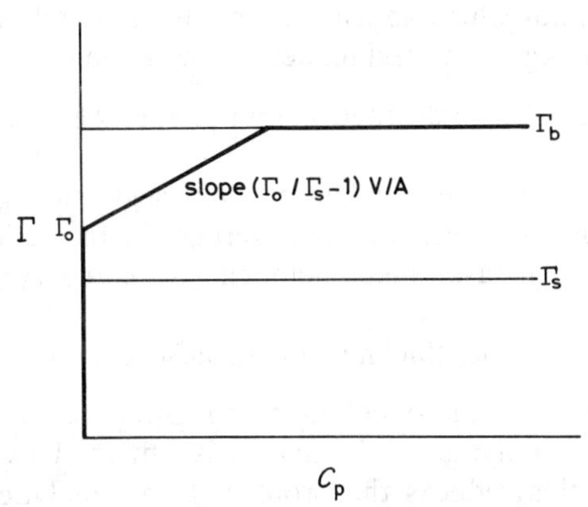

Figure 18 Adsorption isotherm of a mixture of small (s) and big (b) molecules. The (sharp) individual isotherms for the two fractions, with plateau levels Γ_s and Γ_b respectively, are also given.

with the polymer concentration c_p below the limit of detectability, the isotherm coincides with the ordinate axis. In this region, virtually all molecules in the system adsorb and the composition of the adsorbate is identical to the overall composition, i.e. here no fractionation occurs. The vertical portion ascends until the surface is saturated with molecules s and b in the weight ratio w_s/w_b; the adsorbed amount Γ_0 is then given by $\Gamma_0^{-1} = w_s\Gamma_s^{-1} + w_b\Gamma_b^{-1}$.

Beyond this point, addition of more polymer leads to displacement of the smaller molecules by the bigger ones. This results in a linear increase of Γ with c_p. In this region there are virtually no big molecules in the solution; the increase of c_p is only due to small molecules. The slope of this isotherm part is given by $(\Gamma_0/\Gamma_s - 1)V/A$, which is inversely proportional to the surface to volume ratio.

The displacement process is completed if all the small molecules on the surface have been replaced by the big ones. Then the plateau value Γ_b is reached. At still higher c_p the solution contains an increasing fraction of molecules b, although it remains richer in molecules s than the original polymer solution.

In Section VII we shall discuss a few experimental results corroborating the ideas presented in this section.

VI. Experimental Techniques

In this section we shall briefly discuss some experimental techniques that are developed either with special reference to polymer adsorption or for other purposes, but which appear suitable to study adsorbed polymers.

Broadly speaking, these techniques can be divided into three groups:

(a) Those leading to the adsorbed amount, i.e. to the total surface coverage θ.
(b) Those leading to the number of train segments, giving the bound fraction p or the direct surface coverage θ_1. In the latter case combination with a measurement belonging to category (a) is needed to obtain p.
(c) Those leading to the thickness of the adsorbed layer.

Of these techniques, those belonging to group (a) are relatively least characteristic for polymers. Usually, θ is obtained by some depletion procedure and this reduces the problem to an analytical one, which is beyond our scope. We merely note that the sensitivity of the analytical technique should be independent of the molecular weight of the polymer, especially for polydisperse samples, since in that case the molecular weight of the polymer in solution may differ from that on the surface. With respect to procedures (b) and (c), the general remark can be made that the task is always twofold: first, the technique must be made sufficiently sensitive to obtain a high signal/noise ratio; then, the data need some unavoidable physical interpretation. For instance, approaches of group (c) require some thought on what kind of (averaged) thickness is obtained, and in group (b) measurements the question arises whether the proper number of train segments have been "seen". Owing to the constraints of the covalent bonds, train segments can be oriented on the surface in such a way that they do not contribute to the signal, whereas segments close to (but not on) the surface sometimes do give rise to a signal. Basically, the problem is then to distinguish experimentally between train and loop segments. Also theoretically this problem arises, since the decision as to which segments should be considered as train segments is to some extent arbitrary, because of the gradually decaying force field of the surface. Ultimately the issue is how to match as well as possible the experimental and theoretical abstractions from reality.

A. METHODS LEADING TO THE DIRECT SURFACE COVERAGE θ_1 OR THE BOUND FRACTION p

This group can be subdivided into two main sections: electrochemical and spectroscopic methods. A new technique, which has only very recently given the first results, is small-angle neutron scattering. In addition, there are a few less frequently used methods.

The electrochemical procedures are usually based on measurements of some static double-layer property that can be interpreted to yield θ_1.

Double-layer parameters that have been used to find θ_1 are the double-layer charge or capacitance and the shift of the point of zero charge (p.z.c.). Train segments modify the properties of the Stern layer because they replace solvent molecules, thus changing the thickness of this layer, its dielectric permittivity, and the amount of specifically adsorbed ions in it. These three parameters determine the Stern layer capacitance, which is measurable at not too low electrolyte concentration (i.e. when the contribution of the diffuse part is suppressed). Surface charge measurements (at given surface potential) amount to the same because charge–potential curves are integrated capacitance–potential curves. Similar information can be extracted from p.z.c. shifts, since these reflect the alteration of the state of polarization of the solution layer adjacent to the surface. As a note of caution it may be stressed that shifts of the isoelectric point are identical to those of the p.z.c. only if there is no specific adsorption of counterions. If specific adsorption does occur, it can, in turn, be influenced by the presence of trains.

The interpretation can usually not be done by substitution of molecular dimensions and local permittivities because these quantities are subject to considerable uncertainty. Rather, comparison should be sought with small molecules whose adsorption properties are well known, so this technique is not an absolute method.

For polyvinyl alcohol (PVA) adsorbed on silver iodide, θ_1 could be obtained both from surface charge and from p.z.c. measurements, by using as a reference the corresponding properties of CH_2 and CHOH groups which, in turn, were derived from measurements with low molecular weight adsorbates (Fleer *et al.*, 1972; Koopal and Lyklema, 1976). This approach appeared sufficiently sensitive to detect enrichment of acetate groups in trains in a copolymer of vinyl acetate and vinyl alcohol (Koopal and Lyklema, 1979). Mercury has been a classical model for interfacial electrochemistry, but few studies on uncharged non-biological molecules are available. Work on fractionated PVA (Wójciak and Dutkiewicz, 1964), polyethylene glycol (Yoshida *et al.*, 1972a), and polyvinyl pyrrolidone (Yoshida *et al.*, 1972b) suggests that the procedures developed with AgI may also be worked out for Hg. At the present stage, the work of Wójciak and Dutkiewicz (1964) is rather descriptive, whereas Yoshida *et al.* (1972a,b) fail to distinguish properly between trains and loops. Beyond the field of uncharged synthetic macromolecules, there is a fair amount of double-layer work on mercury involving biopolymers (e.g. polarography in the presence of DNA); for reviews see Miller (1971) and Miller and Bach (1973). Changes of surface charge have also been found on oxides but they have not yet led to quantification of θ_1 (Joppien, 1978).

Kinetic electrochemical measurements to obtain θ_1 are generally in an exploratory state. The reduction in the rate of charge transfer through a charged interface may be coupled to θ_1. In this context it may be mentioned that the presence of polymers adsorbed on colloid particles alters the AC and DC conductivity; since train segments alter the conductivity due to ions within the slipping plane and loop and tail segments do the same for ions beyond this plane, perhaps it may in future be possible to discriminate between train and other segments (Dukhin and Dudkina, 1978; Dukhin *et al.*, 1979).

Spectroscopic techniques comprise infrared, electron spin resonance (e.s.r.), also known as electron paramagnetic resonance (e.p.r.), and nuclear magnetic resonance (n.m.r.) spectroscopy. These methods have in common that spectral alterations occurring upon adsorption in specified groups of the polymer are used to count the number of these groups. If the signal of the adsorbed polymer can be separated from that of the dissolved polymer, the bound fraction $p = \theta_1/\theta$ can be obtained. Such a separation can be achieved, for example, by replacing the solution by solvent or by compensating the signal of the dissolved polymer.

E.s.r. procedures can be applied if a suitable spin label is available that can be incorporated in the polymer chain. Loops and tails are more mobile than trains, so are the labels attached to them, and this results in differences in the spectral line width. Resolution of the spectra yields p. This method has been successfully applied to the adsorption of polyvinyl pyrrolidone (PVP) on silica (Robb and Smith, 1974, 1977; Clark *et al.*, 1976). In this work, the reference signals of mobile and immobile labels were obtained by measuring the e.s.r. spectrum of the labelled PVP in a solution of high and low temperature (i.e. of low and high viscosity), respectively. The procedure appeared sufficiently sensitive to assess the influence of various solvents, aqueous and non-aqueous, on p. A general drawback of e.s.r. methods is, of course, that a label is introduced which may affect the mode of adsorption of the polymer under consideration. Even if a labelled monomer does not detectably adsorb, one cannot be sure that the label does not affect the adsorption of a polymer in which it is introduced, since polymers may adsorb under conditions where their monomeric and oligomeric counterparts would hardly adsorb (especially at low χ_s; compare Fig. 9).

For uncharged polymers very few publications have appeared in which n.m.r. is applied. Basically, the method can measure the same as e.s.r., in that the reduction of mobility of adsorbed segments is reflected in the spectral line width so that, in principle, p is measurable. As some line broadening occurs also with segments in loops and tails, the method tends to overestimate the bound fraction. For isotactic polymethyl methacrylate (PMMA) on silica in chloroform the line broadening was found to be so large that the resonance signal was undetectable, suggesting adsorption in a rather flat conformation (Miyamoto and Cantow, 1972). However, as said, p is probably overestimated.

The n.m.r. technique can also be used to observe relaxation phenomena directly with the help of suitable radiofrequency pulses. Some spin-echo methods offer potentialities in finding p, namely when the experiment is carried out in such a way that the first relaxation reflects all polymer, whereas the echo only refers to the loop and tail fraction (Cosgrove *et al.*, 1981). The technical problems of this n.m.r. technique are considerable. Short pulses (high emitting power) and fast detection of the transient signal are required. Moreover, segments of free chains in the solution interfere with the measurement, so the solution must be separated off before measurements are taken. In addition, resonating nuclei must be removed from the solvent, or their signals suppressed. Despite these difficulties the n.m.r. method has led to reliable bound fraction results (Cosgrove *et al.*, 1981; Barnett *et al.*, 1982). An advantage over e.s.r. measurements is that no foreign label needs to be incorporated.

Infrared (i.r.) analysis is based on counting the number of groups in which, upon adsorption, characteristic frequency changes occur. These groups may belong either or both to the adsorbent or to the polymer. For low molecular weight solutions, the subject matter, including its experimental implementation, has been recently reviewed by Rochester (1980), who discusses how to overcome the difficulty (as compared with i.r. analysis of adsorbates from the gas phase) of the presence of a solvent. If the solvent itself does not absorb i.r. radiation there is not a serious problem, but this condition is quite restrictive since it excludes many solvents, including water. It has been suggested to adsorb the polymer from aqueous solution and then change the solvent for a spectroscopically transparent one; although this procedure has been applied for polymeric adsorbates, our recent insight into the reversibility of polymer adsorption (see Section V) renders it subject to doubt, because conformational changes are likely to occur. Such changes are also expected if the liquid is completely removed; the measurement is then relatively simple, but the conformation of adsorbed polymers changes drastically, air being an extremely poor solvent for polymers.

For low molecular weight components three experimental approaches are available for *in situ* measurements: with dispersions, with compressed pellets, and by i.r. reflection from macroscopically flat surfaces. Blanking of the solvent or solution can be achieved by placing a cell, containing the pure solvent or the equilibrium solution, in the reference beam of the spectrophotometer (Rochester, 1980).

Experiments with pellets of polymer-covered particles could be done in principle, but their interpretation is difficult since p will probably be affected by the close proximity of a second particle. Hence, it is doubtful if measurements with pellets are representative for the same on isolated surfaces. To this end, the theory given above in Section IV must be generalized to cover the adsorption in a constrained space between two or more particles.

As said before, the method rests on the availability of a group either in the surface or in the polymer that gives a clear i.r. signal. Its modification due to adsorption must be detectable and interpretable. A suitable surface group is the silanol group on silica, which in vacuum exhibits a sharp absorption maximum around 3740–3750 cm^{-1}. In solvents the position is different, depending on the nature of that solvent. For polymers the absorption peak of the carbonyl stretching vibration around 1650–1700 cm^{-1} in various solvents has been shown to be appropriate. For a quantitative interpretation, the extinction coefficient per adsorbing group must be known, and this information is usually only available by inference. For instance, it could be assumed that the influence of adsorption of a polymer on the extinction coefficient of a silanol group or a carbonyl group is identical to that of a low molecular weight substance with the same group. In this respect the method is not independent.

The direct surface coverage θ_1 can be estimated directly from the frequency shift of a surface group such as silanol. In this way, only the number of segments interacting directly with silanols is counted. In practice there can be segments adsorbed on the surface which do not interact with the silanol groups. If, as usual, θ_1 is considered to be unity if all silanol groups are perturbed, the method tends to overestimate θ_1. We return to this point in Section VII.B, where we shall discuss an illustrative example from the literature.

In principle, the bound fraction p can be obtained from θ_1, using the relation $p = \theta_1/\theta$. A more direct method is to use the frequency shift of a specific group in the polymer, such as carbonyl or phenyl. In this way p is obtainable directly. The possibility that not all train segments are sufficiently perturbed to be distinguishable from train and loop segments might result in an underestimation of p.

Pioneering work by Fontana and Thomas (1961) and Fontana (1963, 1966) proved the viability of these spectroscopic methods. The authors found, for two methacrylates adsorbing from decalin or dodecane on silica, that θ_1/θ (where θ_1 was evaluated from an analysis of the intensity change of the surface silanol group stretching vibration) agreed reasonably well with p obtained from an analysis of the perturbation of carbonyl groups in the polymer. This extent of agreement is a measure of the internal consistency of the method, although it gives no indication of the absolute accuracy of the obtained p value. More systematic work by Korn (1978) showed that the correspondence between the two determinations of p is not always unity. Differences up to a factor of 3 were found, depending on the nature and molecular mass of the polymer and the solvent. Apparently some scrutiny is needed. Insight into the absolute reliability can only be obtained by comparing θ_1 or p determinations by different methods, and such measurements are scanty. An example is the adsorption of polymethyl methacrylate on silica, for which p was found to be ~0.25 from i.r. measurements (Thies *et al.*, 1964; Thies, 1968), whereas Miyamoto and Cantow (1972), using n.m.r., concluded that p could not be far below unity. Accepting relatively minor differences with respect to chemicals and experimental conditions, the difference is far beyond experimental error and suggests that n.m.r. overestimates and/or i.r. underestimates p, both possibilities agreeing with our caveats given above. On the other hand, for polyethylene glycol adsorbing from various organic solvents onto silica (aerosil), Killmann (1976) found agreement within experimental error between i.r. and microcalorimetric measurements; for this system $p =$ 0.20–0.25. Several examples of i.r. analyses of adsorbed polymers have been included in Rochester's review (1980). We will return to the interpretational problems in Section VII.

The newest development in the determination of the bound fraction is small-angle neutron scattering. Neutrons are scattered by the nuclei of atoms, and the intensity of the scattering at a given angle depends on the spatial distribution of scattering centres. Variation of isotopic composition enables molecular regions of interest to be viewed against a weakly scattering background. Scattering from an adsorbed layer can be detected by "contrast matching", in which undesirable contributions such as the scattering from the adsorbent particles are suppressed by choosing a suitable isotopic composition of the experimental system. In principle, the method can give the complete segment density distribution (Barnett *et al.*, 1982).

Obvious drawbacks are the problem of accessiblity to a nuclear reactor, and the high cost.

In addition to the above mentioned methods to determine p and/or θ_1, there are a few that have hitherto not been applied systematically, sometimes because the method is rather restrictive. For instance, Caselli and Traini (1974) found that the adsorption of gelatin on gold reduced its ability to catalyse the decomposition of H_2O_2. Since the reaction proceeds at the surface, this inhibition apparently reflects θ_1 and, in principle, θ_1 could be obtained after appropriate standardization. This is a typical kinetic method, but it is certainly not general since in our laboratory we found that the double-layer relaxation of silver iodide is not (or not markedly) depressed by the presence of polyvinyl alcohol (van Hövell, 1980). Of more general applicability is the microcalorimetric approach developed by Killmann and Eckart (1971) and Killmann and Winter (1975). The heat of immersion of the adsorbent is measured as a function of the polymer concentration, or as a function of θ. Standardization is needed before θ_1 is obtained. One possibility is to assume that in the plateau region $\theta_1 \rightarrow 1$, but this is an overestimation; alternatively, assessments of the partial enthalpy per adsorbed group can be deduced from immersion experiments in a suitable monomeric analogue compound. According to our own experience, the method works reasonably well for the adsorption of polyvinyl pyrrolidone on silica (Cohen Stuart, 1980).

B. METHODS LEADING TO THE THICKNESS OF THE ADSORBED LAYER

Two groups of techniques are currently used: ellipsometry and hydrodynamics. In ellipsometric measurements, the thickness and average concentration of the adsorbed polymeric films are deduced from changes due to the adsorbate in the properties of elliptically polarized light upon reflection. Ellipsometry is thus a static method. On the other hand, hydrodynamic measurements are typically of a kinetic nature; the thickness is inferred from the outward displacement due to the adsorbed polymer of the slipping layer between the solid and liquid phase if the latter moves tangentially with respect to the first. The two methods do not measure the same effective thickness. Ellipsometry observes the refractive index increment of the adsorbate over that of the solution, whereas hydrodynamics "sees" essentially the extent of free drainage of macromolecular adsorbates. Since both the refractive index increment and the extent of drainage will change in some (not necessarily the same) fashion with increasing distance from the surface, depending on the distribution of the segments over loops and tails, there will be some similarity between the outcome of the two procedures, but the exact definition of the effective layer thickness, subject as it is to interpretational abstractions, may well differ quantitatively.

In ellipsometric measurements, the change in the state of polarization of light upon reflection from a surface is measured. If the measuring system consists of a simple substrate (without an adsorption layer) in a medium of known refractive index, the real and imaginary parts of the substrate refractive index can be derived from the change in amplitude and phase of the light beam upon reflection (Passaglia *et al.*, 1964). If there is a homogeneous non-absorbing film on the surface, there are two interfaces at which the light is reflected. From the phase shift and amplitude of the reflected beam with respect to the incident beam, the thickness and refractive index of the film can be calculated (McCrakin and Colson, 1964), provided that the (complex) refractive indices of the substrate and the solution are known.

An adsorbed polymer layer is clearly not homogeneous, but its concentration (and thus its refractive index) is a continuously decreasing function with increasing distance from the surface. The thickness found from an ellipsometric experiment by considering the adsorption layer as a homogeneous film of constant segment concentration is then some average. McCrackin and Colson (1964) carried out calculations to assess what kind of average is obtained, approximating the adsorbed layer by many homogeneous films whose refractive indices are a function of the distance from the substrate. They concluded that the average ellipsometric thickness exceeds the root-mean-square thickness by a factor that depends on the assumed dependence of the refractive index on the distance from the surface, which for polymeric adsorbates is equivalent to the assumed segment density profile. This factor turned out to be about 1.75, 1.46, and 1.66 for a linear, exponential, and Gaussian concentration profile, respectively. Thus, unambiguous information about the meaning of the ellipsometric thickness is only available if the concentration profile in the adsorbed layer is known. In some cases an exponential distribution might be a reasonable approximation, but if the tail fraction is important this is no longer the case (see Section IV). Moreover, in the dilute regions of the adsorbed layer (in which the tails dominate) the assumption of a constant ("smeared-out") refractive index in a layer parallel to the surface breaks down.

As stated above, ellipsometry gives the thickness and refractive index of the adsorption layer. Since, from the refractive index, the average segment concentration follows immediately, the adsorbed amount per unit of surface area is easily found as the product of the layer thickness and the (excess) layer concentration.

Ellipsometric measurements are only feasible for macroscopically flat surfaces. The sensitivity depends on the differences in refractive index between the substrate, the adsorption layer, and the solution. Because in most cases the refractive index difference between adsorbed film and adsorption medium is quite small, reliable measurements can only be obtained if the refractive index of the substrate is high; preferably it should have a relatively high absorption coefficient. For this reason, mainly metals and some metal oxides have been used as solid substrates (Stromberg *et al.*, 1965a, b, 1970; Stromberg and Smith, 1967; Killmann and von Kuzenko, 1974; Killmann, 1976; Killmann *et al.*, 1977).

A method for measuring layer thicknesses which is closely related to ellipsometry is attenuated total reflection. In this technique a light beam is internally reflected several times at the boundary of a prism (used as the adsorbent) and a polymer solution. In principle the method could give some information about the concentration profile in the adsorbed layer, since the light beam penetrates over a certain distance in the adsorption layer before re-entering the prism. However, this feature has not yet been worked out sufficiently. Attenuated total reflection can be applied only to those adsorbents from which a suitable prism can be constructed. To our knowledge, only one paper has been published (Peyser and Stromberg, 1967) in which this technique was applied to the adsorption of uncharged polymers.

A potentially powerful technique is neutron scattering. Small-angle neutron scattering has been used for dispersions with low molecular weight adsorbates (Cebula *et al.*, 1978) and, very recently, also for polymer adsorption (Barnett *et al.*, 1982); see Section VI.A. This technique is probably the only one that can provide reliable information about the concentration profile in an adsorbed polymer layer. From this, the average layer thickness can be inferred.

Of the procedures used for obtaining a hydrodynamic thickness, perhaps the oldest method is to adsorb the polymer at the inner walls of a capillary of the adsorbent and measure the ensuing reduction in flow rate (Öhrn, 1955, 1958; Takeda and Endo, 1956; Huque *et al.*, 1959). This method is, in principle, rather sensitive because according to Poiseuille's law the liquid flux is proportional to the fourth power of the capillary radius, so a small reduction of the free lumen can lead to a substantial decrease in the rate of flow. However, the ratio between the layer thickness and the capillary radius should not be too small because in that case the reduction in the flow time is immeasurable. One of the difficulties with this method is the need for a homogeneous deposition of the polymer inside the capillary wall; obviously local thicker patches act as obstructions for the entire capillary and hence tend to lead to an overestimation of the layer thickness.

Rather popular is viscometry of dispersed particles. The principle is that the effective radius of the particle increases due to polymer adsorption, and this in turn is reflected in an increase of the intrinsic viscosity $[\eta]$. Experiments can be conducted in a simple viscometer. However, the interpretation is not without pitfalls, and several additional viscosity influencing phenomena have to be accounted for or excluded by some extrapolation procedure. Examples of such complications are the primary and secondary electroviscous effects, the effect of electrolytes (if the particles are charged, which is usually the case in aqueous systems), and aggregation of the particles. Because of these complications, extrapolation procedures (to zero sol concentration) are not always straightforward; moreover, as a safeguard, the absence of a shear dependence also deserves verification. Under unfavourable conditions, degradation of polymer coatings during collisions in shear fields has been found to occur (Ahuja, 1976).

Various variants are available for obtaining the hydrodynamic layer thickness. Sedimentation rates of polymer-covered dispersion particles have been used (Garvey *et al.*, 1976). Laser-Doppler scattering is perhaps the most sophisticated technique, but it is applicable only if the particles are spherical and monodisperse (Garvey *et al.*, 1976; Goossens and Zembrod, 1979). Deviations from ideality, especially those leading to aggregation, are readily detectable. If the particles are charged and the polymer is uncharged, the outward displacement of the slipping plane leads to a reduced electrokinetic potential. Quantitative interpretation is not straightforward, because polymer adsorption also modifies the outer Helmholtz potential, but this difficulty can be overcome by analysing the slope of the curve that relates the mobility to the logarithm of the concentration of potential-determining ions in the region of the isoelectric point (Koopal and Lyklema, 1976; Lyklema, 1976). Scholten (1978) developed a variant, applicable to elongated, magnetic, polymer-covered particles. Sols of such particles are birefringent when subjected to a magnetic field. The decay of this birefringence upon removal of the field yields the rotational diffusion coefficient from which, in turn, the hydrodynamic polymer layer thickness can be inferred.

We repeat that the quantity obtained is at best an average hydrodynamic thickness. It is very difficult to say how this averaging is effected for want of sufficiently detailed information on the flow of liquids through assemblies of loops and tails. For irregular, patchwise adsorption of polymers, some elusive averaging will take place, probably leading to an overestimation of the layer thickness, but even for layers that are homogeneous in this respect theoretical interpretation is still in its infancy and has as yet only been applied to an exponential distribution of segments (neglecting tails), as described in Section IV (Varoqui and Dejardin, 1977). It is especially uncertain what the flow reducing effect is of the long extending tails that are expected to be present in most cases. Cohen Stuart (1982) has shown that hydrodynamic thickness is determined by tails only, loops giving a negligible contribution.

Concluding this section on experimental techniques, we mention that as yet there are hardly any methods available for obtaining information about one of the central parameters in polymer adsorption, namely the adsorption energy parameter χ_s. A start was recently reported by Cohen Stuart (1980) who carried out displacement studies whereby polymer was desorbed by low molecular weight competing substances ("displacers"). Since there exists, in principle, a relation between the concentration of displacer necessary to desorb the polymer completely and the adsorption affinities of polymer and displacer, an estimate of χ_s can be obtained. Despite several interpretational difficulties, this method offers potential for a better comparison of experiment and theory.

VII. Some Illustrative Examples

In this section we describe the results of a few experiments that illustrate the principles given before. Within the scope of this book it is impossible to present a complete overview of all experimental work reported in literature. We restrict ourselves to a few examples, selected to represent the main experimental techniques discussed in the preceding section. Since we shall try to compare the results with recent theoretical developments, only experiments on well characterized systems are dealt with.

A. ADSORBED AMOUNT

Recently, Vander Linden and van Leemput (1978a) reported the results of very careful measurements of the adsorption of polystyrene (PS) samples from cyclohexane (at 35°C, corresponding to a θ-solvent) and carbon tetrachloride (at 35°C, a better solvent) on silica. These experiments cover a very wide range of relative molar masses, from $M = 600$ up to 2×10^6, corresponding to degrees of polymerization from 5 up to 20 000. In our opinion these data belong to the best available in the literature, justifying comparison with theory in some detail.

Figure 19 gives the chain length dependence of the adsorbed amount at a polymer concentration of about 10^{-3} g dm^{-3} in the two solvents. Along the abscissa axis the degree of polymerization is given in a logarithmic scale. The experimental points represent Γ in mg m^{-2} (left-hand side ordinate axis). The curves give the theoretical dependence of θ according to the Scheutjens and Fleer theory (in units of monolayer capacity, right-hand side ordinate axis) at a volume fraction of 10^{-3}. The curves for $\chi_s = 1$ are taken from Fig. 9; those for $\chi_s = 0.6$ have not been shown before. The value $\chi = 0.4$ is close to the experimentally determined value 0.396 for PS in CCl_4 at 25°C (Bristow and Watson, 1958).

Qualitatively, and perhaps (semi)quantitatively, the agreement between theory and experiment is excellent. In a θ-solvent, both Γ and θ increase more strongly than linearly with $\log r$ for short chains [the experimental points for $\chi = 0.5$ and $r < 3000$ obey the relationship $\Gamma \propto \sqrt{r}$ (Vander Linden and van Leemput, 1978a)], whereas for long chains the adsorption increases proportionally to $\log r$. In a better solvent, the adsorbed amount is smaller and increases more slowly with increasing chain length. Vander Linden and van Leemput show in their paper that, even qualitatively, no agreement exists with the theories of Roe and of Silberberg.

It is tempting to conclude from Fig. 19 that the monolayer capacity of

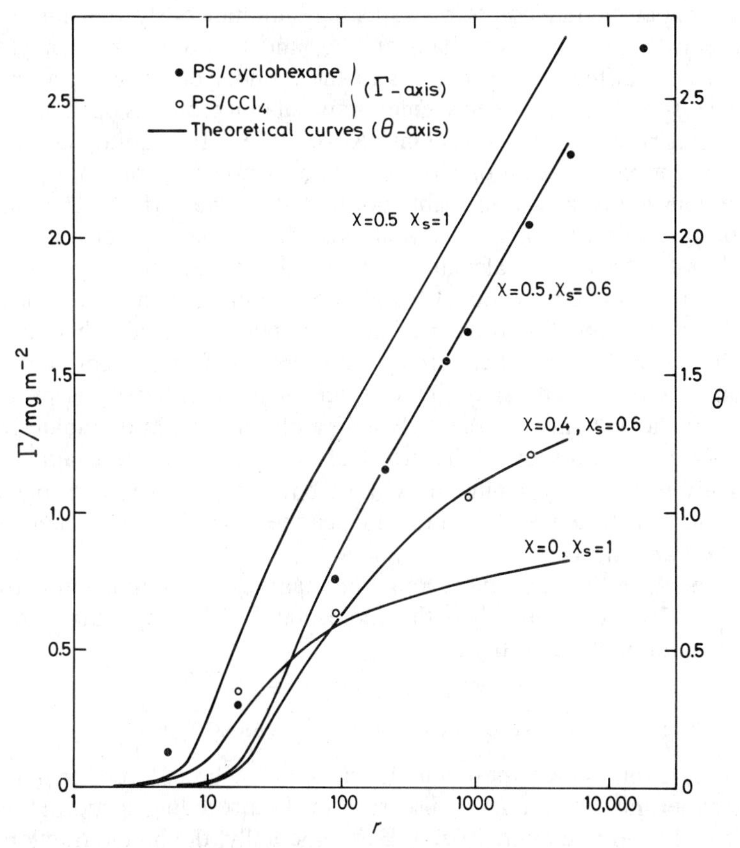

Figure 19 Comparison of the experimental adsorbed amount as a function of chain length with theoretical predictions. The experimental points are for homodisperse polystyrene from cyclohexane (θ-solvent) and carbon tetrachloride on silica (Van der Linden and Leemput, 1978a); the theoretical curves are calculated with the theory of Scheutjens and Fleer. The polymer volume fraction in the solution is 10^{-3}.

PS on silica is 1 mg m^{-2} and that $\chi_s = 0.6$, both for cyclohexane and for CCl_4, but some reservations must be made. First, it is questionable whether one monomer unit of PS may be identified with a theoretical segment in a lattice. Furthermore, there is no *a priori* reason to expect the same value for χ_s in both solvents. Finally, Vander Linden and van Leemput assume that a fully packed monolayer is obtained if each surface silanol group (of which there are 3 per nm^2) interacts with one monomer unit, which would correspond to a monolayer capacity of 0.52 mg m^{-2}. As to the first two points, a variation in the number of segments per chain or in χ_s would not affect the shape of the theoretical curves, but only bring about a shift in the horizontal direction. Hence, it is always possible to find a curve that fits the experimental points by adjusting simultaneously the number of (theoretical) segments per chain and χ_s, and this adjustment might be different for different solvents. As to the monolayer capacity, a one-to-one correspondence between segments and silanol groups is open to doubt, as demonstrated for other systems (Korn, 1978). It is quite conceivable that adsorption will occur on sites where no silanols are present, and some of the benzene rings will probably not lie flat on the surface. Therefore a monolayer capacity might very well exceed 0.52 mg m^{-2}. This is further corroborated by model calculations for a fully packed monolayer of polystyrene; from molecular models a value of around 1 mg m^{-2} is estimated.

At low chain lengths, the experimentally determined adsorbed amount is higher than the theoretical prediction (Fig. 19). There might be several reasons for this discrepancy. The degree of monodispersity is probably lower for short chains, so relatively long molecules might dominate in the adsorbed layer. Moreover, the effect of end groups in the chain can be relatively important for oligomers. Last but not least, the heterogeneity of the silica surface would tend to enhance the adsorption, especially if the adsorbed amount is small.

We conclude that, despite several uncertainties, the agreement between theory and experiment as to the molecular weight dependence of the adsorbed amounts is gratifying.

B. BOUND FRACTION AND DIRECT SURFACE COVERAGE

Figure 20 displays the experimental results for the bound fraction p for the same experimental system as discussed in the preceding section (Vander Linden and van Leemput, 1978a). Experimentally, the bound fraction was obtained using the out-of-plane vibration frequency of the phenyl group around 700 cm^{-1}. The theoretical curves in Fig. 20 are again from the Scheutjens and Fleer theory, using the χ and χ_s values that gave the best fit in Fig. 19. The shapes of the curves around the inflection point at low r are related to the transition of a rather empty surface layer with relatively flat molecules to a more extended layer, and will not be discussed here; the coordinates of the inflection point depend on χ and χ_s.

For a θ-solvent there is (within experimental error) close agreement between theory and experiment, with a steadily decreasing bound fraction if the chain becomes longer. The relatively high p found experimentally for oligomers is probably again due to the experimental problems occurring

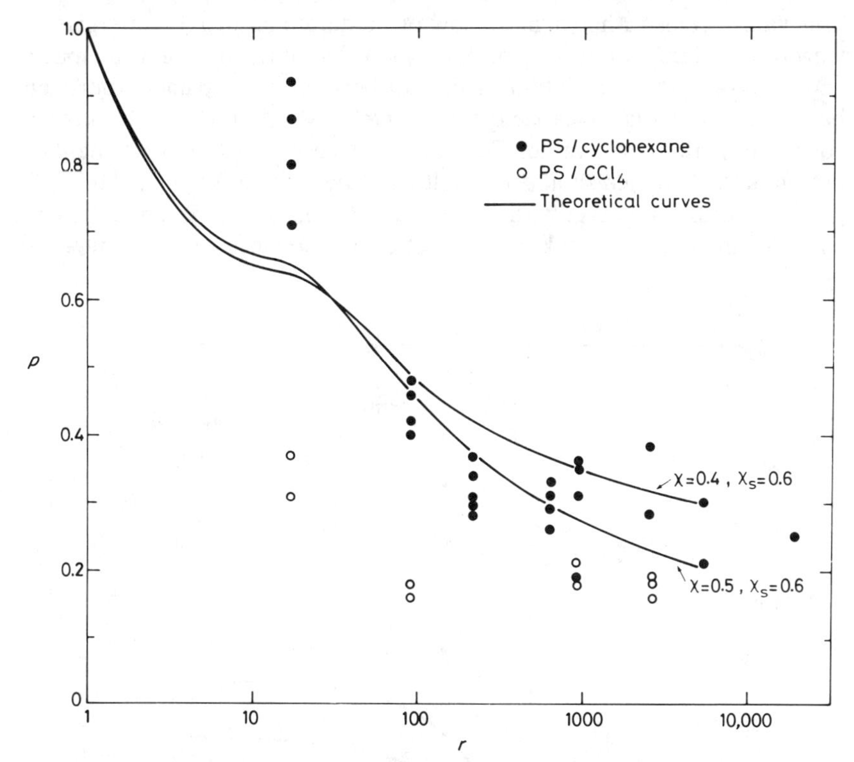

Figure 20 Comparison of the experimental bound fraction (for the same system as in Fig. 19) as a function of chain length with theoretical predictions according to the Scheutjens and Fleer theory.

with short chains, as discussed above. For CCl_4 as the solvent, the general trend of $p(\log r)$ is the same in theory and experiment, but on comparing the two solvents a discrepancy emerges; the experimentally found bound fraction is lower in a better solvent, whereas all theories predict that, at constant ϕ_* and χ_s, p should increase with increasing solvent power. A (much) lower χ_s in CCl_4 as compared with cyclohexane could be responsible for this decrease in p, but a lower χ_s is unlikely in view of the discussion given in connection with Fig. 19. For the moment we can offer no reasonable explanation for this discrepancy. Possibly it is related to different spectroscopic properties in CCl_4 and cyclohexane.

Vander Linden and van Leemput (1978a) have also determined the direct surface coverage θ_1 spectroscopically, using the silanol stretching frequency in the region around 3700 cm^{-1}. Assuming again a one-to-one correspondence between silanol groups and monomer units of PS, they find that in cyclohexane $\theta_1 = 1$ for all except the lowest molecular weights. This value seems rather high for low or moderate adsorption energies. In view of the discussion given above on the value of the monolayer capacity and the gratifying correlation in Fig. 19 between theory and experiment based on a monolayer capacity of 1 mg m^{-2}, we prefer, for the sake of consistency, the latter value. The consequence is that the experimentally determined θ_1 values should be multiplied by a factor of 0.52. Figure 21 gives the results, together with the two theoretical curves. Again the general shape of the experimental and theoretical dependences is the same, with

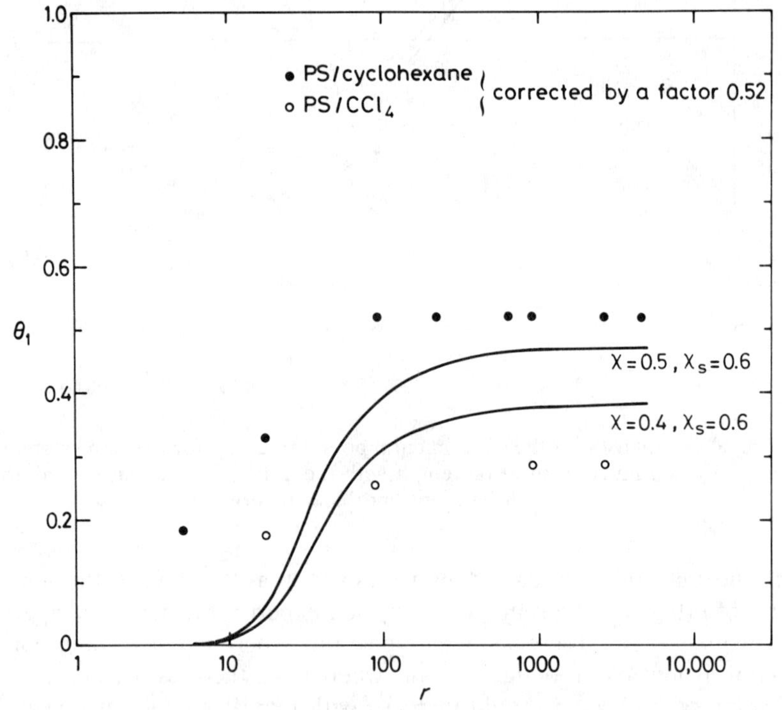

Figure 21 Comparison of the direct surface coverage (for the same system as in Fig. 19) as a function of chain length with theoretical curves from the theory of Scheutjens and Fleer. The experimental values were multiplied by a factor 0.52 (see text).

θ_1 increasing at low r and attaining a platform at high r which is higher in poorer solvents. Quantitatively the agreement is not as perfect as in the previous figures. A better quantitative fit between theory and experiment might possibly be obtained by adjusting the values for χ_s. As discussed before, for oligomers the experimental θ_1 is probably overestimated.

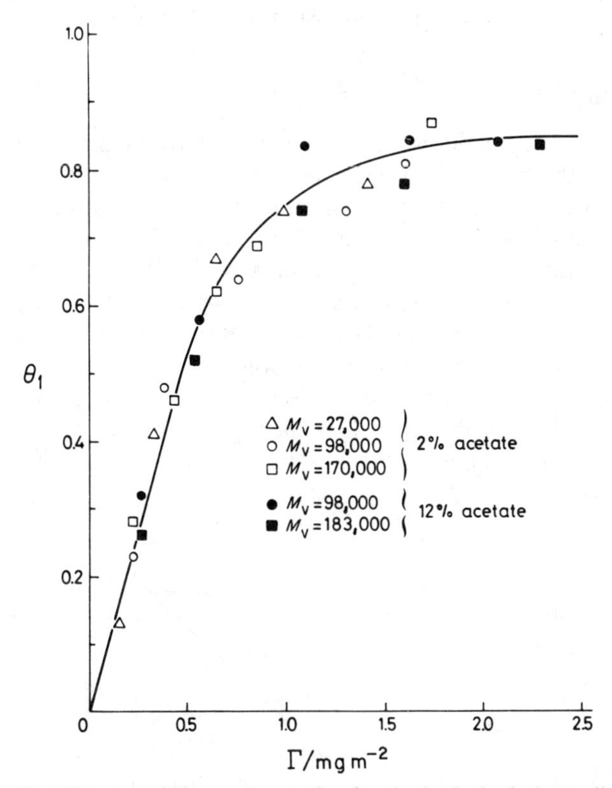

Figure 22 The direct surface coverage θ_1 of polyvinyl alcohol on silver iodide as a function of the adsorbed amount Γ, derived from the shift in the isoelectric point. (Koopal and Lyklema, 1979) The viscosity averaged molecular weight M_v (g mol^{-1}) and the acetate content of the polymer (mol %) are indicated.

We give one example of the direct surface coverage as obtained by an electrochemical method. Figure 22 displays results obtained from the shift of the isoelectric point for polyvinyl alcohol (PVA) adsorbed from water on silver iodide (Koopal and Lyklema, 1979). From the change in the surface charge at fixed pAg a virtually identical curve is obtained (Koopal and Lyklema, 1976). In this case θ_1 is given as a function of the adsorbed amount Γ. Again the results are in qualitative agreement with theoretical predictions. A quantitative comparison is difficult to make since the PVA used, containing some acetate groups, was virtually a copolymer. An interesting feature of Fig. 22 is that θ_1 appears to be a unique function of Γ, regardless of molecular weight and percentage of acetate groups. Also theoretically θ_1 is, at given Γ, approximately independent of M (Scheutjens and Fleer, 1979). In view of the relatively high saturation value of θ_1 one might speculate that in this system also χ_s is rather high.

C. LAYER THICKNESS

Figure 23 gives some ellipsometric results by Killmann (1976) for the average layer thickness of polystyrene from two θ-solvents on metal surfaces. These measurements were confirmed by Smith and Stromberg (1975), thereby correcting older results (Stromberg *et al.*, 1965a) that appeared to be overestimations. Figure 23 shows clearly that the layer thickness is proportional to the square-root of molecular weight, a feature that is also predicted by theory (see Section IV). However, the conclusion by Killmann (1976) that this square-root dependence points to an adsorption of isolated Gaussian coils is not warranted; in Section IV it was shown that the conformation of adsorbed molecules having many relatively short loops and one or two long dangling tails differs greatly from that of a Gaussian coil in solution.

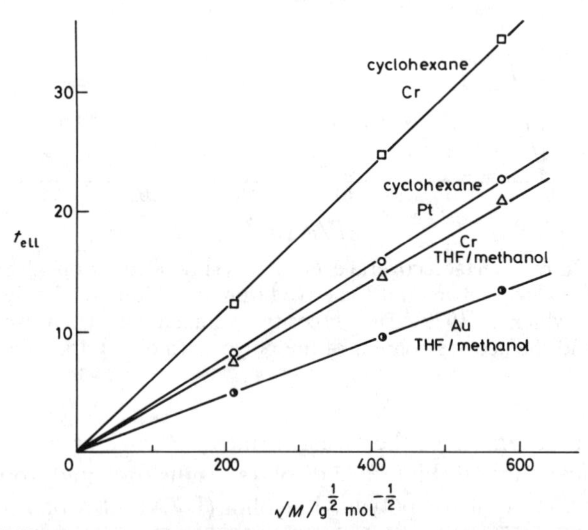

Figure 23 Ellipsometric thickness of polystyrene, adsorbed from two θ-solvents on different metal surfaces, as a function of the square-root of molecular weight. (Killmann, 1976)

A more quantitative comparison as to the value of the layer thickness is difficult to make because the ratio between the ellipsometric thickness and the root-mean-square thickness is only known if an assumption about the distribution of segments in the adsorbed layer is made (see Section VI.B). Moreover, the χ_s values to be used for the systems of Fig. 23 are not known. Nevertheless, some comparison between the layer thickness and the radius of gyration R_g of the free coil in solution is possible. Assuming an exponential concentration profile (which is not completely justified; see Fig. 11), and using the unperturbed dimension of PS coils in θ-solvents (Brandrup and Immergut, 1975), we find a ratio between the r.m.s. layer thickness t_{rms} and R_g of the order of 0.4–1. In Section IV we concluded that, in (semi)dilute solutions and for $\chi_s = 1$, t_{rms} is lower than R_g by a factor of 2 to 3, which is in reasonable agreement with experiment. Theoretically, t_{rms} is not very sensitive to χ_s. The data of Fig. 23 do not allow a definite conclusion about the dependence of t_{rms} on χ_s because of the uncertainty in the relation between the ellipsometric thickness and the root-mean-square thickness.

We mention one example of a hydrodynamic measurement. Following previous investigators (Öhrn, 1955, 1958; Takeda and Endo, 1956; Huque *et al.*, 1959), Priel and Silberberg (1978) measured the reduction in flow time in a glass capillary brought about by the adsorption of polystyrene from toluene. As stated in Section VI.B, an unambiguous interpretation of this type of measurement is not yet possible. The data of Priel and Silberberg can therefore not be interpreted to give a well established thickness–chain length dependence. The viscometric thicknesses found in dilute solutions are of the order of the coil diameter in solution. A recent analysis on the hydrodynamics in polymer-covered capillaries suggests that the root-mean-square thickness is several times lower than the hydrodynamic thickness (Varoqui and Dejardin, 1977). In view of the large uncertainties in the interpretation, the agreement between theory and experiment seems reasonable.

D. INDICATIONS FOR THE OCCURRENCE OF TAILS

We have seen in Section IV that, according to theory, tails play a rather important role in the adsorption of polymers. The only more or less direct technique for the determination of tails is small-angle neutron scattering (see Section VI.A). Recently, the first results with this method were reported (Barnett *et al.*, 1982). These authors found a hump in the concentration profile at some distance from the surface, which is a strong indication of the occurrence of tails, thereby corroborating theoretical predictions. By all other means, only indirect evidence for tails can be obtained.

One of the areas where tails are important is the stabilization of colloidal systems by polymers. In such a steric stabilization the repulsion between two polymer-covered particles is largely determined by the segments in the outermost parts of the adsorption layer. In recent measurements of the thickness of polymer-stabilized free liquid films it was concluded that the high thicknesses obtained could be explained only if long tails, extending much farther into the solution than the r.m.s. thickness, are assumed to be present (Lyklema and van Vliet, 1978). Another example is the work of Sonntag *et al.* (1982), who measured the interaction forces between quartz filaments covered by polymer. Also in this case the range of interaction between the filaments was so large that the authors had to assume the existence of long tails in order to obtain a consistent interpretation. Future work on this type of experimental system might clarify these points further.

Another (indirect) indication for the presence of tails may be found in ellipsometric results where the thickness and adsorbed amount as a function of polymer solution concentration are measured simultaneously. Quite often it is found (see e.g. Killmann and von Kuzenko, 1974) that the thickness increases more strongly with increasing concentration than the adsorbed amount. If the polymer used has a narrow molecular weight distribution, this stronger increase may be ascribed to a higher tail fraction in more concentrated solutions, which is also found theoretically. Hoeve's theory, which considers only trains and loops, predicts that the layer thickness is proportional to the total surface coverage θ [both are proportional to $(-\lambda)^{-1/2}$; see Sections III.B and IV]. Approximately the same result is found in the Scheutjens and Fleer theory if only the thickness due to loop segments is considered. However, according to the latter theory, the overall layer thickness (due to loops *and* tails) increases much more strongly with increasing solution concentration than the adsorbed amount, because of the contribution of tails.

E. HETERODISPERSITY EFFECTS

In Section V we discussed some implications of polymer heterodispersity on adsorption isotherms. One of the conclusions was that the adsorption isotherm of a heterodisperse polymer depends on the surface area/solution volume ratio A/V (see Fig. 17). The reason is that the polymer will distribute itself over a given surface area and a given volume of solution in such a way that virtually only the high molecular weight part of the polymer is on the surface. Let us express the amount of polymer on the surface and that in the solution in the same units, e.g. mg m^{-2}. Then Γ is the amount on the surface and $c_p V/A$ (where c_p is the weight concentration in the solution) is the corresponding amount in the solution (both expressed as weight per unit of surface area). Although the "normal" adsorption isotherm $\Gamma(c_p)$ would depend on A/V, an adsorption isotherm in which Γ is plotted as a function of c_pV/A should be independent of A and V. This is demonstrated in Fig. 24, taken from a recent publication of Koopal (1981). It is gratifying that, notwithstanding the great diversity of the $\Gamma(c_p)$ isotherms, Γ is a unique function of c_pV/A. Other examples of this type have been reported by Cohen Stuart (1980) and Hlady *et al.* (1982). Thus it seems well established that in the adsorption of heterodisperse polymers the total *amount* in solution (per unit of surface area) and not the solution concentration is the relevant variable.

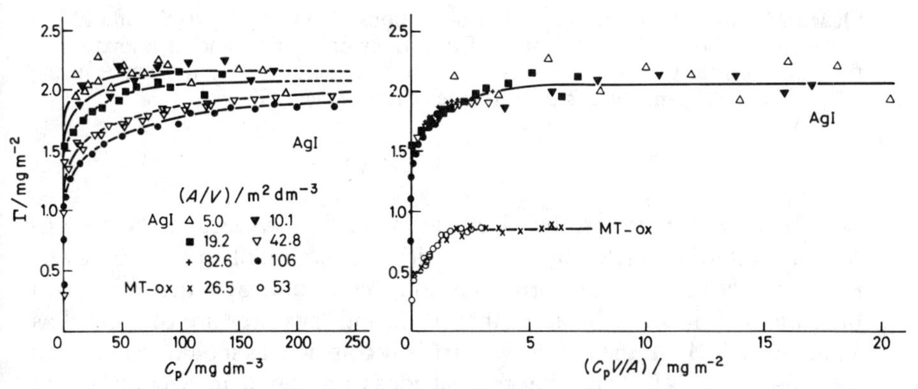

Figure 24 Adsorption isotherms plotted as $\Gamma(c_p)$ (left) and as $\Gamma(c_pV/A)$ (right), for polyvinyl alcohol from water onto AgI and oxidized Sterling MT (Koopal, 1981). The adsorption isotherms $\Gamma(c_p)$ for different surface area/solution volume ratio A/V coincide when Γ is plotted as a function of c_pV/A, the amount of polymer in solution per unit of surface area.

As our last example we consider the adsorption isotherm of a mixture of two relatively narrow fractions of different molecular weight. In Fig. 18 we gave the idealized isotherm as predicted by theoretical considerations. Figure 25 presents the experimental isotherms for two fractionated samples of polyvinyl pyrrolidone ($M_b = 1.0 \times 10^6$ g mol^{-1}, $M_s = 1.44 \times 10^3$ g mol^{-1}) and for a mixture (weight fraction of the small molecules w_s = 0.75) of these two fractions from water onto silica (Cohen Stuart *et al.*, 1980). The isotherm of the mixture was measured at two adsorbent concentrations (i.e. at two A/V ratios). The individual isotherm for the high molecular weight M_b is sharp, as expected. The isotherm for the small molecules is somewhat rounded; this could possibly be attributed to imperfect fractionation and to the fact that in this example M_s was too small to give a sharp isotherm ($M_s = 10^3$ g mol^{-1} corresponds roughly to

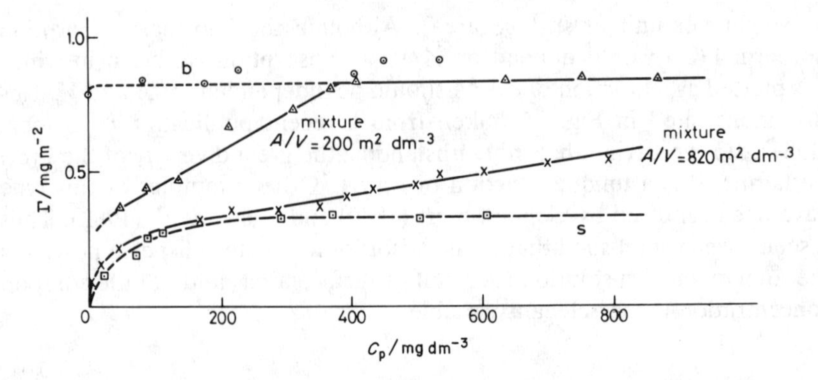

Figure 25 Adsorption isotherms of two fractions of polyvinyl pyrrolidone (M_s = 1440 g mol^{-1}, M_b = 1.0 × 10^6 g mol^{-1}) from water onto silica, and of a mixture of these fractions at two A/V ratios. The weight fraction of the big molecules in the mixture is 0.25. (Cohen Stuart *et al.*, 1980)

r = 10). However, the isotherm of the mixture shows clearly the discontinuous pattern predicted by Fig. 18, with a linear middle part. Moreover, the slopes of these linear portions are in the ratio 3.65, which is close to the ratio 4.07 in A/V. If the isotherms of both mixtures are re-plotted as a function of $c_p V/A$, the two curves virtually coincide, as should be expected (see also Fig. 24). Thus the general ideas presented in Section V are strongly corroborated.

VIII. Applications

Polymer adsorption is a widespread phenomenon and there are many applications of the theory and model experiments discussed above to more complicated situations. Starting from the fundamental side one could, as a first step, envisage a number of extensions of the theory of Section III.

Generalization to copolymers is not a very difficult step. Extra parameters are now needed: a second adsorption energy (because the two types of segment will usually have a differing affinity to the adsorbent); two more χ parameters (for the interaction of the second type of segment with the solvent and for the mutual interaction of both segment types); and a randomness parameter, describing the nature of the distribution (the shorter the blocks the more random). These parameters enter the partition function and, upon maximization, affect the distribution. Extension to proteins might also be feasible, but then many more parameters should be known.

A second generalization is toward polyelectrolytes. The influence of the electric field (which may stem both from the adsorbent and the adsorbate) can only be taken into account by some averaging; a logical way of doing so is to apply again the Bragg–Williams approximation. Since the Stern equation for adsorption of simple ions is also an example of this approximation, the polyelectrolyte adsorption picture is an extension of Stern's model. Polyelectrolyte adsorption is discussed in detail by Hesselink in Chapter 8 of this book.

Generalization to branched polymers (comb-type graft copolymers, crosslinked systems, etc.) requires a different counting procedure of the various conformations, but presents no difficulty in principle.

Apart from these extensions, the influence of the segment volume deserves attention. In all lattice theories, solvent molecules and segments are assumed to occupy just one site, but this assumption clearly breaks down if the segment and solvent molecule sizes differ considerably. This is, of course, a general problem, also applying to polymers in bulk solution. Introduction of a "bulkiness factor" in the Scheutjens and Fleer theory is possible. Provisional studies in our laboratory suggest that introduction of this volume effect may lead to substantially improved agreement with experiment.

A very obvious application is in the field of steric stabilization of colloidal dispersions. The goal can be achieved in two ways. The first and less rigorous approach is to combine the information on the distribution of loops and tails with existing theories on steric interaction for isolated loops and tails. A better approach, of which the first is a limiting case, is to carry out an *ab initio* statistical analysis of polymer adsorbed in the constrained space between two plates. In applying the results of such analyses to real systems the problem of polymer adsorption reversibility emerges again, the crucial question being to what extent a polymeric adsorbate has the capacity to adjust itself during the encounter time of two polymer-covered particles.

Technical and biological examples of polymer adsorption and polymer-stabilized dispersions are abundant and can be encountered in industry (rheology of polymer-stabilized systems, reactions in polymeric adsorbates, drag reduction, adsorbents as crystallization inhibitors, fillers in paints), in pharmacy (stabilization of drugs and monitoring their administration, pharmaceuticals against blood clotting), in medicine and biology (modification of biocompatibility of artificial implants, cell adhesion), in analytical chemistry (chromatography, carrier-fixed enzymes, adsorption of polymers on electrodes), in soil science (flow of water through soils containing microbial debris and adsorbed humic acids), etc. The diversity of these fields underlines the importance of systematic basic studies in the field of polymer adsorption.

References

Ahuja, S. K. (1976). *J. Colloid Interface Sci.* **57**, 438.

Ash, S. G. (1973). In *Colloid Science* (D. H. Everett, ed.), Specialist Periodical Reports, Vol. 1, p. 103, Chemical Society, London.

Barnett, K. G., Cosgrove, T., Crowley, T. L., Tadros, Th. F. and Vincent, B. (1982). In *The Effect of Polymers on Dispersion Properties* (Th. F. Tadros, ed.), p. 183. Academic Press, London.

Benoit, H. (1976). *J. Macromol. Sci.* (*B*) **12**, 27.

Brandrup, J. and Immergut, E. H. (1975). *Polymer Handbook*, Wiley, London.

Bristow, G. M. and Watson, W. F. (1958). *Trans. Faraday Soc.* **54**, 1742.

Caselli, M. and Traini, A. (1974). *J. Electroanal. Chem.* **50**, 387.

Cebula, D., Thomas, R. K., Harris, N. M., Tabony, J. and White, J. W. (1978). *Faraday Discuss. Chem. Soc.* **65**, 76.

Clark, A. T., Robb, I. D. and Smith, R. (1976). *J. Chem. Soc. Faraday Trans. 1* **72**, 1489.

Cohen Stuart, M. A. (1980). Thesis, Agric. Univ. Wageningen, Netherlands.

Cohen Stuart, M. A. (1982), to be published.

Cohen Stuart, M. A., Scheutjens, J. M. H. M. and Fleer, G. J. (1980). *J. Polym. Sci. Polym. Phys. Ed.* **12**, 559.

Cosgrove, T., Barnett, K. G., Cohen Stuart, M. A., Sissons, D. S. and Vincent, B. (1981). *Macromolecules*, **14**, 1018.

DeSantis, R. and Zachmann, H. G. (1977). *Colloid Polym. Sci.* **225**, 729.

Dickinson, E. and Lal, M. (1980). *Adv. Molec. Relax. Process.* **17**, 1.

DiMarzio, E. A. (1965). *J. Chem. Phys.* **42**, 2101.

DiMarzio, E. A. and McCrackin, F. L. (1965). *J. Chem. Phys.* **43**, 539.

DiMarzio, E. A. and Rubin, R. J. (1971). *J. Chem. Phys.* **55**, 4318.

Dukhin, S. S. and Dudkina, L. M. (1978). *Kolloidn. Zh.* **40**, 232.

Dukhin, S. S., Semenikhin, M. N. and Buchko, V. A. (1979). Sbornik VI Konf. po Poverkhnostnukh Silam, Moscow, p. 85.

Everett, D. H. (1964). *Trans. Faraday Soc.* **60**, 1803.

Felter, R. E. and Ray, L. N. (1970). *J. Colloid Interface Sci.* **32**, 349.

Felter, R. E., Moyer, E. S. and Ray, L. N. (1969). *J. Polym. Sci.* (*B*) **7**, 529.

Fleer, G. J., Koopal, L. K. and Lyklema, J. (1972). *Kolloid-Z.* **250**, 689.

Flory, P. J. (1953). *Principles of Polymer Chemistry*, Cornell University Press, Ithaca, N.Y.

Fontana, B. J. (1963). *J. Phys. Chem.* **67**, 2360.

Fontana, B. J. (1966). *J. Phys. Chem.* **70**, 1801.

Fontana, B. J. and Thomas, J. R. (1961). *J. Phys. Chem.* **65**, 480.

Frisch, H. L. and Simha, R. (1956). *J. Chem. Phys.* **24**, 652.

Frisch, H. L. and Simha, R. (1957). *J. Chem. Phys.* **27**, 702.

Garvey, M. J., Tadros, Th. F. and Vincent, B. (1976). *J. Colloid Interface Sci.* **55**, 440.

Goossens, J. W. S. and Zembrod, A. (1979). *Colloid Polym. Sci.* **257**, 437.

Hlady, V., Lyklema, J. and Fleer, G. J. (1982). *J. Colloid Interface Sci.* **87**, 395.

Hoeve, C. A. J. (1965). *J. Chem. Phys.* **43**, 3007.

Hoeve, C. A. J. (1966). *J. Chem. Phys.* **44**, 1505.

Hoeve, C. A. J. (1970). *J. Polym. Sci.* (*C*) **30**, 361.

Hoeve, C. A. J. (1971). *J. Polym. Sci.* (*C*) **34**, 1.

Hoeve, C. A. J. (1977). *J. Polym. Sci.*, *Polym. Symp.* **61**, 389.

Hoeve, C. A. J., DiMarzio, E. A. and Peyser, P. (1965). *J. Chem. Phys.* **42**, 2558.

Hövell, S. van (1980). Unpublished provisional results.
Howard, G. J. and Woods, S. J. (1972). *J. Polym. Sci. (A-2)* **10**, 1023.
Huque, M. M., Fishmann, M. and Goring, D. A. J. (1959). *J. Phys. Chem.* **63**, 766.
I.U.P.A.C. (1972). "Definitions, Terminology and Symbols in Colloid and Surface Chemistry", prepared for publication by D. H. Everett, *Pure Appl. Chem.* **31**, 579.
Jenkel, E. and Rumbach, B. (1951). *Z. Elektrochem.* **55**, 612.
Joppien, G. R. (1978). *J. Phys. Chem.* **82**, 2210.
Killmann, E. (1976). *Croat. Chem. Acta* **48**, 463.
Killmann, E. and Eckart, R. (1971). *Makromol. Chem.* **144**, 45.
Killmann, E. and Kuzenko, M. von (1974). *Angew. Makromol. Chem.* **35**, 39.
Killmann, E. and Winter, K. (1975). *Angew. Makromol. Chem.* **43**, 53.
Killmann, E., Eisenlauer, J. and Korn, M. (1977). *J. Polym. Sci., Polym. Symp.* **61**, 413.
Koopal, L. K. (1981). *J. Colloid Interface Sci.* **83**, 116.
Koopal, L. K. and Lyklema, J. (1976). *Faraday Discuss. Chem. Soc.* **59**, 230.
Koopal, L. K. and Lyklema, J. (1979). *J. Electroanal. Chem.* **100**, 895.
Korn, M. (1978). Thesis, Techn. Univ. Munich, West Germany.
Lane, J. E. (1968). *Aust. J. Chem.* **21**, 827.
Lipatov, Yu. S. and Sergeeva, L. M. (1972). *Adsorbtsiya Polimerov*, Izdatel'stvo Naukova Dumka, Kiev. English transl. *Adsorption of Polymers* (1974), Keter Publ. House, Jerusalem.
Lyklema, J. (1976). *Pure Appl. Chem.* **46**, 149.
Lyklema, J. and Vliet, T. van (1978). *Faraday Discuss. Chem. Soc.* **65**, 25.
McCrackin, F. L. and Colson, J. P. (1964). in *Ellipsometry in the Measurement of Surfaces and Thin Films*, Symposium proceedings (E. Passaglia, R. R. Stromberg and J. Kruger, eds), p. 61, Nat. Bur. Standards, Miscellaneous Publ. No. 256.
Miller, I. R. (1971). In *Recent Developments in Surface and Membrane Science* (J. F. Danielli, M. D. Rosenberg and D. A. Cadenhead, eds), Vol. 4, p. 299, Academic Press, New York and London.
Miller, I. R. and Bach, D. (1973). In *Surface and Colloid Science* (E. Matijević, ed.), Vol. 6, p. 185, Wiley, New York.
Miyamoto, T. and Cantow, H.-J. (1972). *Makromol. Chem.* **162**, 43.
Morawetz, H. (1965). *Macromolecules in Solution*, Interscience, New York.
Motomura, K. and Matuura, R. (1969). *J. Chem. Phys.* **50**, 1281.
Motomura, K., Sekita, K. and Matuura, R. (1971a). *Bull. Chem. Soc. Japan* **44**, 1243.
Motomura, K., Moroi, Y. and Matuura, R. (1971b). *Bull. Chem. Soc. Japan* **44**, 1248.
Öhrn, O. E. (1955). *J. Polym. Sci.* **17**, 137.
Öhrn, O. E. (1958). *Arkiv Kemi* **12**, 397.
Ono, S. and Kondo, S. (1960). In *Handbuch der Physik* (S. Flügge, ed.), Vol. 10, p. 134, Springer, Berlin.
Passaglia, E., Stromberg, R. R. and Kruger, J. (eds) (1964). *Ellipsometry in the Measurement of Surfaces and Thin Films*, Symposium Proceedings, Nat. Bur. Standards, Miscellaneous Publ. No. 256.
Peyser, P. and Stromberg, R. R. (1967). *J. Phys. Chem.* **71**, 2066.
Priel, Z. and Silberberg, A. (1978). *J. Polym. Sci., Polym. Phys. Ed.* **16**, 1917.
Rochester, C. H. (1980). *Adv. Colloid Interface Sci.* **12**, 43.
Robb, I. D. and Smith, R. (1974). *Europ. Polym. J.* **10**, 1005.

Robb, I. D. and Smith, R. (1977). *Polymer* **18**, 500.
Roe, R. J. (1965). *J. Chem. Phys.* **43**, 1591.
Roe, R. J. (1966). *J. Chem. Phys.* **44**, 4264.
Roe, R. J. (1974). *J. Chem. Phys.* **60**, 4192.
Roe, R. J. (1980). *Polym. Sci. Techn.* **12B**, 629.
Rubin, R. J. (1965). *J. Chem. Phys.* **43**, 2392.
Rubin, R. J. (1966). *J. Res. Nat. Bur. Stand.* (*B*) **70**, 237.
Sadakne, G. S. and White, J. L. (1973). *J. Appl. Polym. Sci.* **17**, 453.
Scheutjens, J. M. H. M. and Fleer, G. J. (1979). *J. Phys. Chem.* **83**, 1619.
Scheutjens, J. M. H. M. and Fleer, G. J. (1980). *J. Phys. Chem.* **84**, 178.
Scheutjens, J. M. H. M. and Fleer, G. J. (1982). In *The Effect of Polymers on Dispersion Properties* (Th. F. Tadros, ed.), p. 145. Academic Press, London.
Scholten, P. C. (1978). *Faraday Discuss. Chem. Soc.* **65**, 242.
Silberberg, A. (1962). *J. Phys. Chem.* **66**, 1872, 1884.
Silberberg, A. (1967). *J. Chem. Phys.* **46**, 1105.
Silberberg, A. (1968). *J. Chem. Phys.* **48**, 2835.
Silberberg, A. (1972). *J. Colloid Interface Sci.* **38**, 217.
Simha, R., Frisch, H. L. and Eirich, F. R. (1953). *J. Phys. Chem.* **57**, 584.
Smith L. E. and Stromberg, R. R. (1975). In Conference on Polymers at the Liquid-Solid Interface, Loughborough.
Sonntag, H., Ehmke, B., Miller R. and Knapschinsky, L. (1982). In *The Effect of Polymers on Dispersion Properties* (Th. F. Tadros, ed.), p. 207. Academic Press, London.
Stromberg, R. R. and Smith, L. E. (1967). *J. Phys. Chem.* **71**, 2470.
Stromberg, R. R., Tutas, D. J. and Passaglia, E. (1965a). *J. Phys. Chem.* **69**, 11.
Stromberg, R. R., Tutas, D. J. and Passaglia, E. (1965b). *J. Phys. Chem.* **69**, 3955.
Stromberg, R. R., Smith, L. E. and McCrackin, F. L. (1970). *Symp. Faraday Soc.* **4**, 192.
Takeda, M. and Endo, R. (1956). *J. Phys. Chem.* **60**, 1202.
Tanford, C. (1961). *Physical Chemistry of Macromolecules*, Wiley, New York.
Thies, C. (1968). *Macromolecules* **1**, 335.
Thies, C., Peyser, P. and Ullman, R. (1964). 4th International Congress of Surface Activity, Brussels, 1041.
Truesdell, C. (1945). *Ann. Math.* **46**, 144.
Vander Linden, C. and van Leemput, R. (1978a). *J. Colloid Interface Sci.* **67**, 48.
Vander Linden, C. and van Leemput, R. (1978b). *J. Colloid Interface Sci.* **67**, 63.
Varoqui, R. and Dejardin, P. (1977). *J. Chem. Phys.* **66**, 4395.
Vincent, B. and Whittington, S. (1981) In *Surface and Colloid Science* (E. Matijević, ed.), Vol. 12, Wiley, New York.
Wojciak, W. and Dutkiewicz, E. (1964). *Roczn. Chem.* **38**, 271.
Yamakawa, H. (1971). *Modern Theory of Polymer Solutions*, Harper and Row, New York.
Yoshida, T., Ohsaka, T. and Suzuki, M. (1972a). *Bull. Chem. Soc. Japan* **45**, 3245.
Yoshida, T., Ohsaka, T. and Tanaka, S. (1972b). *Bull. Chem. Soc. Japan* **45**, 326.

Electrolytes

5. Adsorption of Small Ions

J. LYKLEMA

I. General Problems

Ion adsorption is a very common phenomenon. Especially in aqueous solution, adsorbents are, as a rule, charged because there are always some ionic species present (often inadvertently) and, again as a rule, their preferences for a solid surface are unequal. Alternatively, adsorbents may become charged by dissociation of surface groups, i.e. by preferential expulsion of one type of ion. Even if a surface seems to be uncharged, as judged by, for example, electrophoresis, it is likely that there are still ions adsorbed (or desorbed), only to such an extent that the total charge contribution of cations is just balanced by that of anions. At any solid/aqueous solution interface, adsorption of ions is the rule rather than the exception. It is the purpose of this chapter to discuss some basic principles of ion adsorption; clearly, it is impossible to present a complete review in the available space.

Compared with the adsorption of small, uncharged molecules, the adsorption of small ions is strongly, but not exclusively, determined by electric interactions. Since these interactions have a long range, the adsorption behaviour of a given ion is influenced by many more ions than just its nearest neighbours or next-nearest neighbours. As a consequence, ion adsorption laws may differ substantially from those for uncharged species of the same size.

In the description of ion adsorption, several problems recur, some of which are the following:

(i) Accepting the fact that ions can "feel" the presence of an adsorbent even if they do not touch it, how close must an ion be before it can be called "bound"?
(ii) Is it expedient to distinguish different types of adsorbed ions, such as potential-determining, specifically bound, and indifferent ions? Can these types be experimentally distinguished?
(iii) Under what conditions may the electric field of a collection of adsorbed ions be treated as if it were generated by a smeared-out, homogeneous charge?
(iv) Is it possible to separate the electrical from the non-electrical part of the Gibbs energy of adsorption?
(v) How is it possible to treat the transfer of a single ionic constituent in a system that, as a whole, must be electroneutral?
(vi) Are lyotropic sequences (or Hoffmeister series) unexplained deviations from ideality or do they follow certain patterns?
(vii) What is the appropriate way to plot ion adsorption isotherms: adsorbed amount as a function of concentration, activity, or perhaps chemical potential?

These and many other questions will be treated in the coming sections. Anticipating this discussion, it is already noted that the answer to some of the problems raised depends on the point of view from which the question has been asked. For instance, under certain conditions charges may be localized in the statistical-mechanical sense but smeared-out in the electrostatic sense. Or, an ion may be adsorbed in the thermodynamic sense, even if spectroscopically it may be free from the surface. Proper distinctions have not always been made in the literature, and this has sometimes led to considerable confusion. In short, before answering such questions it is mandatory to formulate them appropriately.

Ion adsorption and electrical double layers are closely related, but in this chapter a systematic discussion of double layers will not be given beyond what is necessary for an understanding of the adsorption properties of charged species.

II. Distinction between Bound and Non-bound Ions

Thermodynamically, an ion i is counted as adsorbed (i.e. bound) if it contributes to the surface excess Γ_i of that ion. In the Gibbs convention, Γ_i equals the total amount of i in the system minus the amounts of i in the two adjoining phases, which are assumed to be of constant bulk composition up to the so-called Gibbs dividing plane. The value of Γ_i depends on where this dividing plane is located, and the analysis proceeds as if all excess material is concentrated in that dividing plane, irrespective of the physical position of the adsorbed species i (touching the sorbent surface, or far away from it in the outer periphery of a diffuse layer).

Non-thermodynamic methods measure a different property of an ion and may well produce a quite different answer. Suppose we have some spectroscopic or nuclear resonance technique that can measure alterations in the solvation layer of an ion. By such a method an ion would be classified as "bound" if it had zero, one, or perhaps two solvent molecules between it and the surface. However, the adsorbed amount obtained in this way, if expressed in moles per unit area, would not necessarily agree with the thermodynamic Γ_i based on some agreed dividing plane, since the latter category could well include ions with solvation sheaths indistinguishable from those in the bulk.

Another method of determining ion adsorption is electrokinetics. From the ζ-potential, using diffuse double layer theory, the electrokinetic charge can be calculated, and again can be expressed in moles of ions per unit area. This technique measures ion adsorption beyond the slipping plane, and its outcome does not necessarily coincide with the spectroscopic or thermodynamic value.

The conclusion is inescapable. Different experimental techniques may measure different fractions of the adsorbed ions; the surface excess Γ_i occurring in thermodynamic equations cannot be identified with the surface concentrations obtained by various techniques unless non-thermodynamic assumptions are made. For ion adsorption this dilemma is greater than for adsorption of electroneutral substances, because in the latter case the adsorbate is usually restricted to a monolayer (or little more than that), so the results of different experimental techniques tend to agree mutually and may thus more readily be identified with the thermodynamic excess.

Because of the above difficulty, it is recommended that the nature of the adsorbed amount be specified in any case where confusion may arise (electrokinetic charge, spectroscopically bound ions, etc.).

III. Electroneutrality Matters

Paradoxically, although many investigators have measured adsorption of ions, few if any have measured the adsorption of only *one* ionic species. The reason is that the adsorbent, together with its ionic adsorbate envelope, is electroneutral. So also is the solution, with which it is in equilibrium. Consequently, the transfer of ionic species from bulk to adsorbate proceeds only as electroneutral combinations of several ions together. Analytically, one could measure the depletion of some ion i in solution, say by some spectroscopic or radiotracer technique, and then plot the amount of i adsorbed as a function of the amount of i remaining in the bulk. However, this is not a "complete" adsorption isotherm, because the transfer of i is not the only thing that happens. For instance, if i is a cation, there is co-adsorption of anions, or ions i are exchanged against an equivalent amount of another cation. For this reason it is often instructive, if not desirable, to describe ion adsorption in terms of adsorption of electroneutral entities.

An example may illustrate this. If an insoluble oxide is immersed in an aqueous solution of, say, NaOH, with a pH above the point of zero charge of the oxide, it tends to become negatively charged either by adsorbing OH^- groups or by desorbing protons. Experimentally these processes cannot be distinguished; one merely observes a certain decrease of pH. From the OH^- depletion of the solution one would calculate Γ_{OH^-}, but this is accompanied by co-adsorption of an equal amount of Na^+ ions, so the actual process is adsorption of NaOH molecules, with $\Gamma_{Na^+} = \Gamma_{OH^-} = \Gamma_{NaOH}$. Experimentally indistinguishable from this process is desorption of H^+ ions, followed by the reaction $H^+ + OH^- \rightarrow H_2O$; also in this case each H^+, leaving the surface, is exchanged by a Na^+ ion to maintain electroneutrality. Non-thermodynamic arguments are then needed to locate the excesses of individual ions. In this particular case, it is very likely that the OH^- ions are on the surface and the Na^+ ions in the solution, close to the surface. More specifically, if $\equiv$ROH represents a surface hydroxyl group, this group may react with an OH^- ion to give $\equiv RO^- + H_2O$. Desorption of a proton would yield the same result. Hence, there seem to be good reasons to identify Γ_{OH^-} as the number of $\equiv RO^-$ groups, counted as moles per unit area. However, this identification is of a non-thermodynamic nature and, as we shall show later, it is not entirely correct.

In the example given, the Na^+ ion location cannot be established. To that end one could do additional experiments, for instance electrophoresis. This could yield the part of the Na^+ ions outside the slipping plane. In interfacial electrochemistry such parts of ionic charges are known as "ionic components of charge".

The story is not yet complete, because charged surfaces display the feature of "negative adsorption". The principle is that, say, a negative surface would expel anions; these anions would be enriched in the solution but, again for reasons of electroneutrality, this concentration increase is only measured as an increase of an electroneutral combination of ions. A familiar example of negative adsorption is the Donnan membrane equilibrium; if a charged colloid or dispersion is in equilibrium with a solution free of the colloid, the salt concentration in the dialysate is higher than within the sol. In the above example, if the solution were also to contain NaCl, with $c_{NaCl} \gg c_{NaOH}$, so that the negative adsorption of NaOH may be neglected against that of NaCl, one would observe, in the equilibrium solution, as compared with the situation in the absence of the oxide: (i) a strong depletion of OH^-; (ii) a slightly less strong depletion of Na^+; and (iii) a slight increase of Cl^-. (It will be shown in Section VII.A that negative adsorption is, as a rule, much smaller than positive adsorption of ions; maximally the two are equal.) Actually, because of the electroneutrality argument, only two of the items (i)–(iii) need to be measured; the third can then be calculated. The two measurements yield the two surface excesses Γ_{NaOH} and Γ_{NaCl}, with $\Gamma_{NaOH} > 0$ and $\Gamma_{NaCl} < 0$.

All of this is still in the realm of thermodynamics. In this particular case it would be logical to proceed and identify individual ionic adsorption as follows

$$\Gamma_{OH^-} = \Gamma_{NaOH} \tag{1}$$

$$\Gamma_{Cl^-} = \Gamma_{NaCl} \tag{2}$$

$$\Gamma_{Na^+} = \Gamma_{NaOH} + \Gamma_{NaCl} \tag{3}$$

If things are as presented, $\Gamma_{OH^-} > 0$, $\Gamma_{Cl^-} < 0$, and $\Gamma_{Na^+} > 0$. However, if Cl^- were to have a high affinity to the oxide surface, part of the chloride would be positively and part of it negatively adsorbed (both as NaCl); if the positive adsorption exceeds the negative, Γ_{Cl^-} would also be positive.

These three equations demonstrate the danger of the (often used) method to obtain the surface charge of dispersed particles by determining the counterion adsorption and then equalizing the two excesses Γ_{Na^+} and Γ_{OH^-}. It is now clear that the surface charge $(-F\Gamma_{OH^-})$ and the countercharge $(+F\Gamma_{Na^+})$ differ by an amount $F\Gamma_{NaCl}$. The method works only under conditions where $\Gamma_{NaOH} \gg \Gamma_{NaCl}$.

We can now also see why it is not entirely correct to identify $F\Gamma_{OH^-}$ as the surface charge in the above example of an oxide in a solution of NaOH only; in addition to the positive adsorption on the surface there is some negative adsorption of OH^- from the region close to the surface, and only the sum of the two is measured. For this reason, surface charges in solutions without additional electrolyte can be ill-defined unless supplemented by other measurements. If a carrier electrolyte is present, say NaCl, with

$c_{NaCl} \gg c_{NaOH}$, the negative adsorption is dominated by that of the Cl^- ion, and only then may $F\Gamma_{OH^-}$ be looked at as the surface charge.

The electroneutrality issue recurs in the description of ion adsorption isotherms. Four "logical" ways of presentation offer themselves:

(i) Adsorbed amount of individual species as a function of concentration of that ion in solution, $\Gamma_i(c_i)$.
(ii) Adsorbed amount of individual species as a function of activity of that species, $\Gamma_i(a_i)$.
(iii) Adsorbed amount of electroneutral entity as a function of concentration of that entity, $\Gamma_s(c_s)$.
(iv) Adsorbed amount of electroneutral entity as a function of activity, $\Gamma_s(a_s) = \Gamma_s(\gamma_\pm c_s)$.

Which of these plots should be chosen in a given case depends on the purpose of the representation under consideration, and is to a certain extent a matter of taste. If the purpose is a mere plotting of measured facts, one could use $\Gamma_i(c_i)$. However, if one wished to emphasize that the overall process is that of transfer of electroneutral entities, there may be advantages in the plot $\Gamma_s(c_s)$. In multicomponent mixtures this may be difficult to achieve because there are several (positive and negative) co-adsorption processes going on, but this is not a difficulty of principle. Plots of $\Gamma_i(a_i)$ can only be constructed after making some non-thermodynamic assumption to obtain the single ionic activity; hence such plots give not factual information but *interpreted* factual information. This is a matter of principle. A $\Gamma_s(a_s)$ plot does not have this difficulty.

Plotting adsorbed amounts as a function of activity (as opposed to concentration) brings us to the intrinsic difficulty that not only the bulk but also the adsorbate is highly non-ideal. If we correct for non-ideality in the bulk but not in the adsorbate, an isotherm type is obtained that is clearly internally inconsistent but, since it is not easy to obtain surface activities, one should learn to live with this inadequacy. The issue presents itself whenever such plots are theoretically analysed. Then it must be realized that the non-idealities due to interaction in the bulk are already taken into account, so one has to deal only with the interactions in the adsorbate *in excess* over those in solution. An illustration of this may be found in the Boltzmann equation for ions in a diffuse double layer:

$$c_i(x) = c_i(\infty) \exp[-z_i F\phi(x)/RT] \quad (4)$$

Here, $c_i(x)$ is the concentration of ion i at some location x near the adsorbent, where the (Galvani) potential is $\phi(x)$. The bulk is the reference; at infinite distance from the surface, the potential $\phi(\infty) = 0$ and the concentration $c_i(\infty)$ is the bulk concentration. From $c_i(x)$, $d\Gamma_i$ in a layer of thickness dx, or (after integration) the total value of Γ_i is readily obtained.

It is typical of equation (4) that it assumes that $z_i F\phi(x)$ in the *only* energy difference between an ion at x and an ion at $x = \infty$. For ideally dilute systems this may be correct, but it is no longer so for more concentrated systems. The activity coefficient f_i of an ion in the bulk is determined by the ion atmosphere around that ion, and in the field of the charged adsorbent $f_i(x)$ is likely to differ from the bulk value $f_i(\infty)$, because of the presence of the charged surface. Virtually, equation (4) assumes that all non-idealities due to ion atmospheres are identical in the bulk and in the adsorbate, except for the contribution of the charged surface, expressed through the exponential. Consequently, it would be erroneous to replace $c_i(\infty)$ by $a_i(\infty)$ without simultaneously changing $c_i(x)$ into $a_i(x)$. The latter transition is quite difficult to make, although theories are available (e.g. Kirkwood, 1934). They are quite complex, if only because of the difficult statistical issue of distinguishing between the mean electric potential and the potential of the average force. There is little choice; either equation (4) must be used as its stands, i.e. as an idealization, or it must be replaced by quite a different, complex but rigorous expression. Luckily, in most cases of practical interest, deviations from ideality play only a minor role in the non-diffuse part of a double layer, so equation (4) remains a useful tool. In fact, it is a kind of adsorption isotherm equation.

IV. Electrochemical Potentials

Equalization of the *electrochemical* potential of an ion in the adsorbed state to the same in the bulk formulates the adsorption equilibrium, and this is the basis for the derivation of an ion isotherm equation. Alternatively, for electroneutral combinations of ions, i.e. for electrolytes, equalization of *chemical* potentials is the first step. We have already seen that adsorption of ions can always be written this way. In formulae, using the superscripts S and L to indicate the adsorbed (surface) state and the liquid phase respectively, we have

$$\tilde{\mu}_i^{\mathrm{S}} = \tilde{\mu}_i^{\mathrm{L}} \tag{5}$$

with

$$\tilde{\mu}_i = \mu_i + z_i F\phi \tag{6}$$

and

$$\mu_{\mathrm{s}}^{\mathrm{S}} = \mu_{\mathrm{s}}^{\mathrm{L}} \tag{7}$$

Two facts may be noted. First, the splitting of $\tilde{\mu}_i$ into a purely electrical contribution $z_i F\phi$ and a “chemical” contribution μ_i is a non-thermodynamic operation, because ϕ can only be obtained by bringing an ion i from infinity (the reference state) to the place where ϕ is to be measured, but the isothermal, reversible work of this ion transport also involves a "chemical" contribution. In other words, some assumption must be made to assign a value to ϕ, so that the "chemical" part of the electrochemical potential is (by inference) all that is left after subtraction of $z_i F\phi$.

Secondly, it may seem that working with individual ions magnifies the number of variables, but this is only seemingly so, because electroneutrality conditions must also be obeyed. For instance, if instead of the adsorption of NaCl we wished to study the adsorption of Na^+ and Cl^- separately, we do not have one more degree of freedom because the Gibbs–Duhem relation in the bulk is amplified by the auxiliary electroneutrality condition $c_{Na^+}(\infty) = c_{Cl^-}(\infty)$, or $d\mu_{Na^+}$ and $d\mu_{Cl^-}$ are coupled quantities.

V. Distinguishing Various Types of Ion Adsorption

A. POTENTIAL-DETERMINING AND NERNSTIAN IONS

In the literature there is sometimes confusion as to whether or not a certain ion is potential-determining for a given adsorbent. A great deal of this discussion is due to the fact that there are two ways to define a potential-determining ion:

(a) An ion that obeys Nernst's law, i.e. an ion for which

$$d\Delta\phi = (RT/z_i F)\, d\ln a_i \tag{8}$$

where $\Delta\phi$ is the (Galvani) potential difference between adsorbent and solution. Equation (7) gives rise to the familiar $59/z_i$ mV per decade concentration dependence at 25°C.

(b) An ion that upon adsorption becomes indistinguishable from the solid matrix, so its electrochemical potential in the adsorbed state has no concentration-dependent term

$$\tilde{\mu}_i^S = \mu_i^{S0} + z_i F\phi^S \tag{9}$$

Early in the development of the notion "potential-determining ion" (Lange and Koenig, 1933; Verwey, 1934) the interface between AgI and a solution containing Ag^+ and I^- ions was the foremost example. In fact, for Ag^+ and I^- ions, definitions (a) and (b) both apply, and then by combining equations (5), (6), and (9), with $d\mu_i^L = RT\, d\ln a_i$, equation (8) is immediately obtained.

Later investigations revealed the existence of systems where equation (8) was obeyed, but not equation (9), i.e. there are examples in which the "operational" definition [equation (8)] and the "physicochemical" definition [equation (9)] lead to different conclusions. A notable example is the glass electrode, responding according to equation (8) to H^+, Na^+, K^+, or other ions over several decades, even if these ions, upon uptake in the surface gel layer, would modify its chemical composition. Liquid junction potentials may also obey equation (8), even though no adsorption whatsoever occurs. Clearly, equation (9) is more restrictive than (8) because mechanisms other than those formulated in (9) may lead to (8). One may also ask if, for instance, phosphate ions chemisorbing on oxides should not be called "potential-determining", since they bind strongly and often irreversibly and alter the surface charge and hence $\Delta\phi$. One could say that such ions also determine the potential.

This issue is clearly not settled. In this chapter, we shall therefore henceforth distinguish these types of ions:

(a) *Nernstian ions*, i.e. ions obeying equation (8), whatever the mechanism. This is an "operational" definition.
(b) *Potential-determining ions*, i.e. ions obeying equations (8) and (9), as previously. This is a restricted definition and applies only to such ideal cases as Ag^+ or I^- for AgI, or Ba^{2+} and SO_4^{2-} for $BaSO_4$, usually only for a restricted number of decades because of the occurrence of additional chemical reactions, dissolution and/or complexation processes.
(c) *Surface ions* as ions chemically bound to the surface and contributing to the surface charge σ_0. This definition encompasses a wide range of ion adsorption, since it includes all ions of category (b) but also H^+, OH^-, and chemisorbed phosphates on oxides and chemisorbed complexes and surfactants.
(d) *Ions not chemically bound to the surface*, but contributing to the thermodynamic surface excess Γ_i, which must be considered "adsorbed". To this class belong the counterions in a double layer and all ions of category (c) in so far as they are not chemically but only physically bound. A further discussion of this category follows in Section V.B.

The ions of these four groups are usually easily distinguishable. Nernstian ions must follow equation (8), which can be tested for materials that can act as an electrode in a Galvanic cell. For materials that do not allow such measurements it remains a difficult problem. Most ions of category (b) are readily identified. The distinction between (c) and (d) is virtually a distinction between physisorption and chemisorption, i.e. between binding with and without electron exchange. Practice has shown that, although the difference is not sharp, it is usually not difficult to distinguish, using measurements of adsorption enthalpy, adsorption reversibility, temperature coefficients, etc., in addition to spectroscopic information. That ions contribute to the charge is also usually not difficult to infer. For instance, for phosphates adsorbing on oxides one could measure the so-called R-factor, i.e. the amount of OH^- released per equivalent of phosphate sorbed.

Information on the mode of binding can also be obtained from the general shape of the adsorption isotherm. This will be discussed in Section VII.

B. SPECIFIC AND GENERIC ADSORPTION

There is also much confusion in the literature with respect to the definition of "specific adsorption", apparently because different authors adopt different definitions.

Let us introduce the distinction on the basis of the meaning of the word "specific"; it is a type of adsorption that is different for each type of ion of the same valency. The opposite is "generic adsorption", a term never used but describing precisely what it is, adsorption that is the same for all ions of a given valency, i.e. all ions of the same genus. As an illustration, ion adsorption in the diffuse part of a double layer, as far as it is following Poisson–Boltzmann statistics, is generic because the adsorption energy is $z_i F\phi$ irrespective of the size of the ion or its shape. However, adsorption in the non-diffuse part or Stern layer, leading to "lyotropic sequences" because of differences in adsorption between ions of differing size, is clearly specific.

By this definition, surface ions are also specifically adsorbed ions. However, many authors restrict the word "specific" to physically bound ions because the distinction with chemisorbed surface ions is usually not so difficult. Bearing in mind that surface ions are also specifically adsorbed, we shall now mention some tests to find out whether a given physisorbed ion i is specifically adsorbed. Ions are specifically adsorbed if they are capable of bringing about the following:

(i) *Reversal of the sign of the electrokinetic potential* ζ with increasing a_i at constant or increasing absolute value of the surface charge. This is due to overcompensation of the surface charge by counter charge adsorption in the Stern layer, and clearly it cannot occur if i is not specifically adsorbed, i.e. if the adsorption energy has not a "chemical" contribution in addition to $z_i F\phi$. Overcompensation like this is also known as "superequivalent adsorption". Reversal of ζ-potential leads to the "irregular series" in the coagulation of hydrophobic colloids (see e.g. Overbeek, 1952; Matijević, 1977), so such irregular series are also criteria for specific adsorption.

Reversal of the sign of the electrokinetic potential is often referred to as "charge reversal". This is an easy but sloppy expression. It is true that, if the sign of ζ reverses, the sign of the *electrokinetic* charge σ_{EK} (the charge outside the slipping plane) also reverses. However, the *surface* charge σ_0 retains its sign! Its absolute value may even increase (at fixed concentration in the bulk of the surface ions) because strong specific adsorption in the Stern layer increases the double layer capacitance. For this reason, we had to add "at constant or increasing surface charge" in the first sentence of (i).

(ii) *Shift in the point of zero charge (p.z.c.) or the isoelectric point (i.e.p.)* at otherwise constant conditions. The direction of shift is indicative of the sign of the charge on the ion that is preferentially specifically adsorbing: if it is the anion, Δ(p.z.c.) is positive and Δ(i.e.p.) is negative; if it is the cation, the trends are reversed.

The explanation of these trends is as follows. Consider, by way of example, an adsorbent for which H^+ and OH^- are the surface ions. Hence, a certain value of pH is the p.z.c., and another is the i.e.p. In the absence of specific adsorption the p.z.c. and i.e.p. are identical. Let us now add an electrolyte of which the anion, say Cl^-, adsorbs specifically, whereas the cation, say Na^+, is only generically adsorbed. Chloride ions are then enriched near the surface. This affects the adsorption balance between H^+ and OH^- ions at the p.z.c. Before the Cl^- adsorption, $\Gamma_{H^+} = \Gamma_{OH^-}$, but now $\Gamma_{H^+} > \Gamma_{OH^-}$ because H^+ adsorption is promoted and OH^- adsorption reduced by the presence of the negative charge of the Cl^- ions. In order to restore the p.z.c. condition $\Gamma_{H^+} = \Gamma_{OH^-}$, we must increase the OH^- concentration in the bulk, i.e. we must *increase* the pH. The shift of the isoelectric point is the other way round. By electrokinetic methods, one measures the surface charge plus the Stern charge, up to the slipping plane (or slipping layer). If Cl^- ions adsorb specifically, this would render ζ more negative and the i.e.p. is restored by letting more H^+ ions adsorb, i.e. by *decreasing* the pH. Similar reasoning applies to cation specific adsorption.

This opposite shift of p.z.c. and i.e.p. is a very characteristic criterion for specific adsorption. Using some double layer models one could even assess the amount specifically adsorbed from Δ(p.z.c.) and Δ(i.e.p.). Values of i.e.p. can be obtained by any electrokinetic method, and p.z.c. by (potentiometric) titration of the surface ions, i.e. by potentiometric proton titration.

(iii) *Production of lyotropic sequences in the affinity for a surface.* If, on a given negative surface under otherwise identical conditions, one were to observe that $\Gamma_{Rb^+} > \Gamma_{K^+} > \Gamma_{Na^+} > \Gamma_{Li^+}$ this would be a clear indication of specific adsorption of the cation. In practice, lyotropic series are mostly arrived at indirectly. For instance, the sequence just mentioned would reveal itself in an increase of the negative surface ion adsorption from Li^+ to Rb^+ (i.e. in a double layer capacitance, increasing in this direction) or in a decrease in the same order of the coagulation concentration for negative, electrostatically stabilized sols of the adsorbent material.

Sometimes the question has been asked whether or not one should use the term "specific adsorption" if the difference between different counter-ions is only a matter of distance of closest approach, with no difference of adsorption energy involved. In double layer theory, this is the familiar case of charge-free Stern layers. According to our definition, this is indeed a case of specific adsorption, but it is a rather theoretical case. If different ions are at different, but close, distances from a surface, it is likely that there will also be specific differences in the alteration of their respective solvation shells, so it is also probable that there will be specific adsorption energies in addition to the generic term $z_i F\phi$.

Finally, it may be noted that it is likely also that diffuse double layer models can become specific if they are treated more rigorously than the idealized limiting cases, leading to Boltzmann's law [equation (4)]. It reminds us that generic adsorption is only an idealization of reality, although it is a very useful idealization.

VI. Smeared-out and Discrete Charges

Briefly, a layer of adsorbed ions is smeared-out if we may consider it as a homogeneous layer with everywhere the same charge density. However, if it is necessary for an adequate description to consider the ions as being on specified sites, the layer is discrete. The question arises of when it is permissible to treat them as smeared-out and by what criteria this can be judged. It is first necessary to distinguish between the two main approaches, the electrostatic and the statistical-mechanical.

According to electrostatics, the potential of a given array of fixed point charges, if measured very close to these charges, varies periodically with the locations of these charges, but if measured at some distance the periodicity fades away and the potential can be calculated as if the charged layer were smeared-out. Figure 1 illustrates this. Electrostatically, therefore, the choice between "discrete" or "smeared-out" is in this case a matter of approximation. The further away, the more the individuality of all charge carriers fades. Electrostatics has many more examples of this; as regards the calculation of ϕ, one could treat the charges on a non-conducting sphere as smeared-out, provided that the distance to the surface is large enough. Similarly, a positive charge in combination with a nearby equal negative charge may be looked at as an ideal dipole if considered from sufficient distance. It follows that electrostatically the distinction is a matter of distance and not a matter of principle. The distance at which the discreteness can be ignored is discussed in textbooks of electrostatics, but as an order of magnitude we could say that, if the observation distance is less than or approximately equal to κ^{-1}, surface charges further apart than that must be accounted for individually.

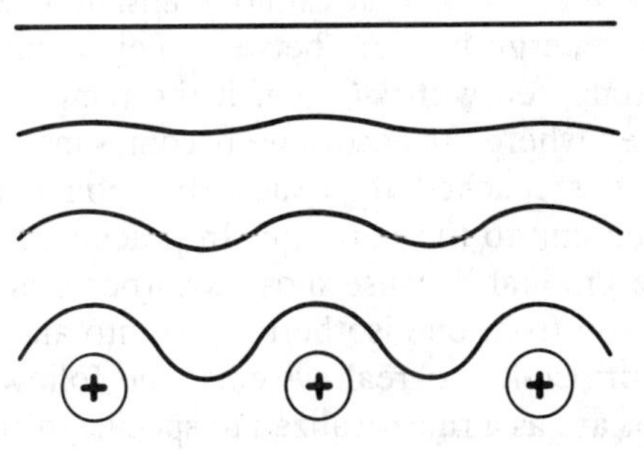

Figure 1 Qualitative picture of equipotential planes near a discrete array of positive charges.

In statistical mechanics distinction is made between "localized" and "mobile" adsorption. Localized adsorption is adsorption on discrete, recognizable sites, and adsorbed molecules can move to another site only by desorbing and re-adsorbing. In mobile adsorption there are no distinguishable sites, the adsorbate behaving as a two-dimensional gas with a high mobility parallel to the adsorbent. Hence, localized adsorption is discrete adsorption, and mobile adsorption is smeared-out. This distinction has a physical nature, and the two canonical partition functions are quite different. For localized adsorption of N ions on N_s sites, if the lateral interaction may be neglected (the Langmuir picture), one has

$$Q_{\text{loc}}(N, N_s, T) = \frac{N_s!}{(N_s - N)!N!} q_{\text{loc}}^N \tag{10}$$

whereas for N mobile ions on an area A, again without lateral interaction,

$$Q_{\text{mob}}(N, A, T) = q_{\text{mob}}^N / N! \tag{11}$$

Typically, in equation (11) there is no number of sites. The molecular partition function q_{mob} is largely translationally determined; it is a function of A and T. On the other hand, q_{loc} is determined by the vibration of the bound ions around an equilibrium state; the permutation factor $N_s!/(N_s - N)!N!$ gives the number of ways in which N indistinguishable ions are accommodated over the N_s distinguishable sites. These two partition functions are needed later for the derivation of adsorption isotherm equations. They are of a different shape for the two cases, and hence offer in principle a means of discrimination, provided that detailed data are available. We shall return to this in Section VII.

In the statistical picture, diffuse ions are clearly mobile (in three dimensions). For ions in contact with the surface, the difference between mobile and localized is the difference between ions translating without any barrier parallel to the surface and ions that cannot translate over the surface but only vibrate. If the energy barriers between neighbouring sites become lower and lower compared with kT (e.g. if the temperature is raised), a stage may be reached where the adsorbate becomes mobile. Theoretically, the transition point is reached if an adsorbed ion completes just one vibration before hopping to the next site. In practice, however, the transition will be more gradual because most adsorbents are heterogeneous, so a gradual transition from one isotherm type into another is observed.

Applying these principles to real systems, the following may be concluded. Surface ions are as a rule localized to specific, distinguishable sites. Diffuse ions are mobile *by definition*. Stern ions assume an intermediate position. They are traditionally treated as localized, but most elaborations are inconsistent (see Section VII).

The strict electrostatic consequence is that the surface potential and Stern layer potential should be periodic functions, whereas potentials in a diffuse double layer part are smeared-out. However, in most cases in the literature all three of these potentials are considered as smeared-out. In view of the foregoing, this can only be approximately correct. More precisely, the kind of experiment that one has in mind determines how good this approximation is. If surface ions move on the surface with a time of residence of, say, 10^{-9} s per site, and electrical measurements are taken with a time scale of, say, 1 s, treatment of the surface ions for the purpose of this particular measurement as smeared-out is fully justified. If, on the other hand, the relaxation time of the surface charge is not negligibly short compared with the time scale of the experiment, this approximation is not permissible.

VII. Ion Adsorption Isotherms

The shapes of adsorption isotherms $\Gamma_i(a_i)$ or $\Gamma_i(c_i)$ for individual ions are very dependent on the part of the double layer where the ion is located (surface, Stern layer, or diffuse layer). Very roughly, adsorption in a diffuse double layer follows the square-root of the concentration, adsorption in a Stern layer follows a (modified) Langmuir or (modified) Volmer type isotherm, whereas surface ion adsorption tends to be semilogarithmic. In principle, this offers the possibility of a more penetrating analysis of ionic adsorbates, but the theories must be used with care because deviations from these basic types may be substantial and are not always easily accounted for. Anyway, the availability of reliable data over a wide concentration range and otherwise well controlled conditions is essential, and it remains desirable to have additional (e.g. spectroscopic) information.

A. DIFFUSE IONS

From the Gouy–Chapman theory (Gouy, 1906, 1910, 1917; Chapman, 1913) one can derive the ionic components of charge of a diffuse double layer, i.e. the contributions of cations and anions to the surface charge (Grahame, 1947). Expressed in SI units and as moles per unit area, for symmetrical (z–z) valent electrolytes, the surface excesses are

$$\Gamma_+ = \sqrt{(\varepsilon\varepsilon_0 cRT/500F^2z^2)}[\exp(-ze\phi_d/2kT) - 1] \qquad (12)$$

$$\Gamma_- = -\sqrt{(\varepsilon\varepsilon_0 cRT/500F^2z^2)}[1 - \exp(ze\phi_d/2kT)] \qquad (13)$$

where ε is the relative dielectric constant, ε_0 the permittivity of free space (8.854×10^{-12} C V^{-1} m^{-1}), F the Faraday, and ϕ_d the diffuse double layer potential. If the surface is negative, $\phi_d < 0$, $\Gamma_+ > 0$, and $\Gamma_- < 0$, meaning that cations are positively adsorbed and anions negatively, although of course both contribute positively to the diffuse charge. (Note that, for very low potentials, where the exponentials may be replaced by the linear terms of their series expansions, $|\Gamma_+| = |\Gamma_-|$, but that for higher potentials the positive adsorption exceeds the negative.)

Equations (12) and (13) indicate that Γ_+ and Γ_- are proportional to $c^{1/2}$, provided that ϕ_d is independent of c. This provision is difficult, because ϕ_d itself is a function of the electrolyte concentration, since the screening of the surface charge is better with higher electrolyte contents (actually, this screening is measured by the Debye length κ^{-1} which is also proportional to $c^{1/2}$). For practical purposes the way out of this difficulty is to replace ϕ_d by ζ. The check for adsorption in a diffuse layer is then the $\Gamma_i(c_i^{1/2})$ proportionality with the proper constant at fixed ζ. Also for negative adsorption the $\Gamma_i(c_i^{1/2})$ dependence can be tested; it is even easier because in the limit of high negative potential $\exp(ze\phi_d/2kT) \rightarrow 0$, so Γ_- becomes independent of potential (see Section VIII).

B. SURFACE IONS

Adsorption isotherms for surface ions tend to be semilogarithmic except if the electrolyte concentration and potential are so low that the diffuse part of the double layer dominates. This latter case is rather exceptional, occurring only at surface potentials of a few tens of mV and concentrations of 10^{-3} molar at most. Under those conditions the isotherm is dictated by the sum of the absolute values of Γ_+ and Γ_- in equations (12) and (13), because of electroneutrality. It leads again to a $c^{1/2}$ dependence, now at constant potential of the surface.

If the potential and/or the electrolyte concentration are not so low, the double layer capacitance C is dominated by the capacitance of the Stern layer, even if this layer is charge-free. This follows from electrostatics. For two capacitances in series, because

$$C^{-1} = C_{\text{diff}}^{-1} + C_{\text{Stern}}^{-1} \tag{14}$$

the total capacitance is determined by the lower of the two. Since C_{diff} increases with $c^{1/2}$ and exponentially with the potential, C_{diff}^{-1} becomes rapidly negligible against C_{Stern}^{-1} if potential and concentration are not very low. Then $C \approx C_{\text{Stern}}$.

For most interfaces, C_{Stern} is not *very* sensitive to the surface charge (i.e. to the adsorbed amount of surface ions), and is independent of bulk concentration. Typically, for dispersed systems such as silver halides and oxides, C_{Stern} varies by less than a factor of 2–3 over >8 decades in concentration of potential-determining ions (i.e. over a range of 500 mV of surface potential in 10^{-1} mol dm^{-3} solutions), and in many cases C_{Stern} may even be almost constant. This constancy or limited variability is due to the fact that C_{Stern} is determined by its geometric part $\varepsilon_0 \varepsilon_{\text{St}}/d$ and by the specifically adsorbed amount. There are also exceptions to this trend; they occur especially if specific processes occur in addition to the electrostatic ones, such as surface reactions after a given surface charge has been reached.

Anyway, considering trends only, if C is not too variable, the amount of surface ions adsorbed is roughly proportional to the surface potential, i.e. to $\Delta\phi$. Then, even if the ions are not Nernstian, the relation between $\Delta\phi$ and a_i is more likely to be of the semilogarithmic type as in equation (8) than linear because, again as a trend, the deviations from Nernstian behaviour are not so drastic that they would completely obscure the basic trend.

The basic trend is that adsorption isotherms for surface ions are semi-logarithmic. Several examples corroborate this. For instance, almost linear or somewhat convex or concave semilogarithmic plots are obtained over several decades for Ag^+ and I^- on AgI, for Ag^+ on Ag_2S, for H^+ and OH^- on many oxides at not too extreme pH, and for phosphate on hematite (α-Fe_2O_3).

Although this semilogarithmic relation is only a trend, it offers a way to discriminate against specific adsorption in the Stern layer, which is of a modified Langmuir or modified Volmer type. In carrying out the test of fitting adsorption equations to experimental data it must be borne in mind that perfectly semilogarithmic dependences appear, over a limited concentration range, as if they are Langmuirian when plotted linearly. It is therefore necessary to compare several ways of plotting, preferably over many decades of the bulk concentration.

C. SPECIFICALLY ADSORBED IONS

The most familiar equation for specific adsorption of ions is that of Stern (1924). This equation has several variations, one of which relates the fraction θ_i of surface sites covered by specifically adsorbed counterions i to the bulk concentration, the potential ϕ_s at the location of adsorption, and the specific adsorption free energy Φ_i:

$$\frac{\theta_i}{1-\theta_i} = \frac{c_i}{55.5}\exp\left(-\frac{z_i e \phi_s}{kT} + \frac{\Phi_i}{kT}\right) \tag{15}$$

In two respects this variant is a simplification of the more complete original equation: (1) it considers no superequivalent adsorption ($\theta_i \leqslant 1$); (2) it considers only *one* type of specifically adsorbing ions (the original equation has *two* ions). Still, equation (15) is very suitable to demonstrate some of the principles and approximations on which it is based.

First, equation (15) at $\Phi_i = 0$ is just the Langmuir equation with adsorption energy $z_i F \phi_s$, so it may be looked at as a Langmuir equation corrected for specific adsorption.

Second, the equation is in principle inconsistent. Langmuir adsorption is *localized* adsorption; in this picture one needs the local potential at the adsorption site (the bottom curve of Fig. 1), but in equation (15) ϕ_s is the smeared-out (i.e. averaged) potential in the layer of adsorption. In other words, a localized adsorption picture is used together with a potential belonging rather to a mobile adsorbate. Essentially, this is the Bragg–Williams approximation, used in statistical thermodynamics to retain separable partition functions in non-ideal systems.

This principal defect has been to a large extent overcome in the "discreteness of charge" pictures. Following older work by Ershler, Grahame, and others, Levine has made substantial contributions to the development of these ideas (e.g. Levine *et al.*, 1967). There are several approaches to this problem. One of them, mainly propounded by Grahame (1958) and Levine *et al.* (1962), is the "cut-off disc" model in which each specifically adsorbed ion is surrounded by a charge-free space beyond which the radial distribution changes in some fashion (depending on model parameters) towards its average value at large distance. A different approach is based on cluster theory (Buff and Stillinger, 1963) and/or on a hexagonal lattice assumption with which various authors have occupied themselves. Two modifications of smeared-out Stern theory ensue: (i) the electrical contribution to the adsorption energy of the Stern ion is altered. Whether or not this is a generic or specific contribution depends on the premises of the model; are the parameters dependent on the size of the ion?; (ii) the potential ϕ_d of the diffuse part of the double layer is altered.

Quantitatively, the differences with smeared-out models increase with increasing θ_i. Usually, they become measurable at Stern charges that are so high that other deviations from ideality occur also, so unambiguous evidence for quantitative correctness is difficult to obtain with dispersed systems. Ionized monolayers, double layers on mercury, or, in general, systems with homogeneous and high surface charges are better test objects. Still, it is one of the features of discrete ion theory that it offers an explanation for some hitherto largely unexplained facts, such as the maximum of ζ as a function of $\Delta\phi$, which has several times been found.

Stern theories accounting for the discrete ion effect cover the most consistent localized pictures that are nowadays available. As far as this author is aware, the other limiting situation, viz. the completely mobile Stern adsorbate, has not been considered at all. The derivation should start from a mobile partition function and not from a localized one. For the canonical case, this starting point is equation (16) and not (10). Here, q_{mob} consists of a translational part parallel to the surface, q_{tr}, a vibrational part q_{v} perpendicular to it, and a contribution of the potential energy q_{pot}. For the translational part we write

$$q_{\text{tr}} = \frac{2\pi mkT}{h^2}(A - N_i a_0) \tag{16}$$

which is the translational partition function for point charges (Hill, 1960) corrected for finite size (N_i ions i of cross-section a_0 on area A), m is the mass of the ion, and h is Planck's constant. Further,

$$q_{\text{v}} = \exp(-h\nu/2kT)/[1 - \exp(-h\nu/kT)] \tag{17}$$

and

$$q_{\text{pot}} = \exp[(-z_i e\phi_s/kT) - (g_{\text{chem},i}/kT)] \tag{18}$$

In equation (17) ν is the classical vibration energy normal to the surface, and in equation (18) ϕ_s is the average potential in the plane of adsorption. Obviously, in this case a smeared-out potential is required. It is a matter of discussion whether in equation (18) also a non-electrical energy is needed. Anyway, if it were high as with chemisorption, the ion would become localized. We have included $g_{\text{chem},i}$ in equation (18), presuming it to be small. It is a molar free energy containing any entropy contribution belonging to the non-electrical term, but not the configurational contribution to the entropy.

With these premises, an adsorption isotherm equation can be derived by standard methods of statistical thermodynamics:

$$q_{\text{mob}} = \frac{2\pi mkT}{h^2}(A - N_i a_0)\frac{\exp(-h\nu/2kT)}{1 - \exp(-h\nu/kT)}$$
$$\times \exp(-z_i e\phi_s/kT) - (g_{\text{chem},i}/kT) = q'_{\text{m}}(A - N_i a_0)$$

$$\ln Q_{\text{mob}} = N_i \ln q'_{\text{m}} + N_i \ln(A - N_i a_0) - N_i \ln N_i + N_i$$

$$\tilde{\mu}_i^{\text{S}} = -RT\left(\frac{\partial \ln Q}{\partial N_i}\right)_{A,T} = -RT\ln\left[q'_{\text{m}}\left(\frac{A - N_i a_0}{N_i}\right)\exp\left(-\frac{N_i a_0}{A - N_i a_0}\right)\right]$$

Since the electrical term $z_i F\phi_s$ enters through q'_{m}, this is an electrochemical potential; the potential inside the solution is the reference. Defining the covered fraction $\theta_i = N_i a_0/A$ and equalizing $\tilde{\mu}_i^{\text{S}} = \mu_i^{\text{L}} = \mu_i^{\text{L}0} + RT\ln x_i$, we finally obtain

$$\frac{\theta_i}{1 - \theta_i}\exp\left(\frac{\theta_i}{1 - \theta_i}\right) = K_{\text{mob}} x_i \tag{19}$$

with

$$K_{\text{mob}} = q'_{\text{m}} a_0 \exp(\mu_i^0/RT) \tag{20}$$

This equation can be somewhat modified to let it more resemble the Stern equation (15). First, $(g_{\text{chem},i} - \mu_i^0)$ is just the specific adsorption free energy Φ_i. Then, also writing the $z_i e\phi_s$ term explicitly,

$$\frac{\theta_i}{1 - \theta_i}\exp\left(\frac{\theta_i}{1 - \theta_i}\right) = k_i\frac{c_i}{55.5}\exp[(-z_i F\phi_s/kT) + (\Phi_i/kT)] \tag{21}$$

with

$$k_i = \left(\frac{2\pi mkT}{h^2}\right)a_0\frac{\exp(-h\nu/kT)}{1 - \exp(-h\nu/kT)} \tag{22}$$

Equation (21) is the mobile counterpart of the localized equation (15), and it is an extension of the Volmer adsorption equation (Volmer, 1925), just as equation (15) is an extension of the Langmuir equation (Langmuir, 1918). Both are of the same order of approximation. The differences are twofold.

(i) In equation (21) there is an exponential factor on the left-hand side that is absent from (15). At low θ_i this has little effect, but at high θ_i it results in a much slower asymptotic approach of saturation. (ii) The proportionality constant of c_i is different and tends to be higher in the mobile case (Hill, 1960), so as an overall picture equation (22) has a steeper initial rise, but then levels off more slowly.

No examples are known to this author where equation (21) has been subjected to experimental tests involving ionic adsorbates.

With allowance for the difference between mobile and localized adsorption, it may now be summarized that, as far as isotherm equations are concerned, the distinction between equation (15) or (22) and semilogarithmic behaviour is one of the means to distinguish between surface ions and specifically adsorbing non-surface ions.

D. LYOTROPIC SEQUENCES

The systematic variation of the adsorbability with the size of a counterion at otherwise fixed conditions is one of the most direct indications of specific adsorption. Double layer capacitances beyond the diffuse double layer range provide the best examples. It is the rule rather than the exception that, at given bulk concentration of surface ions, the capacitance and hence the surface charge varies with the size of the counterion. For instance, on negatively charged AgI, the negative charge (due to adsorbed I^- ions) increases at fixed pI and salt concentration in the direction $Li^+ < Na^+ < K^+ < Rb^+$; on negative hematite (α-Fe_2O_3) at constant pH this sequence is reversed. Differences of several tens of percent between otherwise similar ions such as Li^+ and Rb^+ are easily attained. Lyotropic sequences in coagulation concentrations constitute another class of examples. In the two examples mentioned, the capacitance sequence is the opposite of the coagulation concentration sequence: a high capacitance, i.e. a high charge at given bulk composition, corresponds to a relatively low stability.

Electrostatically, a high capacitance means that the surface charge is well screened by counterions. Good screening can occur if the counterions are close to the surface and/or if they accumulate strongly in the Stern layer (in short, by strong interaction with the surface). In the extreme case, the surface ion and counterion form an intimate, electroneutral entity. Then $C \rightarrow \infty$ because an infinite number of surface ions can adsorb without increasing $\Delta\phi$. The seemingly paradoxical countertrend between C and coagulation concentration rests on the same principle; strong interaction with the surface means that relatively less ions are found in the diffuse double layer part, and this part governs stability.

Thanks to work by Gurney (1953), Pearson (1963, 1966, 1968), and others, the lyotropic series are now largely understood. The basis for the interpretation is that in the bulk the interaction between ions of different size follows broadly the trend of Fig. 2. This trend can, for example, be assessed by judging activity coefficients. Qualitatively, the explanation is that small cations and small anions attract each other very strongly electrostatically, because of the high electric field strengths that they produce; big cations and big anions bind by hydrophobic bonding. This picture can be worked out more precisely by considering the details of the water structure around the ions.

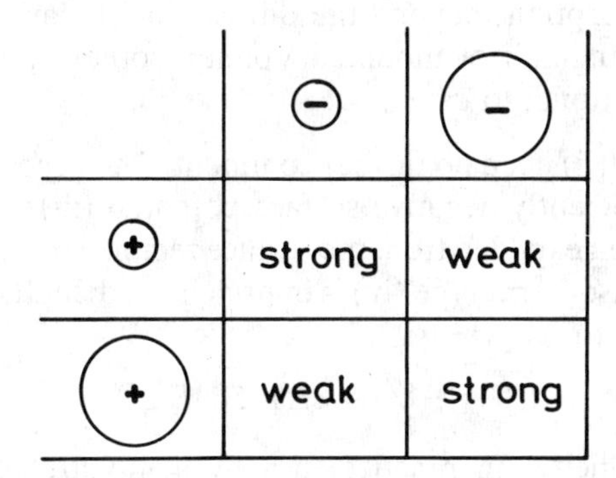

Figure 2 Interaction scheme for ion pairs of different size.

For the purpose of predicting lyotropic sequences in adsorption, one simply has to extrapolate this picture by letting one of the ions of a pair be the surface ion. If a surface is negative, with small anions as the surface ion, the surface will be strongly hydrated (it is then a "high energy surface"), and Li^+ adsorbs more strongly than Rb^+. However, if the surface anions are big, the surface is relatively hydrophobic and "low energy"; then Rb^+ adsorbs better than Li^+. Negative hematite is an example of the first, negative silver iodide of the second. Many more illustrative examples can be offered. Deserving of mention are experiments by Dumont and Watillon (1978) who prepared two different types of rutile (TiO_2) surface, one of low energy and one of high energy, and found that the alkali ion lyotropic sequence on one of these surfaces is the opposite of that on the other.

Similar rules are obeyed by anions, although their behaviour tends to be somewhat less predictable.

VIII. Negative Adsorption

The negative adsorption of electrolytes by charged surfaces, discussed in Section III, can become high enough to be analytically measurable. It is interesting to juxtapose common (or positive) adsorption and negative adsorption:

(i) In common adsorption $\Gamma > 0$; in negative adsorption $\Gamma < 0$.

(ii) Common adsorption can be inferred from a decrease in the bulk concentration; negative adsorption gives rise to an increase in concentration.
(iii) At high $\Delta\phi$, common adsorption of counterions tends to be specific; negative adsorption of co-ions is then generic.
(iv) Common adsorption, beyond the diffuse double layer range, follows a modified Langmuir or modified Volmer isotherm; negative adsorption is proportional to $c^{1/2}$.

Features (i) and (ii) need no further comment. The genericity of negative adsorption at sufficiently negative surface potential (iii) is due to the fact that all ions are then expelled from the surface region, so specific interaction vanishes. The $c^{1/2}$ isotherm type (iv) is in principle attributable to equations of the type (12). For $\phi_d \rightarrow -\infty$,

$$\Gamma_- \rightarrow -\sqrt{(\varepsilon\varepsilon_0 cRT/500F^2z^2)} \tag{23}$$

without any specificity. In practice, usually a mixture of electrolytes is used, at least one to create a sufficient surface potential, another one to be negatively adsorbed. The coefficient of the $\Gamma(c^{1/2})$ relation is then different, but the proportionality with $c^{1/2}$ persists.

Figures 3 and 4 give two illustrations. In Fig. 3 $\sigma_d = -2F\Gamma_-$ is plotted

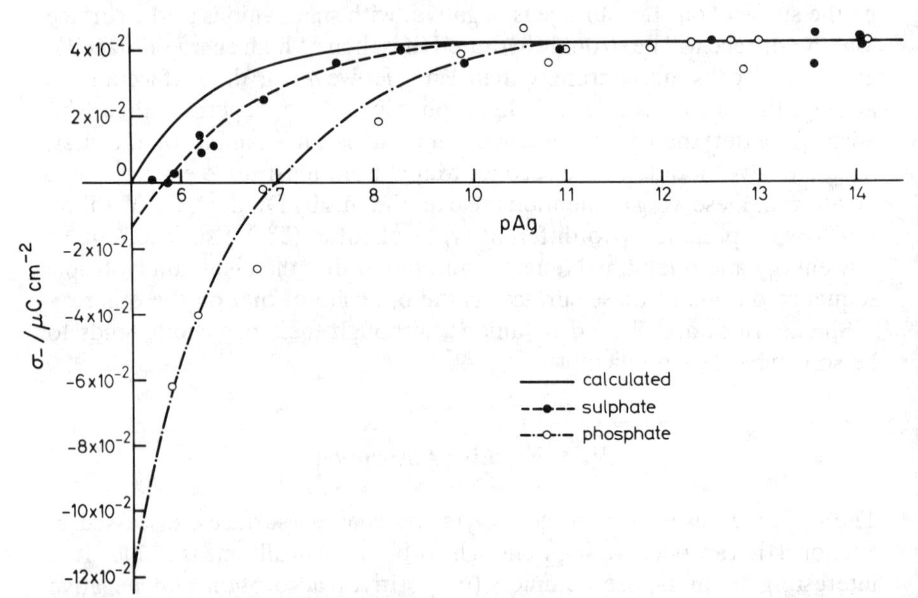

Figure 3 Negative adsorption of sulphate and phosphate on negative silver iodide expressed in μC cm^{-2}; electrolyte concentration 5×10^{-5} mol dm^{-3}; one unit of pAg corresponds to 59 mV.

at fixed c_{salt} as a function of the surface potential. In this case, $\Delta\phi$ and pAg are proportional and the origin is the p.z.c. in the absence of specific adsorption. The drawn curve is theoretical, calculated from a modification of equation (12). There are two electrolytes, KI and K_2SO_4; I^- ions are surface ions and potential-determining, but a small fraction of them is also negatively adsorbed; the electrical factors contain exponentials with $z_i = -1$ and $z_i = -2$ (van den Hul and Lyklema, 1967). It is seen that, if $\Delta\phi$

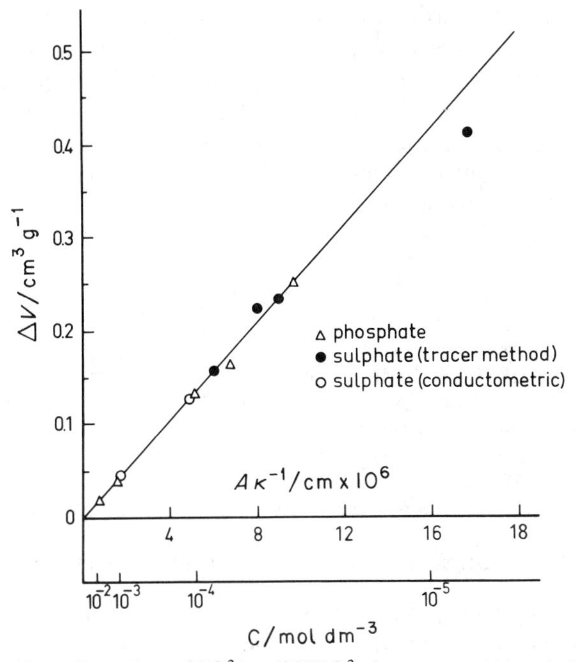

Figure 4 Negative adsorption of SO_4^{2-} and HPO_4^{2-} ions on negative AgI as a function of the electrolyte concentration, expressed as the expelled volume; results of different analytical methods are compared.

becomes more negative, ϕ_d does the same and below *ca.* −150 mV all exponentials have vanished. Negative adsorption is then complete and generic. The experiments show that with SO_4^{2-} and HPO_4^{2-} this level is attained, albeit at a more negative $\Delta\phi$. Specific positive adsorption of these ions is the competing factor, more strongly so for HPO_4^{2-} than for SO_4^{2-}. If $\Delta\phi$ is not very negative, these ions can even give a net positive adsorption leading to the reversal of the electrokinetic charge.

Figure 4 plots the negative adsorption for the same system at high $\Delta\phi$ as a function of electrolyte concentration. On the ordinate axis the "squeezed-out volume" is plotted. It is a direct representation of Γ_-, and identical to the volume of solution adjoining the adsorbent that is entirely devoid of the expelled ion; beyond this volume c_i has its bulk value. On the horizontal axis $A\kappa^{-1}$ is plotted; A is a concentration-independent factor, depending on the valencies of the ions in the mixture. Since κ^{-1} contains $c_i^{-1/2}$, we see that the $\Gamma_-(c_i^{1/2})$ relationship is very well corroborated. Another feature of this way of plotting is that $A\kappa^{-1}$, which has dimension [length], is just the thickness of the squeezed-out volume ΔV. Hence, the slope of the straight line is the surface area of the adsorbent.

Negative adsorption as a tool for surface area determination works only under restricted conditions, but if it does work it is an almost ideal procedure because no assumptions concerning molecular cross-sections are needed. In the field of clay colloid chemistry, where surface potentials and areas tend to be high, this method is more or less routine (see e.g. Schofield and Talibuddin, 1948), but Fig. 4 proves that it can be applied to much lower specific areas, down to the order of $1\ m^2\ g^{-1}$.

References

Buff, F. and Stillinger, F. H. (1963) *J. Chem. Phys.* **39**, 1911.
Chapman, D. L. (1913). *Phil. Mag.* **25**, 475.
Dumont, F. and Watillon, A. (1978). Paper presented at the Euromech 104 Conference in Louvain, Belgium.
Gouy, G. (1906) *Ann. Chim. Phys.* **8**, 291; **9**, 75; (1910) *J. Phys.* **9**, 457; (1917) *Ann. Phys.* **7**, 129.
Grahame, D. C. (1947). *Chem. Rev.* **41**, 441.
Grahame, D. C. (1958). *Z. Elektrochem.* **62**, 264.
Gurney, R. W. (1953). *Ionic Processes in Solution,* McGraw-Hill, New York, Toronto, London.
Hill, T. L. (1960). *Introduction to Statistical Thermodynamics,* Addison-Wesley, Reading (Mass.) and London, pp. 130, 192.
Hul, H. J. van den and Lyklema, J. (1967). *J. Colloid Interface Sci.* **23**, 500.
Kirkwood, J. G. (1934). *J. Chem. Phys.* **2**, 767.
Lange, E. and Koenig, F. O. (1933). In *Handbuch der Experimentalphysik* Vol. 12, part 2, 263.
Langmuir, I. (1918). *J. Am. Chem. Soc.* **40**, 1361.
Levine, S., Bell, G. M. and Calvert, D. (1962). *Can. J. Chem.* **40**, 518.
Levine S., Mingins, J. and Bell, G. M. (1967). *J. Electroanal. Chem.* **13**, 409.
Matijević, E. (1977). *J. Colloid Interface Sci.* **58**, 374.
Overbeek, J. Th. G. (1952). In *Colloid Science* (H. R. Kruyt, ed.), Ch. 8, p. 314, Elsevier, Amsterdam, Houston, New York.
Pearson, R. G. (1963). *J. Am. Chem. Soc.* **85**, 3533; (1966) *Science* **151**, 172; (1968) *J. Chem. Ed.* **45**, 581, 683.
Schofield, R. K. and Talibuddin, O. (1948). *Discuss. Faraday Soc.* **3**, 51.
Stern, O. (1924). *Z. Elektrochem.* **30**, 508.
Verwey, E. J. W. (1934). Thesis, State University of Utrecht, The Netherlands.
Volmer, M. (1925). *Z. Phys. Chem.* **115**, 253.

6. Adsorption of Ionic Surfactants

D. B. HOUGH and H. M. RENDALL

I. Introduction

The adsorption of ionic surfactants at the solid/solution interface is of major technological and commercial importance. Broadly speaking, within the context of this chapter, the surfactant is used to perform one, or sometimes both, of two major functions, namely the control of colloid stability and the control of wetting behaviour (either air–water or oil–water systems on a solid substrate). Colloid stability is of interest, for example, in relation to detergency and anti-redeposition mechanisms, the formulation of pigment and pharmaceutical dispersions, agricultural soil conditioning, emulsion polymerization, flotation (where small particles must be aggregated into a size range suitable for floatability), and other mineral separation processes, such as those based on selective coagulation. Wetting is important, for example, in detergency, dispersibility of powders, dyeing, flotation, the application of pesticides and herbicides, printing, and tertiary oil recovery. It should be remembered, of course, that alteration of the wetting properties of a liquid on a solid by addition of a surfactant is determined not only by its adsorption at the solid/liquid interface but also by its adsorption at the two other interfaces present in the system.

As a consequence of such ubiquitous use of ionic surfactants there is a vast and ever-expanding literature on the subject of their adsorption at solid/liquid (primarily aqueous) interfaces. Because of the complexity of the interactions involved, no single coherent and unifying picture has yet emerged that can provide a universally satisfactory account of the adsorption process for a range of surfactant–solid combinations. Nevertheless, various attempts have been made, particularly during the last two decades, to gain a physical and quantitative understanding of the interactions of ionic surfactants at the solid/aqueous solution interface. In our approach to the subject we have sought to follow two guiding principles.

(i) We have concentrated on the physical and chemical mechanisms that have been reported to influence adsorption, and on results that provide an insight into these underlying mechanisms. This seems more appropriate than seeking a blanket coverage of the published literature which is clearly impossible even if it were desirable.
(ii) We have emphasized a number of subtleties in the measurement and interpretation of adsorption results. Many reported studies are of little value because insufficient thought has been given to the control of important variables, to the capabilities of the experimental technique, or to the limitations imposed by the theoretical framework used in the interpretation. These considerations are not well covered in the existing literature, and we hope that our comments will help to foster a more rigorous approach in future studies.

In reviewing the literature we have in some instances referred to review articles on surfactant adsorption rather than to the widely scattered primary publications. In this way we have aimed not only to achieve a more economical coverage of published work, but also to benefit from considered opinions of experts in the field.

In Section II we deal in some detail with the adsorption process and with the variety of mechanisms (not all of which are independent) that have been deemed to contribute to the adsorption interactions. Underlying principles are identified for certain easily definable situations, for example, low adsorption coverage, very high adsorption coverage, low-energy or high-energy interfaces (as classified in relation to wetting behaviour). While every attempt is made to generalize the theoretical treatment and to achieve a common description of a wide variety of practical situations, the limitations of our current understanding, and of particular theoretical approaches, are also emphasized. In Section III we discuss experimental methods for substrate characterization and adsorption measurements. Here emphasis is placed on the limitations both of experimental methods and of approaches to the interpretation of the results. Illustrative results are presented in Section IV, designed to amplify the theoretical discussions of Section II or, as appropriate, to highlight the gaps in our current understanding.

II. The Adsorption Process

A. SURFACE CHARGE AND THE ELECTRICAL DOUBLE LAYER

It is well established that one of the major factors influencing the adsorption of surfactant ions at any interface is that of electrical interactions within the environment of an electrical double layer. An understanding of charge generation and the structure of the double layer, together with the relevance of measured electrokinetic potentials, is therefore pertinent to our interpretation of the adsorption process.

The generation of surface charge when a solid is in contact with aqueous solution is common to almost all systems. Only under certain solution conditions will the (net) surface charge be zero, i.e. at the point of zero charge (p.z.c.). Overall neutrality in the system demands that an equal and opposite charge exists in the solution (excluding excess charge in the solid phase), the distribution of which, together with the surface charge, constitutes the electrical double layer (see e.g. Overbeek, 1952). Surface charge may arise through one of a number of mechanisms. In the case of inorganic salts such as AgI, $BaSO_4$, and CaF_2 it is a consequence of preferential dissolution of one type of lattice ion over the other. It is usually considered that the adsorption–dissolution process of these lattice ions, which are termed potential-determining ions, does not affect the chemical potential of the solid and that electrochemical equilibrium between bulk solid and bulk solution can be represented (e.g. Smith 1973) by the Nernst equation

$$\Delta\phi = N\Delta\,(\mathrm{pX}) \tag{1}$$

where $\Delta\phi$ is the change of the inner (Galvani) potential of the solid on changing the concentration of potential-determining ion, type X and valency z, and where

$$N = -2.303(kT/ze) \tag{2}$$

with e being the fundamental unit of charge. It is generally assumed that $\Delta\phi$ can be replaced by $\Delta\psi_0$ which refers to the change in surface potential.

Provided that possible changes in the so-called χ-potential, arising in the solid phase and/or the pristine layer of oriented water dipoles adjacent to the solid surface, may be neglected, it follows that

$$d\psi_0/d(pX) = N \tag{3}$$

It is convenient at this point to outline the electrical double layer model that is commonly adopted for solid/liquid interfaces, this being attributed to Grahame (1947) and described here in the form used by Smith (1973, 1976). The model is illustrated in Fig. 1. The Stern region of integral

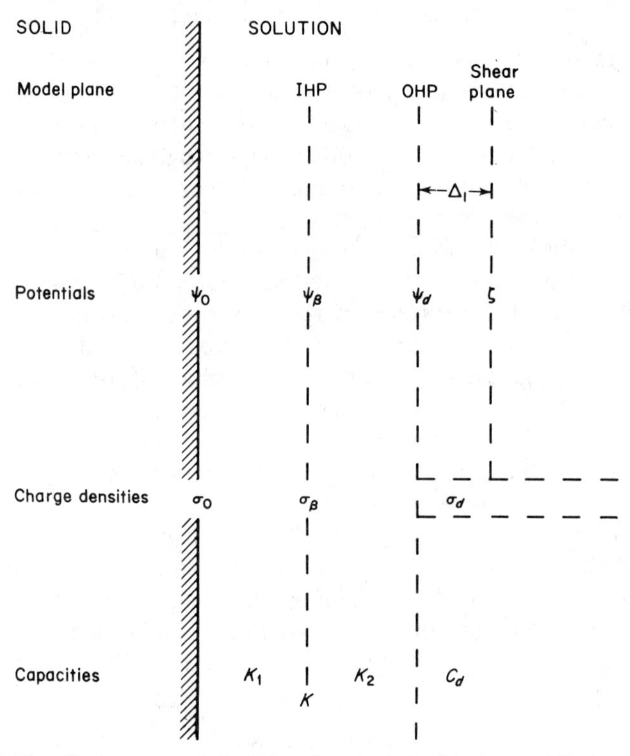

Figure 1 The Grahame model of the electrical double layer. (Grahame, 1947)

capacity K between the surface, of charge density σ_0 and potential ψ_0, and outer Helmholtz plane (OHP or Stern plane) of potential ψ_d, is subdivided by the inner Helmholtz plane (IHP) of charge density σ_β and potential ψ_β. The IHP corresponds to the locus of centres of specifically adsorbed ions. The total charge density at and beyond the OHP is σ_d and the differential capacity of this diffuse region is C_d. The electrokinetic potential ζ at the plane of shear is assumed to refer to a distance Δ_1 into the solution from the OHP. The foregoing parameters are related in the following double layer equations (Smith, 1973).

$$K_1 = \sigma_0/(\psi_0 - \psi_\beta) \tag{4}$$

$$K_2 = -\sigma_d/(\psi_\beta - \psi_d) \tag{5}$$

$$K^{-1} = K_1^{-1} + K_2^{-1} \tag{6}$$

$$C_d = -d\sigma_d/d\psi_d \tag{7}$$

$$\sigma_0 + \sigma_\beta + \sigma_d = 0 \tag{8}$$

For the case of a symmetrical $(z:z)$ electrolyte,

$$\sigma_d = -(2kT\varepsilon\kappa/ze)\sinh(ze\psi_d/2kT) \tag{9}$$

where ε is the absolute permittivity of the bulk solvent and κ is the Debye–Hückel reciprocal double layer parameter, which for the symmetrical electrolyte of concentration C mol dm^{-3} is given by

$$\kappa^2 = 10^3(2z^2e^2CN_A)/\varepsilon kt \tag{10}$$

By making use of the above equations, Smith (1973, 1976) has shown that in the absence of specific adsorption of ions, and in the region of the isoelectric point (i.e.p.), which corresponds to $\psi_d = 0$,

$$(d\psi_0/d\psi_d)^0 = 1 + C_d^0K^{-1} \tag{11}$$

where the superscript 0 refers to i.e.p. conditions. Then, putting

$$s_1 = (d\psi_d/d(pX))^0 = (d\psi_d/d\psi_0)^0(d\psi_0/d(pX))^0 \tag{12}$$

and substituting equations (3) and (11) into (12) gives

$$Ns_1^{-1} = 1 + C_d^0K^{-1} \tag{13}$$

Note here that, from equations (7) and (9), $C_d^0 = \varepsilon\kappa$ and is therefore a function of the supporting electrolyte concentration.

Since ψ_d cannot be measured directly in order to test the validity of equation (13), use is generally made of the experimentally accessible ζ-potential which refers to the potential at some idealized plane of shear between solid and solution phases when tangential motion is induced between them. The replacement of ψ_d by ζ is accompanied by two main problems. The first is that the calculation of ζ-potentials from electrokinetic data, such as from electrophoresis measurements, is uncertain because of difficulties associated with the utilization of double layer relaxation corrections (Wiersema *et al.*, 1966) over certain κa regions (a = particle radius), and with the use of bulk viscosity and permittivity data which may not apply to the high field strength extant in the double layer (Lyklema and Overbeek, 1961; Bijsterbosch and Lyklema, 1978). These difficulties are minimized for small values of ζ. The second problem is the evaluation of Δ_1. For low potentials (say <25 mV) the Debye–Hückel approximation applies for the diffuse double layer, such that

$$\zeta = \psi_d \exp(-\kappa\Delta_1) \tag{14}$$

The need to estimate Δ_1 is removed at i.e.p. conditions when the identity $\psi_d = \zeta = 0$ is exact.

Smith combined equations (13) and (14) with the term

$$s = (d\zeta/d(pX))^0 \tag{15}$$

to give

$$Ns^{-1} = (1 + C_d^0 K^{-1}) \exp(\kappa\Delta_1) \tag{16}$$

A plot of Ns^{-1} against C_d^0 for silver iodide sol particles in solutions of KNO_3 gave an essentially straight line with the expected intercept of unity at zero electrolyte concentration. In view of the approximate constancy of K with electrolyte concentration it was deduced by Smith that Δ_1 was less than about 0.5 nm, so (for small ψ_d at least) ζ and ψ_d may be taken as identical. Similar conclusions were reached by Smith (1976) for SiO_2 and TiO_2 and, using a different approach, by Lyklema (1977) for AgI. This conclusion provides some justification for the commonly assumed equality of ψ_d and ζ when interpreting many colloidal and interfacial phenomena, including that of ionic surfactant adsorption. Nevertheless, the assumption should be reserved for small values of ψ_d and for surfaces that do not possess extensive layers of structured solvent such that Δ_1 may be large. This latter requirement is unlikely to exclude many of the widely studied interfaces.

It should be pointed out that there are numerous examples in the literature (e.g. Bell *et al.*, 1973) where H^+ and OH^- ions are assumed to be potential-determining for various inorganic salts. Using $CaCO_3$ and $Ca_3(PO_4)_2$ as examples, Foxall *et al.* (1979) have shown, with the use of equation (16), that the pH may have a role secondary to that of the fundamental potential-determining lattice ions, simply serving to control the solution equilibria between the free potential-determining ions and those in the form of solution complexes. Similar conclusions may be made about other salts such as various apatites (Somasundaran and Goddard, 1979).

The charging mechanism of inorganic oxides such as SiO_2, TiO_2, and Al_2O_3 relies on the presence of ionizable groups on the solid surface, these being assumed to be amphoteric hydroxylated groups. The charging mechanism is normally represented as

$$MOH_2^+ \overset{H^+}{\leftrightharpoons} MOH \overset{H^+}{\leftrightharpoons} MO^- \tag{17}$$

For such an interface H^+ and OH^- ions are generally described as potential-determining. (Strictly, a case could be made for restricting this term to lattice ions where there is some justification for a Nernst relation.) Although this charging mechanism is rather idealized and may be represented in other ways (Healy and White, 1978), it appears to provide a workable model for oxide surfaces. On the basis of this model, Smith (1976) devised a relationship analogous to the simple Nernst equation (3), viz.

$$\left(\frac{d\psi_0}{d(pH)}\right)_{\sigma_0 \to 0} = N - \left(\frac{kT}{2N_s e^2 \theta_c} \frac{d\sigma_0}{d(pH)}\right) \tag{18}$$

where N_s is the total surface density of ionizable sites; θ_c is the equal fraction of oppositely charged sites at the p.z.c. and is determined by the equilibrium constants for the two surface dissociation reactions. The second term on the right-hand side of equation (18) denotes deviation from the Nernst equation for the system. Smith then went on to show that for conditions around the i.e.p., in the absence of specific adsorption, an equation similar to (16) can be derived:

$$Ns^{-1} \approx (1 + C_d^0 K_A^{-1}) \exp(\kappa\Delta_1) \tag{19}$$

where

$$K_A^{-1} = K^{-1} + K_s^{-1} \tag{20}$$

Here K_s is an effective capacity in series with the Stern capacity and which accounts for the extra term in equation (18). In addition to the deduction that Δ_1 is less than about 0.5 nm the straight-line plots of Ns^{-1} against C_d^0 obtained by Smith (1976) for TiO_2 and SiO_2 provide good indication of the validity of the proposed model for these oxide interfaces, together with an estimate of the fractions of dissociable surface groups charged at the p.z.c.

The dissociation of ionizable surface groups provides the primary charging mechanism for a number of other surfaces in contact with aqueous solution. Thus, for example, the charge on polystyrene latex particles and on other polymer surfaces (both in colloidal and bulk form) may arise as a consequence of the incorporation of particular initiators or other additives into the polymerization process, which on termination of the polymerization are present in the surface as ionizable groups. Typical monofunctional surface groups are carboxylate, sulphate, or amidinium groups (see e.g. Goodwin *et al.*, 1973, 1979). Alternatively, the surface may be amphoteric, such as that of polystyrene latices prepared by Homola and James (1977) which possess both carboxylate and ammonium groups, the ratio of which can be varied so as to control the p.z.c. These latices, as far as charge is concerned, have uses as model oxide systems (James *et al.*, 1977).

A polyamide surface such as nylon 6,6 possesses similar amphoteric properties. The p.z.c. of such a surface will naturally depend both on the relative numbers of each type of grouping and on their respective dissociation constants. A modified Nernst relationship can accordingly be derived (Rendall and Smith, 1978).

The surface charge of carbon blacks is also due to the presence of ionizable groups. Most of these groups are recognized as various oxygen complexes chemically bound to the carbon atoms of the basal planes (Medalia and Rivin, 1976) and which function as weak acids. The majority can be removed by heating to about 1000°C in vacuum. Heating to between 2500 and 3000°C induces crystallization of the carbon, the surfaces then behaving like the inert basal planes of graphite (Saleeb and Kitchener, 1965). On these graphitized surfaces some residual oxygen complexes will probably be present only at the edges of the crystallites. Indeed, it is well established from gas adsorption studies that graphitized carbons present very uniform hydrophobic surfaces, only about one in 1500 surface sites being polar (Young *et al.*, 1954).

A rather different surface charging mechanism is illustrated by clays and micas (van Olphen, 1977). These layer silicates are built up from two-dimensional arrays of silicon–oxygen tetrahedra and aluminium (or magnesium)–oxygen–hydroxyl octahedra superimposed in various ways. Because of the substitution of some Al(III) for Si(IV) in the silica tetrahedra, or Mg(II) for Al(III) in the octahedral layer, the crystals possess a net negative charge which is primarily independent of solution conditions. The edges of clay particles, however, exhibit a pH-dependent surface charge akin to that on oxide surfaces and which is due to hydroxylation and subsequent ionization of broken Si–O and Al–O bonds (van Olphen, 1977).

B. THE ADSORPTION ISOTHERM

The adsorption of surfactant ions can be considered as a special case of the specific adsorption of ions referred to in Grahame's treatment of the electrical double layer. The ionic headgroup is, by definition, located at the IHP. It is sometimes convenient at very low coverage to assume that the IHP and OHP coincide (i.e. Stern's original model). This assumption does not invalidate the model as long as it is remembered that σ_β then refers to charges located in the OHP while σ_d refers, of course, to the diffuse layer charge *outside* this plane.

In general, the number of moles per unit area (Γ_1) of surfactant (1) adsorbed depends on the bulk solution concentration C_1. The relationship between Γ_1 and C_1 is the equation for the adsorption isotherm. The choice of the correct isotherm is not straightforward. A good fit between experimental measurements and the chosen isotherm is not always a good criterion for assessing the validity of that isotherm (see e.g. Kitchener, 1965; Rendall and Smith, 1979). Useful procedures to test the self-consistency of a given treatment may be obtained from consideration of conditions of "congruence".

The adsorption is said to be charge congruent if, as far as electrostatic interactions are concerned, the amount adsorbed is a function of potential only at constant charge. It is potential congruent if the amount adsorbed is a function of charge only at constant potential. If we take the example of AgI as the adsorbent, following de Keizer and Lyklema (1980), the adsorption at constant ionic strength and temperature can generally be described with four variables, Γ_1, C_1, σ_1, pAg (≡ potential), only two of which are independent. Then an adsorption isotherm can be written as

$$F'(\text{pAg}, \Gamma_1)\, C_1 = f'(\Gamma_1) \tag{21}$$

or

$$G'(\sigma_0, \Gamma_1)\, C_1 = g'(\Gamma_1) \tag{22}$$

The adsorption isotherm is said to be congruent in charge (or potential) provided that the function F' (or G') is separable into the product of two functions, one depending on only the electrical parameter, pAg (or σ_0), and the other depending on Γ_1 only. In this case equations (21) and (22) can be rewritten as

$$F(\text{pAg})\, C_1 = f(\Gamma_1) \tag{23}$$

and

$$G(\sigma_0)\, C_1 = g(\Gamma_1) \tag{24}$$

Thus, a test of charge (or potential) congruence is that plots of Γ_1 against C_1 at different σ_0 (or pAg) values should be mutually parallel. Then, taking the standard free energy of adsorption ΔG^0_{ads} as the part of the adsorption free energy that is independent of surface coverage, we may write either $F(\text{pAg})$ or $G(\sigma_0)$ as $\exp(-\Delta G^0_{\text{ads}}/kT)$. Adsorption is congruent in charge (or potential) if the electrical component of the free energy term, written in terms of charge (or potential), is constant. Thus equations (23) and (24) become

$$f(\Gamma_1) = C_1 \exp(-\Delta G^0_{\text{ads}}/kT) \tag{25}$$

and

$$g(\Gamma_1) = C_1 \exp(-\Delta G^0_{\text{ads}}/kT) \tag{26}$$

All components of the adsorption free energy that depend on surface coverage, such as those arising from lateral interactions between the adsorbate ions, are included in the functions f or g and do not occur in ΔG^0_{ads}.

It seems unlikely that perfect congruence of either type exists in most ionic surfactant/solid systems, because (i) the electrical contribution to the adsorption free energy is likely to depend both on potential, in the form of a term $ze\psi_\beta$, and on charge because of, for example, changes in dipole orientations at the interface, which are controlled by the surface charge dependent electric field across this region, (ii) σ_0 and ψ_β are generally not independent variables, and (iii) components of the adsorption free energy (in f or g) that are dependent on surface coverage may also depend on the charge (or potential).

de Keizer and Lyklema (1980) have employed the Frumkin–Fowler–Guggenheim (FFG) isotherm as part of some investigations into the type of congruence that best describes the adsorption of tetraalkylammonium ions on AgI:

$$\frac{\theta}{1-\theta}\exp(A\theta) = \frac{C_1}{55.51}\exp\left(-\frac{\Delta G^0_{ads}}{kT}\right) \tag{27}$$

where the fractional surface coverage θ is given by Γ_1/N_s, and N_s is the total number of adsorption sites (in moles) per unit area for monolayer saturation adsorption. Adsorption is assumed to be reversible and localized, and the surface is taken to be energetically homogeneous. Lateral interactions are accounted for by A, the value of which, according to equation (27), can be estimated from the maximum slope $(\mathrm{d}\theta/\mathrm{d}\ln C_1)_{max}$ of the isotherm which occurs at $\theta = 0.5$. Furthermore, at $\theta = 0.5$ substitution of A into equation (27) gives the value of ΔG^0_{ads}.

Although the above treatment based on the FFG isotherm represents a novel and interesting alternative to the forms of adsorption isotherm more conventionally applied to surfactant adsorption, there are two major limitations. It is implicitly assumed that the factor A is a constant throughout the range of adsorption coverage. In reality, this term could change in sign, as well as in magnitude, as θ is increased. At low coverages, A would reflect (repulsive) electrostatic interactions between adsorbed ions. For surfactant ions, however, attractive chain–chain lateral interactions are widely believed to become important at higher coverage. It is frequently reported that the apparent adsorption energy becomes more favourable as coverage increases. This behaviour is sometimes ascribed to the formation of "hemimicelles" (see Section II.C). Secondly, electrostatic interactions are strongly influenced by the level of supporting electrolyte. Congruence of the isotherms could therefore be expected only when all measurements are made at a single ionic strength. An alternative approach to congruence tests is reviewed in Section II.F.

It is instructive to compare equation (27) with the widely used Stern–Langmuir isotherm, which may be written (e.g.. Rendall *et al.*, 1979) in the form

$$\frac{\theta}{1-\theta} = \frac{C_1}{55.51} \exp\left(-\frac{\overline{\Delta G_{\text{ads}}}}{kT}\right) \tag{28}$$

Here the adsorption free energy $\overline{\Delta G_{\text{ads}}}$ is taken to depend on the mean potential ψ_β in the adsorption plane, and on other factors. Possible subdivisions of this adsorption energy are discussed below. In contrast, equation (27) incorporates only the intrinsic electrostatic properties of the "bare" interface into ΔG^0_{ads}, and allows for variation in the electrostatic term with coverage in the factor $\exp(A\theta)$.

Implicit in the derivation of the above equations are the following assumptions:

(i) Only one type of ion, the surfactant ion, is specifically adsorbed, so $\sigma_\beta = z_1 e \Gamma_1$. This assumption is generally reasonable at low surface coverage in the presence of electrolyte which may plausibly be considered as "indifferent".

(ii) One surfactant ion replaces one solvent molecule, i.e. ion and solvent molecules are of the same size (this is especially true of equation (29) below). Even though a better fit over a wide range of coverage may be obtainable with an equation modified to take account of the relative sizes of the species, there is always some doubt about the form of a surfactant adsorption isotherm beyond the low-coverage region.

(iii) The surface is taken as homogeneous and, in the Stern treatment, allowance can be made, through a mean potential term, for electrical interactions.

(iv) Dipole terms and self-atmosphere potentials, as well as lateral chain–chain interactions, are, in particular, excluded in the Stern treatment.

The Stern–Langmuir isotherm should therefore be reserved for extremely low coverage by surfactant ions, or for comparison between different surfactants at identical coverage (i.e. to obtain comparative information). It is interesting that isotherms of the type given by equation (28) are frequently used to interpret adsorption and electrokinetic data, and other more "indirect" data such as those from contact angle measurements, with some disregard for the restrictions of the assumptions employed in their derivation. For example, those authors interested in surfactant adsorption at mineral/water interfaces are concerned with conditions of surface coverage where lateral interactions between adsorbed surfactant ions are possible. Widespread use is made of an isotherm in the form

$$\Gamma_1 = 2rC_1 \exp(-\Delta G_{ads}/kT) \tag{29}$$

where r is the radius of the adsorbed ion. This expression is an analogue of equation (28) at low coverage ($\theta \rightarrow 0$) suggested by Grahame (1947) (see also Sparnaay, 1972). An important part of the interpretation of the adsorption process (see e.g. Fuerstenau and Healy, 1972; Somasundaran and Goddard, 1979) involving the use of equation (29) hinges on the expected occurrence of lateral interactions between the hydrophobic moieties of the adsorbed ions (see Section II.C). Thus, although the general form of the Stern isotherm has been retained, a fundamental modification of the original concept has been introduced. In the following sections we will discuss factors that may be considered to influence adsorption interactions, and it will become apparent that the less than rigorous application of equations such as (28) and (29) often does provide a useful semi-quantitative analysis of complex adsorption mechanisms.

C. MECHANISMS OF ADSORPTION

1. *Contributions to the Adsorption Energy*

Much attention has been devoted to elucidating the various contributory mechanisms to the adsorption process for a wide variety of surfactants and adsorbents. To this end ΔG_{ads} is usually assumed to comprise a number of *additive* contributions. In the first instance we may write

$$\Delta G_{ads} = \Delta G_{elec} + \Delta G_{spec} \tag{30}$$

where ΔG_{elec} accounts for electrical interactions and ΔG_{spec} is a specific adsorption term which contains all other contributions to the adsorption free energy that are dependent on the "specific" (non-electrical) nature of the system. Although convenient, any such subdivision of ΔG_{ads} is questionable in that the electrical and chemical states of a molecule are not necessarily independent quantities.

A semantic point to be noted here is that ΔG_{spec} is, in the terminology of some authors, referred to as a chemical free energy. In this chapter we reserve the term "chemical" to account for chemical bonding (as it is normally understood) between adsorbent and adsorbate. Furthermore, it will be implicitly understood that ΔG_{ads} refers to a mean adsorption energy as in equation (28).

2. *Electrical Interactions*

Usually ΔG_{elec} is ascribed totally to coulombic interactions. Following de Keizer and Lyklema (1980), however, we may include a dipole term such that

$$\Delta G_{\text{elec}} = \Delta G_{\text{coul}} + \Delta G_{\text{dip}} \tag{31}$$

where

$$\Delta G_{\text{coul}} = z_1 e \psi_\beta \tag{32}$$

and

$$\Delta G_{\text{dip}} = \sum_j \Delta n_j \mu_j E_s \tag{33}$$

The ΔG_{dip} term is taken to arise through the exchange process occurring between the surfactant ion in bulk solution and in the adsorbed state, with accompanying desorption of n water molecules. In equation (33), Δn_j is the change in the number of adsorbed dipoles j of moment μ_j, and E_s is the electric field strength across the plane of the adsorbed species (taken here as the IHP). Surfactant dipoles as well as water dipoles are accounted for in equation (33), and n will be related to Δn_j according to the mode of adsorption of the surfactant ion, especially with respect to the possible replacement of orientated water dipoles by segments of the hydrocarbon chain. In criticism of equations (31)–(33), it might be said that ΔG_{coul} and ΔG_{dip} are not mutually independent, for the same reasons outlined in the arguments against the existence of pure charge or potential congruence in adsorption isotherms (see Section II.B).

If we neglect ΔG_{dip} (as in the original Stern equation), the basic interpretation of ΔG_{elec} is much simplified. Three examples may be considered:

(i) When the surfactant ions are counterions to the net charge density $(\sigma_0 + \sigma_\beta)$, then z_1 and ψ_β are of opposite sign, so $z_1 e \psi_\beta < 0$ and electrical interaction promotes the adsorption process. In the absence of other specifically adsorbed ions (such that σ_β is initially zero) this situation will exist at very low surface coverages ($|\sigma_\beta| < |\sigma_0|$, where σ_0 and σ_β are of opposite sign) for cationic surfactant/negatively charged surface and anionic surfactant/positively charged surface combinations.

(ii) If $(\sigma_0 + \sigma_\beta)$ is of the same sign as the surfactant ions, then z_1 and ψ_β are of like sign and $z_1 e \psi_\beta > 0$, i.e. electrical interactions oppose adsorption. In the absence of specifically adsorbed ions this situation will exist for anionic surfactant/negatively charged surface and cationic surfactant/positively charged surface combinations. It will also exist at higher surface coverage for the combinations cited in the first example, i.e. sufficient surfactant ions, initially as counterions, adsorb until $|\sigma_\beta| > |\sigma_0|$ (allowed for in terms of a ΔG_{spec} contribution). That is, the i.e.p. is traversed and ψ_β is reversed in sign, so $z_1 e \psi_\beta$ becomes positive and opposes the favourable ΔG_{spec} term.

(iii) Under i.e.p. conditions referred to above, ΔG_{elec} will be zero (neglecting ΔG_{dip}) and adsorption is governed by the ΔG_{spec} term.

That ΔG_{elec} may be the dominating term in ΔG_{ads}, at least at low surface coverage, is inferred from examples in the literature where ionic surfactants are shown not to adsorb to any appreciable extent on certain surfaces of like charge. An example is depicted in Fig. 2 for the adsorption of sodium

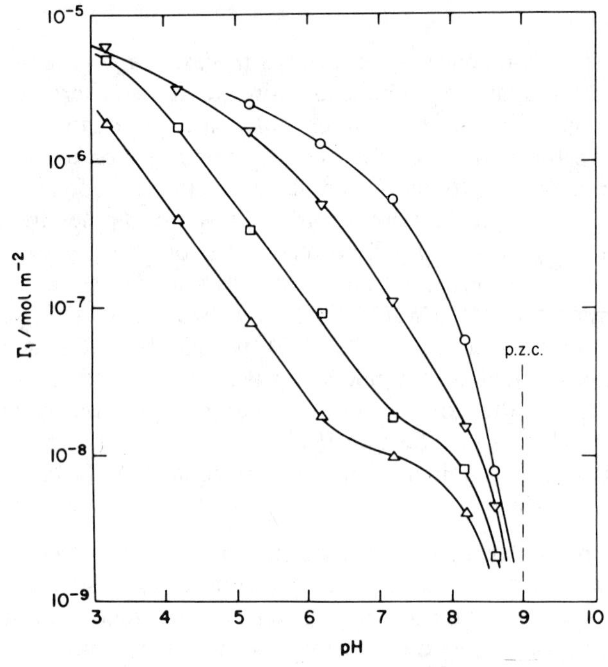

Figure 2 The amount of sodium dodecyl sulphonate adsorbed on alumina as a function of pH, at 2×10^{-3} mol dm^{-3} ionic strength. Surfactant concentrations, in mol dm^{-3}: △, 1×10^{-5}; □, 3×10^{-5}; ▽, 1×10^{-4}; ○, 2.5×10^{-4}. (Fuerstenau, 1971)

dodecyl sulphonate on alumina (p.z.c. = pH 9.1) as a function of pH (Fuerstenau, 1971). Adsorption, as expected from the above considerations, is seen to decrease as the surface charge becomes progressively less positive. Above the p.z.c. of the original surface the adsorption is immeasurably small. The effect is also illustrated for this system by electrokinetic measurements as a function of pH at different surfactant concentrations (Fuerstenau, 1971) shown in Fig. 3. At pH values above the p.z.c. the ζ-potential curves coincide, indicating zero adsorption.

Before progressing to ΔG_{spec} a final word should be said about ΔG_{elec}. The treatment outlined above ignores self-atmosphere or "discreteness of charge" effects (Levine *et al.*, 1967) and takes account only of the mean potential ψ_β in the adsorption plane. The latter term is frequently assumed to be known, and values have been calculated for ΔG_{spec} that automatically incorporate the self-atmosphere correction. This correction, being an electrical quantity, will be dependent on σ_β; indeed, variations in the calculated ΔG_{spec} with σ_β have been explained on this basis (Levine *et al.*, 1967). However, under conditions of very low surface coverage, i.e. low σ_β, in accordance with the strict application of equations (28) and (29), self-atmosphere effects can probably be ignored.

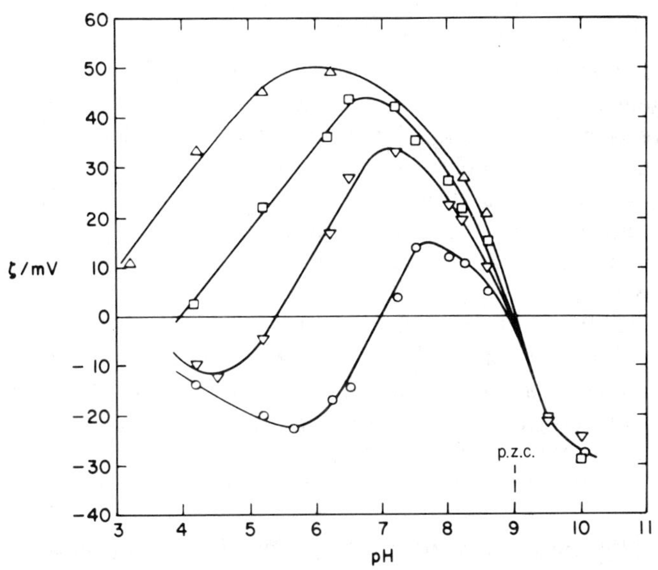

Figure 3 ζ-Potentials of alumina as a function of pH, at various concentrations of sodium dodecyl sulphonate, in mol dm^{-3}: △, 1×10^{-5}; □, 3×10^{-5}; ▽, 1×10^{-4}; ○, 2.5×10^{-4}. Ionic strength 2×10^{-3} mol dm^{-3}. (Fuerstenau, 1971)

3. *Specific Interactions*

(a) *Contributions to* ΔG_{spec}. The subdivision of ΔG_{spec} into supposedly separate independent interactions can be made in a number of slightly different ways according to whether one chooses to group together interactions of a particular type, e.g. van der Waals interactions, or interactions between particular species, e.g. hydrocarbon chain–surface interactions. Neither subdivision can be perfect and the choice is primarily one of convenience according to the system under study and the experimental variables employed (e.g. Fuerstenau and Healy, 1972; Healy, 1974; Somasundaran and Hanna, 1977). In this chapter we will adopt the following breakdown:

$$\Delta G_{\text{spec}} = \Delta G_{\text{cc}} + \Delta G_{\text{cs}} + \Delta G_{\text{hs}} + \ldots \tag{34}$$

The first term on the right-hand side of equation (34), ΔG_{cc}, is identified as the free energy change due to cohesive chain–chain interactions between the hydrophobic moieties of the adsorbed ions, whereas ΔG_{cs} and ΔG_{hs} account for chain–substrate and headgroup–substrate interactions respectively. In the reviews of Fuerstenau (1971) and Healy (1974) it may be noted that the former describes ΔG_{cs} in terms of hydrophobic bonding whereas the latter author attributes ΔG_{cs} to van der Waals dispersion interactions. Both authors explain ΔG_{cc} in terms of hydrophobic interaction, albeit indirectly by Fuerstenau. This rather mixed terminology, which also appears in other reviews, needs clarification.

The general consensus is that hydrophobic interactions refer to the general tendency shown by apolar portions of ions or molecules to aggregate in an aqueous environment, thus partially or completely removing them from contact with water molecules. The driving free energy is principally entropic, arising from the destruction of short-lived structures of water molecules organized around the apolar moieties (see e.g. Clark *et al.*, 1977). It seems reasonable to suppose therefore that both ΔG_{cc} and ΔG_{cs} will involve types of hydrophobic interaction, the latter being highly dependent on the nature of the solid surface and any associated structured water, and on whether or not this structure is disrupted by the hydrophobic chains. Dispersion interactions, as well as providing an implicit contribution to hydrophobic interactions (van Oss *et al.*, 1980), will always be operative whether or not hydrophobic interaction, as defined above, is deemed to occur.

Other physical components of ΔG_{cs} may also be envisaged depending upon the specific structures of adsorbent and adsorbate. For example, the adsorption of a surfactant possessing an alkylaryl hydrophobic residue may involve ion(or dipole)–induced dipole interactions as a result of π-electron polarization of the aromatic group on close approach to the surface (Snyder, 1968; see also the discussion below on headgroup effects). Surfactants with fluorocarbon chains are spectacularly more strongly adsorbed than their hydrocarbon analogues onto fluorocarbon polymers, although van der Waals interactions would predict the opposite. This may well reflect configurational compatibility of the fluorocarbon residue for the twisted PTFE chain (Rance, 1976; Ottewill, 1979).

(b) *Chain–chain interaction and hemimicellization.* One of the principal concepts that have been employed to explain the adsorption of surfactants onto strongly hydrated mineral surfaces, usually in the form of oxides, is that of "hemimicellization" initially postulated by Gaudin and Fuerstenau (1955) and developed and used by Fuerstenau (see e.g. Fuerstenau, 1970, 1971; Fuerstenau and Healy, 1972) and other authors (e.g. Somasundaran and Goddard, 1979). Briefly, for such surfaces the dominant contribution to ΔG_{spec} is attributed to ΔG_{cc}. This term, the basis of which is explained above, arises from the tendency of the hydrophobic moieties of the adsorbing surfactant to remove themselves from the aqueous environment by forming two-dimensional aggregates on the adsorbent surface. The expected similarities between these structures and bulk micellar units and the driving free energy of formation in each case explain the term hemimicelle. The experimental data on which this postulate depends are typified by the much quoted sodium dodecyl sulphonate/alumina system, in the form of the plots shown in Fig. 4 of the amount adsorbed and ζ-potential against surfactant concentration, obtained by Wakamatsu and Fuerstenau

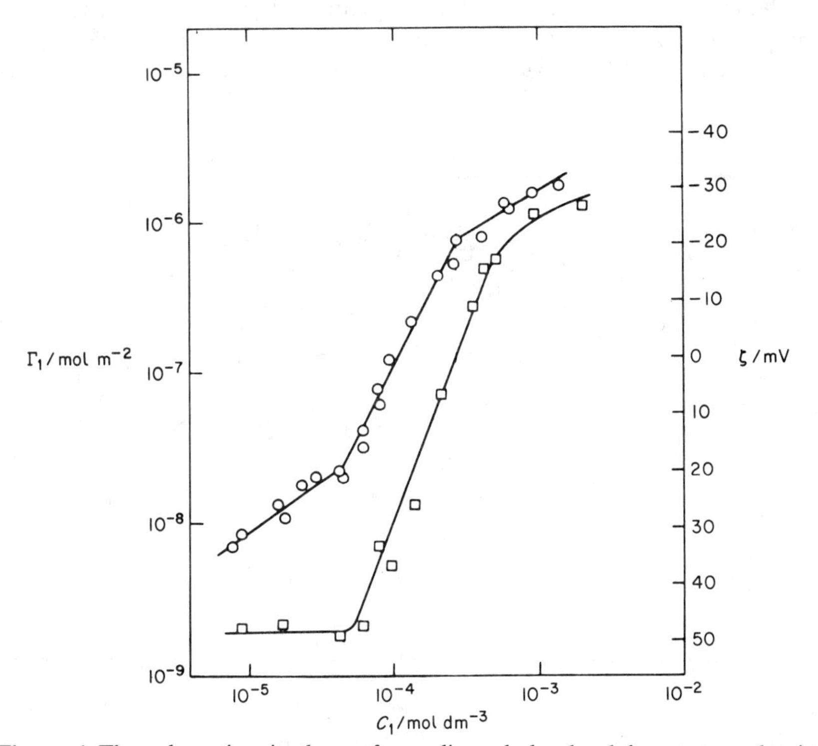

Figure 4 The adsorption isotherm for sodium dodecyl sulphonate on alumina (○), and corresponding ζ-potentials of the alumina particles as a function of the equilibrium surfactant concentration (□), at pH 7.2 and 2×10^{-3} mol dm^{-3} ionic strength. (Fuerstenau, 1971)

(1968). According to these authors the initial rise in adsorption, which occurs with no accompanying change in ζ-potential, indicates simple exchange of surfactant ions (acting as indifferent ions) with the indifferent counterions of the supporting electrolyte in the double layer. Then, at a particular concentration known as the hemimicelle concentration (h.m.c.), the adsorption increases dramatically as hemimicelles form on the adsorbent; a concomitant change in ζ-potential occurs and, although not shown in the diagram, sharp changes are observed in other interfacial properties such as contact angle (Wakamatsu and Fuerstenau, 1973) and sedimentation rate of the Al_2O_3 particles (Somasundaran and Goddard, 1979). At a particular surfactant concentration in the hemimicellization process the i.e.p. is exceeded and, as explained earlier, the adsorption process is thereafter hindered by the positive $z_1 e \psi_\beta$ term, so the slope of the adsorption isotherm is reduced. Further arguments in favour of the hemimicellization concept have been put forward by the various authors named above and may be illustrated using electrokinetic and flotation data for the homologous series of cationic alkylammonium acetates on negatively charged quartz, shown in Figs 5 and 6 (Somasundaran *et al.*, 1964; Fuerstenau *et al.*, 1964, respectively), and also by adsorption isotherms for

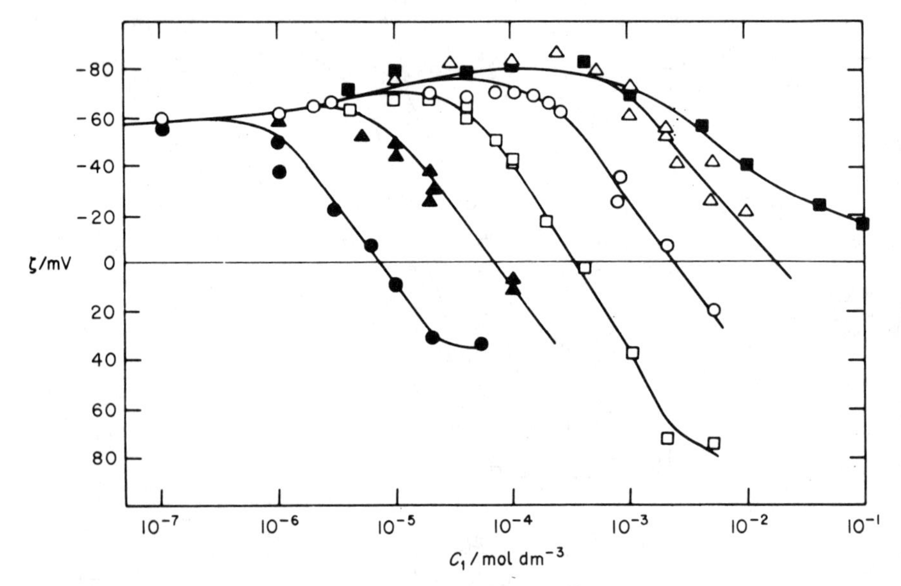

Figure 5 ζ-Potentials of quartz at neutral pH as a function of the concentration of alkylammonium acetates of varying hydrocarbon chain length: ●, C_{18}; ▲, C_{16}; □, C_{14}; ○, C_{12}; △, C_{10}; ■, ammonium acetate; from data of Somasundaran *et al.*, 1964. (Fuerstenau, 1971)

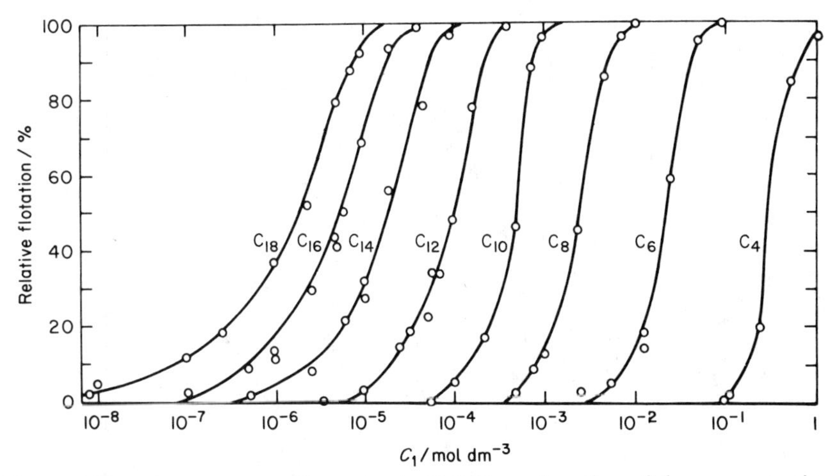

Figure 6 Flotation response of quartz at pH 6–7 as a function of the concentration of alkylammonium acetates of varying hydrocarbon chain length, as shown on curves. (Fuerstenau *et al.*, 1964)

sodium alkyl sulphonates on alumina as the continuing example (Wakamatsu and Fuerstenau, 1968) shown in Fig. 7. The sharp changes in adsorption, ζ-potential, and flotation depicted in these diagrams (and attributed to hemimicellization) are highly dependent on alkyl chain length as expected from the analogy with the bulk micellization process. Moreover, the absence of sharp changes in ζ-potential and adsorption for surfactants of chain length C_8 and less is, according to Fuerstenau (1970), further support for this analogy. This author does not refer to the corresponding flotation data which apparently contradict this inference in that the C_4, C_6, and C_8 surfactants do indeed cause flotation of the quartz.

If the hemimicellization concept is adopted, it is reasonable to suppose that the free energy change associated with the process can be given by $\Delta G_{cc} = n\phi$, where ϕ is the free energy involved in removing a CH_2 group from the aqueous environment and n is the number of CH_2 (and CH_3) groups in the alkyl chain. (Strictly, $\Delta G_{cc} = \text{const} + n\phi$ would be preferable; this does not affect the following argument.) Then, at the i.e.p., when $\Delta G_{elec} = 0$ and $\Delta G_{ads} = \Delta G_{spec} = \Delta G_{cc}$, it follows from equation (29) that

$$\ln(2rC_1^0/\Gamma_1^0) = n\phi/kT \tag{35}$$

where superscript 0 refers to conditions at the i.e.p. Alternatively, at the h.m.c.,

$$\ln(2rC_1^h/\Gamma_1^h) = n\phi/kT + z_1 e\psi_\beta^h/kT \tag{36}$$

where superscript h refers to conditions at the h.m.c. Equation (35) has

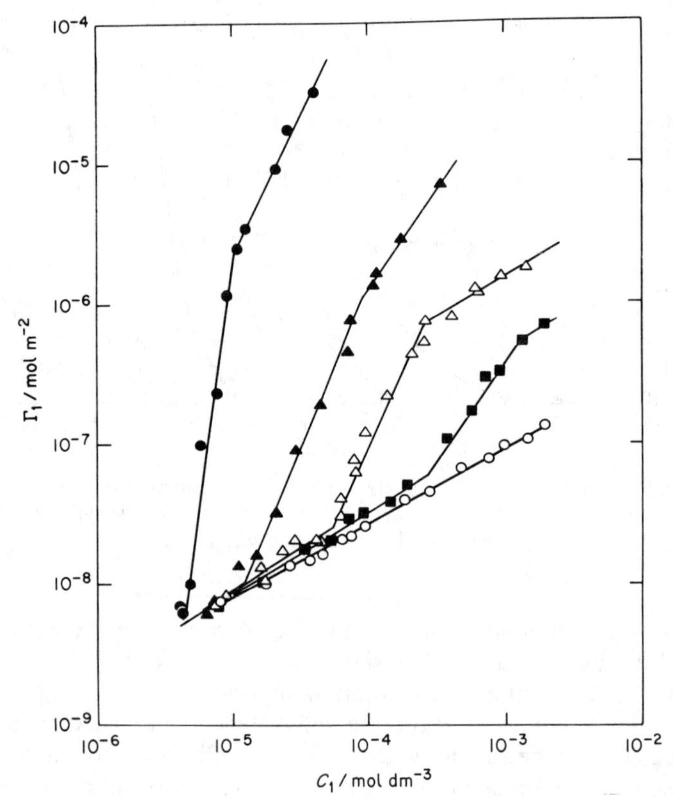

Figure 7 Adsorption isotherms for sodium alkyl sulphonates of varying hydrocarbon chain length on alumina, at pH 7.2 and 2×10^{-3} mol dm^{-3} ionic strength: ●, C_{16}; ▲, C_{14}; △, C_{12}; ■, C_{10}; ○, C_8. (Wakamatsu and Fuerstenau, 1968)

been used by Somasundaran *et al.* (1964) to obtain values of ϕ, on the reasonable assumption that r and Γ_1^0 are constant for a particular homologous series of a surfactant. Lin and Somasundaran (1971) have estimated values of ϕ from (36), taking r and ψ_β^h as constant for a surfactant series; the approximate constancy of ψ_β^h was inferred from such data as those shown in Fig. 5. On this basis plots of C_1^0 or C_1^h against n should display slopes of ϕ/kT, where C_1^0 is obtained from the electrokinetic data at $\zeta = 0$ and C_1^h can be obtained either directly from the adsorption isotherm or from the electrokinetic data or flotation data, by extrapolating the sharply descending electrokinetic curves back to the curve for the indifferent ammonium ion and the sharply ascending flotation curves back to zero flotation. These plots for the data of Figs 5–7 are shown in Fig. 8. Good straight lines are obtained, the similar slopes of which give values of ϕ of about 1 kT, which is in the same order of magnitude as for the micellization of straight alkyl chain surfactants (see Lin and Somasundaran, 1971) and is in accordance with Traube's rule (Traube, 1891; see Adamson, 1976).

Thus the analogy between micellization and hemimicellization appears to be further corroborated, at least for the mineral systems chosen by the

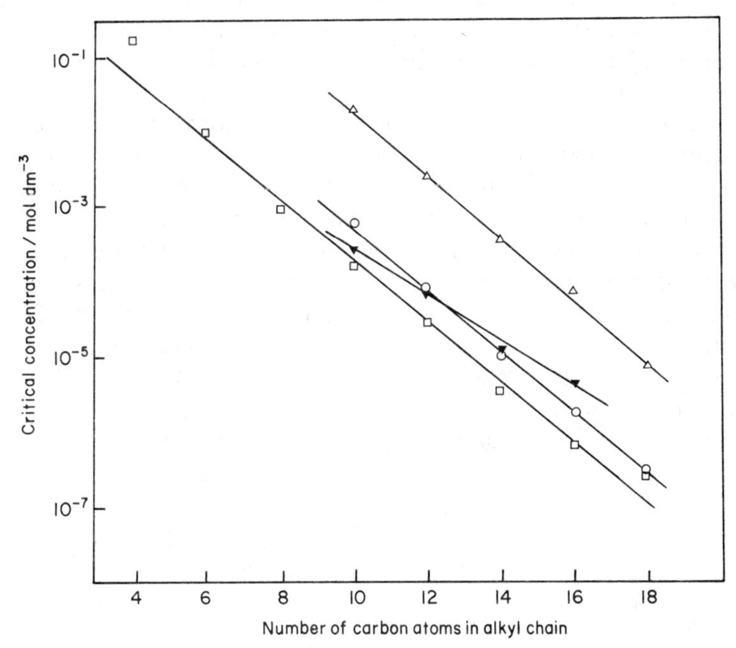

Figure 8 Effect of hydrocarbon chain length on critical concentrations of alkylammonium ions on quartz (open symbols) and for alkyl sulphonate ions on alumina (filled symbols). ○, C_1^h (electrokinetics), from data of Somasundaran *et al.*, 1964 (Lin and Somasundaran, 1971). △, C_1^0 (electrokinetics) (Fuerstenau, 1971). □, C_1^h (flotation), from data of Fuerstenau *et al.*, 1964 (Fuerstenau and Healy, 1972). ▼, C_1^h (adsorption), from data of Wakamatsu and Fuerstenau, 1968 (Lin and Somasundaran, 1971)

aforementioned authors. Nevertheless it seems to us that the details of the hemimicellization model should be regarded with some caution. Whilst it is highly reasonable to attribute, for example, the increased slope of adsorption isotherms (e.g. Fig. 7) to lateral interactions between adsorbed ions, the details of the mechanism and the appropriate form for the adsorption isotherm require further consideration. Thus, for example, in equations (27) and (62) lateral interactions are deemed to give to the effective adsorption energy a contribution that is proportional to coverage;

in the hemimicellization literature cited above, ΔG_{cc} is implicitly treated as coverage-independent above the h.m.c. Furthermore, it is not immediately obvious that the highly localized adsorption implicit in the formation of hemimicelles should demonstrate as marked a response to the mean potential as may be observed at the i.e.p. in Fig. 4. Finally, the chain length dependence of ΔG_{cc} referred to above was in fact derived from conditions (e.g. constant adsorption density) where any plausible model for the adsorption process would be expected to follow Traube's rule. We therefore judge that the appropriate form for the chain–chain term remains an open question.

(c) *Chain–solid interactions.* In contrast to their apparent insignificance on oxide surfaces, chain–solid interactions would be expected to occur for less polar surfaces, i.e. those that do not generally cause significant structuring of interfacial water (e.g. by hydrogen bonding or dipole orientation), other than perhaps in the first layer of water molecules. Typically, paraffinic surfaces of many polymers and waxes, carbonaceous surfaces, and other low energy "hydrophobic" surfaces will fall into this category. For such systems ΔG_{cs}, in contrast to ΔG_{cc}, would be expected to govern the first stages of adsorption, so the initial slopes of the isotherms will be much higher than those for hydrated mineral surfaces. An example is given in Fig. 9 for the adsorption of alkyl quaternary ammonium ions on negatively

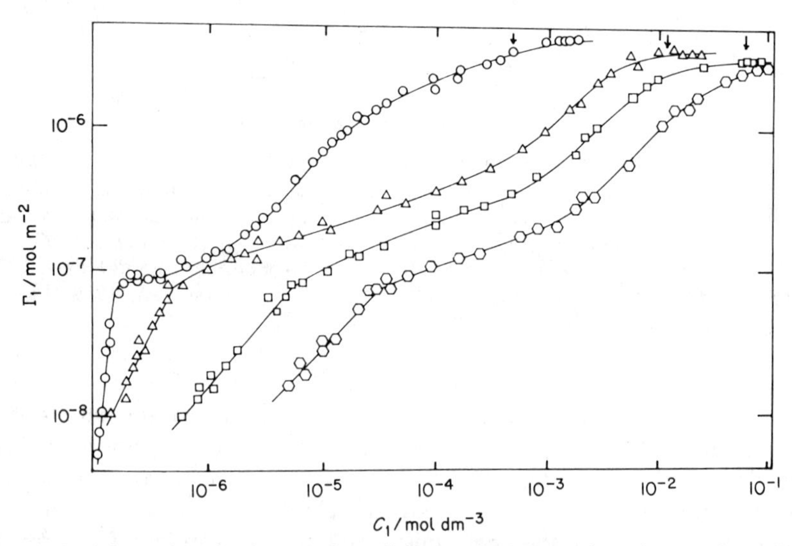

Figure 9 Adsorption isotherms for alkyltrimethylammonium ions, of varying hydrocarbon chain length, on polystyrene latex particles at pH 8 in 10^{-3} mol dm^{-3} KBr solution. ○, C_{16}; △, C_{12}; □, C_{10}; ○, C_8; arrows indicate c.m.c. values. (Connor and Ottewill, 1971)

charged polystyrene latex particles (Connor and Ottewill, 1971). For each of these isotherms the inflection at the lower surfactant concentration can be accounted for by the reversal of sign of ΔG_{elec} as the i.e.p. is traversed. The inflection at the higher surfactant concentration may reflect the orientation of the alkyl chains from a horizontal orientation of saturation coverage into a vertically oriented monolayer as the surfactant concentration is increased. In this way solid–water and surfactant chain–water interactions would be minimized at high coverage. One might expect, for surfaces of hydrocarbon character, that ΔG_{cs} per CH_2 group will be of the same order of magnitude as ΔG_{cc}, that is about $1\,kT$. The electrophoresis data of Rendall *et al.* (1979), for a series of anionic surfactants on positively charged nylon sols, which are interpretable in terms of chain–solid interactions, bear out this prediction (see Table 1 and accompanying text).

(d) *Headgroup effects.* The term ΔG_{hs} is introduced to cover all contributions not explicitly accounted for as chain–chain, chain–solid, and electrostatic interactions. This might, for example, include chemical interactions (e.g. hydrogen bonding, covalent bonding) and accompanying or separate solvation terms. These factors are obviously closely analogous to those involved in the specific adsorption of inorganic ions. There is, however, little obvious value in the practice of setting these factors down as separate contributions to ΔG_{spec} (see Fuerstenau and Healy, 1972; Healy, 1974; Somasundaran and Hanna, 1977).

A certain amount of confusion may arise in discussion of the effect of headgroup size. For example, the apparently strong binding of alkylaryl sulphonates to proteins has been attributed to charge delocalization over the benzene sulphonate headgroup (Steinhardt and Reynolds, 1969). However, it is not necessarily appropriate to consider the benzene ring as part of the headgroup. For micellization (Shinoda *et al.*, 1963) and for adsorption both at hydrophobic interfaces (Rendall *et al.*, 1979) and at hydrophilic surfaces (Dick *et al.*, 1971) the benzene ring gives a contribution equivalent to 3–4 CH_2 groups in the alkyl chain. The same problem arises in assessing the role of headgroup size in the adsorption of cationic headgroups such as ammonium, trimethylammonium, pyridinium, and quinolinium attached to the same long hydrocarbon chain. The adsorption of ions of this type onto negatively charged AgI sols has been investigated by electrophoresis measurement by Ottewill and Rastogi (1960), and attempts have been made to relate headgroup size to biological activity in such a series of drugs (Barlow *et al.*, 1971). Nevertheless the reliable definition of headgroup, let alone headgroup size, becomes increasingly difficult in progressing through the series.

A more clear-cut example of the importance of ΔG_{hs} occurs in the case of dominating chemical interaction. This may be of great importance, for example, in achieving specificity in flotation. Examples include xanthate adsorption on sulphides, and oleate on haematite, calcite, apatite, and fluorite (see e.g. Somasundaran, 1975; Fuerstenau, 1971).

D. ADSORPTION AT LOW SURFACE COVERAGES

It was pointed out in Section II.B that one of the restrictions placed on the use of adsorption isotherms, such as the Stern–Langmuir equation (28), is that the fractional coverage by the adsorbate must be low in order to avoid lateral interactions and discreteness of charge effects. Under these conditions we can rewrite equation (28) using the substitutions given in equations (30)–(32) and ignoring ΔG_{dip}, viz.

$$\ln\left(\frac{\theta}{1-\theta}\right) = \ln\left(\frac{C_1}{55.5}\right) - \frac{z_1 e\psi_\beta}{kT} - \frac{\Delta G_{\mathrm{spec}}}{kT} \tag{37}$$

On the basis of the electrical double layer model depicted in Fig. 1, Rendall *et al* (1979) have derived expressions for conditions of low surface coverage relating the Stern potential ψ_{d} to the concentration of surfactant (given here as $\mathrm{pS} = -\log_{10} C_1$) by combination of equations (4)–(9) with equation (37). Using the notation developed above, and making the reasonable assumption mentioned in Section II.B that the IHP and OHP are coincident such that $\psi_\beta = \psi_{\mathrm{d}}$ (the identity of σ_β being retained) and K_2 tends to infinity, their expressions simplify to

$$\frac{\mathrm{d(pS)}}{\mathrm{d}\psi_{\mathrm{d}}} = \frac{1}{N} - \frac{K + C_{\mathrm{d}}}{2.303\sigma_\beta(1-\theta)} \tag{38}$$

at constant ψ_0, and

$$\frac{\mathrm{d(pS)}}{\mathrm{d}\psi_{\mathrm{d}}} = \frac{1}{N} - \frac{C_{\mathrm{d}}}{2.303\sigma_\beta(1-\theta)} \tag{39}$$

at constant σ_0. In an analogous fashion to the definitions of s and s_1 in equations (12) and (15), the following quantities are defined:

$$s_2 = \left(\frac{\mathrm{d}\zeta}{\mathrm{d(pS)}}\right)_{\zeta\to 0} \tag{40}$$

and

$$s_3 = \left(\frac{\mathrm{d}\psi_{\mathrm{d}}}{\mathrm{d(pS)}}\right)_{\psi_{\mathrm{d}}\to 0} \tag{41}$$

so that at low potentials, from equation (14),

$$s_2 = s_3 \exp(-\kappa\Delta_2) \tag{42}$$

where Δ_2 is the distance between the OHP and the plane of shear in the presence of the adsorbed layer. Then, at the i.e.p., where $\sigma_{\mathrm{d}}^0 = 0$ so that

$\sigma_\beta^0 = -\sigma_0^0 = -K\psi_0^0$, equations (38) and (39) become

$$Ns_3^{-1} = 1 - \frac{kT}{z_1 e}\left(\frac{K + C_d^0}{K}\right)\frac{1}{\psi_0(1-\theta^0)} \qquad (43)$$

at constant ψ_0 ($=\psi_0^0$), and

$$Ns_3^{-1} = 1 - \frac{kT}{z_1 e}\frac{C_d^0}{\sigma_0(1-\theta^0)} \qquad (44)$$

at constant σ_0 ($=\sigma_0^0$). For the case of an interface of low initial Stern potential ψ_d^i, and for conditions of low adsorption at the i.e.p. such that $(1-\theta^0) \approx 1$, by substitution of the approximate identity of $\sigma_0 \approx -\sigma_d \approx \varepsilon\kappa\psi_d^i = C_d^0\psi_d^i$, equations (43) and (44) simplify to a common approximation,

$$Ns_2^{-1} \approx \left(1 - \frac{kT\exp(-\kappa\Delta_1)}{z_1 e\zeta^i}\right)\exp(\kappa\Delta_2) \qquad (45)$$

where s_3 has been substituted by s_2 and ζ^i is the initial ζ-potential of the system before adsorption of surfactant.

The important conclusion from equation (45) is that a plot of Ns_2^{-1} against $kT/z_1e\zeta^i$ should have unit intercept and slope, the latter being independent of the type of surface, of the total ionic strength, and of the surfactant ions adsorbed, provided that (i) $\Delta_1 \approx \Delta_2$, which may be a reasonable approximation at very low surface coverages when the plane of shear is likely to be displaced negligibly by the isolated hydrocarbon chains, and (ii) the electrical double layer model and the theory developed therefrom are not grossly in error. This approximate expression therefore provides a useful test of the self-consistency of the treatment, similar in nature to the congruence tests discussed elsewhere in this chapter. A further feature of the above analysis is that, at constant ionic strength and potential-determining ion concentration, the slope s_2 of the ζ-potential against concentration curves should be independent of the specific adsorption potential (i.e. of ΔG_{spec}). This conclusion was made earlier by Ottewill *et al.* (1960) on the basis of a simplified adsorption model in which changes in the electrostatic term $z_1e\psi_\beta$ were ignored in calculating the slopes. However, these authors reported slopes exceeding the Nernst gradient N, the maximum attainable in any circumstances with the Stern–Langmuir theory. This observation might reflect the importance of favourable lateral interactions, especially when later results are taken into account (Osseo-Asare and Fuerstenau, 1973; Osseo-Asare *et al.*, 1975). This point is raised again in Section IV.D.

Equation (37) shows that (relative) values of ΔG_{spec} for a series of surfactants at a particular interface may be obtained from C_1^0, the concentration of each surfactant required to attain i.e.p. conditions. Rendall *et al*. (1979) conducted electrophoresis measurements on sols of a polyester, nylon 6,6, and AgI, in the presence of various anionic and cationic surfactants. Plots of ζ-potential against pS are shown in Fig. 10 for a positively charged nylon sol at pH 2 and at constant ionic strength in the presence of a number of anionic surfactants. Values of ΔG_{spec} were obtained by substitution of the calculated θ^0 for each surfactant into equation (37),

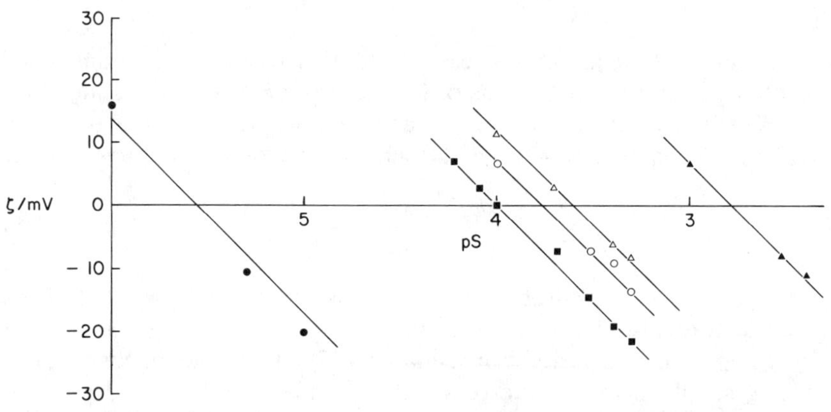

Figure 10 ζ-Potential against pS for nylon sol at pH 2 in the presence of the sodium salts of: ■, dodecyl sulphate; ○, octylbenzene sulphonate; ●, hexadecyl sulphonate; △, dodecyl sulphonate; ▲, decyl sulphonate. (Rendall *et al*., 1979)

taking $\psi_\beta \equiv \psi_d = 0$. Values of θ^0 were determined on the basis that $\sigma_\beta^0 = -\sigma_0^0$ at the i.e.p. A value of $\sigma_0^0 = 0.62$ μC cm^{-2} was used, as calculated from earlier electrophoresis results (Rendall and Smith, 1978) assuming a constant σ_0 system. This value of σ_β^0 corresponds to about 3.9×10^{12} adsorbed surfactant ions per cm^2 (~26 nm^2 per ion). Rendall *et al*. (1979) assumed the total number of adsorption sites to be 2×10^{14} cm^{-2} (i.e. one site per 0.5 nm^2), so $\theta^0 = \Gamma_1/N_s = -\sigma_\beta/z_1 e N_s = 0.0194$. Values of ΔG_{spec} obtained in this way from the data of Fig. 10 are shown in Table 1. It is interesting

Table 1 Values of ΔG_{spec} for anionic surfactants on nylon sol.

Surfactant	$-\Delta G_{spec}/kT$	c.m.c./mol dm^{-3}
C_{10} sulphonate	6.4	4.4×10^{-2}
C_{12} sulphonate	8.4	9.8×10^{-3}
C_{16} sulphonate	12.9	8.0×10^{-4}(50°C)
C_{12} sulphate	9.3	8.1×10^{-3}
C_8 benzene sulphonate	8.8	1.1×10^{-2}

ΔG_{spec} values from Rendall *et al*. (1979); c.m.c. values from Mukerjee and Mysels (1971).

to note the incremental change of about 1 kT per CH_2 group in these values for the homologous series of alkyl sulphonates. However, in contrast to the similar values discussed in Section II.C that were ascribed to ΔG_{cc}, these values probably reflect chain–solid interactions, ΔG_{cs}, at such low fractional coverages. Incremental changes of about 1 kT would be expected in that the nylon surface is, with the exception of its charged sites, principally composed of CH_2 groups which would undergo hydrophobic interactions with the alkyl chains in their expected horizontal mode of adsorption.

The value chosen for N_s (Rendall *et al.*, 1979) was clearly somewhat arbitrary. However, provided that N_s does not vary dramatically with chain length at the low coverages (low θ^0) involved, equation (37) at the i.e.p. reduces, at constant κ and potential-determining ion concentration, to

$$\Delta G_{spec}/kT = \text{const.} - 2.303\,\text{pS}^0 \qquad (46)$$

As was pointed out above, the Stern–Langmuir treatment should be used to obtain comparative, rather than absolute, information about ΔG_{spec}. Within the limits of its derivation, equation (46) demonstrates that the required information may be obtained from a single i.e.p. determination for each surfactant under chosen conditions of κ and potential-determining ion concentration. According to Rendall *et al.* (1979) such determinations therefore provide a sensitive technique for the identification of small differences in affinity of surfactants for a solid/liquid interface. The effect of structural differences in surfactants (other than simple chain length variation) can also be seen by this method. For example, the ΔG_{spec} values for dodecyl sulphate and dodecyl sulphonate ions in Table 1 indicate the stronger binding of sulphate over sulphonate headgroups on to nylon. Also, the presence of the benzene ring in the octylbenzene sulphonate ion appears to influence adsorption in much the same way as it affects the c.m.c. Moreover, the octylbenzene moiety resembles the dodecyl chain in each respect. Effects of aromatic groups in the hydrophobic chain as observed by numerous authors have been catalogued by Rosen (1978).

Rendall *et al.* (1979) have also tested the predicted variation of s_2 under different conditions of ionic strength and potential-determining ion concentrations for the surfactant/sol combinations of sodium dodecyl sulphate/nylon (varying ionic strength), dodecyltrimethylammonium bromide/polyester (varying ionic strength and pH), and sodium dodecyl sulphate/AgI (varying ionic strength and pAg), the latter system being expected to approximate to the "constant ψ_0" condition. The fits were acceptable in most cases. Using these data, a plot of Ns_2^{-1} against $kT/z_1e\zeta^i$ was constructed as shown in Fig. 11. The experimental points from the three quite different surfaces over a range of surfactant concentrations, ionic strength, and potential-determining ion concentrations are seen to lie close to the line predicted by equation (45), taking $\Delta_1 = \Delta_2$. The agreement between the results is not sensitive to any detailed model of the Stern region, and lends good support for the generality of the model and the theory developed by these authors.

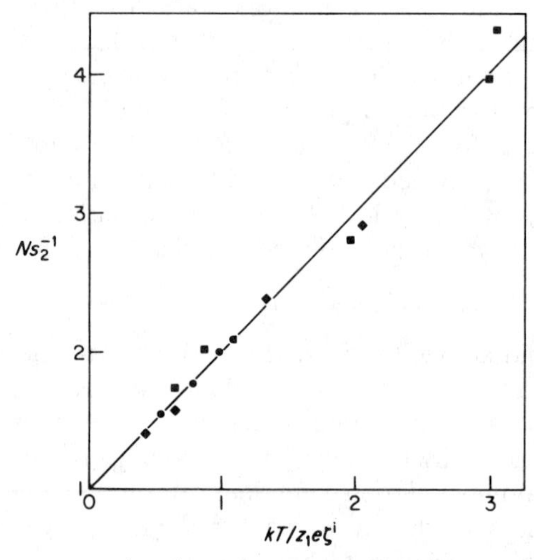

Figure 11 Ns_2^{-1} against $kT/z_1e\zeta^i$ for sodium dodecyl sulphate on nylon (●) and on AgI (◆), and for dodecyltrimethylammonium bromide on polyester (■), at various ionic strengths and pH (or pAg). (Rendall *et al.*, 1979)

E. ADSORPTION AT HIGH COVERAGE

Surfactant adsorption tends to reach a plateau value around the c.m.c. of the surfactant. Further increases in solution concentration will predominantly increase the concentration of micelles, which are not generally surface active. A number of workers have reported adsorption maxima in the region of the c.m.c. (Fava and Eyring, 1956; Mukerjee and Anavil, 1975; Saleeb and Kitchener, 1965; Day *et al.*, 1967; Sexsmith and White, 1959). Many of these observations have been identified as experimental artefacts, as exemplified by the following cases.

Heterogeneity of the substrate can lead to experimental adsorption maxima. Natural cotton and cellulosic materials, for example, contain waxes which provide adsorption sites below the c.m.c. of the surfactant. However, these materials may well be solubilized above the c.m.c., giving rise to an apparent reduction in surfactant adsorption. Careful washing of the substrate with a suitable solvent prior to the adsorption studies will frequently remove the maximum for a variety of adsorbents (Ginn *et al.*, 1961; Day *et al.*, 1967; see also Fig. 16).

A further possible cause of apparent adsorption maxima is the failure to achieve separation from the solution, prior to analysis, of a small amount of "fines"—of disproportionately high surface area. The surfactant adsorbed by these particles will be analysed as though present in bulk solution. Such stabilization is of course more likely to be a problem above the c.m.c. (Day *et al.*, 1967).

Adsorption maxima may well be found in systems containing mixed chain lengths or mixed surfactant type. The micellar composition of such systems will almost invariably differ from the bulk composition of the surfactant mixture. Above the c.m.c., therefore, the free surfactant, in equilibrium with the adsorbed layer, will differ in composition from that in the sub-micellar region, and will vary with total surfactant present. (Even below the c.m.c. the amount adsorbed in equilibrium with a given bulk concentration may vary with surface/solution ratio through fractionation of the different components of the mixture.) These considerations are of great importance, for example, in studies of typical commercial surfactants, which normally contain a range of isomers.

Adsorption maxima for pure adsorbates on homogeneous surfaces are not necessarily prohibited on thermodynamic grounds (Hall, 1980a). However, there is no case known to us in which an adsorption maximum has been unequivocally demonstrated for a pure ionic surfactant at a smooth homogeneous interface.

The adsorption achieved at the "plateau" is frequently assumed to correspond to monolayer coverage. The factors that promote surfactant adsorption at the solid/liquid interface, whether chain–chain or chain–hydrophobic surface interaction, are obviously closely akin to those that promote micelle formation. It therefore seems highly likely that the bulk concentration at which a complete monolayer is formed would be very similar to the c.m.c. Indeed, a complete close-packed monolayer of surfactant molecules in a vertical orientation on a hydrocarbon type surface might be considered as a low-curvature analogue of a micellar surface. Complications, however, arise in detailed quantitative interpretation of plateau adsorption results. The available surface area of the substrate is not in general well known, leading to considerable uncertainty as to the true area per molecule at "saturation". Furthermore, the asymmetry of surfactant molecules allows for a wide range of possible molecular orientations at the interface, including the possibility that bilayers rather than monolayers might be formed. Many authors quote an area per molecule at the plateau, and state without further justification that this corresponds to a close-packed monolayer.

In the case of high energy (highly polar) interfaces it is not obvious that there should be a strong driving force for the adsorption of hydrocarbon chains. Thus it has been reported that little adsorption of ionic surfactants onto substrates such as alumina and silica will take place in the absence of favourable charge interactions (see Sections II.C and IV.C), and that a minimum negative charge density on clay minerals is necessary before a complete adsorbed layer (for surface area determination) can be guaranteed (Kivel *et al.*, 1963; Greenland and Quirk, 1963). In these circumstances it is often postulated that a first layer is adsorbed with the charged head-groups towards the interface, and that a second layer of surfactant may then arise through interaction with the exposed hydrocarbon chain. Support for this model is advanced on the basis of the calculated areas occupied per molecule at complete coverage, from the existence of two distinct plateau regions in some adsorption isotherms, and from the assumption that the postulated hydrocarbon-rich surface of the adsorbed layer at intermediate concentrations can explain the "hydrophobic" nature of the surface, with regard to wetting, in this region. Recently, direct force–distance measurements on a mica substrate in the presence of cationic surfactant have been reported (Pashley and Israelachvili, 1981). These studies indicate strong interactions at distances consistent with the existence of an adsorbed bilayer in the micellar region. Perhaps more surprisingly, bilayers of cationic and nonionic surfactants on polycarbonate surfaces have been inferred from capillary flow resistance measurements (Bisio *et al.*, 1980).

Interfaces of intermediate polarity might be expected to show behaviour intermediate between those of graphitized carbon (essentially monolayer coverage at the c.m.c.) and clay or oxide systems (adsorption requiring strong favourable electrostatic interaction). The effect of interfacial polarity on adsorption has been studied in relation to the soap titration technique of Maron *et al.* (1954) and to the stabilizing effect of surfactants in emulsion polymerization. There appears to be a direct correlation between the area per adsorbed molecule at the c.m.c. and the interfacial tension of the "bare" interface (Vijayendran, 1979; Piirma and Chen, 1980). This observation seems highly reasonable. The interactions leading to close-packed monolayer adsorption on low energy (i.e. non-polar) surfaces, and to micelle formation, are essentially similar. Depending on the strength of the monolayer–substrate interaction, the surfactant concentration that would be in equilibrium with a close-packed adsorbed layer might be expected to lie below, at, or above the c.m.c. of the surfactant. In the last case, a close-packed monolayer might not be achieved, because micelle formation would take place first. It should be noted, however, that the adsorption plateau is often quoted as occurring slightly above the c.m.c., possibly indicating an increase in monomer activity above the c.m.c. (Saleeb and Kitchener, 1965). This effect would not be detected by the surface tension measurements commonly employed in the soap titration technique. In addition, it is important to remember that the state of charge of the interface, as well as interfacial polarity, may influence the adsorption. Thus, on a polycarbonate membrane with a net negative charge, adsorbed bilayers of cationic and nonionic surfactants above the c.m.c. were inferred from the increased flow resistance of the membrane. However, an anionic surfactant, although adsorbed to some extent, did not affect the flow, indicating a much lower adsorbed layer density (Bisio *et al.*, 1980). In this case electrostatic interaction, rather than interfacial polarity, appears to be the determining factor.

For a given surfactant–substrate combination the area per molecule at the plateau varies with the nature and amount of the supporting electrolyte present (e.g. Connor and Ottewill, 1971; Saleeb and Kitchener, 1965). These results may be rationalized in terms of a reduction in headgroup–headgroup repulsion as the total ionic strength is increased, and when specific "counterion binding" of the ion is increased by a change of ion type. This parallels the effect of electrolytes on adsorption at the solution/air interface, and the factors discussed by Tanford (1972, 1974) in relation to micelle shape.

Surfactant adsorption isotherms close to the c.m.c. often have an apparently Langmuirian shape (see e.g. Saleeb and Kitchener, 1965; Arai and Yoshizaki, 1971), but this masks an extremely complex series of underlying factors. In this region, for example, surfactant ions may rearrange from a horizontal to a vertical orientation, and electrostatic interactions may be heavily modified by varying degrees of "counterion binding". These complexities may be seen from the form of the adsorption curve when plotted on a log(adsorption) against log(concentration) basis, as exemplified by the results of Connor and Ottewill (1971) in Fig. 9.

F. AN ALTERNATIVE THERMODYNAMIC APPROACH TO ADSORPTION

The theories outlined in the preceding sections, describing effects of surfactant ion adsorption on electrical double layer properties such as electrokinetic potentials, rely to varying extents on a detailed model of the inner layer, this usually being the Grahame model. According to Hall (1980b) the application of such models to interpret electrokinetic data is not only cumbersome but also involves certain assumptions whose theoretical status is not secure. This author has sought to develop a rigorous thermodynamic approach to the problem which does not require the use of such a detailed model (Hall, 1980b; Hall *et al.*, 1980). Concepts such as surface potential and inner-layer capacitance, and the need to subdivide the Stern region by the inner Helmholtz plane, do not appear in this treatment. However, the notion of a Stern potential and its identity with ζ-potential is preserved and the diffuse double layer is assumed to be described by the Poisson–Boltzmann equation.

The basis of Hall's theory is a theorem that was previously developed (Hall, 1978) to describe the electrochemical contribution to the interaction between charged plates, a limiting case of which refers to an isolated interface. In brief, one considers a solid surface in contact with an electrolyte solution, there being a dividing surface beyond which, on the solution side, the ion distribution conforms to the Poisson–Boltzmann equation. Then, at equilibrium and at constant T and p, an exact differential quantity dL is given by

$$\mathrm{d}L = \sum_i \Gamma_i \,\mathrm{d}\mu_i \tag{47}$$

where, for strongly adsorbed ions of chemical potentials μ_i, the Γ_i quantities are essentially the surface excesses which appear in the Gibbs adsorption isotherm. The dividing surface can be conveniently placed at the origin of the diffuse layer, the potential there being ψ_d. The summation extends over all independent solution species on both sides of the dividing surface, the μ_i at the dividing surface being given by

$$\mu_i = \mu_i^\mathrm{b} - z_i e \psi_\mathrm{d} \tag{48}$$

where μ_i^b is the chemical potential of species i in bulk solution, and is assumed to be given by

$$\mu_i^\mathrm{b} = \mu_i^0(T, p) + kT \ln n_i^\mathrm{b} \tag{49}$$

where n_i^b is the bulk number density of that species. Wherever possible the choice of independent species is made such that

$$\sum_i \Gamma_i z_i e = -\sigma_\mathrm{d} \tag{50}$$

Thus, in terms of the Grahame model, the summation in equation (50) is the total charge density arising from both the surface charge and any specifically adsorbed ions. The variable μ_i is more suitable than μ_i^b (i.e. in effect, pX) as a basis for congruence tests.

Two of the situations considered by Hall that are relevant to this chapter refer to the adsorption of surfactant ions on both uncharged and charged surfaces. The treatments are based on virial expansions of the surface pressure Π in terms of amounts adsorbed. Thus, on an uncharged surface under conditions of constant ionic strength maintained by an indifferent electrolyte, the Gibbs adsorption equation takes the form

$$\mathrm{d}\Pi = \Gamma_1 kT \,\mathrm{d} \ln n_1^\mathrm{b} \tag{51}$$

For small Γ_1 it is supposed that Π can be expressed as a virial expansion in integral powers of Γ_1 so that

$$\Pi = kT\Gamma_1 + B_{11}\Gamma_1^2 + \ldots \tag{52}$$

The second term on the right of equation (52) allows for interaction between adsorbed surfactant ions. Since such interactions at low coverages can be expected to be largely electrostatic, B_{11} is likely to depend primarily on κ. Combining equations (51) and (52) gives

$$kT\,\mathrm{d}\ln n_1^{\mathrm{b}} = kT\,\mathrm{d}\ln\Gamma_1 + 2B_{11}\,\mathrm{d}\Gamma_1 \tag{53}$$

from which one obtains

$$kT\left(\frac{\mathrm{d}^2\ln n_1^{\mathrm{b}}}{\mathrm{d}\Gamma_1\,\mathrm{d}\kappa}\right)_{\Gamma_1} = 2\frac{\mathrm{d}B_{11}}{\mathrm{d}\kappa} \tag{54}$$

Using equation (9) for σ_{d} in the approximate form for small potentials, such that $\psi_{\mathrm{d}} = -\sigma_{\mathrm{d}}/\varepsilon\kappa$, together with equation (50) we obtain the identity

$$\Gamma_1 = \varepsilon\kappa\psi_{\mathrm{d}}/z_1 e \tag{55}$$

Then, from the theorem expressed through equations (47)–(49), at constant T, p, and Γ_1, and for small Γ_1 (such that ψ_{d} is small), it can be shown that

$$kT\ln n_1^{\mathrm{b}} - z_1^2 e^2\Gamma_1/\varepsilon\kappa = \text{constant} \tag{56}$$

from which, on differentiation, one obtains

$$kT\left(\frac{\mathrm{d}^2\ln n_1^{\mathrm{b}}}{\mathrm{d}\Gamma_1\,\mathrm{d}\kappa}\right)_{\Gamma_1} = -\frac{z_1^2 e^2}{\varepsilon}\left(\frac{\mathrm{d}(1/\kappa)}{\mathrm{d}\kappa}\right) \tag{57}$$

Combining equation (57) with equation (54) and integrating gives

$$B_{11} = z_1^2 e^2/2\varepsilon\kappa + B_{11}^0(T,p) \tag{58}$$

where $B_{11}^0(T,p)$ is the integration constant. Now, integrating equation (53) at constant T, p, and κ gives

$$kT\ln n_1^{\mathrm{b}} = kT\ln\Gamma_1 + 2B_{11}\Gamma_1 + g(T,p,\kappa) \tag{59}$$

where $g(T,p,\kappa)$ is the integration constant. However, since $(kT\ln n_1^{\mathrm{b}} - z_1 e\psi_{\mathrm{d}})$ depends only on Γ_1, it follows that g cannot depend on κ, so from equations (58) and (59),

$$kT\ln n_1^{\mathrm{b}} - z_1 e\psi_{\mathrm{d}} = g(T,p) + kT\ln\Gamma_1 + 2B_{11}^0\Gamma_1 \tag{60}$$

On a charged surface under conditions similar to those above, except that a potential-determining ion species 2 is present, it is reasonable to suppose that Π can be written as

$$\Pi = \Pi_0 + \Pi_1 + \Pi_2 + B_{12}\Gamma_1\Gamma_2 \tag{61}$$

where Π_0 is the surface pressure at the p.z.c. in the absence of surfactant, Π_1 is the surface pressure of surfactant ions when $\Gamma_2 = 0$ which can be expanded as shown in equation (52), and Π_2 is the surface pressure of potential-determining ions when $\Gamma_1 = 0$ which is a function of Γ_2 and the indifferent electrolyte concentration. The term B_{12} accounts for interactions between adsorbed ions of 1 and 2, and again will be expected to depend on κ. Then, by arguments similar to those applied for an uncharged surface, for the surfactant ions one obtains

$$kT \ln n_1^{\mathrm{b}} - z_1 e \psi_{\mathrm{d}} = g(T, p) + kT \ln \Gamma_1 + B_{12}^0 \Gamma_2 + B_{11}^0 \Gamma_1 \qquad (62)$$

and for the potential-determining ions

$$kT \ln n_2^{\mathrm{b}} - z_2 e \psi_{\mathrm{d}} = f(T, p, \Gamma_2) + B_{21}^0 \Gamma_1 \qquad (63)$$

where $B_{12}^0 = B_{21}^0$ is a function of T and p only. It is worth noticing that equation (62) is similar in form to the Stern–Langmuir equation (28), but with lateral interaction accounted for in the last term. Hall (1980b) has shown that B_{12} describes the effect of surfactant adsorption on the concentration of potential-determining ions for which the surface charge ($\equiv\Gamma_2$) has a given value. If this effect disappears at high κ, then B_{12}^0 is zero. If one makes the simplifying assumption that both B_{11}^0 and B_{12}^0 are zero, which is implicit in the treatment of Rendall *et al.* (1979) discussed in the preceding section, then for small initial surface charges it can be shown (Hall 1980b), following from equations (62) and (63), that

$$\frac{kT}{z_1 e}\left(\frac{\partial \ln n_1}{\partial \psi_{\mathrm{d}}}\right)_{n_2, \kappa} = 1 - kT\left[\frac{1}{z_1 e \psi_{\mathrm{d}}^{\mathrm{i}}} + \frac{z_2^2}{z_1^2}\left(\frac{\partial^2 \Gamma_2}{\partial \mu_2^2}\right)\frac{z_1 e \psi_{\mathrm{d}}^{\mathrm{i}}}{2\Gamma_2}\right] \qquad (64)$$

Equation (64) is a generalization of the corresponding expressions derived by Rendall *et al.* (1979). Its particular form depends on the charging mechanism of the surface under consideration. Hall provides simple arguments to show that within experimental error the final term is usually insignificant, such that at the i.e.p.,

$$\frac{kT}{z_1 e}\left(\frac{\partial \ln n_1}{\partial \psi_{\mathrm{d}}}\right)_{n_2, \kappa}^0 = 1 - \frac{kT}{z_1 e \psi_{\mathrm{d}}^{\mathrm{i}}} \qquad (65)$$

which is identical to equation (45) taking $\zeta = \psi_{\mathrm{d}}$.

Thus, from quite general thermodynamic concepts, and without the use of a detailed model for the inner region, general expressions can be derived to relate ζ-potentials and amounts adsorbed to surfactant and potential-determining ion concentrations; furthermore, these expressions under certain limiting conditions, are essentially identical to the equations derived by Rendall *et al.* (1979), the latter having already been shown to be consistent with experimental results. Tests of the validity of the thermodynamic expressions using experimental data (see Hall *et al.*, 1980) are therefore also identical to those made by Rendall *et al.* (1979) as depicted, for example, in Fig. 11 for a number of different combinations of surfactant, surface, and solution conditions. Moreover, the general conclusions made regarding the nature of the adsorption process in terms of electrical and specific interactions, and the utility of i.e.p. data in describing the affinity of surfactant ions for a particular surface, are common to both approaches.

Application of Hall's approach to a study of the primary charging mechanism of interfaces that conform to the minimal requirements of the model (Hall *et al.*, 1980; Hall and Rendall, 1980) has confirmed predictions that (a) for all points of identical Γ_2 (i.e. at constant σ_0), ψ_d follows a Nernst relation with pX, and (b) plots of σ_0 against μ_2 as defined by equation (48) are congruent. Furthermore, these tests provide a simple, rapid diagnostic indication of the breakdown of model assumptions.

The theorem that dL in equation (47) is an exact differential leads to a generalization of these conditions to multi-component systems, and hence to the suggestion that there may be better choices of experimental conditions to test theories of surfactant adsorption than those conventionally adopted. The two above conditions (a) and (b) apply to a general system provided that the concentrations of all ions concerned are varied in such a way that the quantities $(n_i^b)^{z_j}/(n_j^b)^{z_i}$ are constant for all i, j pairs. Thus, for example, in the study of an anionic surfactant S^- adsorbing onto AgI, the surfactant concentration should be varied in such a way that $[Ag^+][S^-]$ = constant, or equivalently that $[S^-]/[I^-]$ = constant. The congruence test would then indicate the region within which it is justifiable to assume, for example, that counterion binding is absent and that displacement of the plane of shear is negligible.

III. Methods

A. CHARACTERIZATION OF SUBSTRATES

1. *Surface Area*

(a) *Gas adsorption.* Surface areas are frequently determined by gas or vapour adsorption, generally in conjunction with the BET isotherm (Brunauer *et al.*, 1938). The major disadvantage of this method is that the sample is studied in a dry state, which might be substantially different from that in contact with the solution of interest. Thus, precipitated materials may lose a significant fraction of surface area on drying (van den Hul and Lyklema, 1968). Especially for granular solids, problems may arise either from incomplete wetting of the substrate under certain conditions, or from dispersal of the substrate to create fresh surface area. A number of materials such as clays and cellulose can swell in contact with polar liquids. The apparent surface area of cellulose can vary by orders of magnitude depending on the nature of the adsorbed layer (Stamm and Millett, 1941). In this latter case, it is doubtful whether the concept of a surface area is relevant to the adsorption of small molecules, including surface active agents, from solution.

(b) *Adsorption of dyes and surfactants.* A possible method to obtain a "wet" surface area is to study the adsorption from solution of a suitable molecule. However, molecules that might be expected to give "monolayer" adsorption from solution at the solid/liquid interface are very much larger and more complex than the N_2 molecules commonly employed in gas adsorption. There is therefore always some doubt as to whether a complete, close-packed monolayer will be formed. Parts of the surface may be inaccessible to large molecules. The area per molecule will depend on the orientation of the adsorbed molecule at a given interface, and, particularly with surface active agents, there may be doubt as to whether a monolayer or a bilayer is adsorbed. Amongst the molecules that have been employed are cationic surfactants for negatively charged mineral surfaces (Greenland and Quirk, 1963; Kivel *et al.*, 1963), methylene blue (Pham Thi Hang and Brindley, 1970), and *p*-nitrophenol (Giles and Tolia, 1964; Giles *et al.*, 1970). It has been suggested (Giles and Tolia, 1964) that extrapolation to zero time of the amount of *p*-nitrophenol "adsorbed" by fibres can give a reliable measure of the external specific surface area.

(c) *Soap titration.* A method widely used in assessment of the surface areas of latex dispersions is the "soap titration" technique of Maron *et al.* (1954). This consists essentially of determining the displacement of the c.m.c. of a surfactant in the presence of a known amount of latex. The c.m.c. is identified from the break in a surface tension–log(concentration) plot, or from conductance measurements. The specific surface area of the dispersion may then be obtained provided that the area occupied per molecule at the c.m.c. is known. This method is therefore primarily suited to the characterization of different batches of the same latex type.

(d) *Electron microscopy.* Provided that the substrate is of well defined geometry and fairly uniformly sized, the geometric surface area may be obtained from electron microscopy. Obviously monodisperse spherical particles are particularly readily characterized in this way. Comparison of BET and geometric areas can give an indication of the extent of surface roughness of the sample. The problem of obtaining reliable values for the area per molecule, even for well defined substrates, may be illustrated with reference to the results of van den Hul and Vanderhoff (1971). They observed a plateau adsorption of sodium dodecyl sulphate on a polystyrene latex of 2.6×10^{-4} mol g^{-1}. This corresponds to ~ 0.50 nm^2 per ion on the basis of the BET area, or to ~ 0.41 nm^2 per ion on the basis of electron microscope size. Obviously, much larger discrepancies are to be expected with the asymmetric and polydisperse samples generally available as adsorbents.

(e) *Double layer measurements.* Two essentially independent methods for obtaining surface areas from double layer measurements were suggested by van den Hul and Lyklema (1968). The first makes use of the double layer capacity, and the analytically determined excess charge on the substrate. The second makes use of the non-specific nature of the "negative adsorption" or exclusion of co-ions at high charge densities. The negative adsorption may be inferred from a detailed analysis of titration measurements. Obviously these methods are applicable primarily to non-swelling and non-porous substrates which (for negative adsorption) are capable of attaining a high surface charge density. A significant feature of this work was the substantial loss in surface area experienced by the silver halide dispersions on drying (van den Hul and Lyklema, 1968).

A recently proposed method of obtaining surface areas from the application of a congruence test to double layer measurements (Hall and Rendall, 1980) has not yet been tested experimentally. In principle the method could be more rigorous and more generally applicable than the capacitance measurement; it should be more direct than the negative adsorption method and obviates the need for high charge densities. However, it is applicable only to surfaces that conform to the double layer model outlined above.

2. *Charge and Potential Measurements*

The electrical state of the substrate/solution interface in terms of surface charge and double layer potential profile can be assessed with the aid of techniques such as conductometric and potentiometric titrations and electrokinetic measurements. Total surface charges can be obtained from titration data (e.g. for AgI see Lyklema, 1961, 1966; for oxides see Yates and Healy, 1976; Parks and de Bruyn, 1962; for polymer latex particles see van den Hul and Vanderhoff, 1968, 1970; Goodwin *et al.*, 1973). In the case of low surface area solids, for which high levels of suspended material may be necessary for analytical accuracy, problems may arise from the suspension effect (Bolt, 1957). Furthermore, the accuracy of surface charge densities obtained from the titration data is obviously also dependent on the reliability of the chosen specific surface area of the solid. The problems inherent in the conversion of electrophoretic mobilities to ζ-potentials have been referred to in Section II. Ambiguities also arise in the application of other electrokinetic techniques to obtain ζ-potentials. For example, doubt has been expressed concerning the validity of all electrokinetic measurements conducted on packed beds (Biefer and Mason, 1959).

B. TECHNIQUES FOR ADSORPTION MEASUREMENT

1. *Direct Measurements*

(a) *Radiotracers.* Direct studies of amounts adsorbed have been attempted with radiolabelled compounds. The technique has been reviewed by Muramatsu (1973). Adsorption at the air/solution interface has been studied by direct counting through a thin window (Sekine *et al.*, 1970; Tajima, 1971; Nilsson, 1957; Weiss *et al.*, 1974). Even in the case of adsorption studies using "soft" radiation, e.g. tritium (Muramatsu *et al.*, 1973), a contribution from the underlying solution may not be insignificant. "Harder" radiation sources such as ^{14}C and ^{35}S give considerably higher counts from the underlying solution. This method is, in principle, applicable to surfactant adsorption at the solid/liquid interface but we are unaware of any published study. The main problem is that radiation hard enough to penetrate the substrate would undoubtedly give rise to a substantial count rate from the sub-solution as well as from the adsorbed layer.

An alternative approach to the study of adsorption at the solid/solution interface is to count after removal of the solid (e.g. a glass slide). However, the inevitable passage of the slide through the solution/air interface may cause fresh deposition, or removal, of surfactant, and additional surfactant may be deposited through the evaporation of a finite amount of solution adhering to the slide. Even when the slide emerges visibly dry (Ter-Minassian-Saraga, 1975a), i.e. when there is a reasonably high solution/substrate contact angle, there is no guarantee that the amount of surfactant retained on the slide is equal to the amount in adsorption equilibrium with the solution. Thus, for example, in a study of the effect of sodium dodecyl sulphate and Aerosol OT (sodium diethylhexylsulphosuccinate) on the wetting of paraffin wax (Padday, 1967), a significant amount of adsorption at the solid/liquid interface was postulated to explain the contact angle behaviour. Nevertheless, after withdrawal of the surface from contact with the solution, wetting with pure water gave the behaviour typical of an uncontaminated fresh wax surface, implying that no surfactant had been left there. This may be contrasted with the Langmuir–Blodgett technique of depositing insoluble monolayers by withdrawal of the substrate through an air/solution interface (Blodgett, 1935; Blodgett and Langmuir, 1937). Thus, the extent of surfactant transfer to or from the air/solution interface is generally unknown. The method of radiocounting after withdrawal is principally suited to the study of molecules that irreversibly adhere to the substrate and that are not surface-active at the air/solution interface.

A similar method has recently been proposed for the measurement of adsorption at high concentration, or low surface area, where conventional solution depletion measurements are of inadequate sensitivity, or for multicomponent mixtures where it may be desirable to minimize the composition changes that result from substantial adsorption. The "slurry method" (Nunn *et al.*, 1981) requires the measurement of the increased amount of the component of interest in the separated solids, rather than the depletion of the equilibrium solution. It is necessary to have an analytical method that can detect the adsorbed material in the presence of the substrate (e.g. a sufficiently penetrating radiation). Adsorbate present in the liquid within interstices may be allowed for following careful weighing. Attempts to remove this liquid, however, would give rise to the uncertainties discussed above.

(b) *Infrared studies of adsorbed surfactants.* Although multiple reflection infrared techniques are widely used to study, for example, hydrogen bonding of adsorbed molecules to the substrate, such techniques are unsuitable for applications involving aqueous solutions because of the high infrared absorption by water. The technique of obtaining an infrared spectrum for a powdered substrate (dispersed within a KBr disc) subsequent to its equilibration with the surfactant solution, followed by a drying process, has been used to provide evidence for the operation of chemisorption mechanisms for certain surfactant–substrate combinations such as oleic acid–oleate adsorption on haematite (Peck *et al.*, 1966) and hydroxamic acid on oxidized iron ores (Raghavan and Fuerstenau, 1975). However, identification of a particular bonding type in this way does not necessarily imply the existence of such a bond between substrate and surfactant in the aqueous environment, it being possible that some alteration in the chemical state of the surfactant headgroup may occur during sample preparation. Connor (1968) used a polystyrene film with adsorbed alkyl quaternary ammonium ions to demonstrate the absence of chemisorption processes for this system. This technique is subject to the same ambiguities as the KBr disc method.

(c) *Neutron scattering.* Neutron scattering techniques have recently been developed which allow detailed studies of adsorbed layers. The scattering cross-sections for hydrogen and deuterium are very different, so by using appropriately deuterated molecules in H_2O–D_2O mixtures the contrast can be varied in order that selected features may be studied. Harris *et al* (1979) have studied the adsorption of sodium deuterododecanoate onto polystyrene latex. On the assumption of a uniform adsorbed shell, they were able to calculate shell thickness and density, and hence obtain an adsorption isotherm. The results correlate well with an isotherm obtained by Connor (1968) at a slightly different pH and ionic strength. The saturation adsorption value varied very sharply with pH, probably reflecting the weak acid character of the headgroups (see Fig. 12).

The technique is still not fully developed as far as surfactant adsorption is concerned. It remains to be established whether any information is available that could not be obtained more conveniently by other means.

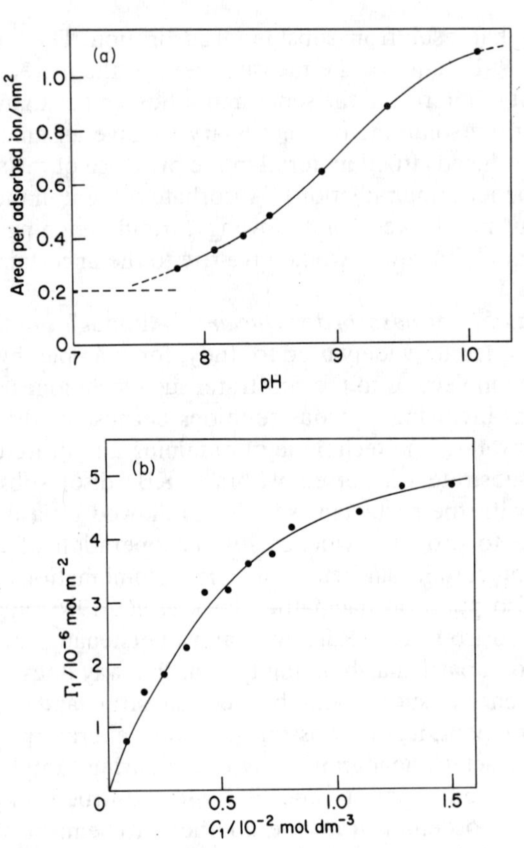

Figure 12 Adsorption of sodium deuterododecanoate onto polystyrene latex particles; data from neutron scattering measurements. (Harris *et al.*, 1979) (a) Saturation adsorption density against pH (broken line indicates value corresponding to crystallographic area of ion). (b) Adsorption isotherm at pH 8.1; total ionic strength, 2×10^{-2} mol dm^{-3}.

2. *Solution Depletion Methods*

(a) *Background.* The most widely used approach to the direct determination of adsorption is to study the depletion of the species of interest from solution in equilibrium with the adsorbent. The problem then reduces to the determination of the solution concentration of the species in the presence of sufficient interface per unit volume to cause a measurable change in concentration. The method is therefore applicable only when the specific surface area of the substrate is not too small. It must be stressed that only an apparent adsorption isotherm is obtained by these direct measurements. Difficulties inherent in the application of theoretical expressions to interpret these results are discussed in Section III.C. Some of the principal methods for measuring surfactant concentrations are discussed below.

(b) *Radiotracers.* Provided that a suitably labelled sample can be prepared, solution concentrations may be obtained by counting a suitable sample, using a simple Geiger–Müller tube or a liquid scintillation counter. The method has the advantage of being highly specific and highly sensitive (Connor, 1968). It is therefore particularly useful either for very low concentrations, or for the study of single components of a mixed system. The main disadvantages are the necessity to prepare special (labelled) samples, and the need to adopt appropriate procedures for handling radiolabelled compounds. Examples of the application of radiotracers are given by Connor (1968), Pearson and Wade (1972), Dixit and Biswas (1975), and Kulkarni and Somasundaran (1980).

(c) *Dye extraction.* A number of techniques have been developed that rely on the extraction into a solvent (e.g. chloroform) layer of an electroneutral complex of the surfactant with a dye molecule of opposite charge. The concentration is deduced either visually in a titration experiment, or spectrophotometrically after a suitable extraction procedure (Epton, 1948; Longwell and Maniece, 1955; Abbott, 1962; Mukerjee, 1956; Barr *et al.*, 1948; Few and Ottewill, 1956). Such methods are rapid, simple, and require little in the way of specialized equipment. Clearly, these measurements are not highly specific, and indeed find widespread application in the monitoring of total ionic surfactant levels in environmental studies.

(d) *Direct spectrophotometric analysis.* Where the surfactant contains a suitable chromophore (e.g. alkylbenzene sulphonates, pyridinium ions), surfactant concentrations may be determined directly by spectrophotometric measurements (Dobias, 1977, 1978; Rosen and Nakamura, 1977; Greenland and Quirk, 1963).

(e) *Interferometry.* The refractive index of aqueous solutions is affected to a similar extent by simple electrolyte as by surfactant ions. Consequently interferometry is not generally useful for adsorption studies, particularly in the presence of supporting electrolyte. This technique has been used mainly with zwitterionic and nonionic surfactants (e.g. Hough, 1973). However, there are a few reports of interferometric analysis applied to ionic surfactant adsorption (Robb and Alexander, 1967; Connor, 1968).

(f) *Surface tension.* Surface tension, like refractive index, is more widely used to assess concentrations of nonionic surfactants (e.g. Ottewill and Walker, 1974) than of ionic surfactants, for which more sensitive techniques are generally available. However, surface tension measurements may prove useful in some parts of the isotherm (Saleeb and Kitchener, 1965), and are widely used in the "soap titration" method to identify the c.m.c. (Maron *et al.*, 1954). Of course, surface tension measurements cannot be used to examine whether adsorption increases above the c.m.c., except after dilution of the equilibrated solution.

(g) *Ion selective electrodes.* Ion selective electrodes have been used to determine the uptake of, for example, Na^+ or Br^- ions in conjunction with oppositely charged surfactant ions (e.g. Sexsmith and White, 1959). Electrodes selective to surface-active ions are now available, and have been used in studies of surfactant binding to soluble polymers and proteins (Birch *et al.*, 1974; Oakes, 1974).

C. INTERPRETATION OF ADSORPTION MEASUREMENTS

1. *Calculation of Amount Adsorbed*

(a) *Solution depletion methods.* It is normally assumed that the amount of ionic surfactant calculated from the change in solution concentration, usually after separation of the solution from the solid, may be identified with the quantity σ_β, i.e. that the molecules "lost" from solution are those having a direct and "specific" interaction with the substrate. It is important to note that solution depletion (and most other methods used to deduce surfactant adsorption) cannot, strictly, provide this information directly. The quantity calculated from the change in solution concentration will contain a contribution from the diffuse layer charge σ_d as well as from σ_β. Equation (8) expresses the condition for electrical neutrality, which must always be obeyed for any interface at equilibrium. The values of σ_0 and σ_β depend directly on the bulk concentrations of potential-determining ions and of specifically adsorbed (surfactant) ions, although their magnitudes may also be modified indirectly by other ionic species, and by each other, through electrostatic factors. The interactions governing the balancing charge σ_d, however, are non-specific, and all ions of the same charge type will contribute to σ_d in proportion to their bulk concentrations. While σ_d is composed of an excess of ions of charge opposite to $(\sigma_0 + \sigma_\beta)$ and a deficiency of ions of like charge, the former contribution becomes increasingly important as $|\psi_d|$ increases. Thus, depending on the magnitude of ψ_d and on the level of supporting electrolyte, the amount of a particular ion in the diffuse layer may vary from virtually 100% of that required to balance $(\sigma_0 + \sigma_\beta)$, through zero, to a deficiency as compared with the concentration in bulk solution. In the presence of a swamping concentration of indifferent electrolyte, the contribution to σ_d of the surfactant species will be small, and the measured adsorption in this case may approximate closely to σ_β. On the other hand, at low levels of supporting electrolyte the amount of surfactant in the diffuse layer may become significant when surfactant and substrate have opposite charge. Indeed, it has been suggested that the measured adsorption of ionic surfactants at low concentrations onto oppositely charged oxides may be entirely within the diffuse layer (Section II.C). Nevertheless, this contribution is frequently ignored when surfactant adsorption is discussed. This is particularly likely to cause confusion when the study is extended to consider the overall charge balance at the interface. The results of Sexsmith and White (1959) for the uptake of cationic surfactants, and simultaneously Br^- ions, onto cellulosic substrates typify the outcome of such a study. Below a certain level of surfactant adsorption they found no change in bulk Br^- ion concentration, whereas above this level a Br^- ion was "adsorbed" for every additional surfactant cation adsorbed. The authors inferred a two-stage adsorption mechanism, "ion-exchange adsorption" at low adsorption densities and "ion-pair adsorption" at higher densities. Similar inferences underlie many published discussions of surfactant adsorption. However, observations such as those of Sexsmith and White (1959) do not present *a priori* evidence of a change in adsorption mechanism. The electrical neutrality condition requires the presence of a balancing charge in the diffuse layer. When surfactant and substrate have opposite charge the onset of (in this case) Br^- ion "adsorption" need indicate only the point at which $|\sigma_\beta| > |\sigma_0|$. Indeed, in the literature on hemimicellization (see Section II.C) the surfactant concentration corresponding to a dramatic increase in adsorption, which is identified with the formation of hemimicelles, generally refers to a condition where $|\sigma_\beta| < |\sigma_0|$. The postulated change in adsorption mechanism is not likely therefore to be attended in this case by any dramatic uptake of surfactant counterions.

The following points are also worthy of note.

(i) Use of the term "ion-exchange" as an adsorption *mechanism* is potentially misleading. It is regularly introduced in situations where electrical neutrality is the dominant factor. As we have seen, the distinction between loss of a counterion or addition of a co-ion to the primary charge is, in one sense, almost coincidental. In contrast, however, other authors (e.g. Allen and Matijević, 1970) seem to imply that "ion-exchange" involves a highly ion-specific interaction with the substrate. It is our view that, in general, use of the term "ion-exchange" to describe an adsorption process obscures rather than clarifies the underlying mechanisms.

(ii) The substrate is normally separated from the dispersion prior to analysis, to prevent interference in the analysis by the substrate or by the adsorbed surfactant. In many cases, centrifugation or filtration methods are applied to remove small particles. Some thought should be given to possible concomitant errors arising from (a) removal of loosely bound material, especially multilayers, in a strong shear field, (b) alteration of the adsorption balance through double layer overlap in the settled solid, (c) loss of surface area through aggregation of the compressed solid, and (d) adsorption of surfactant onto (or leaching of surface active material from) the membrane filter. The latter point emphasizes the need for suitable pre-treatment of the filter.

(b) *Electrokinetic methods.* Unlike solution depletion methods, electrokinetic measurements may allow the determination of a charge density within the plane of shear, close to or even identical with the OHP. The question of whether "adsorbed" ions are specifically adsorbed or within the diffuse layer does not arise here. However, even within the range of κa values and potentials where quantitative interpretation is possible, the results still do not provide a simple direct measure of the amount adsorbed, as is frequently supposed. Changes in the electrokinetic charge density represent changes in the *net* charge density within the plane of shear. Such changes include not only those arising from the adsorption of the surfactant ions but also those from the rearrangement of potential-determining ions (so affecting σ_0), and from any other ionic species that is specifically adsorbed, including, in particular, co-adsorbed counterions.

The effect of the rearrangement of potential-determining ions may be significant. For example, the theoretical isotherms shown in Figs 13–15 (discussed in more detail below) display a factor of 2 between the amounts adsorbed at the i.e.p. for interfaces of the same initial primary charge density, depending on whether it is assumed that ψ_0 or σ_0 remains constant throughout the adsorption process. Of course, for a generalized charging system such as that discussed by Levine and Smith (1971), it is not necessarily reasonable to assume constancy of either ψ_0 or σ_0.

Co-adsorption of counterions cannot be neglected at high surfactant coverage (see e.g. Saleeb and Kitchener, 1967; Bibeau and Matijević, 1973). Moreover, in addition to the uncertainties in calculating ζ-potentials at high potentials from electrokinetic data (as mentioned previously), ζ becomes an insensitive function of charge at high charge density [see equation (9)]. For these reasons quantitative information about amounts of surfactant adsorbed can only be obtained from the region at and around the i.e.p. (see Section II) where surfactant coverage is low and ζ is low or zero. As discussed in Section II, measured shifts in the i.e.p. upon surfactant addition provide a reliable guide to relative adsorption. Of course, measured ζ-potentials, including those away from the i.e.p. region, may be of direct interest in their own right in relation to, for example, studies of electrical interactions between interfaces.

2. *Analysis of Results*

Once a set of reliable adsorption data has been obtained the conclusions reached may well depend significantly on the way in which these data are handled. We have seen the complexity of adsorption at charged interfaces, where several interacting factors influence the adsorption versus concentration relationship. This means that simple analytical expressions are rarely available to analyse the results. Most workers therefore rely on qualitative conclusions based on isotherm shape. Extreme care is necessary if unsupported and often wholly misleading explanations are to be avoided here.

Such pitfalls may readily be demonstrated by plotting typical "results", generated by the application of a theoretical isotherm, in different ways. Figures 13–15 show different representations of the simple Langmuir and Stern–Langmuir isotherms. The Langmuir isotherm may be written in the form

$$\theta/(1-\theta) = C_1/C_{1(\frac{1}{2})}$$

where θ is the fractional coverage of the surface, C_1 is the solution concentration, and $C_{1(\frac{1}{2})}$ is the concentration at half-coverage which is a direct measure of the adsorption free energy. For low coverages the Langmuir isotherm can therefore be shown in an essentially dimensionless form by plotting θ against $C_1/C_{1(\frac{1}{2})}$. For direct comparison the Stern–Langmuir isotherm, equation (28), can be depicted in terms of the same variables for adsorption, both at constant charge and potential. In this case $C_{1(\frac{1}{2})}$ is the concentration that would produce half-coverage by a species whose Langmuir adsorption free energy was numerically equal to the specific adsorption free energy of the ion. Thus the two Stern–Langmuir plots for constant charge and constant potential will intersect the Langmuir plot at their respective i.e.p. corresponding to $\psi_\beta = 0$ in each case. Other parameters used to generate Figs 13–15 are given in Table 2.

In order to show results over a wide range of surfactant concentrations, many authors have chosen to plot the amount adsorbed against the logarithm of the concentration. Figure 13 (a) and (b) show that both for the Langmuir and for the Stern–Langmuir isotherms this procedure gives rise to a curvature that might immediately suggest cooperative effects in the adsorption process. However, such effects are, by definition, absent from the theoretically calculated results. Figure 14 shows the three isotherms plotted on a log-log basis. The Stern–Langmuir plots could each be represented, well within the limits of experimental precision, by two essentially straight-line portions with a distinct break-point. To the unwary such an isotherm shape might well be taken to indicate a change in adsorption mechanism, although such a conclusion would be false in the case of the theoretically calculated adsorption values. Figure 15 compares the same isotherms plotted in a linear form for a surface of fixed monolayer capacity (see Table 2). In this case, if results conforming to the Stern–Langmuir

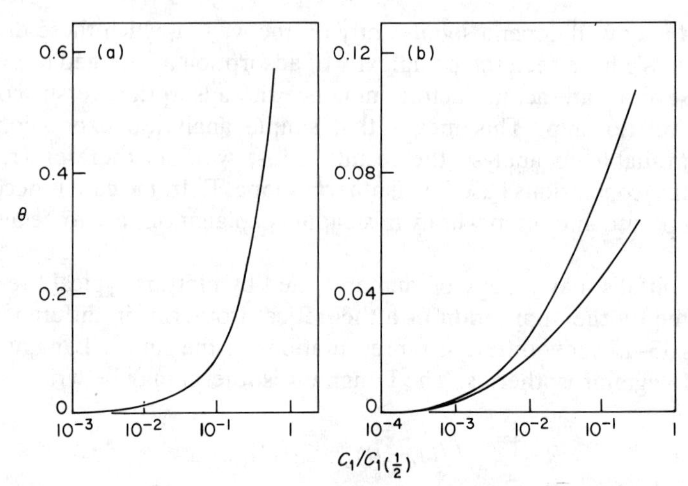

Figure 13 Fractional surface coverage against logarithm of equilibrium concentration for (a) Langmuir isotherm, and (b) Stern–Langmuir isotherm, for parameters given in Table 2 (upper curve, constant ψ_0; lower curve, constant σ_0).

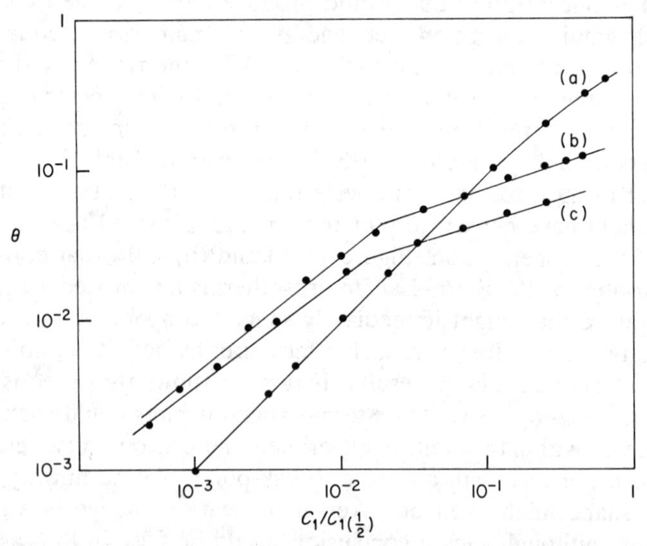

Figure 14 Logarithm of fractional surface coverage against logarithm of equilibrium concentration for (a) Langmuir isotherm, (b) Stern–Langmuir isotherm at constant σ_0, and (c) Stern–Langmuir isotherm at constant ψ_0. Parameters as in Table 2 and accompanying text.

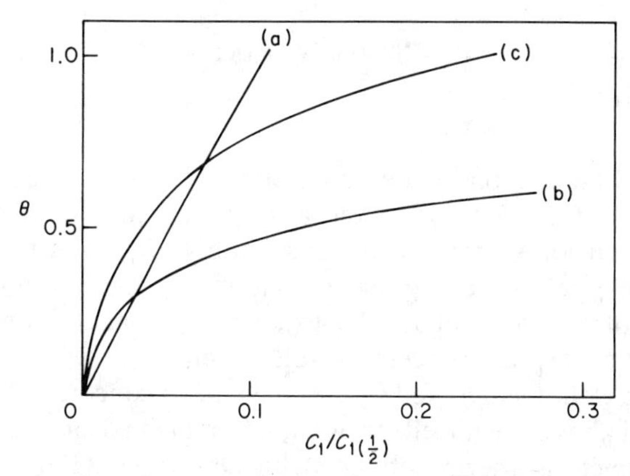

Figure 15 Fractional surface coverage against equilibrium concentration for (a) Langmuir isotherm, (b) Stern–Langmuir isotherm at constant σ_0, and (c) Stern–Langmuir isotherm at constant ψ_0. Parameters as in Table 2 and accompanying text.

Table 2 Parameters used in theoretical Stern–Langmuir calculations.

Ionic strength	10^{-2} mol dm^{-3}
K	30 μF cm^{-2}
σ_0 (no adsorption)	1.0 μC cm^{-2}
ψ_d (no adsorption)	40 mV
ψ_0 (no adsorption)	73.3 mV
N_s	2×10^{14} cm^{-2}

isotherm were wrongly assumed to be describable by the simple Langmuir isotherm, the erroneous conclusion would be that adsorption had occurred on a surface possessing a restricted number of adsorption sites (approximately 1/10 of monolayer capacity in this case).

Unfortunately, many published works that claim to have advanced our understanding of ionic surfactant adsorption on solid substrates have relied either on the use of ambiguous (or even erroneous) experimental data or on misinterpretations of acceptable data, many of which have been highlighted in this section. It is strongly to be recommended that future work should be carried out wherever possible in such a way that quantitative inferences from the shapes of accurately obtained isotherms can be more rigorously and objectively tested. Attention is particularly drawn to the discussions on congruence in Section II.B, and to Fig. 11 and accompanying text in Sections II.D and II.F.

IV. Illustrative Results

A. CARBON

The pure carbon/aqueous solution interface is, in principle, a model hydrophobic surface for surfactant adsorption studies. However, the properties of carbon surfaces vary considerably according to their origin and treatment prior to adsorption measurement (e.g. washing, graphitization). It is not therefore yet possible to advance a clear and comprehensive picture of the adsorption process for this system.

Most of the well documented studies have been concerned with adsorption up to and above the micellar region, i.e. at high adsorption coverages. Figure 16 depicts isotherms obtained by Day *et al.* (1967) for sodium

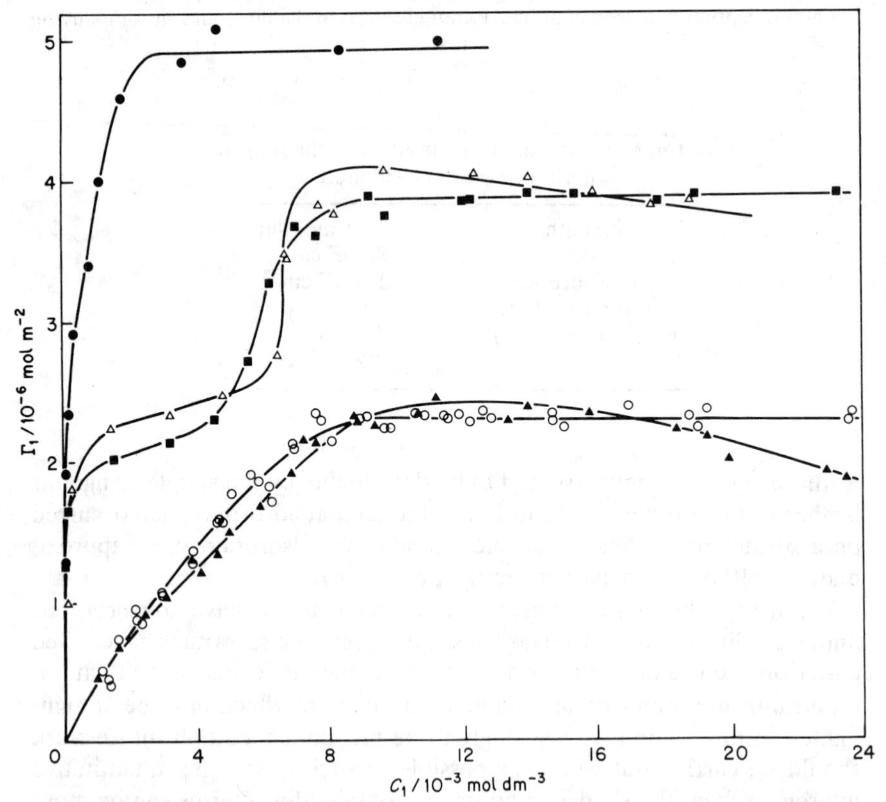

Figure 16 Adsorption isotherms for sodium dodecyl sulphate on carbon substrates. Graphon in 10^{-1} mol dm^{-3} NaCl (●), and without added electrolyte (■). Spheron 6 (▲), and after washing (○), and after heat treatment at 2700°C (△). (Greenwood *et al.*, 1968; Day *et al.*, 1967)

dodecyl sulphate (SDS) on a carbon black (Spheron 6) before and after various washing processes. The disappearance of the adsorption maximum on washing suggests the removal of matter from the carbon black which only becomes extractable at higher surfactant concentrations (Day *et al.*, 1967). The principal effect of graphitization of such materials is to remove the hydrophilic ionizable oxygenated sites from the surface to produce surfaces of a more uniform and hydrophobic character (see Section II.A), which would be expected to exhibit adsorption properties different from those of the original carbon black surfaces. It might be hoped that the graphitized surface would be more reproducible, and more suitable as a model surface, than the untreated surfaces. The results of Day *et al.* (1967) shown in Fig. 16 demonstrate an increased surfactant adsorption on Graphon compared with its ungraphitized form, Spheron 6. For both types of surface the onset of maximum adsorption occurred in the c.m.c. region. The limiting area per molecule, as calculated on a monolayer basis, decreased from about 0.70 nm^2 on the Spheron 6 (washed) to 0.42 nm^2 on the Graphon. Isotherms obtained for samples of Spheron 6 (unwashed), after heat treatments at various temperatures up to and including that used industrially for graphitization, showed a progressive change in shape towards that for Graphon.

The isotherm for Graphon in Fig. 16 shows a marked point of inflection corresponding to an area per molecule of about 0.70 nm^2. A similarly shaped isotherm was originally reported by Skewis and Zettlemoyer (see Zettlemoyer, 1968) for SDS on Graphon and was interpreted by these authors, with the aid of calorimetric evidence, to represent a two-stage adsorption process. At low surfactant concentrations the alkyl chains adsorb parallel to the surface, presumably to maximize favourable hydrophobic interactions between the alkyl chains and the hydrophobic surface. The point of inflection on the isotherm is then deemed to correspond to saturation coverage in this orientation. A further increase in the surfactant equilibrium concentration results in a progressive reorientation of the alkyl chain towards the perpendicular with respect to the surface, the polar head groups being directed towards the solution phase. Saturation adsorption then corresponds to a vertically orientated monolayer. Above the point of inflection, associative chain–chain as well as chain–solid hydrophobic interactions would be expected to supply the driving free energy for adsorption. In contrast to the above, Zettlemoyer *et al.* (1957) found no comparable evidence for a two-stage adsorption process for sodium dodecylbenzene sulphonate on Graphon. Adsorption was concluded to occur with the aromatic ring in juxtaposition to the graphitic surface, with the sulphonate group and part of the alkyl chain extending into the solution. The reported limiting area per molecule of 0.60 nm^2 at plateau adsorption, however, would neither prove nor disprove this suggested orientation. Isotherms exhibiting no point of inflection below the plateau adsorption value have also been obtained by Saleeb and Kitchener (1965) for the anionic surfactant Aerosol OT (sodium diethylhexylsulphosuccinate) and the cationic surfactant cetyltrimethylammonium bromide (CTAB) on a variety of carbon black surfaces and their graphitized counterparts. As in the aforementioned studies, graphitization resulted in an increase in surfactant adsorption in accordance with the production of a more hydrophobic, non-ionogenic surface; plateau adsorption commenced in the region of the surfactant c.m.c.

There appears to be no significant published work for ionic surfactant adsorption on graphitized carbons in which the effect of alkyl chain length has been specifically studied (contrast other types of surface, including carbon blacks, discussed below). It is predictable that the chain–solid contribution to the free energy of adsorption would increase by about 1 kT for each additional CH_2 group, as would the chain–chain contribution. The expected outcome of such a study is therefore that the isotherm shape would be preserved, and displaced towards lower concentrations as the chain length increased. Nevertheless, the calculated areas per molecule at plateau adsorption and at any inflection points in the isotherms might well give a good supportive indication of the surfactant orientation, and hence of the type of interaction predominating at different stages of the adsorption process. For example, we might obtain, for a homologous series of sodium alkyl sulphates, adsorption isotherm shapes similar to that obtained by Zettlemoyer (1968) and Day *et al.* (1967) for SDS as shown in Fig. 16. If these authors' interpretations are correct, the area per molecule at the point of inflection would increase as the alkyl chain length is increased, and at plateau adsorption would be essentially independent of chain length. This prediction has apparently been verified for carbon black (Koganovskii *et al.*, 1977), as discussed below.

A marked increase in the adsorption of SDS on Graphon on the addition of 10^{-1} mol dm^{-3} of sodium chloride was measured by Greenwood *et al.* (1968) and is shown by the isotherm in Fig. 16. Plateau adsorption increased from an area per molecule of 0.42 nm^2 (no salt) to 0.33 nm^2 (10^{-1} mol dm^{-3} NaCl) and to 0.28 nm^2 in 0.5 mol dm^{-3} NaCl. The disappearance of the point of inflection at higher salt concentrations was considered by Greenwood *et al.* (1968) to be a consequence of the much steeper rise in adsorption at low surfactant concentrations. Saleeb and Kitchener (1965) found that the presence of 10^{-2} mol dm^{-3} concentrations of Li, Na, Ag, and Cs chlorides increased the adsorption of the respective metal salts of Aerosol OT on graphitized carbon blacks in the order Li $<$ Na $<$ Ag $<$ Cs; the decrease in the limiting area per molecule for the original sodium salt of Aerosol OT was from 0.70 nm^2 to 0.66 nm^2 on the addition of NaCl. Electrophoretic mobility curves of the graphite particles in these electrolytes as a function of Aerosol OT concentration showed qualitatively similar shapes to the adsorption isotherms (Saleeb and Kitchener, 1967). In view of the observed trend of the hydrated ionic radii and the c.m.c. of the corresponding metal salts of Aerosol OT in the order Li >Na > Ag > Cs, it was concluded by Saleeb and Kitchener that the increase in surfactant adsorption is a result of the increased counterion binding (or Stern adsorption) of the metal ions in the surfactant ion adsorbed layer as the hydrated ionic radius decreases. That is, the more specifically adsorbed smaller ions provide more efficient electrostatic screening between the adsorbed surfactant ions. In other words, the *repulsive* ΔG_{elec} term is reduced as the degree of counterion binding increases. From the derived ζ-potentials and with the aid of Gouy–Chapman theory, Saleeb and Kitchener (1967) deduced the net charge density σ_c within the electrokinetic plane of shear, i.e. the excess charge of surfactant ions over bound counterions, and compared it with the total charge density σ_a of surfactant ions, as obtained from the adsorption isotherms. At very low surface coverages the ratio σ_c/σ_a was close to unity, as might be expected. As the surface coverage was increased the ratio decreased to about 0.16, 0.12, and 0.11 for, respectively, the Li, Cs, and Ag forms of Aerosol OT. These results, together with surface conductance measurements made by Saleeb and Kitchener (1967), indicate that at higher surface coverages only about 12% of the surfactant counterions are in the diffuse double layer. They provide further evidence for the existence and interpreted effects of surfactant counterion binding in the adsorbed layer. Furthermore, as Saleeb and Kitchener (1967) pointed out, the results highlight the error made by some authors in interpreting adsorbed amounts directly from electrokinetic data (see Section III.C).

Zettlemoyer *et al.* (1962) have demonstrated the even greater effect of specifically bound calcium ions in dodecyl sulphate adsorbed layers on Graphon. Calcium ion binding was demonstrated to occur up to a limiting stoicheiometric ratio of one calcium ion to two dodecyl sulphate ions. The limiting area per molecule at saturation adsorption at this binding ratio was accordingly reduced from 0.46 nm^2 (Ca^{2+} absent) to 0.28 nm^2 (see Zettlemoyer, 1968), which indicates a highly close-packed vertically orientated monolayer in which all lateral charge repulsion has been screened out.

The adsorption process of ionic surfactants on the less well defined carbon black surfaces is more difficult to interpret. Day *et al.* (1967) have suggested that the absence of a point of inflection in the isotherm for SDS on Spheron 6 (Fig. 16) below the c.m.c. may be a consequence of the heterogeneous nature of such surfaces prohibiting the formation of an ordered and uniform layer of adsorbed ions in the parallel orientation. It is, presumably, coincidental that the plateau value on carbon black corresponds closely to the adsorption density at the inflection for graphitized carbon (Fig. 16). Saleeb and Kitchener (1965) found that the addition of electrolyte induced dramatic changes in the originally Langmuir-shaped isotherms of Aerosol OT on various carbon blacks. Adsorption continued to increase dramatically above the c.m.c. and maxima were observed at areas per molecule that were suggested to correspond to bimolecular layers. The authors proposed that a first layer of surfactant might be induced, by hydrophilic sites on the surface, to adsorb with headgroups towards the surface, and that a second layer would form on the exposed alkyl chains. Arai and Yoshizaki (1971), on the other hand, obtained smooth Langmuirian isotherms for a series of sodium alkyl sulphates. On the questionable assumption (see Section III.C) of simple Langmuir adsorption they deduced a limiting area per molecule of 0.64 nm^2, independent of chain length, and hence suggested a vertically oriented monolayer at saturation adsorption. This, however, contrasts with the area per molecule of 0.42 nm^2 attributed to adsorption in this orientation onto Graphon, as discussed above. The free energy of adsorption of the surfactants increased in the expected order with increasing alkyl chain length.

From a number of adsorption studies of alkyl sulphates and sulphonates on acetylene black, the Russian workers Koganovskii and Klimenko and their collaborators purport to have adequately explained the adsorption process. However, the poor presentation of their data (e.g. see Koganovskii *et al.*, 1977) prohibits a critical analysis of these claims. In brief, they have deduced that adsorption up to the c.m.c. region occurs via the alkyl chains in parallel orientation to the surface (Koganovskii *et al.*, 1977). In this region the alkyl chains become close-packed; the area per molecule increases by predictable increments as the length of the alkyl chain is increased (Koganovskii *et al.*, 1977). Above the c.m.c. further adsorption occurs (Koganovskii *et al.*, 1977, 1979), accompanied by rapid changes in the differential heat of adsorption (Klimenko *et al.*, 1979). These changes are explained by a form of three-dimensional association of surfactant on the surface akin to bulk micellization (Klimenko, 1979; Klimenko *et al.*, 1979; Koganovskii *et al.*, 1979). For fractional coverages up to about 0.6 (with respect to monolayer adsorption in the c.m.c. region) the adsorption data were shown to fit an equation analogous to the Hill equation (Koganovskii *et al.*, 1975, 1977), which accounts for lateral interaction between adsorbed species as well as for adsorbent–adsorbate interactions. The calculated adsorption free energies of the alkyl sulphates were greater than for the corresponding sulphonates, and for each homologous series an incremental increase in adsorption energy of about 1 kT per CH_2 was obtained. It may be noted that electrical interactions do not appear explicitly in this treatment and are apparently ignored by the authors.

The above mentioned Russian authors cite Fuerstenau's concept of hemimicellization (Section II) as similar or equivalent to their view of surfactant association in the adsorbed layer (Koganovskii *et al.*, 1975). Thus, for example, the sudden change in the measured heats of adsorption (Klimenko *et al.*, 1979) at a surface coverage corresponding to about 2 nm^2 per molecule, as derived from the data given by Koganovskii *et al.* (1977) and Klimenko *et al.* (1979), was deemed to represent the onset of the hemimicellization process. An attempt by Koganovskii *et al.* (1975) to quantify the association using a two-dimensional equation of state analogous to the van der Waals gas equation, however, is inadequate as presented. Properly interpreted, the treatment would indicate a first-order phase transition in the adsorbed layer, as has been predicted for surfactants with more than eight carbon atoms, on the basis of a statistical mechanical analysis (Cases *et al.*, 1975; Cases, 1979).

Electrical interactions between the ionic headgroups of surfactants and any fixed surface charges on the surface do not appear to play an obvious role in the adsorption process on carbon. Isotherms obtained for cationic and anionic surfactants are generally quite similar in form, even at low surface coverages (contrast, for example, oxide surfaces). Graphitized surfaces are usually considered to carry a negligible charge which would explain these observed similarities. However, significant ζ-potentials of about -65 mV, derived from electrophoresis data by Saleeb and Kitchener (1967) for graphitized Sterling MT particles and from the ζ-potential data of Parfitt and Picton (1968), are difficult to ignore in this context, even though the charge on these surfaces may reside at the edges of the graphitic sheets (Kitchener, 1967). The apparent absence of headgroup–surface charge interactions on the more polar carbon black surfaces is even more difficult to explain. One minor piece of evidence in favour of such interactions does, in fact, emerge from the work of Saleeb and Kitchener (1965) who reported that CTAB showed a slightly greater affinity for carbon black than for graphitized surfaces, whereas Aerosol OT had the opposite effect. These results could be explained by a favourable ΔG_{elec} term for the cationic surfactant and an unfavourable (i.e. repulsive) ΔG_{elec} term for the anionic surfactant during their initial stages of adsorption on negatively charged carbon black surfaces. Graphitization, with the concomitant reduction in surface charge density, would then lead to a decrease in the mean free energy of adsorption for CTAB and an increase for Aerosol OT, as observed. Saleeb and Kitchener (1965) then suggested that the much higher limiting areas per molecule for CTAB on Spheron 6 ($\sim$0.6 nm^2) compared with Graphon ($\sim$0.40 nm^2) might also be explained by charge interactions. If some of the surfactant ions were adsorbed on Spheron 6 in a reversed orientation in the monolayer, with cationic headgroups in juxtaposition to the anionic surface charges, the close packing in the monolayer might be restricted. However, the results of Connor and

Ottewill (1971) show that the adsorption density of cationic surfactants on polystyrene latex increases as the density of primary charge increases. Their observation casts doubt on the above explanation of the differences between Spheron 6 and Graphon.

B. POLYMERS

For surfactant concentrations near the c.m.c. the adsorption of ionic surfactants on hydrophobic polymers might be expected to resemble that on carbon surfaces. Although the distribution of primary charge on the two types of surface may be different—surface charges are assumed to be evenly and randomly distributed on a polymeric material, and may be absent from major areas of the adsorbing surface of graphitized carbon (Kitchener, 1967)—the total primary charge, even on a typical latex particle, is generally more than an order of magnitude lower than the surfactant density typical of monolayer coverage. Consistent with this statement, it is noted that both cationic and anionic surfactants are regularly reported to attain high adsorption coverages on identical polymer surfaces. Polymer surfaces of widely differing polarity and charge density are available, which makes it possible to test the picture of surfactant adsorption onto hydrophobic surfaces that emerges from the studies on carbon. Furthermore, the low coverage region of the adsorption isotherm on polymer surfaces has been studied in some detail by a number of authors.

Connor and Ottewill (1971) have reported the adsorption of a range of quaternary ammonium surfactants onto polystyrene surfaces bearing carboxylic acid groups. The area per molecule at "saturation" coverage (in the region of the c.m.c.) decreased with chain length, with ionic strength, and with surface charge (by pH change). Thus for dodecyltrimethylammonium bromide (DTAB) a limiting area per molecule of 0.68 nm^2 at pH 4.6 reduced to 0.56 nm^2 at pH 8. For CTAB an area per molecule of 0.59 nm^2 at pH 4.6 reduced to 0.47 nm^2 at pH 8. Increasing the bromide ion concentration from 10^{-3} to 5×10^{-2} mol dm^{-3} decreased the latter area to 0.36 nm^2. The limiting areas for CTAB on polystyrene are similar to that (0.40 nm^2 per molecule) on Graphon (Saleeb and Kitchener, 1965). Furthermore, the general trends with ionic strength and chain length follow the pattern for carbon adsorbents. Increasing the surface charge, however, increases the maximum adsorption at saturation, contrary to the postulate of Saleeb and Kitchener (1965).

The effects of surfactant chain length on saturation adsorption at the polymer/solution interface have been studied by a number of authors. Thus, for example, Maron *et al.* (1954) showed that the areas occupied on styrene–butadiene latices by soap molecules at the c.m.c. varied steadily with chain length from 0.41 nm^2 for laurate to 0.23 nm^2 for stearate. The latter value closely approaches that ($\sim$0.21 nm^2) for the cross-section of a hydrocarbon chain obtained from X-ray studies, and for close-packed insoluble monolayers at the air/water interface. Robb and Alexander (1967) obtained isotherms for a range of cationic surfactants on polyacrylonitrile latices. Based on the geometric areas (a less satisfactory choice than usual because the authors concluded that their particles were slightly porous) we have calculated saturation areas per surfactant ion of 0.65 nm^2 (C_{12} chain), 0.40 nm^2 (C_{14} chain), and 0.35 nm^2 (C_{16} chain). While the absolute areas are in considerable doubt, the trend with chain length is clearly established.

Piirma and Chen (1980) showed that for alkyl sulphonates on polystyrene at 47°C (chosen to be above the Krafft point of the C_{16} salt) the limiting areas per molecule varied from 0.52 nm^2 for the C_{12} salt to 0.25 nm^2 for the C_{16} salt. At 22°C the limiting area for the C_{12} compound decreased to 0.47 nm^2, and further to 0.35 nm^2 on the addition of 0.9 mol dm^{-3} NaCl. These figures not only confirm the widely reported effect of added electrolyte, but also indicate that the limiting area per molecule showed a systematic increase with temperature.

A major influence on the limiting area for surfactant adsorption at the c.m.c. is reported to be the polarity of the interface (Vijayendran, 1979). A linear relationship was found between the logarithm of the area per molecule of laurate ion and the monomer/water interfacial tension (an approximate guide to the polarity of the polymer). Where it was possible to calculate the interfacial tension, or the polarity, of the polymer/solution interface, the above correlation seems to be established. The extreme values quoted were for polystyrene (monomer/water interfacial tension 40–43 mN m^{-1}; area per laurate ion 0.45–0.59 nm^2) and methylacrylate (monomer/water interfacial tension 13–14 mN m^{-1}; area per laurate ion 1.46–1.75 nm^2). While the surface charges of these substrates are not explicitly known in most cases, it is clear that the variations in area per molecule are much larger than those recorded by Connor and Ottewill (1971) for a single surface type. It therefore seems reasonable to assume that a more fundamental difference in surfactant–substrate interaction is involved. The results are consistent with the idea that chain–substrate interactions will become weakened as the interface becomes more polar, and that the free energy for monolayer formation will become increasingly less favourable as compared with that for micelle formation. It has been suggested (Vijayendran, 1979) that these findings might explain the relative difficulty of stabilizing more polar polymer systems during emulsion polymerization, and that the limiting area per molecule may give a reasonable guide to the surface composition of copolymer latices.

While the effects of primary surface charge are clearly detected at saturation adsorption, charge effects are much more dramatically demonstrated in studies at low coverage. Particularly in the case where surfactant and substrate are of opposite charge, considerable success has been achieved in the explicit description of experimental results in terms of the theory outlined in Section II.

The isotherms of Connor and Ottewill (1971) for cationic surfactant adsorption on polystyrene latex show a pronounced "knee" when plotted logarithmically, in the vicinity of the i.e.p. (see Fig. 9). As we have shown in Section III.C, this is the shape expected if the Stern–Grahame model holds. Furthermore, the equilibrium concentration for the C_{14} salt doubled at the i.e.p. when the primary surface charge was doubled (through pH change). This is also consistent with the Stern–Grahame model developed in Section II. Although these items of evidence strongly suggest that the Stern–Langmuir treatment gives a good semi-quantitative description of the results of Connor and Ottewill (1971), some quantitative difficulties remain. The slopes of the logarithmic adsorption plots below the knee are unexpectedly high (Connor and Ottewill, 1971). Correspondingly, the gradients $d\zeta/d \log C_1$ at the i.e.p. (Connor and Ottewill, 1971; Connor, 1969) are rather high and may exceed the Nernst value, which is the maximum permitted by the simple Stern–Langmuir model (Section II.A). This indicates that the adsorption may be more complex than was assumed in the theoretical treatment. Thus, in the work of Connor and Ottewill (1971), although the chain length dependence of the adsorption followed Traube's rule, and infrared studies as well as solvent desorption studies provided evidence that specific chemical-type interactions between head-groups and surface were not operative, the above mentioned discordances do indeed raise the interesting possibility of adsorption effects dependent on surfactant headgroup type.

The results of Rendall *et al.* (1979) were discussed in some detail in Section II.D. It was found that adsorption very close to the i.e.p. of nylon and polyester dispersions, in the presence of anionic and cationic surfactants respectively, could be described using the Stern–Grahame model. The chain length dependence of adsorption was consistent with chain–substrate interactions, and the response to changes in ionic strength and primary surface charge confirmed the importance of electrostatic interactions. Results for a range of anionic surfactants on a positively charged nylon dispersion, in particular, could be reproduced quantitatively on the basis of the theoretical treatment.

The transition from the low-coverage isoelectric region, where the simple Stern–Grahame model is substantially successful, to high coverage regions, where other factors have to be considered, is a matter of considerable interest. Studies such as those of Connor and Ottewill (1971) and of Bibeau and Matijević (1973) in which adsorption and electrokinetic measurements were made are potentially of great value here. It is clear, from the results of Connor and Ottewill (1971), that the net charge within the shear plane, as deduced from the ζ-potentials (see Section III.C), levels off at a surfactant concentration where the adsorption isotherm is still rising steeply, and that the maximum electrokinetic charge is equal to only a small fraction of the maximum surfactant adsorption. Bibeau and Matijević (1973) found that the electrokinetic charge density ($\equiv \sigma_0 + \sigma_\beta$) of a PVC latex could be resolved into additive components of the primary surface charge and adsorbed surfactant ions at low adsorption densities, i.e. the equality $\sigma_\beta = z_1 e \Gamma_1$ is inferred to hold. For higher coverages, however, the electrokinetic charge reflected only a fraction of the adsorbed surfactant ion charge. For an adsorbed layer density chosen to correspond to that in a surfactant micelle, the estimated proportion of bound counterions within the surfactant adsorbed layer was ~75% (Bibeau and Matijević, 1973), which is similar to a corresponding value of ~70% obtained for a surfactant micelle (Stigter and Mysels, 1955). This result indicates the potential value of the analogy between high adsorption coverage and micellization. In the presence of magnesium ions, a much smaller fraction of the adsorbed surfactant ions was detected in the electrokinetic measurements of Bibeau and Matijević (1973), consistent with the expected stronger binding of Mg^{2+} ions than Na^+ ions to the adsorbed layer. These results therefore show similar trends to those of Saleeb and Kitchener (1967) for Aerosol OT on carbon surfaces (Section IV.A).

Adsorption studies on PTFE provide an interesting contrast to results on hydrocarbon polymer surfaces. The adsorption of octanoate, dodecanoate, and perfluorooctanoate ions onto a PTFE latex has been reported by Rance (1976). No adsorption of octanoate was detected, whereas the perfluorinated analogue reached an adsorption plateau corresponding to an area per molecule of 0.54 nm^2, i.e. of the order of monolayer coverage. It was suggested that the twisted chain structure of the perfluoro ion allowed it to fit into the molecular architecture of the PTFE surface, whereas the straighter hydrocarbon chain would not find such a fit. On the other hand, Owens *et al.* (1978), in a detailed study of the wetting behaviour of aqueous (hydrocarbon) surfactant solutions on polyethylene and PTFE plates, concluded that adsorption at the two polymer surfaces was similar. This conclusion was supported by some electrokinetic evidence. Other than being ascribed to variations between samples according to their source and preparation, there is no obvious explanation for this major difference in findings. The issue awaits clarification.

C. OXIDES

In Section II.C a distinction was drawn between physical mechanisms of adsorption of ionic surfactants on hydrophilic surfaces such as oxides, and on surfaces possessing some hydrophobic character such as carbon, polymers, and various inorganic salts. The distinction resides primarily in the apparent absence, or at least much smaller significance, of attractive interactions between the hydrophobic moiety of the surfactant ion and the hydrated oxide surface, in contrast to their existence on hydrophobic surfaces. With minor reservations it was concluded that the essential features of the hemimicelle concept developed by Fuerstenau and coworkers appear to provide a physically acceptable description of adsorption on clean hydrated oxide surfaces. It is worth noting that the development of this concept has been based, by and large, on adsorption studies on silica (in the form of quartz) and alumina only. It is true to say, however, that these substrates are usually considered to show surface chemical behaviour typical of oxides in general. In view of the large corpus of literature that has emerged from Fuerstenau's group, a summary of some of its key contributions to the hemimicellization theme may be useful to the reader.

The original postulate of an adsorption mechanism akin to a bulk micellization process was first made by Gaudin and Fuerstenau (1955) based on electrokinetic measurements on the dodecylammonium chloride/quartz system. More extensive electrokinetic measurements reported by Fuerstenau (1956) for a homologous series of alkylammonium acetates on quartz demonstrated the chain length dependence of the adsorption process and also showed that, in accordance with the hemimicelle hypothesis, an increase in the negative surface charge on the quartz leads to a lowering of the h.m.c. Flotation data obtained by Fuerstenau *et al.* (1964) for a homologous series of alkylammonium acetates on quartz, as illustrated in Fig. 6, were used to strengthen the hemimicelle model. A more quantitative treatment of the adsorption process was provided by Somasundaran *et al.* (1964) based on electrokinetic data for the alkylammonium acetate/quartz system (see Fig. 5), these being almost identical to the data of Fuerstenau (1956) referred to above. Surprisingly, comprehensive adsorption data for this homologous series of surfactants on quartz appear not to have been obtained by Fuerstenau and his coworkers. Adsorption and electrokinetic data reported by Ball and Fuerstenau (1971) for dodecylammonium acetate on quartz at neutral pH do show the sort of trends referred to in Section II.C, as illustrated by Fig. 4 for the sodium dodecylsulphonate/alumina system. However, the three regions of the isotherm are not as distinct as those shown in Fig. 4, and the amounts adsorbed are about an order of magnitude smaller than values reported by de Bruyn (1955) for the same system. Furthermore, values of the adsorption free energy derived from the data obtained by Ball and Fuerstenau (1971) at different temperatures did not show the abrupt changes expected of a hemimicellization process. Healy and Dick (1971) suggested that, in contrast to alkylsulphonate adsorption on alumina, the hydration of the amine headgroup may play an inhibiting role in the adsorption process on quartz, or possibly that this headgroup may be able to structure interfacial water such that the chain–chain hydrophobic interaction term cannot operate effectively.

Adsorption isotherms for DTAB and CTAB on silica in the pH range 8–10 have been reported by Bijsterbosch (1974). At pH 8 and pH 9 the isotherms for the DTA^+ ion displayed a two-step character, whereas at pH 10 the isotherm was smooth up to the final plateau value. The isotherms for the CTA^+ ion were similarly smooth at every pH. In all cases measurable Br^- ion adsorption did not occur until final plateau adsorption was approached. According to Bijsterbosch (1974) these results, together with isotherms obtained in the presence of butanol and with electrophoresis data, support a model proposed by Rupprecht (1971). At low surfactant concentrations adsorption occurs as a result of electrostatic attraction between surfactant ions and surface without the need for attendant adsorption of Br^- ions. At higher surfactant concentrations a bilayer is formed as a result of hydrophobic interactions between chains of the first and second layers. As expected, co-adsorption of Br^- ions also occurs in this region. On the other hand, adsorption measurements made by Ter-Minassian-Saraga (1966) for DTAB on silica were partly explained by this authoress in terms of the adsorption of surfactant dimers. This mechanism was also offered (Ter-Minassian-Saraga, 1975b) to explain the different isotherm shapes encountered in Bijsterbosch's work. The latter author subsequently raised a number of objections to this explanation (Bijsterbosch, 1975).

Adsorption studies on alumina reported by the Fuerstenau school parallel those made on quartz surfaces and present a similar picture of the hemimicellization model of adsorption. Fuerstenau and Modi (1959) demonstrated from electrokinetic measurements that dodecylammonium cations only adsorb on alumina when they are counterions to the primary surface charge, so indicating the insignificance of specific chain–substrate and headgroup–substrate interactions. Below the p.z.c. of the alumina the surfactant ions acted as indifferent co-ions to the surface charge in that they behaved identically to Na^+ ions. More recently, this effect has been illustrated (see Fuerstenau, 1971) with data for the sodium dodecylsulphonate/alumina system, in the form of amounts adsorbed and ζ-potential as a function of pH (Figs. 2 and 3). Correlations made between changes in the ζ-potential and the shape of the adsorption isotherm for sodium dodecyl sulphonate on positively charged alumina (Somasundaran and Fuerstenau, 1966; Wakamatsu and Fuerstenau, 1968), as illustrated in Fig. 4 and discussed in Section II.C, were also demonstrated for a homologous series of alkyl sulphonates by Wakamatsu and Fuerstenau (1968); their isotherms appear in Fig. 7. Adsorption of the C_{14} and C_{16} homologues was found to continue above monolayer coverage. Presumably surfactant ions in the first layer are predominantly adsorbed with headgroups directed towards the surface while those of the second layer assume the opposite orientation. The expected hydrophobizing effect of the first layer of ions was demonstrated by Wakamatsu and Fuerstenau (1973) for the homologous series, from contact angle measurements. A sharp increase in contact angle occurred at each surfactant concentration corresponding to the h.m.c. Perhaps unexpectedly, no subsequent decrease in contact angle was observed at higher surfactant concentrations, as is observed for cationic surfactant/silica systems (e.g. Elton, 1957) in which contact angle returns to zero.

Further evidence in favour of the hemimicelle concept has been supplied in the form of thermodynamic data for the sodium dodecyl sulphonate/alumina system: by Roy and Fuerstenau (1968) from heats of immersion studies; by Somasundaran and Fuerstenau (1972) from adsorption data at different temperatures; and by Healy and Dick (1971) who used adsorption data previously reported by Somasundaran *et al.* (1966).

The adsorption data of Tamamushi and Tamaki (1959) for sodium dodecyl sulphate on alumina as a function of pH show, in accordance with the hemimicelle hypothesis, that adsorption does not occur when surfactant and surface are of like charge. However, dodecyl ammonium ions were found to adsorb on positively charged alumina surfaces. Electrokinetic measurements made by Doss (1976) for the DTA^+/alumina system also contradict the postulate that only ΔG_{elec} and ΔG_{cc} terms contribute to the overall adsorption free energy. Curves of ζ-potential versus pH for alumina in different concentrations of the surfactant cation did not converge at the p.z.c. of the oxide (as would be expected for the hemimicelle model) but instead converged at about 2 pH units lower than the original p.z.c. of the surface. This sort of result for oxides has been used to suggest the operation of a chemisorption mechanism (under the heading of ΔG_{hs} as discussed below) but could also be explained by the presence of hydrophobic sites on such surfaces (arising from impurities, for example), such that a ΔG_{cs} term is operative. Balzer and Lange (1979) have also found, for a series of alkylbenzene sulphonates on alumina, that adsorption still occurs to a small extent when the primary surface charge is made negative.

As a further test of the hemimicelle theory, Dick *et al.* (1971) have explored the role in adsorption of an aromatic nucleus in the hydrophobic moiety of the surfactant, and the effects of branching of the accompanying paraffinic chain. Adsorption isotherms were obtained for five isomers of sodium dodecylbenzene sulphonate on alumina. The dodecyl chain was located *para* to the headgroup and isomerism resided in the attachment of the hydrocarbon chain to the benzene group by carbon atoms 2 to 6 inclusive—chain branching therefore increasing in this order. Individually the isotherms for the surfactants displayed trends similar to those for alkyl sulphonate ions. An increase in chain branching resulted in decreased adsorption and a corresponding increase in the h.m.c., concomitant with a decreasing contribution of lateral chain–chain interactions. A comparison of derived values of ΔG_{cc} for the five isomers with those of the linear alkyl sulphonates showed that the aromatic nucleus contributed a hydrophobic interaction energy to the adsorption process equivalent to about 3 CH_2 groups, as observed in micelle formation (Shinoda *et al.*, 1963).

Rosen and Nakamura (1977) have made adsorption and electrokinetic measurements on a homologous series (C_4 to C_9) of α,ω-bis(sodium *p*-sulphophenoxy)alkanes on alumina. These symmetrical surfactants comprised two benzene sulphonate groups separated by a linear alkyl chain. In general the adsorption and ζ-potential behaviour of these systems were similar to those for the alkyl sulphonates. However, the distinctiveness of the three commonly identified regions of the adsorption isotherms was found to depend upon surfactant structure and ionic strength. Calculated values of ΔG_{cc} were about 0.55 kT per CH_2 group, which is almost half the value obtained for the alkyl sulphonates (Section II.C). Rosen and Nakamura (1977) concluded that the symmetrical ions are adsorbed horizontally on the alumina surface such that their alkyl chains are only partly removed from the aqueous environment.

Relatively little attention has been directed towards studies of ionic surfactant adsorption on other oxide surfaces. A few examples are worth citing. Vilcu and Olteanu (1975) have studied the adsorption of alkylpyridinium and alkyloxyethylenepyridinium ions on a silica–alumina cracking catalyst at pH 7 (above the measured p.z.c. of pH 6). Adsorption and electrokinetic data were qualitatively similar to those for the alkyl sulphonate/alumina and alkylammonium/quartz systems reviewed above. Calculated values of ΔG_{cc} showed that the ethylene oxide groups (either one or two per ion) *increased* the effective carbon chain length by between 0.6 and 1.2 CH_2 groups per ethylene oxide unit, in accordance with the trend of the solution c.m.c. The adsorption and electrokinetic results of Contreras and Watillon (1977), of Shergold and Kitchener (1975), and of Han *et al.* (1973) for SDS on ferric oxide (in the second case, as haematite) also provide reasonable measures of agreement with the hemimicelle theory of adsorption. As the positive charge on the oxide was reduced towards the i.e.p., the surfactant adsorption decreased towards zero. Differences in isotherm shape obtained by Contreras and Watillon (1977) for SDS and CTAB, and accompanying differences in sol stability, were explained by these authors by differences in the structuring properties of the ferric oxide interface under different pH conditions.

In a study of CTAB adsorption onto titanium dioxide, for the purpose of extracting this oxide into organic liquids, Stratton-Crawley and Shergold (1981) found that, although the general shapes of isotherms were indicative of the operation of some form of hemimicellization, adsorption nevertheless occurred, albeit only at high surfactant concentrations, when the charge on the oxide surface was rendered positive. Contact angle measurements and plateau adsorption values indicated bilayer formation on the negatively charged surfaces, the first layer being orientated with headgroup towards the substrate and the second layer in the reverse orientation.

The existence of some sort of strong "chemical" interaction between certain surfactant headgroups and particular oxide surfaces can be accounted for within the term ΔG_{hs} (Section II.C). Such chemisorption is indicated, but not necessarily proven, by the use of infrared spectroscopic measurements (Section II.B), and possibly from adsorption, electrokinetic, or flotation studies. The observation that adsorption of a surfactant ion (whether observed directly or inferred from electrokinetic or flotation data) may continue to occur, when the primary charge on the oxide is reversed such that the surfactant ions become co-ions in the diffuse layer, is not strong evidence for chemisorption of the headgroup, as is frequently assumed, unless chain–substrate or other "physical" headgroup–substrate interactions are known to be absent. Then, by elimination, the operation of a chemisorption term may be tentatively deduced. These deductions would be expected to be supported by a significant degree of irreversibility in the adsorption process.

Electrokinetic measurements have been used as evidence for chemisorption in a number of adsorption systems, including laurate and oleate on alumina and ferric oxide (Han *et al.*, 1973), dodecyl sulphate on natural haematite (Shergold and Mellgren, 1969), and oleate on titanium dioxide (Purcell and Sun, 1963). Infrared studies have indicated the occurrence of chemisorption for oleate on haematite (Peck *et al.*, 1966) and on rutile (see Fuerstenau, 1970), and for dodecyl sulphate on alumina (Yoon and Salman, 1976). The last cited work is an example of the ambiguous nature of the infrared technique in that adsorption data for an SDS/alumina system, as referred to previously (Tamamushi and Tamaki, 1959), suggest the absence of a chemisorption term.

Undoubtedly, the existence of a chemisorption mechanism is plausible for surfactant–substrate combinations in which metal soap formation is likely to occur. However, for some long chain ions such as fatty acids and amines, this adsorption mechanism may be somewhat obscured by the complicated solution behaviour of these hydrolysable species. For example, oleate ions in solution show a tendency to form a number of association complexes, and depending upon solution properties such as ionic strength, pH, and temperature, some of these highly surface-active species can be present in sufficient proportions to play a dominating role in the adsorption process. The importance of such solution equilibria in influencing the adsorption of oleate on haematite and the corresponding flotation of the mineral has been investigated by Kulkarni and Somasundaran (1980). The subject has also been reviewed by Somasundaran and Goddard (1979).

D. IONIC SOLIDS

At first sight it might seem that ionic solids would resemble hydrophilic oxide surfaces rather than the non-polar carbon and hydrocarbon surfaces reviewed in Sections IV.A and IV.B. However, such solids, and in particular the classical model surface of silver iodide, are generally classified as hydrophobic, both in the sense that they approximately conform to currently accepted theories of hydrophobic colloid stability, and that such interfaces may display significant contact angles with aqueous solutions (see e.g. Billett *et al.*, 1976).

There is indeed good evidence that chain–substrate interactions play a significant role in surfactant adsorption on AgI. The results of Rendall *et al.* (1979) for AgI support the application of the Stern–Langmuir adsorption isotherm, together with the Grahame model of the interface, for low coverages and for low potentials (i.e. near the i.e.p.). For SDS on AgI the slopes $d\zeta/d \log C_1$ responded to pAg and to ionic strength in accordance with the predictions of equations (43) and (45), as depicted in Fig. 11.

Extensive electrokinetic studies of the adsorption of cationic surfactants onto negatively charged AgI, and of anionic surfactants onto positively charged AgI surfaces, have been reported by Ottewill and coworkers (Ottewill and Rastogi, 1960; Ottewill and Watanabe, 1960; Osseo-Asare *et al.*, 1975). The experiments of Ottewill and Watanabe (1960) and of Ottewill and Rastogi (1960) were carried out at pAg 3 and pAg 12 respectively—rather far removed, especially in the latter case, from the i.e.p. of the substrate (pAg 5.6). It cannot automatically be assumed that under these conditions electrokinetic measurements will provide a sensitive monitor of adsorption. Also, even near the i.e.p. the simple theory developed in Section II might be inadequate here, in that the slopes $d\zeta/d \log C_1$ reported in these papers at i.e.p. conditions for both anionic and cationic surfactants substantially exceed the Nernst value, which is the maximum attainable in any circumstances when the Stern–Langmuir isotherm applies. The observed slopes could be explained if, at higher coverage, a coverage-dependent attractive energy could be incorporated into the isotherm to oppose the coverage-dependent electrical interaction energy term. Lateral chain–chain interactions, whether or not visualized as a hemimicellization process, could provide such an effect.

Support for this suggestion, in the case of anionic surfactant adsorption, might be found in some later studies (Osseo-Asare and Fuerstenau, 1973; Osseo-Asare *et al.*, 1975). In the presence of long-chain sulphonates at low concentrations, the ζ-potential against pAg curves were displaced towards lower pAg in a manner consistent with specific adsorption of the chain–substrate type. At higher concentrations, the ζ-potentials passed through a maximum, and eventually became more negative, as the pAg was decreased. Similar observations for oxide systems are widely quoted in support of the concept of hemimicellization (see e.g. Fuerstenau, 1970). The earlier study of Ottewill and Watanabe (1960) at pAg 3 lies well beyond this maximum in ζ-potential and hence refers to a region where chain–chain interactions could be deemed to occur. By analogy, the same would be true for the cationic surfactant/AgI system studied by Ottewill and Rastogi (1960).

Adsorption isotherms for CTAB and cetyl pyridinium bromide (CPB) onto precipitated hydroxyapatite and calcium carbonate, and onto a natural carbonate-fluoroapatite (francolite), have been reported by Hanna and Saleeb (1980). For CTAB on hydroxyapatite, the adsorption increased approximately linearly with concentration to reach a plateau, in the vicinity of the c.m.c., at an adsorption density of 0.49 nm^2 per molecule, close to the values quoted in Section IV.A for graphitized carbon, and so suggesting the formation of close-packed monolayers. The isotherm for CTAB on calcium carbonate, however, was S-shaped and reached a maximum adsorption density of 2.83 nm^2 per molecule. The natural ore showed adsorption behaviour similar to that of the calcium carbonate, suggesting a carbonate-like surface layer. Calcination of the francolite up to 1000°C shifted the adsorption behaviour towards that for hydroxyapatite, suggesting the removal of carbonate species from the interface.

The reasons for the difference in saturation adsorption between the phosphate and carbonate systems are not at all clear. Hanna and Saleeb (1980) calculated the total density of negative charges at the interfaces, based on unit cell dimensions, and found that the negative site density for calcium carbonate was 5.2 times lower than for hydroxyapatite. This compares closely with the ratio (5.7) between the adsorption densities of CTAB on the two substrates. It was therefore suggested that the adsorption depends on localized interactions between surfactant cations and negative surface sites. However, the authors' calculations also show that the total number of negative sites on apatite is about 5 times larger than the saturation adsorption density of CTAB ions. This difference may reflect the much larger molecular area of CTAB than of the lattice ions. There is therefore no reason to predict that CTAB adsorption would decrease proportionally if the negative site density at each interface was reduced. Based on the above figures, calcium carbonate would have just the number of adsorption sites required to sustain a close-packed monolayer of surfactant ions. In support of the hypothesis that surface charged sites are important in determining the maximum adsorption, the authors showed that adsorption is increased by the addition of SO_4^{2-} ions (which might adsorb and increase the density of negative sites) and reduced by the addition of Na^+ and Mg^{2+} ions (which might be expected to compete for adsorption sites). This effect may be contrasted with the effects of electrolyte on saturation adsorption on carbon and polymer substrates, where increased ionic strength leads to increased saturation adsorption (Sections IV.A and IV.B).

In marked contrast to the above observations, Cases *et al.* (1975) have reported adsorption of alkylammonium ions onto calcite, which extends at least to a level corresponding to bilayer coverage. It is not clear whether the presence of 0.4% of the adsorbing species as the alkylamine molecule at the pH (~8.2) of measurement enhances adsorption of the alkylammonium ions, as compared with alkyltrimethylammonium ion adsorption. An interesting feature of these studies (Cases *et al.*, 1975; Cases, 1979) is that isotherms for apatite, calcite and silicate substrates, plotted on a θ versus log C_1 basis, are not smooth but are composed of a series of steps. The authors have interpreted their results on the basis of a model in which two-dimensional condensation of the surfactant ions occurs (through chain–chain interactions) on a series of uniform domains within each of which the adsorption sites are identically energetic. Physically, such a heterogeneous surface could represent the different crystal planes exposed to the solution by a crystalline material. The adsorption results may then be analysed to give an energy distribution curve for adsorption sites on the substrate. Grinding of the mineral substrate gave rise to an almost smooth and step-free adsorption isotherm which was displaced towards higher equilibrium concentrations. Correspondingly, the energy distribution curve shows a reduction in high-energy sites and the production of a much more energetically homogeneous surface characterized by the lowest energy sites (Cases, 1979).

The model of a patch-wise heterogeneous solid surface has been developed in some detail for adsorption at the solid–gas interface (see e.g. House and Jaycock, 1978). Its application to adsorption at the solid/aqueous solution interface would allow surface characterization under the conditions appropriate to many applications. The prime requirement is a much more extensive set of reliable adsorption data than is usually obtained. The results may be of particular interest, for example, in understanding the inhibition of homogeneous nucleation by surfactants (Goujon *et al.*, 1976).

A further point to consider is that the process of grinding a crystalline material to achieve a sufficiently high surface area for adsorption studies may, at the same time, drastically alter its surface properties. Wherever possible, therefore, studies aimed at elucidating technological issues should be conducted on unground samples.

Finally, mention is made of systems in which adsorption of the surfactant ion is considered to be dominated by interactions of a "chemical" nature between headgroup and surface. These include soaps, such as oleate, on fluorite, barite, calcite, and apatite minerals, dodecylbenzene sulphonate, dodecyl sulphonate, and Aerosol OT on calcium carbonate and calcium phosphate, and xanthate on sulphide minerals. The reviews of Fuerstenau (1971) and of Somasundaran and Goddard (1979) include some discussion of this topic, and the interested reader is referred to these articles.

E. CLAYS AND SILICATES

Silicates and clays are fundamentally more complex and heterogeneous than the systems so far described, and are therefore less suitable as model substrates to test theories of adsorption. Nevertheless surfactant adsorption onto silicate and clay minerals has been widely studied because of its technical and commercial importance. Even for a given mineral type, chemical composition can be widely variable. Thus, for example, the variable nature and proportions of cations associated with particular clay type minerals may strongly influence the adsorption of anionic surfactants by those minerals (Ralston and Kitchener, 1975; Schott, 1967; Read and Manser, 1972). Furthermore, the swelling behaviour of certain clays, such as the montmorillonites, when contacted with water, is strongly dependent on the nature of the cations present between the aluminosilicate sheets—available surface areas for surfactant adsorption being affected accordingly. Such minerals, in fact, may be intercalated with surfactants, to give regular interlayer spacings which may be measured by X-ray diffraction, so giving the surfactant adsorbed layer thickness (e.g. Weiss, 1980; Greenland and Quirk, 1962).

Although the face charges of clay platelets are fixed and essentially independent of solution conditions, the edge charges can show oxide type pH dependence and may therefore give rise to pH dependent adsorption. Dissolution and hydrolysis processes, which can substantially change the surface properties of clays and silicates, may also be responsible for variations in surfactant adsorption with solution conditions (e.g. ionic strength), with time, and with pre-treatment of the substrate (e.g. Hanna and Somasundaran, 1979).

The mineral mica, however, occupies a special place in fundamental studies. It is possible to cleave mica in such a way as to provide a clean molecularly smooth surface. Thus a uniquely suitable substrate is obtained for contact angle measurements and for studies of interfacial forces. A recent study of interactions between mica surfaces possessing adsorbed layers of CTAB (Pashley and Israelachvili, 1981), together with contact angle measurements, has provided good information on adsorption at this interface. Measurements of intersurface separations showed that a surfactant monolayer was formed on each mica surface at concentrations above about 6×10^{-5} mol dm^{-3}, and a bilayer at about 10^{-3} mol dm^{-3} (i.e. in the vicinity of the c.m.c.). This adsorption would be promoted by the high (fixed) primary charge on the surface (for the first monolayer) and by chain–chain interactions between the two layers in the bilayer. Evidence was also obtained for hydration effects at the mica/aqueous solution interface, and for their disruption by the adsorbed surfactant.

Pearson and Wade (1972) reported the adsorption of CTAB on kaolinite. They noted an inflection in the isotherm at a solution concentration of about 5×10^{-4} mol dm^{-3}, which was interpreted as due to a changeover to adsorption (by van der Waals attraction) onto the positively charged crystal edges. Equally, and perhaps more plausibly, the inflection could be taken to indicate the onset of bilayer formation on the crystal faces.

Flegman and Ottewill (1967) measured the adsorption of hexadecyl sulphate ions onto kaolinite (at pH 3). Saturation adsorption corresponded either to a monolayer of 0.35 nm^2 per molecule over both the negatively charged faces and positively charged edges, or, more likely in the authors' opinion, to bilayer formation on the positive edges, with adsorption of the anions on the similarly charged faces being absent. This interpretation was reinforced by electrophoresis and flocculation measurements. Hanna and Somasundaran (1979) studied the adsorption of SDS and sodium dodecylbenzene sulphonate (DOBS) onto kaolinite. The adsorption showed some complex time-dependent behaviour. Saturation adsorption for DOBS at low pH corresponded to 0.34 nm^2 per molecule, similar to values obtained by the same authors for DOBS on calcium carbonate and calcium phosphate. Nevertheless, it would seem more likely, by analogy with the above work, that a bilayer of DOBS was formed on the positively charged edges of the clay platelets. In agreement with this hypothesis, the adsorption was strongly pH-dependent, with saturation adsorption in the alkaline region being about 12% of that at low pH (Hanna and Somasundaran, 1979). Throughout the pH range, adsorption was enhanced by the addition of salt. It was also found by Hanna and Somasundaran, however, that the shape of the isotherm was dependent on both the cation and the anion used (see Somasundaran and Goddard, 1979). Differences in adsorption capacity and the presence of adsorption maxima were explained in terms of differences in ion size and hydration properties.

Significant features of adsorption onto silicate minerals are exemplified by studies of CTAB adsorption onto amosite asbestos (Ralston and Kitchener, 1975) and of dodecylammonium ions onto chrysotile asbestos (Atkinson, 1973). CTAB gave an adsorption plateau at solution concentrations around 5×10^{-5} mol dm^{-3} for both pH 5.5 and pH 10.5, with a slightly higher adsorption density in the latter case. This was taken to represent saturation of the negatively charged sites. A steep rise in adsorption observed around 10^{-4} mol dm^{-3} CTAB was attributed to chain–chain association (Ralston and Kitchener, 1975). Chrysotile asbestos showed a marked reduction in the uptake of dodecylammonium ions as a consequence of ageing in HCl prior to adsorption, the effect developing over several days. This loss in adsorption capacity was postulated to be due to silicic acid condensation reactions.

Acknowledgements

Work on this chapter was started while one of us (D.B.H.) was a member of staff in the Department of Pure and Applied Chemistry, University of Strathclyde, Glasgow. Library services and other facilities made available by this department are gratefully acknowledged. We are particularly grateful to Dr T. L. Whateley of the Department of Pharmaceutical Chemistry of the university for his help in the collection and collation of some of the literature cited in this review and for many useful discussions on its subject matter.

References

Abbott, D. C. (1962). *Analyst* **87**, 286.
Adamson, A. W. (1976). *Physical Chemistry of Surfaces*, 3rd edn, pp. 90, 389. John Wiley and Sons, New York, London.
Allen, L. H. and Matijević, E. (1970). *J. Colloid Interface Sci.* **33**, 420.
Arai, H. and Yoshizaki, K. (1971). *J. Colloid Interface Sci.* **35**, 149.
Atkinson, R. J. (1973). *J. Colloid Interface Sci.* **42**, 624.
Ball, B. and Fuerstenau, D. W. (1971). *Discuss. Faraday Soc.*, **52**, 361.
Balzer, D. and Lange, H. (1979). *Colloid Polym. Sci.* **257**, 292.
Barlow, R. B., Lowe, B. M., Pearson, J. D. M., Rendall, H. M. and Thompson, G. M. (1971). *Mol. Pharmacol.* **7**, 357.
Barr, T., Oliver, J. and Stubbings, W. V. (1948). *J. Soc. Chem. Ind.* **67**, 45.
Bell, L. C., Posner, A. M. and Quirk, J. P. (1973). *J. Colloid Interface Sci.* **42**, 250.
Bibeau, A. A. and Matijević, E. (1973). *J. Colloid Interface Sci.* **43**, 330.
Biefer, G. J. and Mason, S. G. (1959). *Trans Faraday Soc.* **55**, 1239.
Bijsterbosch, B. H. (1974). *J. Colloid Interface Sci.* **47**, 186.
Bijsterbosch, B. H. (1975). *J. Colloid Interface Sci.* **51**, 212.
Bijsterbosch, B. H. and Lyklema, J. (1978). *Adv. Colloid Interface Sci.* **9**, 147.
Billett, D. F., Hough, D. B. and Ottewill, R. H. (1976). *J. Electroanal. Interfacial Electrochem.* **94**, 107.
Birch, B. J., Clarke, D. E., Lee, R. S. and Oakes, J. (1974). *Anal. Chim. Acta* **70**, 417.
Bisio, P. D., Cartledge, J. G., Keesom, W. H. and Radke, C. J. (1980). *J. Colloid Interface Sci.* **78**, 225.
Blodgett, K. B. (1935). *J. Am. Chem. Soc.* **57**, 1007.
Blodgett, K. B. and Langmuir, I. (1937). *Phys. Rev.* **51**, 964.
Bolt, G. H. (1957). *J. Phys. Chem.* **61**, 1166.

Brunauer, S., Emmett, P. H. and Teller, E. (1938). *J. Am. Chem. Soc.* **60**, 309.
de Bruyn, P. L. (1955). *Trans AIME* **202**, 291
Cases, J. M. (1979). *Bull. Minéral.* **102**, 684.
Cases, J. M., Goujon, G. and Smani, S. (1975). *AIChE Sym. Ser.* **71**, 100.
Clark, A. H., Franks, F., Pedley, M. D. and Reid, D. S. (1977). *J. Chem. Soc. Faraday Trans. 1* **73**, 290.
Connor, P. (1968). Ph.D thesis, University of Bristol.
Connor, P. (1969). 5th International Congress of Surface Active Substances, **2**, 469.
Connor, P. and Ottewill, R. H. (1971). *J. Colloid Interface Sci.* **37**, 642.
Contreras, S. and Watillon, A. (1977). *Croat. Chem. Acta* **50**, 1.
Day, R. E., Greenwood, F. G. and Parfitt, G. D. (1967). 4th International Congress of Surface Active Substances, **2B**, 1005.
Dick, S. G., Fuerstenau, D. W. and Healy, T. W. (1971). *J. Colloid Interface Sci.* **37**, 595.
Dixit, S. G. and Biswas, A. K. (1975). *AIChE Sym. Ser.* **71**, 88.
Dobias, B. (1977). *Colloid Polym. Sci.* **255**, 682.
Dobias, B. (1978). *Colloid Polym. Sci.* **256**, 465.
Doss, S. K. (1976). *Inst. Min. Metall. Trans.* (*C*) **85**, 195.
Elton, G. A. H. (1957). 2nd International Congress on Surface Activity, **3**, 161.
Epton, S. R. (1948). *Trans Faraday Soc.*, **44**, 226.
Fava, A. and Eyring, H. (1956). *J. Phys. Chem.* **60**, 890.
Few, A. V. and Ottewill, R. H. (1956). *J. Colloid Sci.* **11**, 34.
Flegman, A. W. and Ottewill, R. H. (1967). 4th International Congress of Surface Active Substances, **2B**, 1271.
Foxall, T., Peterson, G. C., Rendall, H. M. and Smith, A. L. (1979). *J. Chem. Soc. Faraday Trans. 1* **75**, 1034.
Fuerstenau, D. W. (1956). *J. Phys. Chem.* **60**, 981.
Fuerstenau, D. W. (1970). *Pure Appl. Chem.* **24**, 135.
Fuerstenau, D. W. (1971). In *The Chemistry of Biosurfaces* (M. L. Hair, ed.), Vol. 1, p. 143, Marcel Dekker, New York.
Fuerstenau, D. W. and Healy, T. W. (1972). In *Adsorptive Bubble Separation Techniques* (R. Lemlich, ed.), p. 91, Academic Press, New York and London.
Fuerstenau, D. W. and Modi, H. J. (1959). *J. Electrochem. Soc.* **106**, 336.
Fuerstenau, D. W., Healy, T. W. and Somasundaran, P. (1964). *Trans. AIME* **229**, 321.
Gaudin, A. M. and Fuerstenau, D. W. (1955). *Trans. AIME* **202**, 958.
Giles, C. H. and Tolia, A. H. (1964). *J. Appl. Chem.* **14**, 186.
Giles, C. H., D'Silva, A. P. and Trivedi, A. S. (1970). *J. Appl. Chem.* **20**, 37.
Ginn, M. E., Kinney, F. B. and Harris, J. C. (1961). *J. Am. Oil Chem. Soc.* **38**, 138.
Goodwin, J. W., Hearn, J., Ho., C. C. and Ottewill, R. H. (1973). *Br. Polym. J.* **5**, 347.
Goodwin, J. W., Ottewill, R. H. and Pelton, R. (1979). *Colloid Polym. Sci.* **257**, 61.
Goujon, G., Cases, J. M. and Mutaftschiev, B. (1976). *J. Colloid Interface Sci.* **56**, 587.
Grahame, D. C. (1947). *Chem. Rev.* **41**, 441.
Greenland, D. J. and Quirk, J. P. (1962). Proc. Nat. Conf. Clays Clay Minerals, *Clays Clay Minerals*, **9**, 484.
Greenland, D. J. and Quirk, J. P. (1963). *J. Phys. Chem.* **67**, 2886.

Greenwood, F. G., Parfitt, G. D., Picton, N. H. and Wharton, D. G. (1968). *Adv. Chem. Ser.* No. 79, 135.
Hall, D. G. (1978). *J. Chem. Soc. Faraday Trans. 2* **74**, 1757.
Hall, D. G. (1980a). *J. Chem. Soc. Faraday Trans. 1* **76**, 386.
Hall, D. G. (1980b). *J. Chem. Soc. Faraday Trans. 2* **76**, 1254.
Hall, D. G. and Rendall, H. M. (1980). *J. Chem. Soc. Faraday Trans. 1* **76**, 2575.
Hall, D. G., Rendall, H. M. and Smith, A. L. (1980). *Croat. Chem. Acta* **53**, 147.
Han, K. N., Healy, T. W. and Fuerstenau, D. W. (1973). *J. Colloid Interface Sci.* **44**, 407.
Hanna, H. S. and Saleeb, F. Z. (1980). *Colloids Surf.* **1**, 295.
Hanna, H. S. and Somasundaran, P. (1979). *J. Colloid Interface Sci.* **70**, 181.
Harris, N. M., White, J. W. and Ottewill, R. H. (1979). Institut Laue-Langevin Report, Grenoble, Experiment 09-05-249.
Healy, T. W. (1974). *J. Macromol. Sci., Chem.* **A8**, 603.
Healy, T. W. and Dick, S. G. (1971). *Discuss. Faraday Soc.* **52**, 378.
Healy, T. W. and White, L. R. (1978). *Adv. Colloid Interface Sci.* **9**, 303.
Homola, A. and James, R. O. (1977). *J. Colloid Interface Sci.* **59**, 123.
Hough, D. B. (1973). Ph.D thesis, University of Bristol.
House, W. A. and Jaycock, M. J. (1978). *Colloid Polym. Sci.* **256**, 52.
van den Hul, H. J. and Lyklema, J. (1968). *J. Am. Chem. Soc.* **90**, 3010.
van den Hul, H. J. and Vanderhoff, J. W. (1968). *J. Colloid Interface Sci.* **28**, 336.
van den Hul, H. J. and Vanderhoff, J. W. (1970). *Br. Polym. J.* **2**, 121.
van den Hul, H. J. and Vanderhoff, J. W. (1971). In *Polymer Colloids* (R. M. Fitch, ed.), Plenum Press, New York.
James, R. O., Homola, A. and Healy, T. W. (1977). *J. Chem. Soc. Faraday Trans. 1* **73**, 1436.
de Keizer, A. and Lyklema, J. (1980). *J. Colloid Interface Sci.* **75**, 171.
Kitchener, J. A. (1965). *J. Photogr. Sci.* **13**, 152.
Kitchener, J. A. (1967). 4th International Congress of Surface Active Substances, **2B**, 139.
Kivel, J. Albers, F. C., Olsen, D. A. and Johnson, R. E. (1963). *J. Phys. Chem.* **67**, 1235.
Klimenko, N. A. (1979). *Colloid J.* (U.S.S.R.) (Engl. transl.) **41**, 666.
Klimenko, N. A., Polyakov, V. E. and Permilovskaya, A. A. (1979). *Colloid J.* (U.S.S.R.) (Engl. transl.) **41**, 913.
Koganovskii, A. M., Klimenko, N. A. and Permilovskaya, A. A. (1975). *Colloid J.* (U.S.S.R.) (Engl. transl.) **37**, 588.
Koganovskii, A. M., Klimenko, N. A. and Chobanu, M. M. (1977). *Colloid J.* (U.S.S.R.) (Engl. transl) **39**, 307.
Koganovskii, A. M., Klimenko, N. A. and Chobanu, M. M. (1979). *Colloid J.* (U.S.S.R.) (Engl. transl.) **41**, 851.
Kulkarni, R. D. and Somasundaran, P. (1980). *Colloids and Surfaces* **1**, 387.
Levine, S. and Smith, A. L. (1971). *Discuss. Faraday Soc.* **52**, 290.
Levine, S., Mingins, J. and Bell, G. M. (1967). *J. Electroanal. Chem. Interfacial Electrochem.* **13**, 280.
Lin, I. J. and Somasundaran, P. (1971). *J. Colloid Interface Sci.* **37**, 731.
Longwell, J. and Maniece, W. V. (1955). *Analyst* **80**, 167.
Lyklema, J. (1961). *Kolloid-Z.* **175**, 129.
Lyklema, J. (1966). *Discuss. Faraday Soc.* **42**, 81.
Lyklema, J. (1977). *J. Colloid Interface Sci.* **58**, 242.
Lyklema, J. and Overbeek, J. Th. G. (1961). *J. Colloid Sci.* **16**, 501.

Maron, S. H., Elder, M. E. and Ulevitch, I. N. (1954). *J. Colloid Sci.* **9**, 89.
Medalia, A. I. and Riven, D. (1976). In *Characterisation of Powder Surfaces* (G D. Parfitt and K. S. W. Sing, eds), p. 279, Academic Press, London.
Mukerjee, P. (1956). *Anal. Chem.* **28**, 870.
Mukerjee, P. and Anavil, A. (1975). In *Adsorption at Interfaces*, ACS Symposium Ser. **8**, 107.
Mukerjee, P. and Mysels, K. J. (1971). *Critical Micelle Concentrations of Aqueous Surfactant Systems*, NBS Publication, NSRDS-NBS 36, Nat. Bur. Stand., Washington.
Muramatsu, M. (1973). In *Surface and Colloid Science* (E. Matijević, ed.), Vol. 6, p. 101, John Wiley and Sons, New York.
Muramatsu, M., Tajima, K., Iwahashi, M. and Nukina, K. (1973). *J. Colloid Interface Sci.* **43**, 499.
Nilsson, G. (1957). *J. Phys. Chem.* **61**, 1135.
Nunn, C., Schechter, R. S. and Wade, W. H. (1981). *J. Colloid Interface Sci.* **80**, 598.
Oakes, J. (1974). *J. Chem. Soc. Faraday Trans. 1* **70**, 2200.
van Olphen, H. (1977). *An Introduction to Clay Colloid Chemistry*, 2nd edn, Wiley-Interscience, New York, London.
van Oss, C. J., Absolom, D. R. and Neumann, A. W. (1980). *Colloid Polym. Sci.* **258**, 424.
Osseo-Asare, K. and Fuerstenau, D. W. (1973). *Croat. Chem. Acta* **45**, 149.
Osseo-Asare, K., Fuerstenau, D. W. and Ottewill, R. H. (1975). In *Adsorption at Interfaces*, ACS Symp. Ser. **8**, 63.
Ottewill, R. H. (1979). In *Surface Active Agents*, p. 1, Society of Chemical Industry, London.
Ottewill, R. H. and Rastogi, M. C. (1960). *Trans. Faraday Soc.* **56**, 880.
Ottewill, R. H. and Walker, T. (1974). *J. Chem. Soc. Faraday Trans. 1* **70**, 917.
Ottewill, R. H. and Watanabe, A. (1960). *Kolloid-Z.* **170**, 132.
Ottewill, R. H., Rastogi, M. C. and Watanabe, A. (1960). *Trans. Faraday Soc.* **56**, 854.
Overbeek, J. Th. G. (1952). In *Colloid Science* (H. R. Kruyt, ed.), Vol. 1, p. 115, Elsevier, Amsterdam.
Owens, N. F., Richmond, P., Gregory, D., Mingins, J. and Chan, D. (1978). In *Wetting, Spreading and Adhesion* (J. F. Padday, ed.), p. 127, Academic Press, London.
Padday, J. F. (1967). 4th International Congress of Surface Active Substances, **2B**, 299.
Parfitt, G. D. and Picton, N. H. (1968). *Trans Faraday Soc.* **64**, 1955.
Parks, G. A. and de Bruyn, P. L. (1962). *J. Phys. Chem.* **66**, 967.
Pashley, R. M. and Israelachvili, J. N. (1981). *Colloids Surf.* **2**, 169.
Pearson, J. T. and Wade, G. (1972). *J. Pharm. Pharmacol.* **24**, Suppl. 132P.
Peck, A. S., Raby, L. H. and Wadsworth, M. E. (1966). *Trans. AIME* **238**, 301.
Pham Thi Hang and Brindley, G. W. (1970). Proc. Nat. Conf. Clays Clay Minerals, *Clays Clay Minerals*, **18**, 203.
Piirma, I. and Chen, S.-R. (1980). *J. Colloid Interface Sci.* **74**, 90.
Purcell, G. and Sun, S. C. (1963). *Trans. AIME* **226**, 6, 13.
Raghavan, S. and Fuerstenau, D. W. (1975). *J. Colloid Interface Sci.* **50**, 319.
Ralston, J. and Kitchener, J. A. (1975). *J. Colloid Interface Sci.* **50**, 242.
Rance, D. G. (1976). Ph.D thesis, University of Bristol.
Read, A. D. and Manser, R. M. (1972). *Inst. Min. Metall. Trans.* (*C*) **81**, C69.

Rendall, H. M. and Smith, A. L. (1978). *J. Chem. Soc. Faraday Trans. 1* **74**, 1179.
Rendall, H. M. and Smith, A. L. (1979). In *Surface Active Agents*, p. 37, Society of Chemical Industry, London.
Rendall, H. M., Smith, A. L. and Williams, L. A. (1979). *J. Chem. Soc. Faraday Trans. 1* **75**, 669.
Robb, D. J. M. and Alexander, A. E. (1967). Monograph No. 25, Society of Chemical Industry, London.
Rosen, M. J. (1978). *Surfactants and Interfacial Phenomena.* John Wiley and Sons, New York.
Rosen, M. J. and Nakamura, Y. (1977). *J. Phys. Chem.* **81**, 873.
Roy, P. and Fuerstenau, D. W. (1968). *J. Colloid Interface Sci.* **26**, 102.
Rupprecht, H. (1971). *Kolloid-Z.* **249**, 1127.
Saleeb, F. Z. and Kitchener, J. A. (1965). *J. Chem. Soc.* 911.
Saleeb, F. Z. and Kitchener, J. A. (1967). 4th International Congress of Surface Active Substances, **2B**, 129.
Schott, H. (1967). *Kolloid-Z.* **219**, 42.
Sekine, K., Seimiya, T. and Sasaki, T. (1970). *Bull. Chem. Soc. Japan* **43**, 629.
Sexsmith, F. H. and White, H. J. (1959). *J. Colloid Sci.* **14**, 598.
Shergold, H. L. and Kitchener, J. A. (1975). *Int. J. Miner. Process.* **2**, 249.
Shergold, H. L. and Mellgren, O. (1969). *Inst. Min. Metall. Trans. (C)* **78**, C121.
Shinoda, K., Tamamushi, B., Nakagawa, T. and Isemura, T. (1963). *Colloidal Surfactants*, Academic Press, New York and London.
Smith, A. L. (1973). In *Dispersions of Powders in Liquids* 2nd edn (G. D. Parfitt, ed.), p. 86, Applied Science Publishers, London.
Smith, A. L. (1976). *J. Colloid Interface Sci.* **55**, 525.
Snyder, L. R. (1968). *J. Phys. Chem.* **72**, 489.
Somasundaran, P. (1975). *AIChE Sym. Ser.* **71**, 1.
Somasundaran, P. and Fuerstenau, D. W. (1966). *J. Phys. Chem.* **70**, 90.
Somasundaran, P. and Fuerstenau, D. W. (1972). *Trans. AIME* **252**, 275.
Somasundaran, P. and Goddard, E. D. (1979). *Mod. Aspects Electrochem.* **13**, 207.
Somasundaran, P. and Hanna, H. S. (1977). In *Improved Oil Recovery by Surfactant and Polymer Flooding* (D. O. Shah and R. S. Schechter, eds), p. 205, Academic Press, New York, London.
Somasundaran, P., Healy, T. W. and Fuerstenau, D. W. (1964). *J. Phys. Chem.* **68**, 3562.
Somasundaran, P., Healy, T. W. and Fuerstenau, D. W. (1966). *J. Colloid Interface Sci.* **22**, 599.
Sørenson, E. (1973). *J. Colloid Interface Sci.* **45**, 601.
Sparnaay, M. J. (1972). In *The International Encyclopedia of Physical Chemistry and Chemical Physics* (D. D. Eley and F. C. Tompkins, eds), Topic 14, "Properties of Interfaces" (D. H. Everett, ed.), Vol. 14, "The Electrical Double Layer", p. 104, Pergamon Press, Oxford.
Stamm, A. J. and Millett, M. A. (1941). *J. Phys. Chem.* **45**, 43.
Steinhardt, J. and Reynolds, J. A. (1969). *Multiple Equilibria in Proteins*, Academic Press, New York.
Stigter, D. and Mysels, K. J. (1955). *J. Phys. Chem.* **59**, 45.
Stratton-Crawley, R. and Shergold, H. L. (1981). *Colloids Surf.* **2**, 145.
Tajima, K. (1971). *Bull. Chem. Soc. Japan* **44**, 1767.
Tamamushi, B. and Tamaki, K. (1959). *Trans. Faraday Soc.* **55**, 1007.
Tanford, C. (1972). *J. Phys. Chem.* **76**, 3020.

Tanford, C. (1974). *J. Phys. Chem.* **78**, 2469.
Ter-Minassian-Saraga, L. (1966). *J. Chim. Phys.* **63**, 1278.
Ter-Minassian-Saraga, L. (1975a). *AIChE Sym. Ser.* **71**, 68.
Ter-Minassian-Saraga, L. (1975b). *J. Colloid Interface Sci.* **51**, 211.
Traube, I. (1891). *Annalen* **265**, 27.
Vijayendran, B. R. (1979). *J. Appl. Polym. Sci.* **23**, 733.
Vilcu, R. and Olteanu, M. (1975). *Rev. Roum. Chim.* **20**, 1041.
Wakamatsu, T. and Fuerstenau, D. W. (1968). *Adv. Chem. Ser.* **79**, 161.
Wakamatsu, T. and Fuerstenau, D. W. (1973). *Trans. AIME* **254**, 123.
Weiss, A. (1980). *Chem. and Ind.* (London), No. 9, 382.
Weiss, J., Zografi, G. and Simonelli, A. P. (1974). *J. Pharm. Sci.* **63**, 381.
Wiersema, P. H., Loeb, A. L. and Overbeek, J. Th. G. (1966). *J. Colloid Interface Sci.* **22**, 78.
Yates, D. E. and Healy, T. W. (1976). *J. Colloid Interface Sci.* **55**, 9.
Yoon, R. H. and Salman, T. (1976). In *Colloid and Interface Science* (M. Kerker, ed.), Vol. 3, p. 233, Academic Press, New York.
Young, G. J., Chessick, J. J., Healy, F. H. and Zettlemoyer, A. C. (1954). *J. Phys. Chem.* **58**, 313.
Zettlemoyer, A. C. (1968). *J. Colloid Interface Sci.* **28**, 343.
Zettlemoyer, A. C., Schneider, C. H. and Skewis, J. D. (1957). 2nd International Congress on Surface Activity, **3**, 472.
Zettlemoyer, A. C., Skewis, J. D. and Chessick, J. J. (1962). *J. Am. Oil Chem. Soc.* **39**, 280.

7. Adsorption of Dyes

C. H. GILES

Robert Boyle (1664) published the first scientific treatise on colour, and in it wrote: "The Liquors that Dyers imploy to tinge are qualified to do so by multitudes of little corpuscles of the Pigment or Dying stuff, which are dissolved and extracted by the Liquor, and swim to and fro in it, those Corpuscles of Colour . . . insinuating themselves into, and filling all the Pores of the Body to be Dyed, . . .". It will be seen later that this was a most percipient observation, and if translated into modern terms expresses quite closely our present understanding of the process of dye adsorption by porous solids.

I. Outline of Nature of Dyes and Substrates

In this chapter we are mainly concerned with the dyeing of textile fibres, and firstly therefore an outline of their chemical nature and of some of the morphological properties of the fibres will be helpful to an understanding of the mutual interactions involved.

A. DYES

To be technically useful, dyes must have (i) intense colour, (ii) fastness, i.e. resistance to the various chemical and mechanical stresses they meet in the manufacture and use of the finished coloured product, (iii) solubility in the medium of application, which is almost always water, and (iv) the ability to be adsorbed and retained by the fibre (substantivity) or to be chemically combined with it (reactivity) (Allen, 1971; Abrahart, 1977).

1. *Colour*

This depends on the presence in the molecule of a long conjugate chain of single and double bonds—the chromophore—with one or more polar substituents—the auxochromes. Detailed discussion of the subject is outside the scope of this chapter, and reviews or monographs should be consulted (e.g. Reynolds, 1943, 1944; Lewis, 1945; Ferguson, 1948; Peters and Sumner, 1956; Murrell, 1964; Coates, 1967; Griffiths, 1976).

The most important chromophore is the aromatic azo system, which accounts for by far the largest proportion of commercial dyes, followed by that of anthraquinone and other polycyclic quinones; triphenylmethane and related systems, heterocyclic systems, e.g. azine and thiazine, and other miscellaneous systems are of lesser importance. The chromophore itself absorbs radiation mainly in the ultraviolet region and as a substance appears only very weakly coloured. The function of the auxochromes is to intensify the colour of the chromophore, by their ability to move the absorption bands of the chromophore to longer wavelengths by stabilizing resonance forms of the structure; thus visible colour appears. The intensity of the colour of dyes is among the highest found in any type of substance, the molar extinction coefficients being of the order of 10^4. In practice this means that only some 10 to 100 multilayers of dye are required just to be visible to the eye.

2. *Fastness*

Ability to withstand various wet treatments, especially washing, and exposure to light are two of the most important properties of dyes in fibres. Fastness to light depends on the complex inter-relation of many factors internal and external to the dyed fibre, and is outside the scope of this chapter, but fastness to washing is closely related to the mechanism by which the dye is adsorbed, as we shall later observe. Dyes that are covalently bonded to the fibre molecule, i.e. reactive dyes, and those that form water-insoluble particles in the fibre, during the dyeing process, have the highest wash fastness. Water-soluble dyes often have lower wash fastness, which varies greatly from one dye to another but is aided by high molecular weight of the dye and a minimum number of solubilizing groups in the molecule.

3. *Solubility*

Solubility in water is conferred by ionogenic or polar groups; these may have to operate only during the actual dyeing operation, and afterwards they may be inactivated so that the wet fastness of the finished material is as high as possible. The principal solubilizing groups used for commercial dyes are shown in Table 1.

Table 1 Solubilizing groups used with various classes of dyes

Group	Dye classes
	Permanent
Ionogenic	
Sulphonate, $-SO_3^-Na^+$	Acid wool, direct cotton, chrome mordant, 1:1 dye–metal complex wool dyes
Amino and quaternary ammonium, $-\overset{+}{N}H_2HCl^-$, $-N^+R_3Cl^-$	Cationic dyes for cellulose, wool, silk, and acrylic fibres
Polar	
Hydroxy, amino, and sulphamido, $-OH$, $-NH_2$, $-SO_2NH_2$, etc.	Disperse dyes for synthetic polymer fibres and 2:1 dye–metal dyes for wool and nylon
	Temporary
Sodium phenolate, $-O^-Na^+$	Insoluble azo (azoic) dyes for cellulose (naphthols coupled on the fibre with diazo compounds); vat dyes, later insolubilized by oxidation to quinone groups
Isothiouronium, $-CH_2SC(=\overset{+}{N}R_2)NR_2\ Cl^-$	Phthalocyanine dyes for cellulose

4. *Typical Dye Structures*

Some typical structures of commercial dyes, which illustrate the above general principles are as follows:

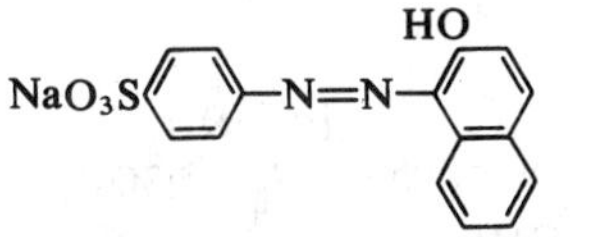

NaO₃S–C₆H₄–N=N–(2-hydroxy-1-naphthyl) [HO]

C.I. (Colour Index, 1975) 15510 (Orange II); azo dye for wool

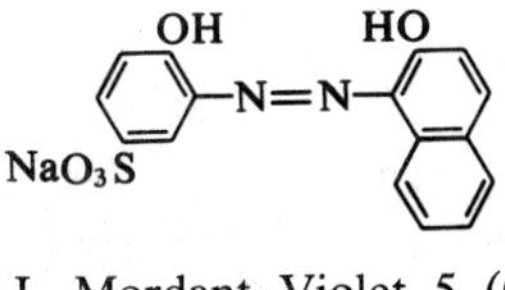

NaO₃S, OH–C₆H₃–N=N–(2-hydroxy-1-naphthyl) [HO]

C.I. Mordant Violet 5 (C.I. 15670); the colour is developed by chelation with chromium in the dyeing process

C.I. 61530; blue anthraquinone dye for wool

C.I. 61500; blue disperse dye for synthetic polymer fibres

Blue cationic dye for acrylic fibres

C.I. 14880; 1:1 dye–metal complex, blue dye for wool

2:1 dye–metal complex; bordeaux shade dye for protein fibres and nylon

Pyranthrone, C.I. 59700; orange vat dye for cellulose

C.I. 24410; blue direct dye for cellulose

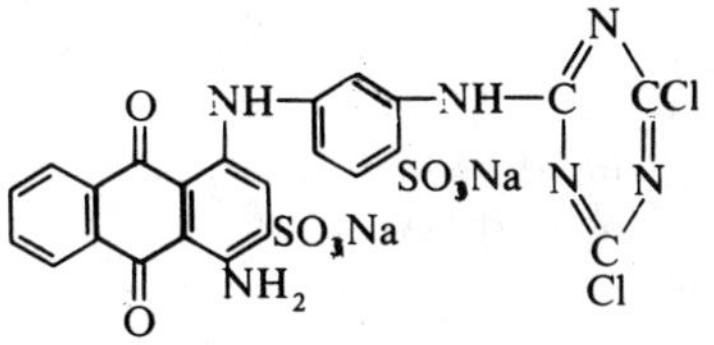

Reactive dye for cellulose; reddish blue shade, with triazine reactive group

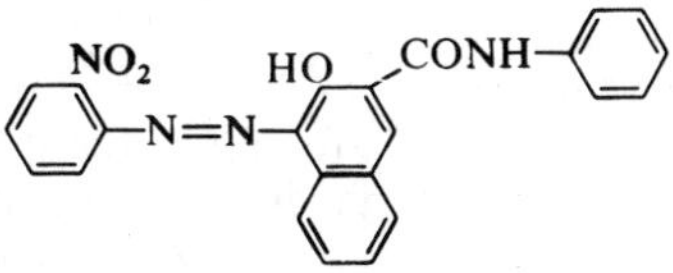

Orange dye for cellulose, formed in the fibre by diazo coupling

B. FIBRES

The principal repeat units of the molecular structures of the important textile fibres are as follows:

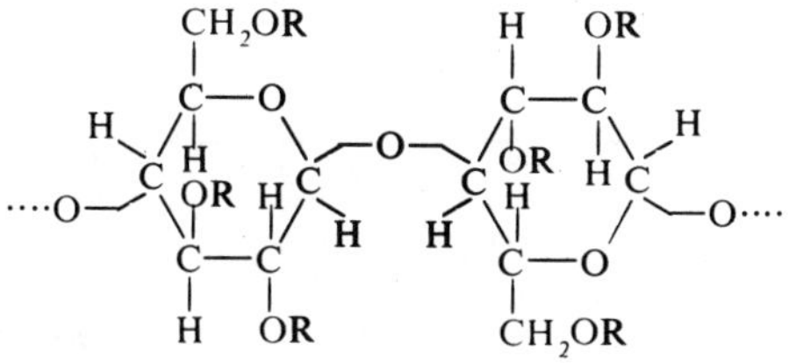

Cellulose (R = H) or cellulose triacetate (R = COMe); secondary cellulose acetate has a proportion of free hydroxy groups, in the ratio of about two hydroxy to five acetyl groups

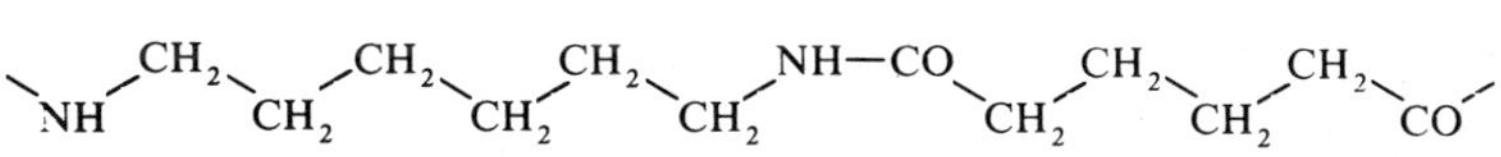

$$\text{—NH—CH}_2\text{—CH}_2\text{—CH}_2\text{—CH}_2\text{—CH}_2\text{—CH}_2\text{—NH—CO—CH}_2\text{—CH}_2\text{—CH}_2\text{—CH}_2\text{—CO—}$$

Nylon 6,6

$$\cdots\text{—O—CH}_2\text{—CH}_2\text{—O—CO—C}_6\text{H}_4\text{—CO—}\cdots$$

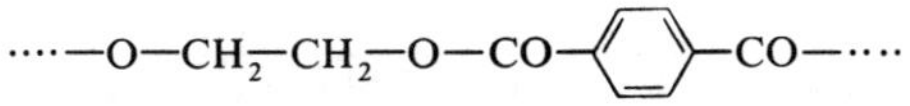

Polyester, e.g. Terylene

$$\cdots\text{—CH(CN)—CH}_2\text{—CH(CN)—CH}_2\text{—}\cdots$$

Polyacrylonitrile; some sulphonate groups are also present

NH–CHR–CO–NH–CHR–CO

Protein; amongst the side-chains R operative in dyeing are those containing carboxyl or amino groups

It is obvious that all these fibres contain many potential centres of attraction for the polar or ionic groups of dye molecules. How far these attractive forces operate in actual dyeing is the question discussed below.

Apart from the chemical structures of the fibres, their morphology, especially their pore structure, is of considerable interest in enabling us to understand their dyeing properties. All the above-mentioned fibres have some form of porosity, most likely that due to intermolecular spaces

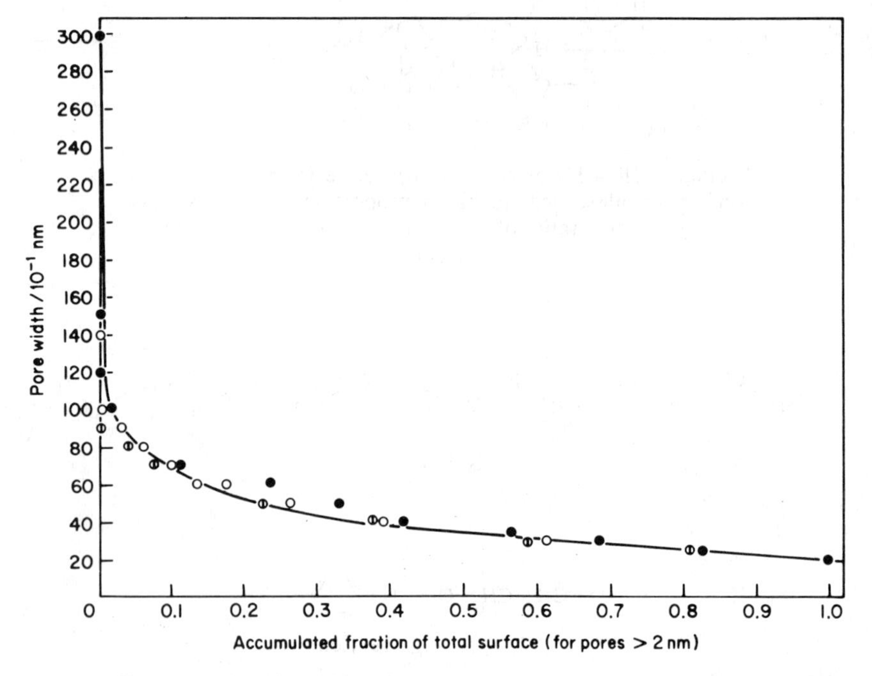

Figure 1 Change with pore width, of water-accessible internal surface in cellulose. (◐) Normal dried regenerated film (max. pore width, 14 nm); (○) Viscose rayon fibre (max. pore width, 30 nm); (●) "Never dried" regenerated film (max. pore width, 30 nm). Total internal surface in water-accessible pores (diam. >2 nm), 325 m^2 g^{-1}. Plotted from data of Stone *et al.*, 1969. (Giles and Haslam, 1978)

between the micelles or bundles of polymeric chains that make up the fibre. Little quantitative investigation on the size and distribution of pores in fibres appears to have been carried out, except the researches of Stone *et al.* (1968, 1969). These authors developed an interesting and simple method for measuring the pore size distribution of various forms of cellulose and regenerated cellulose, by a solute exclusion technique. One set of their data is shown in Fig. 1.

II. Methods of Investigation of Dye Adsorption

By far the most used method of determining the amount of dye adsorbed by a substrate is spectroscopy in the visible region. Because of their intense colour it is easily possible to follow the change in concentration of dyes in solution by any standard absorptiometer, using either hand recording or automatic pen recording. By this means the effect of change in the common variables of interest (time, temperature, concentration of dye or other reagent, pH of solution) can readily be studied. This method has been favoured almost exclusively since the modern electronic absorptiometers and spectrophotometers became available some four decades ago. The intensity of colour of the dye solutions of interest in this type of study usually requires the employment of very short-path optical cells in the instrument. The normal 5 or 10 mm cells may be used if the original solutions are diluted before measurement, but this is a time-consuming operation and also it introduces possibilities of error. To avoid these undesirable features 1.0 mm cells may be used, if desired with spacers to reduce the effective path length to 0.1 mm or less. Specially optically ground and standardized spacers may be obtained, but they are expensive. It has been found that ordinary thin glass plates, such as are used as microscope cover glasses, are quite satisfactory, provided that the cell length when they are used is calibrated with solutions of known concentration. Glass spacers however are fragile and very liable to breakage; we have found that in their place polyester film can be used. This may be obtained in the requisite gauge and each piece may be discarded after use. This is desirable because some dyes may slightly stain the film, but these may often be rinsed off and the spacer re-used.

III. Basic Factors Determining Dye Adsorption by Porous Solids

Three principal factors determine the adsorption: the interfacial tension between dye solution and substrate, dye–substrate forces, and the porosity of the substrate.

A. INTERFACIAL TENSION

Several investigations are on record, of measurement of dye solution tensions against air, and against polymer models for fibres spread in monolayers. The literature on the surface tension of dye solutions is sparse, but the facts recorded are consistent. Milicević (1964) measured the surface tensions of aqueous solutions of various sulphonated and unsulphonated benzeneazo-2-naphthol dyes and found that alkyl substitution of the dye reduces the surface tension of its solutions, as would be expected, but a dye having no alkyl substituent, Orange II (see first formula above), has very little lowering effect on the surface tension of water. Giles and Soutar (1971) measured the surface tensions of aqueous solutions of a number of anionic and cationic dyes, and found that every one lowered the tension of water against air, but that traces of impurity in normally recrystallized dyes have a marked effect in accentuating the reduction. Probably these impurities are intermediates remaining from manufacture, and the most effective way of removing them was found to be foaming them off from the surface. Some typical results are shown in Fig. 2. It was clear that even a very high degree of sulphonation, viz. four sulphonate groups to two

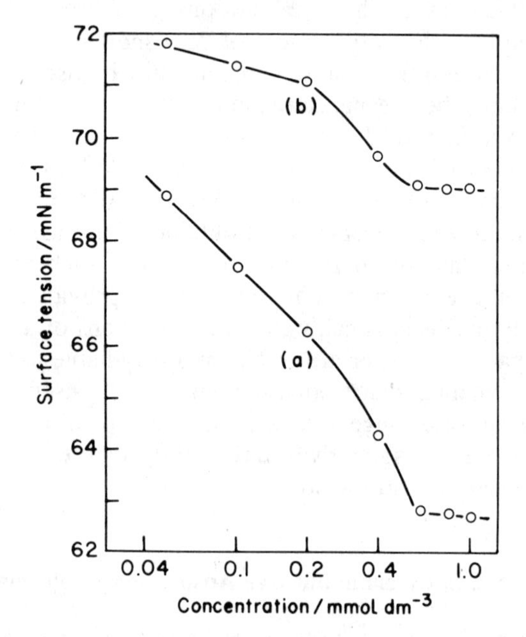

Figure 2 Surface tension of aqueous solutions of highly purified anionic dyes, at 18.5 ± 1.25°C. (a) C.I. (Colour Index) Acid Red 88. (b) C.I. Direct Blue 1. (Giles and McIver, 1977)

naphthalene nuclei, a much higher ratio than in most commercial dyes, gives a dye that reduces water surface tension. This means that most, if not all, dyes in pure water will tend to concentrate at the air/solution interface.

The above situation is of course not quite the same as that in dyeing a textile fibre, where the dye solution is in contact with the fibre, not with air. Fibre polymers or model compounds of very similar chemical properties can however be spread as monolayers on dye solutions and the interfacial

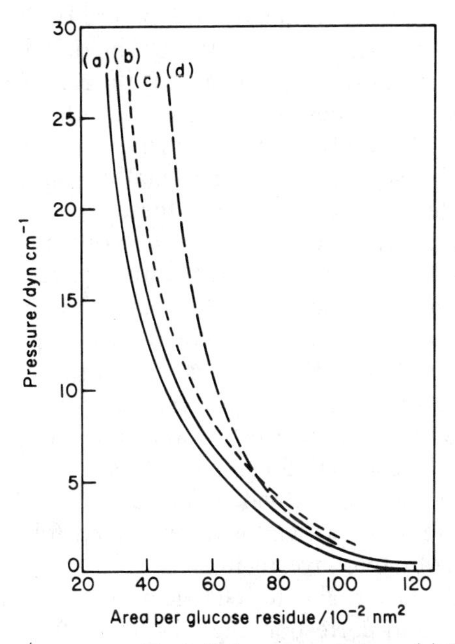

Figure 3 Pressure/area curves for cellulose monolayers. (a) Control, on water alone; (b)–(d) on solutions (10^{-5}–10^{-4} mol dm^{-3}) of azo direct cotton or acid wool dyes, at room temperature. (b) C.I. Direct Yellow 12, Red 2, Blue 1, Acid Orange 7 (Orange II), Reds 1, 88, in water alone; (c) as (b) in 0.1 mol dm^{-3} NaCl; (d) C.I. Direct Green 26 in water alone. (Giles, 1975)

tension then measured. In these conditions there is found to be a reduction of the interfacial tension when dye is dissolved in the underlying aqueous layer (i.e. an increase in the surface pressure of the film, which is actually a measure of the air/water tension less the monolayer/water tension). An example is shown in Fig. 3. The monolayer method as it can be applied to the present subject is discussed below.

B. DYE–SUBSTRATE FORCES

1. *Methods of Identification*

The following types of force singly or together contribute to the adsorption affinity of dyes for solid substrates: ionic and ion-dipole, hydrogen bond, van der Waals, hydrophobic, and covalent bond. A variety of methods of identifying the operative forces in particular adsorption systems can be used, and their application will be apparent as the appropriate results are discussed below. Some methods however require a more detailed discussion.

In many cases it is possible to identify the nature of these forces from the change in adsorption when certain factors are modified, e.g. the change in adsorption with pH in wool dyeing can identify the action of ion-exchange and other forces, as will be described below. It is however possible to identify specific forces by studying the interaction of model compounds. Two methods of study used here have been that of continuous variations and the monolayer technique. In both of these cases the complicating action of the substrate structure, especially the porosity of fibres, which introduces kinetic retardation and other disturbing effects, is eliminated and direct observation of dye–substrate force can often be made.

(a) *Method of continuous variations*. In this method some physical property, e.g. spectral absorption, viscosity, dielectric constant, refractive index, is measured for a range of binary solutions each containing as solutes the two components A and B under investigation, in a range of proportions from 100%A to 100%B, each solution having the same total molarity. The method seems to have been described as far back as 1911 (see Arshid *et al.*, 1955, for earlier references), and is often named Job's method, from its extensive early use by that investigator.

The plots are usually linear, and any change of slope indicates a complex of A with B, in the ratio corresponding with the inflection (Fig. 4). In many cases more than one complex is so indicated. For the theoretical basis of the method, see e.g. Arshid *et al.* (1955, 1956), Giles *et al.* (1961), and earlier work referred to therein. Quantitative measurements may be made by the method, for example, thermodynamic parameters of reactions in solution have been determined (Giles *et al.*, 1972; Giles and McIntosh, 1973), and these can throw light on some types of dye–substrate interaction, as will be described later.

A particular advantage of the method, apart from its rapidity and simplicity, which enables large numbers of systems to be tested, is that it can be used as readily in aqueous as in non-aqueous media; this can be said of few, if any, of the other methods for detecting hydrogen bonds. This advantage is obviously important in any study of the behaviour of dyes and fibres in aqueous baths. Some examples of complexes detected by this means using, for example, esters as models of cellulose acetate or polyethylene terephthalate, and glucose, ketones, or hydroxy compounds as models of cellulose, are given in Table 2. The method also enables intramolecular bonding to be detected, by the reduced ability of groups in a molecule to engage in intermolcular bonding, e.g. the intramolecular bond in *o*-hydroxyazobenzene derivatives reduces their ability to bond intermolecularly with phenols.

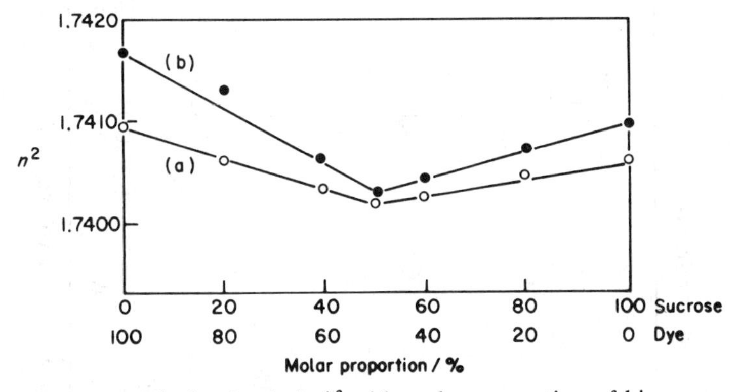

Figure 4 Change of (refractive index)^2 with molar proportion of binary aqueous solutions of sucrose with C.I. Acid Blue 25 at 14°C. (a) 0.5 mmol dm^{-3}; (b) 1.0 mmol dm^{-3}. (Giles and McIntosh, 1973)

Examination of results with a given pair of solutes in two solvents shows that in only a few cases is intermolecular bonding prevented by the solvent; notable exceptions are some systems in water, e.g. hydroxy groups in simple carbohydrates appear unable to form intermolecular hydrogen bonds with aliphatic or aromatic hydroxy groups in a second solute. This fact has an important bearing on any interpretation of the mechanism of dye–cellulose bonding. Phenolic, amino, azo, and ester groups bond mutually in water, indicating that such a mechanism probably operates in dye–ester bonding with synthetic fibres.

It is also possible as stated above to use the method to measure thermodynamic parameters for the bonding reaction, from the displacement of the inflection in the curve with change in temperature. Table 3 gives some results obtained thus. The method was originally developed by Yoshida *et al.* (1964). A related method, the solute excess method, was used and found to give similar results; one component in the mixture of compounds A and B in a binary solution is greatly in excess of the other, and it is assumed that at equilibrium the reaction has moved to completion.

Table 2 Complexes detected by refractometry

Type of bond	Solutes: a	Solutes: b	Solvent	Total mol. concn.	Temp./°C	Mol. ratio of Complex, a:b
Alk—OH + Ar—OH	D-Glucose	Phenol	W	0.25	20	0
			EG	0.1	20	1:6
>C=O + Ar—OH	Acetone	Phenol	W	0.1	20	0
	Diisobutyl ketone		D	0.25	20	1:2
	Ethyl methyl ketone		CCl_4	0.1	14.83	1:2, 1:2
–N=N– + Ar—OH	Azobenzene	Benzyl alcohol	CCl_4	0.1	20	1:1
		Phenol	B	0.25	19	1:1, 1:2
–COOR + Ar—OH	Ethyl acetate	Phenol	CCl_4	0.1	13.11	1:1
–COOR + Ar—Cl	Ethyl acetate	Chlorobenzene	CCl_4	0.1	14.85	1:1
			EW	0.1	14	1:1
	Phenyl acetate		B	0.1	21	1:1
			EW	0.25	14	1:1
–COOR + NO_2	Ethyl acetate	*p*-Nitrophenol sulphuric ester	W	0.04	20	2:1
	Glycerol triacetate		W	0.04	20	1:1

B = benzene; D = dioxan; EG = ethylene glycol; EW = ethanol–water (1:1 vol./vol.); W = distilled water.

Table 3 Thermodynamic parameters for bonding reactions in binary solution

Solutes	Solvent	Method*	Temp./ °C	$-\Delta G$/ kJ mol^{-1}	$-\Delta H$/ kJ mol^{-1}	$-\Delta S$/ J K^{-1} mol^{-1}
Phenol, Acetonitrile	CCl_4	i.r.	22	8.4	17.6	29
	xylene	c.v.	28	8.0	15.5	25
Phenol, Chlorobenzene	CCl_4	c.v.	14.8 28.0	2.1 1.7	13	38
Ethyl acetate, Chlorobenzene	H_2O + EtOH	s.e.	21.0 39.0	4.6	14	34
Sodium naphthalene-sulphonate, sucrose	H_2O	c.v.	20.5 39.0	7.1 5.9	28	71
C.I. Acid Blue 25, sucrose	H_2O	c.v.	14.0 37.0	9.6 2.5	101	314

* i.r. = infrared spectroscopy (Flett, 1952); c.v. = continuous variations; s.e. = solvent excess (Giles *et al.*, 1972; Giles and McIntosh, 1973).

The parameter used was refractive index in this method (Chandra and Basu, 1960; Giles *et al.*, 1972; Giles and McIntosh, 1973).
(b) *Gas–liquid chromatography*. This method has been used by de Vries *et al.* (1967, 1968) and de Vries and Smit (1967) to measure the heat of reaction of cellulose acetate, as stationary phase, with a number of compounds that are models of non-ionic dyes as used for non-ionic synthetic fibres. Some of their results are given in Table 4.

Table 4 Heats of adsorption for various solutes on secondary cellulose acetate (CA) and cellulose triacetate (CTA). (de Vries *et al.*, 1967, 1968; Smit *et al.*, 1967)

Solute	Temperature range/°C	$\Delta H°$(CA)/ kJ mol^{-1}	$\Delta H°$(CTA)/ kJ mol^{-1}
Alcohols			
Methanol	60–80	36.4	34.3
Ethanol	60–90	46.9	39.8
Propanol	60–80	52.3	41.9
n-Heptanol	120–140	53.6	44.0
Phenols			
Phenol	150–171	56.5	55.3
p-Ethylphenol	150–171	64.9	63.6
p-Chlorophenol	150–171	63.6	73.6
Hydrocarbons			
n-Hexane	20–40	3.4	3.8
Benzene	30–50	22.2	26.0
Toluene	30–50	30.1	32.6
Halogen compounds			
Chloroform	60–80	27.2	24.7
Bromoform	110–130	31.0	31.0
Carbon tetrachloride	30–50	23.0	19.3

(c) *The monolayer method* (Giles, 1978). When a polymer is spread as a monolayer on water, or an aqueous solution, its polar or ionic groups orientate as closely as possible to the water surface, not necessarily in an extended straight line, but possibly intermingled randomly with neighbouring polymer chains. When the monolayer is forced to contract, the chains condense and tend to cover the whole of the water surface, unless prevented by solute molecules from the water below. The ease with which this condensation takes place has been shown to have a direct relation to the interaction between the monolayer and the solute in the solution (sub-phase) beneath the monolayer (Giles and McIver, 1975). The surface pressure/area curve for the polymer film may be altered in slope in three general ways by the presence of the solute, and the nature of the alteration can be used to diagnose the nature of the reaction between the polymer and the solute. The principle of the diagnosis is illustrated in Fig. 5. Examples of the results of this method are given later. An additional use for the monolayer method is with surface viscosity measurement, combined with the continuous variations procedure, which can give information on the nature of the polymer film–solute bonds.

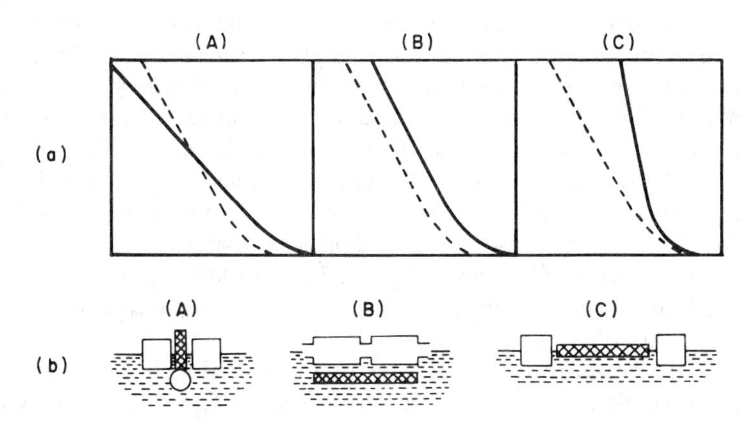

Figure 5 (a) The three general types of change of slope of the surface pressure/area curve for a monolayer of a polymer when a solute is introduced into the aqueous sub-phase. The broken line represents the curve for the monolayer on pure water. (b) The suggested relative orientations of polymer molecules (open rectangles) and solute molecules (hatched rectangles) in the monolayer on aqueous solutions, corresponding to the three respective curve slope changes of (a) above. (A) shows the penetration of the monolayer by a monoionic solute, where the ionic group remains in the water and the hydrophobic portion of the molecule penetrates the monolayer. (B) shows the orientation of a long solute molecule with ionic groups at each end (e.g. a direct cotton dye), beneath the monolayer. (C) shows crosslinking of monolayer polymer chains by a bifunctional solute, whereby the viscosity of the film is greatly increased. In (A) and (C) the polymer chains are shown in cross-section, in (B) in side elevation. (Giles and McIver, 1975)

The great advantage of the monolayer method in interpreting phenomena of fibre dyeing is that the interference of the pore structure of the fibre is eliminated, and only dye–fibre polymer reactions are involved in the observed effects. As will be mentioned below, pores have a twofold effect on dyeing: they retard dye diffusion in the fibre, and they can facilitate internal aggregation of the dye, which may mask the action of dye–fibre reactions.

(d) *Titration methods*. These involve the study of the variation of amount of dye adsorbed by a fibre, with change in pH of the dyebath; the variation can offer clear clues to the action of ion-exchange in the dyeing. The method has been used with protein, nylon, and acrylic fibres. Details are given later.

2. *Identification in Various Systems*

(a) *Man-made ester fibres.* These are dyed with the so-called disperse dyes, which are characterized by absence of ionic groups in the molecule. Water solubility, which is low, and affinity for the fibres are ensured by the presence of one or more polar substituents on the main chromophore, which is in most cases an azobenzene or an anthraquinone structure. The low solubility in water, e.g. 1–100 mg dm^{-3}, even at the boil, requires that dispersions of the dyes are used. Dyeing takes place from the saturated aqueous solution, from which individual dye molecules are adsorbed by the fibre, to be replaced by dissolution of others from the suspended dye particles (Vickerstaff, 1954). The three principal types of ester fibre differ in their resistance to penetration by dye in accordance with their degree of crystallinity. Secondary cellulose acetate is the most easily penetrated, and is dyed at about 80°C, cellulose triacetate is dyed at about 100°C, but polyester (polyethylene terephthalate) is the most highly crystalline and requires very high temperatures, up to 120°C for effective dyeing within a normal period of about 1 hour (Waters, 1950). Dyeing can be accelerated and lower temperatures used if so-called carriers are added to the dyebath; these are aromatic compounds, e.g. biphenyl derivatives, which appear to act by making the polymer molecules more flexible.

A glance at the constitution of the fibre molecules and the disperse dyes, and consideration of the data in Tables 3 and 4, suggests that the forces involved in the dye adsorption are polar, most probably hydrogen bond in nature. This suggestion is reinforced by various quantitative observations. Thus, for example, the maximum amount of dye (in a given structural class) adsorbed by (sec.) cellulose acetate rises with the number of hydrogen atoms in the dye molecule available for intermolecular bonding (Fig. 6). Also the maximum adsorption rises with the water solubility of the dye (Fig. 7), and since the latter property depends on attraction by water molecules for the polar substituent groups in the dye molecule, it is reasonable to attribute the adsorption to a similar attraction by the polymer.

It is clear from Fig. 7 that the dyes are more soluble in the fibre than in water, which implies that they can penetrate further into the fibre than merely into its water-filled pores. This is clearly illustrated by the behaviour of polyester fibres. These have a water content in the air-dry condition of 50–100 mmol kg^{-1}, but they can adsorb up to 180 mmol kg^{-1} of simple anthraquinone disperse dyes (at 100°C) (Schuler and Remington, 1954), which in view of the difference in molecular size of the water and the dye

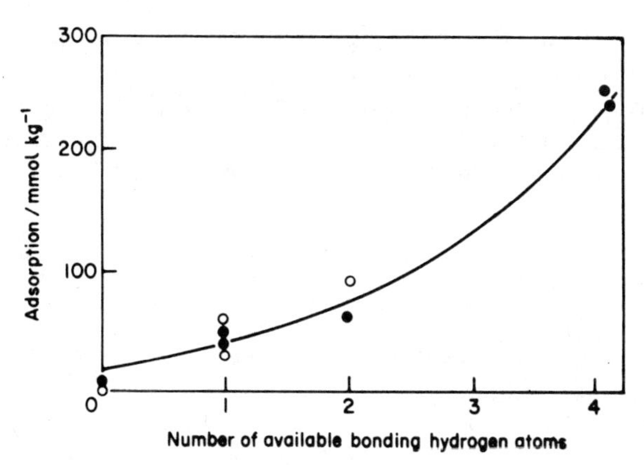

Figure 6 Illustrating the dependence of maximum adsorption of multifunctional azobenzene disperse dyes (no 2- or 2′-substituents) on cellulose acetate at 80°C, and the number of hydrogen atoms in their molecules available for intermolecular bonding. Data from Bird and Harris, 1957. (Giles, 1961)

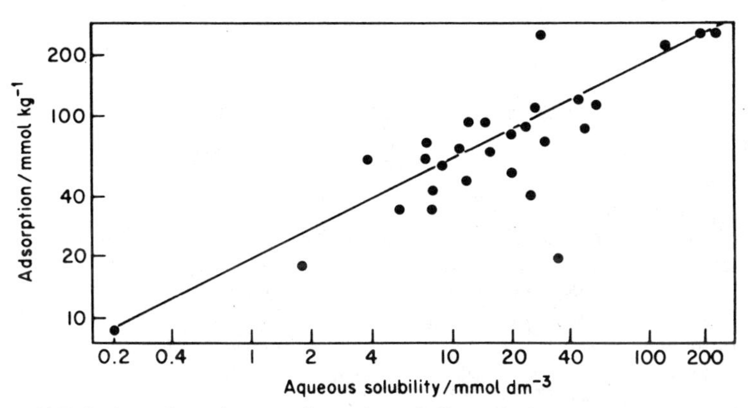

Figure 7 Relation of maximum adsorption of disperse dyes by cellulose acetate at 80°C and their water solubility at 80°C. Data from Bird and Harris, 1957. (Giles, 1961)

molecules, represents penetration of a very much greater amount of the polymer structure than is available even to the small molecule of water vapour. The high temperature of the dyeing would reduce adsorption compared with that achievable (in due course of time to reach equilibrium) at the room temperature used for water vapour adsorption measurement, so the true comparison here is even more favourable to the dyes.

Another corollary postulated from the above data is that, since adsorption rises with number of bonding groups (Fig. 6), it should rise with the ability of such groups in the dye molecule to juxtapose with polar groups in the polymer substrate. That this situation does in fact occur can be seen from results of investigations by Daruwalla *et al.* (1960) (Fig. 8), who found that adsorption of a disperse dye with a planar molecule is much greater than that of a dye with a similar but non-planar structure.*

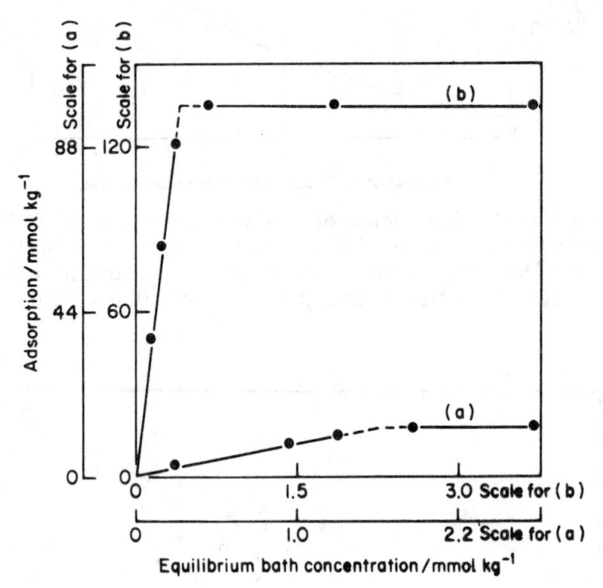

Figure 8 Adsorption isotherms for non-ionic disperse dyes on cellulose acetate at 80°C. Typical data from Daruwalla *et al.*, 1960.

Dye	Saturation adsorption/ mmol kg^{-1}	Affinity/ kJ mol^{-1}
(a) Non-planar	3.1	−2.55
(b) Planar	135	−19.97

Further evidence of the penetration of the dyes is given by the nature of the adsorption isotherms, which are linear (C type) (Giles *et al.*, 1974b), as in Fig. 8. This is a feature of a number of systems in which the solute has greater power of penetration of the substrate than the solvent has. The C type isotherm condition can be considered a special case of the well known Langmuir adsorption system. If the solute molecules are likened to the gas molecules in the Langmuir (1916, 1918) hypothesis, the amount of dye adsorbed, y, is

$$y = by_{\mathrm{m}}c/(1 + bc) \tag{1}$$

* Juxtaposition is confirmed by dichroism measurements (Vickerstaff, 1954, p. 328).

where y_m is the amount adsorbed in a completed monolayer, c is the dye concentration in solution, and b is a constant which is a function of temperature. If, in accordance with the above observations, the dye penetrates deeply into the polymer structure, we may assume that the surface available for adsorption increases as each molecule of dye adsorbed breaks open a little more of the fibre structure. This action may be likened to that of a zip fastener, the slider representing the entering dye molecule. Then we have

$$y_m = k(1 + \alpha c) \tag{2}$$

where k is a constant and α can be considered a coefficient of expansion of the available internal surface. Under C type conditions $y = \beta c$, where β is another constant. It then follows that

$$b = \beta/(k + k\alpha c - \beta c) \tag{3}$$

This equation is valid only up to a critical value $c = c^* = k/(\beta - k\alpha)$, i.e. $k + k\alpha c - \beta c = 0$, because if c is greater than c^* the sign of b would change from positive to negative, but this is impossible because the components of b are positive. Thus at $c = c^*$, $b = \infty$, i.e. the rate of escape of adsorbed dye from the fibre is zero. It follows that, when the concentration c^* is reached, the value of y will remain fixed, provided that no fresh internal surface is formed, and the curve will change abruptly to the horizontal. In some C type curve systems, in which amino acids are adsorbed by the clay montmorillonite in the calcium form, the actual basal spacing of the clay lattice, measured by X-rays, was found to increase with the amount of amino acid adsorbed (Greenland *et al.*, 1962).

It is clear from the above discussion that the most likely orientation of dye relative to fibre molecule is face-to-face, in which as many polar groups as possible engage in dye–fibre interaction, the ester groups acting as proton donors or acceptors. This hypothesis can be verified by use of the monolayer procedure. Figure 9 shows force/area curves for cellulose triacetate films on water, alone and mixed with two disperse dyes. The width of the smaller dye (red) molecule is approximately the same as that of the cellulose triacetate (CTA) polymer chain, and the absence of any expansion of the film when this dye is present is consistent with the face-to-face contact of dye and polymer in the plane of the water surface; when the wider dye (yellow) molecule is present however the film expands by the amount consistent with this mutual orientation.

The continuous variations method applied to these mixed CTA + dye films gives clear evidence of a 1:1 complex between the hexaacetylcellobiose repeat unit of the CTA chain and a molecule of each dye (Fig. 10) (Giles *et al.*, 1970). The increase in viscosity in this way confirms the above hypothesis of face-to-face polar attraction. Other systems, described below,

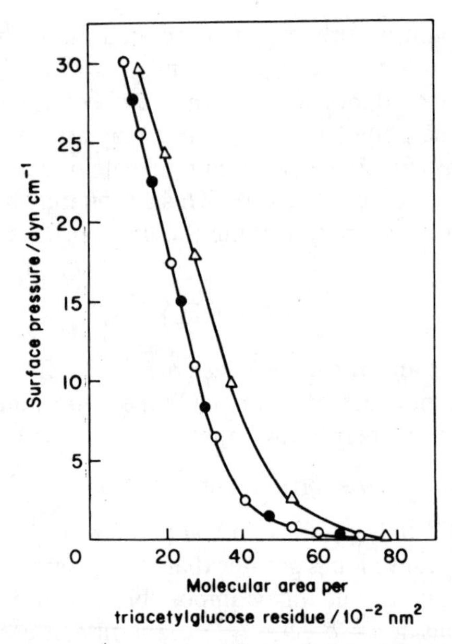

Figure 9 Surface pressure/area curves for cellulose triacetate monolayers (CTA) alone and mixed with disperse dye (one dye molecule to one hexaacetylcellobiose unit). (Giles *et al.*, 1970). ○ CTA alone; ● CTA plus C.I. Disperse Red 13. △ CTA plus C.I. Disperse Yellow 13.

in which dye–polymer hydrogen bonds are probably not formed, do not show such viscosity increase.

(b) *Additivity effects.* A rather remarkable effect shown by some disperse dyes on the man-made ester fibres is that of additivity. This means that a particular dye may have a maximum adsorption on a given fibre of this class, but if dyed in the presence of another dye the first may be adsorbed to the same maximum, but in addition the second may be adsorbed. This property enables deeper shades to be obtained than would otherwise be possible with the particular dyes used singly, but it is very selective—dye A may be additive with B but not with C. The cause seems to be related critically to the geometry of the dye molecules; dyes with very similar molecular shapes are non-additive, but those with different shapes are additive, and each gives adsorption isotherms independent of the presence of the other (Johnson *et al.*, 1964) (Fig 11). The additive dyes, when examined in the pure solid form, are found to crystallize as eutectic mixtures, whereas non-additive dyes form mutual solid solutions in all proportions in the crystal phase. In other words, the additive dyes are mutually incompatible, which suggests that they are able to enter different regions of the substrate without mutual interference (Giles, 1961, 1971).

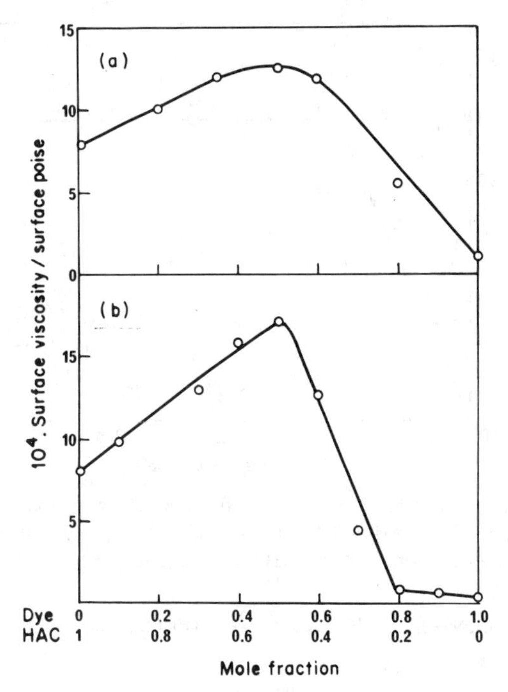

Figure 10 Continuous variation changes of monolayer viscosity of mixtures of cellulose triacetate and disperse dye. (a) C.I. Disperse Red 13; (b) C.I. Disperse Yellow 13; HAC = hexaacetylcellobiose residue. (Giles *et al.*, 1970)

Perhaps the dyes that form solid solutions do so in the dispersion in the dyebath, then enter the fibre in monodisperse form in aqueous solution, and finally re-aggregate in the voids that exist in all the fibres of the present type.

(c) *Disperse dyes on other fibres.* These dyes are used on nylon, on which, unlike the case of the acid wool dyes described below, they have the advantage of not revealing physical differences in the fibre by differences in shade depth, but their fastness to washing is not as good as that of the acid dyes. (Note that disperse dye adsorption is completely reversible.) There seems no reason to doubt that the adsorption mechanism is the same as on the ester fibres. Some data on the relative adsorption of these dyes on cellulose acetate and on nylon (Fig. 12), which show a linear relation between the two sets of adsorption values, seem to confirm this belief.

These dyes merely stain wool and cellulose fibres to very weak shades, and are of no practical interest for them. Bird and Firth (1960) and Aspland and Bird (1961) examined both systems quantitatively and confirmed the very low maximum adsorption for both fibres. The adsorption isotherms however were interesting, C type (linear) curves. This indicates that the

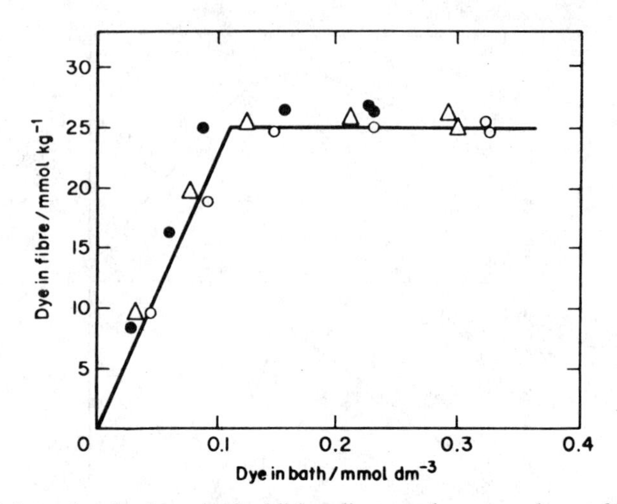

Figure 11 Adsorption isotherms of additive disperse dyes on nylon at 90°C. Data from Johnson *et al.*, 1964. (Giles, 1971). (—), Dye I; ○, dye I in presence of dye II; ●, dye I in presence of dye III; △, dye I in presence of dye IV.

O_2N–C₆H₃(Cl)–N=N–C₆H₄–N(Et)C_2H_4OH (I)

MeCOHN–C₆H₄–N=N–(Me)(OH) (II)

OMe, O (III)

HO O NHMe, MeHN O OH (IV)

dyes penetrate into some of the regions of the fibres not readily penetrated by water, an indication that is also consistent with the linear relation between saturation adsorption on regenerated cellulose fibres and water solubility of the dyes (Aspland and Bird, 1961; Giles, 1975), as for the acetate fibres discussed above.

(d) *Protein, nylon, and acrylic fibres.* These fibre classes are here classified together because their main dyeing processes involve ion-exchange.

Wool is the most important protein fibre in terms of amount used. The dyes used for it are anionic types, either "acid" dyes, e.g. C.I. 15510 and

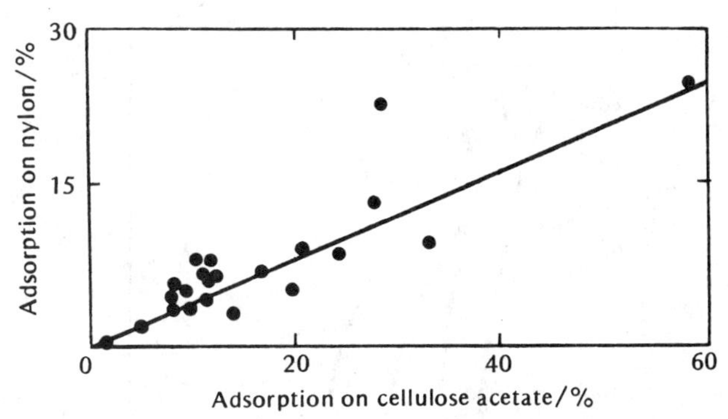

Figure 12 Relation between saturation adsorption of disperse dyes on nylon and on cellulose acetate, at 85°C. Data from Vickerstaff, 1954. (Giles, 1966b)

61530, or mordant dyes, e.g. C.I. 15670, which are combined with chromium during the technical dyeing procedure to form a dye–chromium chelate complex, or metal complex dyes in which the chelation has taken place during the manufacture of the dye. The latter type comprises the sulphonated 1:1 dye–metal complexes, e.g. C.I. 14880, and the 2:1 dye–metal complexes, usually unsulphonated; the metal is in most, but not necessarily all, cases chromium.

In water the amino and carboxyl groups of a protein fibre (W) can be considered as zwitterionic pairs, and entering hydrogen ions from an acid HX then neutralize the carboxyl groups by back-titration:

$$^{-}OOC—W—NH_3^{+} \xrightarrow{H^{+}X^{-}} HOOC—W—NH_3^{+}X^{-}$$

Thus the capacity of protein or nylon fibres for acids or dye colour anions should be equivalent to their free amino group content. That this is so, quite closely, can be seen from the curves of Fig. 13. It is also confirmed by studies of the temperature dependence of the wool titration curve, which reveal that the heat of combination of hydrogen ions is almost zero in acid solution and about $-40\ \text{kJ mol}^{-1}$ in alkaline solutions, values close to the heats of ionization of many organic acids and the heats of dissociation of various substituted ammonium ions respectively (Vickerstaff, 1954). At very low pH values, excess of acid is adsorbed, probably because of protonation or hydrolysis of the peptide or amide groups (Fig. 13). Hydrolysis is implied by the weakening of nylon fibres dyed at these low pH values.

The importance of the ion-exchange reaction in nylon dyeing with sulphonated dyes is also shown by the linear dependence of the dye adsorption on the content of amine end-group in the fibre (Fig. 14) (Burdett, 1975).

Nevertheless there is something of a mystery in the apparent equivalence between the uptake of anions and the amino group content of wool. It is

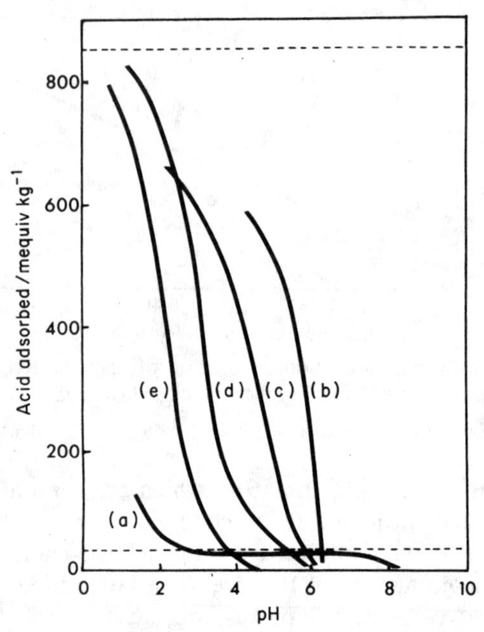

Figure 13 Titration curves for nylon 6,6 (Peters, 1945), and wool (Vickerstaff, 1954). The broken lines represent the values corresponding to the respective free amino group contents of the fibres. Nylon 6,6: (a) dye C.I. 18050 (disulphonated). Wool (monobasic free acids): (b) dye C.I. 26660; (c) C.I. 15510; (d) naphthalene-2-sulphonic acid; (e) hydrochloric acid. (Giles, 1966b)

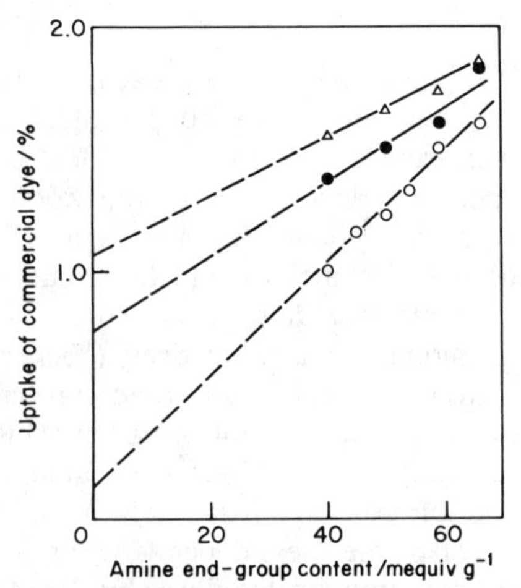

Figure 14 Relationship between dye uptake and amine end-group content of nylon. (△) Solacet Fast Scarlet BS; (●) Solway Purple RS; (○) Naphthalene Scarlet 4RS. (Burdett, 1975)

not clear how the anions can reach all the amino groups in the fibre. The wool fibre consists of the protein keratin in several levels of organization. Firstly, the individual protein chains are grouped, parallel and in contact, in small numbers, possibly two or three, to form the protofibrils; these have probably a coiled structure, with alternating sections of non-helical chain. Numbers of protofibrils are embedded in a matrix, probably also consisting of helical and non-helical protein chains, to form microfibrils, and these in turn form larger macrofibrils. Next the macrofibrils make up the cells of the cortex, the bulk of the fibre, which is enclosed by the scale structure. The cortex in crimped wool has two segments running parallel with the fibre length, the orthocortex and the paracortex. There are also complex intercellular materials and membranes. The surface consists of the scales, each composed of endocuticle and exocuticle. Finally, the scales are covered with a very thin (~10 nm) epicuticle, a layer of protein exposing an outer surface of largely hydrocarbon groups, linked on the inner side to the scale surface by means of cystine groups. This epicuticle offers resistance to penetration of aqueous solutions, and if it is removed chemically, e.g. by alkali, penetration is facilitated.

This description of the complex structure of the protein is difficult to reconcile with the apparent equivalence of anion adsorption and amino group content. There is moreover evidence that at high dye contents adsorption of dye occurs either on peptide groups, with breakdown of the protein chains and collapse of the fibre into powder (Vickerstaff, 1954; Skinner and Vickerstaff, 1945), or as aggregates in cavities in the fibre (Speakman and Smith, 1936; Royer and Maresh, 1943). Thus the data must be regarded as open to re-interpretation. The present most likely explanation is that the dye may be concentrated in voids in the fibre structure, and that the correspondence between the dye anion capacity and amino group content of the fibre may be partly fortuitous.

The titration curves (Fig. 13) show that anions differ in affinity for the fibre; affinity rises with anionic weight, as shown by the shift of pH needed for half-saturation. The forces responsible for this increase in affinity with anionic or molecular weight are difficult to identify exactly, but obviously they must include one or more of the polar and non-polar or van der Waals forces. Measurements of thermodynamic parameters cannot give much useful information in this case because water is desorbed to give place to adsorbed dye, which itself may associate after adsorption, so that heat etc. measurements give only a net result, which is difficult to interpret (Peters and Lister, 1954; Coward and Nursten, 1964; Peters, 1955; Meggy, 1950; Zollinger, 1965).

Investigations made some years ago showed that, either in aqueous solution or in the vapour phase, small organic molecules have higher affinity for wool, silk, and nylon when proton donors than when proton acceptors (Chipalkatti *et al.*, 1954, 1955). In ability to bind with proteins or nylon, substituent groups rank roughly in the order $\mathrm{ArOH} \gg \mathrm{ArNHCOMe} > \mathrm{ArNH_2} > \mathrm{AlkOH} >$ proton acceptors; the phenolic group (but not the phenolate ion) appears to have much higher affinity than other groups.

These conclusions are supported by some tests in monolayers. Thus the two dyes C.I. 63010 and C.I. 58610, both of which have hydrogen atoms (*) free to form intermolecular bonds, were dissolved in the water beneath a protein monolayer. C.I. 58610 was found to have a more marked effect on the rigidity of the film—producing a change of type C (Fig. 5)—than C.I. 63010 (Giles and MacEwan, 1959; Cameron *et al.*, 1958).

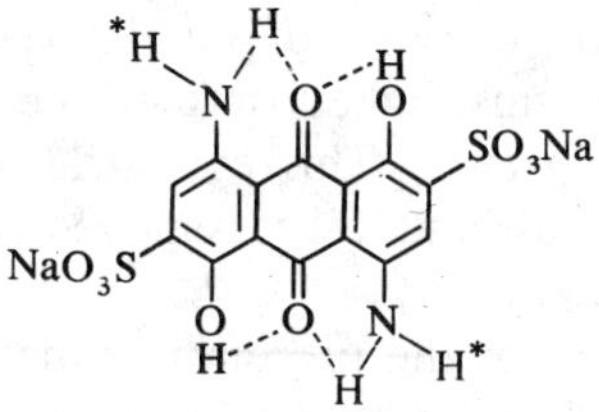

C.I. 63010

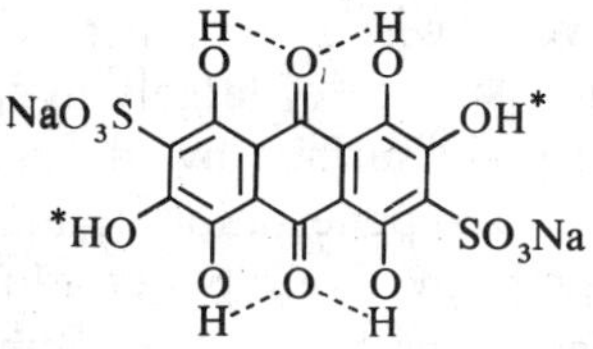

C.I. 58610

The above observations suggest that hydrogen bonds and other polar bonds do not play an important part in the affinity of dyes for wool; they would only do so in dyes whose molecules contain hydroxy groups free to form intermolecular bonds. Few commercial dyes have this characteristic, because it produces undesirable sensitivity to alkali.

This leaves the non-polar forces (van der Waals forces and hydrophobic bonds) as the remaining candidates to account for non-ionic affinity of dyes for wool. There is little firm experimental evidence in support; the increased fastness to washing conferred on wool dyes by increased molecular weight, e.g. substitution by long alkyl chains, which is sometimes thought to indicate the operation of non-polar forces, may be attributed to kinetic effects (decelerated diffusion rate in the fibre pores, as will be referred to later). Possibly the behaviour of wool dyes towards protein monolayers throws some light on the matter. Monosulphonated dyes produce the type A change (Fig. 5), showing that the unsulphonated part of the dye molecule tends to associate with the less polar regions of the protein which constitute the upper surface of the monolayer. Disulphonated dyes, with sulphonate groups at opposite ends of the molecule, produce the type B change, and therefore are oriented flatwise beneath the protein, with their ionic groups in contact with the ionogenic groups of the protein. This rather suggests that the ion–ion forces take precedence over the non-polar ones.

The 1:1 metal–complex dyes behave like the normal dyes without chelating metal, in monolayer reactions; thus monosulphonated 1:1 dyes produce type A (Fig. 5) changes in protein monolayers, and disulphonates produce type B changes (Giles and McIver, 1975). There is no evidence from these investigations that the chelating metal enters into any bonding with the protein (Fig. 15).

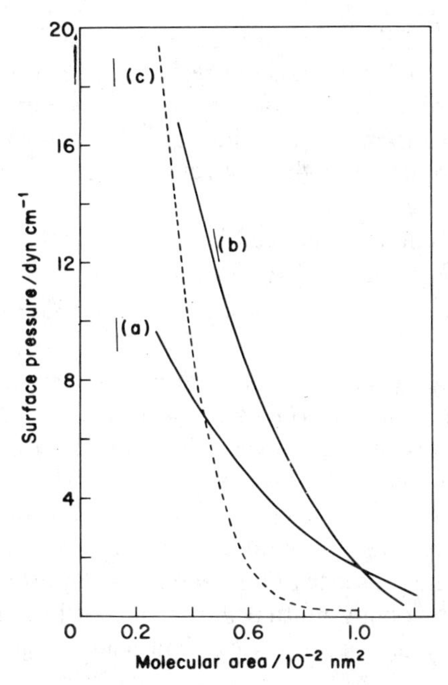

Figure 15 Pressure/area curves for protein monolayers on pure water (c), and on buffered (pH 4.0) 10^{-4} mol dm^{-3} solutions of monosulphonated acid wool dyes. (a) C.I. Acid Yellow 99; (b) C.I. Acid Orange 74. (Giles *et al.*, 1974a)

The 2:1 metal complex unsulphonated dyes show a variety of effects on protein monolayers (Giles and McEwan, 1959), according to the nature and position in the molecule of the polar substituent groups; hydroxy groups have a marked effect, in accordance with the observations discussed above. Many of these dyes have a type of spiro structure, in which the two dye molecules are oriented at right-angles to each other, on opposite sides of the chelating metal atom. When some of these dyes associate, which they appear to do by an interlocking of the planar individual dye molecules, they cause a marked increase in the rigidity of the protein film, which is indicative of non-polar bonding between dye aggregates and protein.

The 2:1 dye–metal complex dyes give adsorption isotherms on wool, nylon, silk, and acetylated wool or nylon that are indicative of a combination of ion-exchange adsorption in water-accessible regions of the fibre, and hydrogen-bond or non-polar adsorption in water-inaccessible regions, especially in the acetylated fibres, in which the ionic groups have been lost by the acetylation (Zollinger, 1956, 1958; Back and Zollinger, 1958; Giles and MacEwan, 1959). Thus some evidence is here given of the operation of the non-polar forces.

(e) *Mordant dyeing*. This is one of the oldest of all dyeing processes, and gives colorations of high fastness. In present-day operation, mordant dyes (e.g. C.I. 15670) are applied, mainly to wool, together with potassium or sodium dichromate as mordant. The mordant may be applied before, during, or after dyeing, though only relatively few dyes are suitable for mordanting during the dyeing operation.

The dichromate anion is reduced by oxidation of disulphide bonds in the protein fibre, forming the Cr(III) cation. The action occurs in steps:

$$\mathrm{Cr(VI)} \xrightarrow{\text{reduction}} \mathrm{Cr(IV)} \xrightarrow{\text{reduction}} \mathrm{Cr(II)} \xrightarrow{\text{oxidation (air)}} \mathrm{Cr(III)}$$

Cr(III) reacts finally by successive addition of dye molecules to give 1:1 and 2:1 dye–metal complexes (Giles, 1944; Hartley, 1969; Dobozy, 1973). Some dyes appear to form the 1:1 and others the 2:1 complex, the evidence being qualitative, based mainly on the colour of the dyed fibre, because the complexes cannot be removed unchanged. In a few cases analysis of the dyed and mordanted fibre after removal of unreacted chromium appeared to show the presence of only the 2:1 complex (Race *et al.*, 1946). Further reactions involving mutual interaction of hydroxy groups on the chromium atom may also occur, giving olation and oxolation products, e.g.

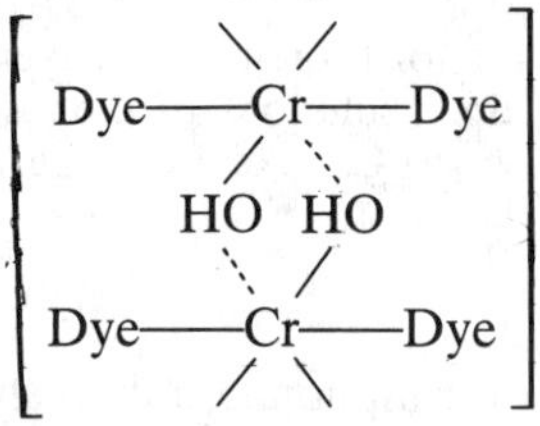

```
[      \ /          ]
[ Dye——Cr——Dye      ]
[      /  `.        ]
[    HO   HO        ]
[      `.  /        ]
[ Dye——Cr——Dye      ]
[      / \          ]
```

for which there is experimental evidence (Merry, 1936; Arshid *et al.*, 1954; Rollinson, 1956). In view of these observations and those on the metal complex dyes discussed above, it appears that the mordanting metal atom probably does not take part in any reaction with the fibre molecule, after the initial reduction, and that therefore the superior fastness to washing of the dyed and mordanted fabrics can be accounted for by the development of large molecular aggregates in the fibre, favoured by the high molecular weight of the smallest units first formed.

(f) *Acrylic fibres*. These are dyed with cationic dyes, large numbers of which have been introduced during the past two decades as a result of intensive research. Many of these new dyes have excellent fastness properties and bright shade. Some modified forms of nylon and polyester fibre have also been developed, containing anionic sites for the adsorption of cationic dyes. Examples of the new type of cationic dyes are given above.

The adsorption isotherms for cationic dyes on these fibres are of the normal Langmuir or L type (Giles *et al.*, 1974b), and by re-plotting these on reciprocal scales, and measuring the intercept, the number of adsorption sites can be determined. The values for this parameter, for various acrylic fibres, agree well with those obtained by other methods (Balmforth *et al.*, 1964) (Tables 5 and 6). These facts provide excellent confirmation of the

Table 5 Acidic end-group content of polyacrylonitrile. (Balmforth *et al.*, 1964)

Before ion exchange (Na^+)	15
(K^+)	5
(H^+)	21
total =	41
After ion exchange (H^+)	41.5
Sulphur analysis	40.5
Dye adsorption (pH 5.4)	40
Reciprocal of number-average mol. wt ($1/\bar{M}_n$)	40

End-group content measured in mequiv. kg^{-1}.
Polyacrylonitrile has a low specific viscosity (=0.15).

hypothesis that the dyeing mechanism is a simple one of ion-exchange whereby the dye cation displaces H^+, Na^+, or K^+. Potentiometric titration shows that most acrylic fibres contain not only the strong acidic groups thus revealed, but additional weak groups, which can be determined by a special method of potentiometric titration. At low pH, adsorption takes place mainly on the strongly acidic groups (Fig. 16). Some of the strong sites appear to be sulphate and sulphonate groups, or phosphoric acid introduced by the redox polymerization catalyst used in the manufacture of the original polymer. The weaker sites have an ionization constant similar to that of carboxylic acids (Burdett, 1975).

(g) *Cellulose fibres* (cf. Giles and Hassan, 1958; Iyer, 1974). There is also with the dye–cellulose system, as with the dye–protein one, something of a mystery regarding the nature of the forces that cause such strong adsorption of dyes, when the action of no very powerful dye–fibre forces seems at first sight to be obvious. Several theories have been proposed, discarded, and later re-proposed to account for the facts, but none has received

Table 6 (Strong) acidic dye sites of polyacrylonitrile fibres (Balmforth *et al.*, 1964)

Fibre	Dye sites/mequiv. kg^{-1} from Langmuir isotherm intercept (pH 5.4 dyebath)	Dye sites/mequiv. kg^{-1} by non-aqueous conductance titration (after ion-exchange)*
Acrilan 16 (Chemstrand)	36	41
Orlon 42 (DUP)	54	54
Dralon (BAY)	44	44
Dralon Neu (BAY)	55	58

* Before removal of Na^+ and K^+ by ion-exchange these values were approximately half those shown here, respectively.

universal acceptance. The present position is that the very earliest theories, dating back to the mid-eighteenth and the mid-nineteenth centuries, are being reconsidered in the light of recent information.

Prior to 1885 cotton had to be dyed using some form of mordant to fix the colouring matter, because the available dyes, both natural and synthetic, had insufficient affinity for the fibre to be useful without this assistance. In that year however the first of the so-called direct cotton dyes—Congo Red—was patented by Böttinger and the Agfa organization, and thence forward very large numbers of dyes with direct dyeing properties for cellulose have continued to be introduced.

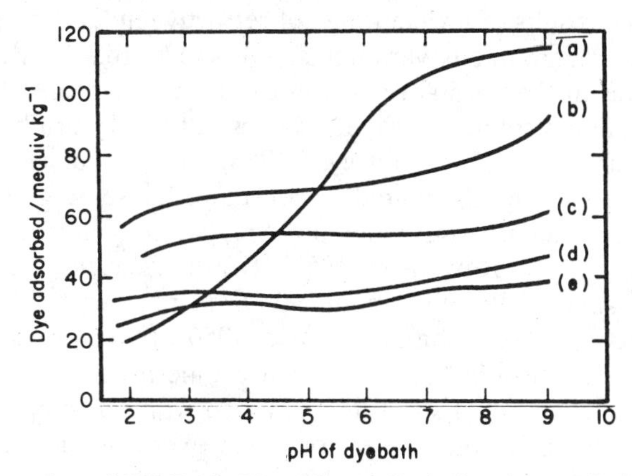

Figure 16 Adsorption of C.I. Basic Blue 22, a cationic dye, by acrylic fibres, against the pH of the dyebath, at 100°C. (a) Courtelle E; (b) Beslon; (c) Orlon 42; (d) Acrilan 16; (e) Acribel. Data from Balmforth *et al.*, 1964. (Giles, 1971)

Congo Red and most of its followers are bis-azo dyes based on benzidine, and possibly the first important suggestion in modern times regarding the requirements for cellulose substantivity was made by Hodgson (1933) when he demonstrated that planarity in the dye molecule is essential. Thus this

NH_2 —N=N— R R′ —N=N— NH_2
SO_3Na R′ R NaO_3S

structure having methyl groups at both R positions is planar and the dye is highly substantive to cellulose, but the dye with methyl groups at both R′ positions has low substantivity and its molecule is non-planar because of steric interference of the methyl groups. If however the 3,3′ positions in the molecule are bridged by, for example, O, NH, or S, both planarity and high substantivity are restored. These observations however do not explain the source of the affinity, they merely emphasize that it probably depends on close face-to-face contact of dye molecule and cellulose chain. The nature of the force that holds the two together has still to be sought.

A variety of suggestions have been made as to the nature of this force (for summaries see e.g. Giles, 1966b, 1975), and include hydrogen bonds, π-bonds, ion–dipole forces, and van der Waals forces.

Several authors (e.g. Vickerstaff, 1954, and others quoted therein) suggested that hydrogen bonds are formed when the dye and the cellulose molecule are aligned closely and in contact face-to-face; the alignment is confirmed by results of dichroism measurements (quoted by Vickerstaff). However, the results of a wide range of refractive index measurements by the method of continuous variations have shown that it is unlikely that carbohydrates in the normal ring-form can form intermolecular hydrogen bonds with simple solutes in water, because of a very firmly bound layer of water which they hold (Arshid *et al.*, 1956). Other evidence also supports this hypothesis of firmly bound water in cellulose, e.g. (i) in water, glucose–water, and sucrose–water bonding is stronger than water–water bonding (Taylor and Rowlinson, 1955), (ii) the freezing point of water in cellulose is abnormally low, about −21°C in regenerated cellulose with 40% water content (Preston and Tawde, 1956), (iii) monolayer tests with a surface-active benzidine dye, of similar geometry to a direct cotton dye, but with long-chain alkyl substituents, agree with the supposition. Thus, the area per dye molecule in the monolayer over cellobiose solution rises to the value needed to accommodate a single layer of water molecules between each dye and cellobiose molecule (Allingham *et al.*, 1954).

A form of hydrogen bonding in which the cellulosic hydroxy groups are attracted by the π-electron system of the dye molecule has been suggested (Bamford, 1954; Lead, 1957; Zollinger, 1954, 1959, 1960), and examined by Yoshida *et al.* (1964). The latter authors used the method of continuous variations, with refractometry, and found that sucrose, glucose, and cellobiose form 1:1 complexes, in water, with simple aromatic sulphonate ions but not with an aliphatic sulphonate. The bond formation is therefore confirmed, and it appears unlike other hydrogen bonds in being formed even in the presence of the water "atmosphere" around the carbohydrate molecule. Yoshida's results were later confirmed with sucrose solutions, using both sodium naphthalene 1-sulphonate and a sulphonated anthraquinone dye (C.I. Acid Blue 25) as second solute, respectively (Giles and McIntosh, 1973) (Sucrose was used in preference to cellobiose because, unlike celloboise, it cannot form a reactive aldehyde tautomer.) Also, in one test methyl-β-D-glucopyranose was used as a better model for cellulose. It would be expected that this form of bond would require a close face-to-face alignment of the carbohydrate and the planar dye molecules; the high entropy value of the bonding (Table 4) seems to show that this situation does occur. The bond thus seems to involve multiple adjacent protons in the carbohydrate and the π-electron system of the dye, and to be effective even in the presence of water.

When sucrose was used with azo dyes it was found, rather surprisingly, that no complex is formed (Campbell *et al.*, 1957; Giles and McIntosh, 1973). This result was obtained both with monoazo dyes of small molecular size, and with a bis-azo direct cotton dye. It is difficult to account for this non-reactivity, when the other aromatic compounds are reactive. Possibly the attraction of the azo group reduces the electron density of the aromatic nuclei on either side of it and prevents their bonding with the carbohydrate hydroxy groups.

There is therefore evidence that dyes without azo groups are assisted in their adsorption to cellulose by the above postulated form of π-bond. The anthraquinone class of vat dyes fall into this category (e.g. C.I. 59700). When their affinity values for cellulose are examined it is seen that they are in fact higher than those for dyes with azo groups (Fig. 17). (They are also higher than those for vat dyes with the CONH group as part of their

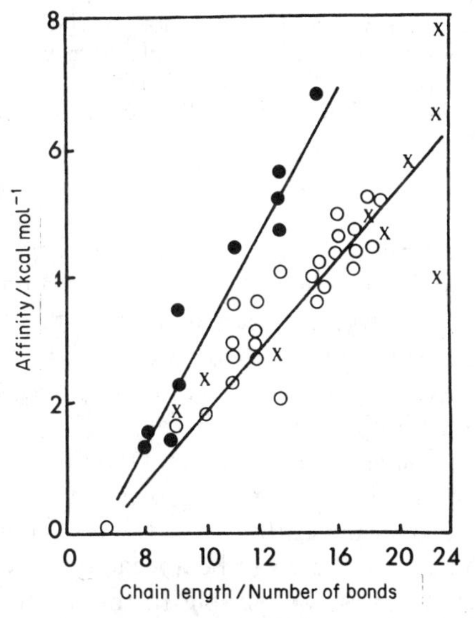

Figure 17 Affinity of dyes for regenerated cellulose against their molecular conjugate chain length. (○, ●) Affinities at 40°C of planar vat dyes and related compounds with respectively one and no –NHCO– group in the conjugate chain, on cuprammonium rayon. (×) Affinities at 50°C of planar sulphonated azo dyes, on viscose rayon. Data from Peters and Sumner, 1955; (×) data from Giles and Hassan, 1958. (Giles and Hassan, 1958)

conjugate system; however, we have no information on the π-bonding properties of this class of dye.)

Most of the direct cotton dyes are azo compounds, and therefore forces other than the π-bond must be responsible for their adsorption. Recent investigations using solutions of cellulose in the cadmium-ethylenediamine (Cadoxen) solvent, and consideration of the possible effect of the fibre pores, have thrown some light on this matter.

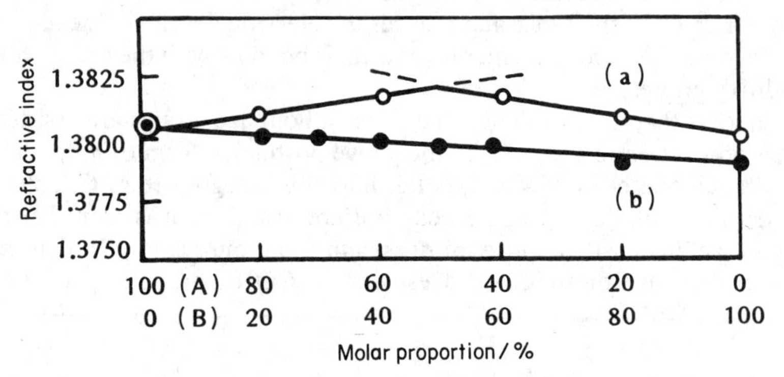

Figure 18 Use of refractive index, by continuous variation method, to detect complex formation in water. (A) Cellulose, calculated as cellobiose residues. (B) (a) C.I. Direct Blue 1; (b) C.I. Direct Blue 10. Blue forms a 1:1 complex with each cellobiose residue in the cellulose chain; Blue 10 forms no complex. (Agnihotri and Giles, 1972)

The method of continuous variations using refractometry and surface viscosity has been applied (Figs 18 and 19). Urea, and also the dye C.I. 24410, by refractometry, are seen to form a 1:1 complex with the cellobiose residue in cellulose, but the dye C.I. 24340 does not, yet both dyes are direct cotton dyes. It would appear that a bond is formed between the cellulosic hydroxy groups and amino groups. However, when tests are made with mixed monolayers of cellulose and the dye C.I. 24410 (Fig. 19), no evidence of a complex is revealed; this is in contrast to the result with cellulose acetate and a disperse dye shown above (Fig. 10). The direct dye–cellulose complex must therefore be a weak one, possibly some form of acid–base attraction. It is comparable in this respect to protein–dye ion-exchange complexes, which similarly are not revealed by film viscometry. The π-electrons in the above-mentioned bond thus have weak basic properties.

The operation of this acid–base attraction is consistent with a series of affinity values for direct cotton dyes on cellulose, given by Vickerstaff

(1954). These show that, in the eleven quoted cases (Giles, 1975), affinity is aided by the presence of basic nitrogen-containing groups. Wegmann's postulate of ion–dipole reaction between dye and cellulose (Giles, 1966) may also be relevant to any enquiry into the source of dye–cellulose affinity, but exact data by which it can be tested seem to be lacking.

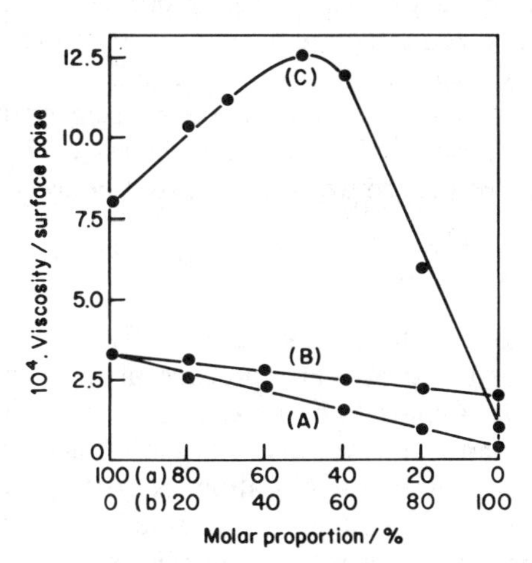

Figure 19 Use of monolayer viscosity, by continuous variation method, to detect presence or absence of complex formation, between dyes and monolayer polymers. (a) Polymer; (b) dye. (A) Cellulose on aqueous solution of Methylene Blue (cationic dye). (B) Cellulose on aqueous solution of C.I. Direct Blue 1 (anionic dye) in 0.01 mol dm^{-3} NaCl. (C) Mixed film of cellulose triacetate (plotted as hexaacetylcellobiose units) with a non-ionic disperse dye (C.I. Disperse Red 13) on water. All measurements at room temperature. (Agnihotri and Giles, 1972)

The face-to-face alignment of direct dye and cellulose molecule is also indicated by pressure/area curves of cellulose monolayers on solutions of direct cotton dyes. These show the type B (Fig. 5) slope change due to the action of the dye on the monolayer. This indication of a face-to-face alignment, with dye molecules located just beneath the cellulose monolayer, is reinforced by the observation that the film expands in proportion to the *width* of the dye molecule (Giles and McIver, 1977). This again points to some dye–polymer force, probably of a weak nature.

Thus at present it appears that the initial operation in attachment of dye to cellulose is due to a combination of several types of weak force: van der Waals non-polar forces, acid–base attraction if the dye molecule contains basic groups, and π-bonds if it has no azo group. These however are in this author's opinion inadequate to account for the normal deep shades obtained in technical operations. Porosity effects, to be described later, must be invoked to give a full explanation of cellulose dyeing.

IV. Adsorption of Dyes by Inorganic Solids

This is important in at least three processes: photographic emulsion sensitization, dyeing anodized aluminium, and the measurement of specific surface and porosity of industrial powders. The first of these is a highly specialized subject, outside the scope of the present chapter; there is a considerable literature covering it (see, e.g. Herz, 1977, and references therein). The other two procedures are discussed below.

A. DYEING ANODIZED ALUMINIUM

This is practised commercially on a wide scale. There is a very extensive literature coverage of the theory and practice of the anodizing process, and on the practice of its coloration, but comparatively little on the theory of the coloration (for an extensive review of the whole subject, see Giles, 1975; and for a shorter review of coloration theory, see Giles, 1979).

The metal is coated with an integral film oxide by electrolysis in an acid solution. The film is highly porous, and in this respect is comparable to textile fibres. The pores are of the order of 10 nm in diameter and run outwards from the metal surface and perpendicular to it, in a pattern, when viewed by electron microscopy, somewhat resembling a honeycomb. Because of its high porosity the film has a large internal surface, and because of its positive charge in water it readily adsorbs anionic dyes or inorganic ions. Coloration by inorganic oxides or salts is effected by electrolytic means or by application of metal ions from aqueous baths. Thus a gold shade very resistant to outdoor exposure is produced by the incorporation of ferric ions from, for example, ferric ammonium oxalate solution. Anionic dyes are readily applied from low temperature baths.

The film is essentially Al_2O_3 with a hydrated surface. In water OH^- ions are released from the surface and the cationic centres remaining are the source of the attraction for anionic dyes. O'Connor *et al.* (1956) showed that the surface can ionize in either of two ways, depending on the nature of the bond holding the hydroxy group to the surface:

$$Al(OH)_3 \rightleftharpoons [Al(OH)_2]^+ + OH^-$$

$$AlO.OH \rightleftharpoons AlO.O^- + H^+$$

Pretreatment of the film with an acid H^+X^- covers its surface with acid anions that may be either covalently bound, or ionized, or both together, depending upon the nature of the anion:

$$\rangle Al\text{—}OH + H^+X^- \rightleftharpoons \rangle Al\text{—}X + H_2O \rightleftharpoons \rangle Al^+X^- + H_2O$$

The anion, whether covalently bound or ionized, may be replaced by dye anions, and it is this reaction that is the basis of the technical dyeing process.

Confirmation of the operation of the anion exchange process when the anodic film is dyed with sulphonated dyes is given by a number of observations (Giles and Datye, 1963) of the effects of degree of sulphonation of dye and the nature of any inorganic salt present in the dyebath. Thus, it was found that (i) sodium chloride and potassium iodide do not influence the dyeing, (ii) all sulphates are very effective in retarding dyeing, (iii) trisodium phosphate is much more effective than sulphates in retarding dyeing, (iv) the effectiveness of sodium sulphate in retarding dyeing increases with the number of sulphonate groups in the dye anion.

These facts appear to show that the action of inorganic salts is one of repulsion between the dye anion and the inorganic anion, at the film surface, and competition between them for ionic sites in the film, and not merely due to the size of the inorganic anion. It was shown that aggregation of dye by the salts is not an important factor. Table 7 and Fig. 20 illustrate the above observations.

Table 7 Effect of sodium sulphate on dye uptake by anodic alumina, with dyes of different degrees of sulphonation

		Dye and amount adsorbed/g kg^{-1} film		
Sample	Na_2SO_4/ g dm^{-3}	C.I. 15620 (monosulphonated)	C.I. 16045 (disulphonated)	C.I. 16255 (trisulphonated)
1	0	302	22.0	9.7
2	2	149	8.5	0.0
3	5	123	3.7	0.0
4	10	—	1.2	0.0
5	20	92	1.2	0.0
6	40	57	0.0	0.0

The operation of the alternative covalent bond mechanism simultaneously with anion exchange was also demonstrated. For example, sodium sulphate solutions desorb dye from the film, but only partially, the desorbable proportion probably being that present as free anions, and the non-desorbable being covalently attached. Further, sulphate ions in the dyebath

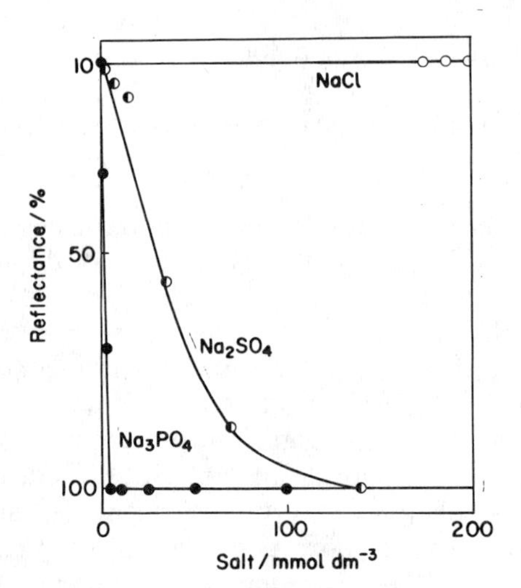

Figure 20 Effect of inorganic ions of different valency upon adsorption of an acid wool dye (C.I. Acid Yellow 3) by sulphuric acid anodic oxide film on aluminium in neutral baths. The reflectance (y-scale) is a measure of the amount of dye adsorbed, 100% reflectance being attained when no dye is adsorbed. (Giles, 1974)

replace dye anions present in the free state in the film, but not those covalently bonded. Therefore when sulphate is present in the dyebath a higher proportion of the dye anions become covalently bonded to the film than when the salt is absent, the free dye anions being partially displaced by sulphate ions. This situation can be inferred from the relative amounts of dye desorbed from dyeing made in the presence and absence of sulphate (Table 8).

Table 8 Effect of sodium sulphate solutions in desorbing C.I. Acid Yellow 3 from anodic film

Dyebath addition	Condition	% Dye desorbed by Na_2SO_4 soln*
None	Unsealed	50 ± 5
None	Sealed	8 ± 2
2 g dm^{-3} Na_2SO_4	Unsealed	20 ± 5
2 g dm^{-3} Na_2SO_4	Sealed	1.5 ± 0.5

* 20 g dm^{-3} Na_2SO_4 at 60°C for 30 min.

After the dyeing operation the film is "sealed" by treatment in boiling water or steam. This produces crystalline aluminium hydroxide which blocks the pore apertures at the surface and improves the resistance of the dyeing to washing and light. The effect is well seen in scanning electron micrographs (see examples and references in Giles, 1974) and by use of the electron probe microanalyser (Lamb, 1979).

The general conclusion reached from the above observations is that sulphonated dyes are adsorbed by the anodic film by both of the alternative mechanisms outlined above, the proportions of the two forms of attached dye being varied according to the dyebath composition, the nature of the dye, and the method by which the film has been formed (which determines the nature of the anions originally present in the undyed film).

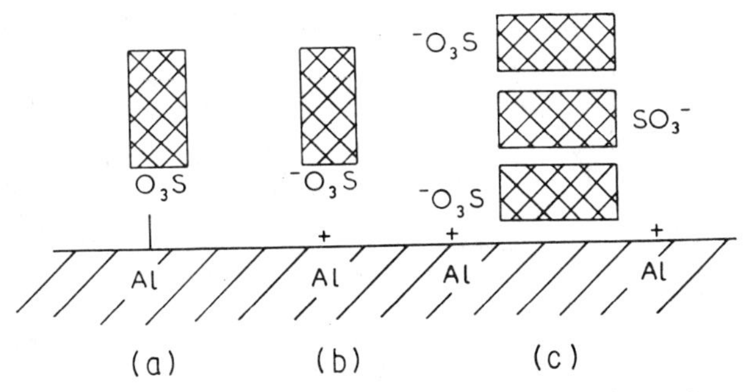

Figure 21 The probable nature of dye–alumina bonds with sulphonated dyes on anodic alumina film. (a) Covalent bond; (b) ion-exchange with individual dye anions; (c) ion-exchange with anionic micelles of dye. (Giles, 1974)

The dye present as free ions may however be in either or both of two forms: single monodisperse anions, or small ionic micelles containing only a few individual dye anions, usually from two to four, arranged in sandwich and anti-parallel forms. (It will be clear from the data in Table 7 that only mono-ionic dyes are suitable for this material; the evidence for the micellar size is given below.) The situation at the dyed film surface is illustrated by Fig. 21.

B. SPECIFIC SURFACE AND POROSITY MEASUREMENT

This can readily be carried out on inorganic powders, using monoionic dyes. These should be either anionic or cationic, depending on the sign of charge on the powder in water. Most powders of technical interest, e.g. silica, charcoal, titania, normally carry a negative charge in water, and consequently cationic dyes are suitable for them. Alumina normally carries a positive charge, for reasons given above, and anionic dyes may be used for it, but if it is pretreated with alkali (the quality supplied for chromatography is usually in this form), cationic dyes may be used.

Certain precautions must be observed to ensure satisfactory results. Thus, dyes that form specific bonds with the surface of the powder, e.g. hydrogen bonds from free hydroxy groups in the dye molecule, or covalent bonds, as in the case of alumina discussed above, should not be used. Neither should dyes be used that may be precipitated by ions from the powder, e.g. Fe^{3+} or Ca^{2+} which may react with anionic dyes. The pH of the dye solution also must be correctly adjusted, if necessary by buffering; for example, alkali from chromatographic alumina powder may partially decolorize the cationic dye Crystal Violet.

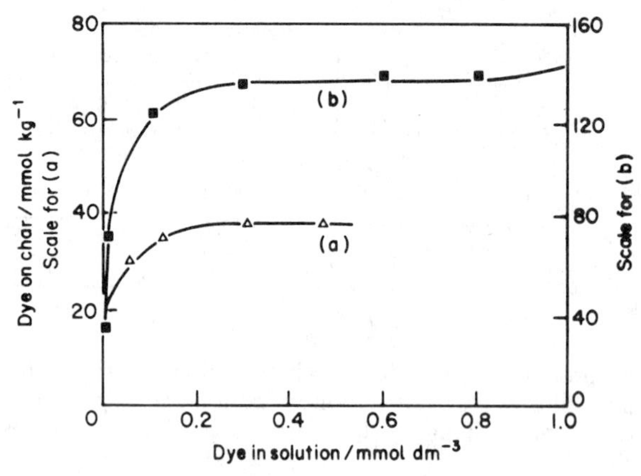

Figure 22 Isotherms for adsorption of typical cationic dyes on bone char, at 18°C. (a) Victoria Pure Lake Blue BO; (b) Methylene Blue BP. (Giles *et al.*, 1969)

The chosen dye may be applied cold and an adsorption isotherm determined. This will usually have a plateau, which is taken to represent the completion of a condensed adsorbed monolayer (Fig. 22). If then the cross-sectional area of the adsorbed dye at the surface is known, the specific surface is readily calculated. It is found that the dyes are adsorbed in the form of sandwich micelles (see Fig. 21), with an average aggregation number that rises with the cube of the weight of the colour ion of the dye. For example, Methylene Blue (ionic wt. 302) has an aggregation number, N, of 2.0, the dye being adsorbed as a dimer, and Victoria Pure Blue BO (i.w. 477) has $N = 9$. The micelles appear to be adsorbed flat, so the area that each covers is the same as that of a single dye molecule. The N values have been obtained by several authors, by different methods, including direct measurement in monolayers (Table 9). Figure 23 illustrates the relation between aggregation number and weight of the colour ion of the dye.

Table 9 Data on relation between ionic weight of monoionic dyes and aggregation number *(N)*

		Approximate molecular area/nm^2		N, in aqueous solution, by		N, by			
Dye	Ionic wt	end-on	flat	diffusion	polarography	adsorption on carbon	streaming potential	Molecular area in monolayer (b)/nm^2	N, by monolayer (a/b)
I	377	0.50	1.50			3.6		0.32(?)	4.7
II	393	0.80	1.43	2.2–2.5	*ca.* 1–5	2.8		0.44	3.3
III	284	0.64	1.20	1.6–1.7	1.6–2.0	1.95–2.0	2.4	0.57	2.1
IV	315	0.90	1.50			2.4–2.6		0.60	2.5
V	372	0.90	2.25		1.5–5.0	3.6	2.6	0.55	4.1
VI	477	0.90	2.70(?)	8.9–9.9		9.0	8.9	0.38	7.1

I = C.I. 15620 (monosulphonate); II = C.I. 62055 (monosulphonate); III = C.I. 52015 (cationic; Methylene Blue); IV = C.I. 50240 (cationic; Safranine T); V = C.I. 42555 (cationic; Crystal Violet); VI = C.I. 42595 (cationic).

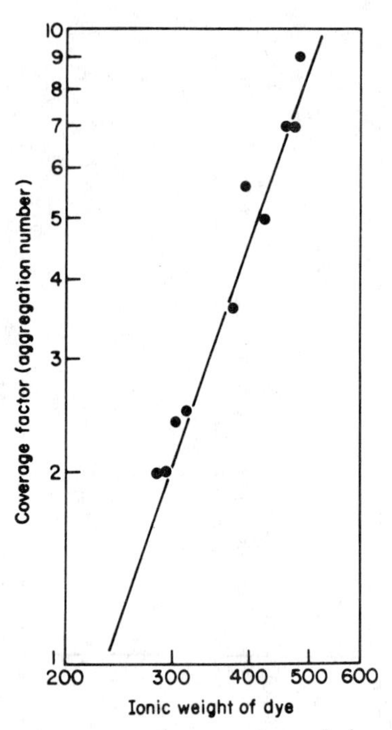

Figure 23 Relation between aggregation number of dyes adsorbed on various non-porous solids, and the weight of the colour ion. (Giles *et al.*, 1969)

The pores in many industrial powders are comparable in size to dye molecules, and consequently can be used as molecular sieves whereby, by selective adsorption, estimates of the pore structure can be obtained. Results so far obtained in this way agree with those from other methods (Giles and D'Silva, 1969; Giles *et al.*, 1969, 1979).

V. Effect of Porosity on Dyeing of Fibres

It will be clear from the cellulose porosity observation (Fig. 1), and the description of wool morphology, that dyes must have access to an extensive pore structure in fibres; intermolecular voids also exist in the synthetic polymer fibres. In such spaces dyes may aggregate. There have been many experimental observations of the presence of discrete aggregates of dye in textile fibres dyed under normal conditions (for a review see Daruwalla, 1974). These have been made, for example, by optical microscopy on wool (Speakman and Smith, 1936; Royer and Maresh, 1943), by X-ray examination on the same fibres (Astbury and Dawson, 1938), by electron microscopy on direct cotton-dyed regenerated cellulose films (Weissbein and Coven, 1960), by low-angle X-ray examination of dyed cellulose acetate (Kratky *et al.*, 1964), by microspectrophotometry and scanning electron microscopy on several varieties of man-made fibres (Ohtsu *et al.*, 1974), and by more indirect means, e.g. by adsorption of organic vapours on dyed wool (Giles *et al.*, 1964), and by adsorption of a solute (*p*-nitrophenol) on dye in cellulose (Giles and Haslam, 1977). (For a fuller discussion see e.g. Giles, 1975.)

Many authors indeed have considered that aggregation of dye is an essential part of the adsorption process, especially of cellulose dyeing, an idea that can be traced back for well over a century, and that has been revived, with good experimental evidence, in recent years by Bach *et al.* (1963).

Little attention however has been given to the mechanism by which the internal aggregation occurs. Dyeing in most of the above systems takes place at or near the boil, and at these temperatures dye aggregates in solution are broken down to the monodisperse condition. The dye must therefore enter the fibre as single molecules or at most in very small aggregates. Why then should it later appear in some cases as aggregates so large that they are visible quite readily in the electron microscope, and even sometimes in the optical microscope? This cannot, at least in those cases where examination has been made of wet dyeings, be due simply to crystallization during drying after dyeing, because these wet dyeings give spectroscopic evidence of the presence of associated dye. A clue to the sequence of events can perhaps be obtained from investigations of the dyeing of cellulose with direct cotton dyes, some examples of which are as follows.

Daruwalla and D'Silva (1963) measured the heat of dyeing of direct dyes on regenerated cellulose film for various dye concentrations. One of their sets of results is illustrated in Fig. 24, in comparision with results obtained here with the same type of film and the same dye, using a solute (*p*-nitrophenol), which by its preferential adsorption on the dye itself rather than on the cellulose, from water, enables estimates to be made of the sizes of dye particles in the film or fibre (for details of the method and the interpretation, see Giles *et al.*, 1971; Giles and Haslam, 1977). The suggestions regarding the sequence of events shown in Fig. 24, and illustrated in Figs 25 and 26, show that three-dimensional aggregates probably form as a result of the break-up of a multilayer structure at high concentrations, this in turn being developed from a true monolayer first formed at very low dye concentrations.

The formation of aggregates is encouraged by the normal addition of inorganic salt (usually sodium chloride) to the dyebath. (A small amount of salt is essential to eliminate the negative potential of cellulose in water,

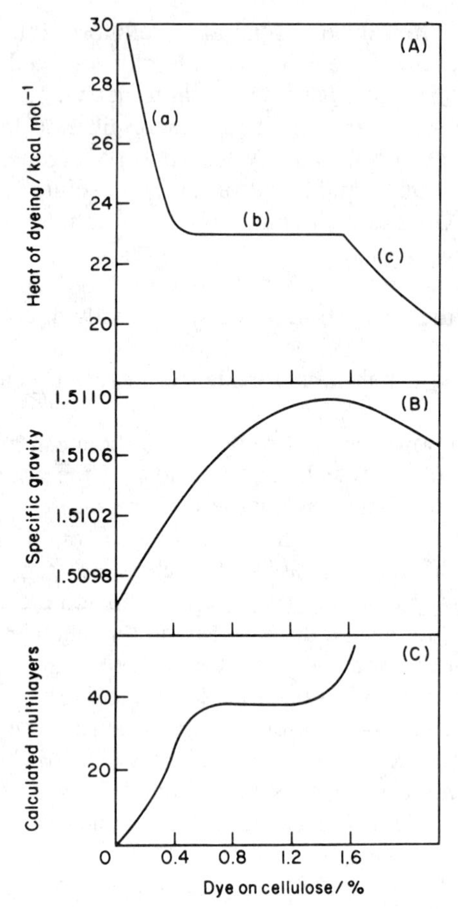

Figure 24 Comparison of three parameters of dye (C.I. Direct Blue 1) on regenerated cellulose, with dye adsorbed; (A) and (B) based on data from Daruwalla and D'Silva, 1963. (Giles and Haslam, 1978)

which tends to repel the anionic dye.) However, there may be a more fundamental reason for aggregation in any type of fibre pore. This may be illustrated by a modification of the Gibbs adsorption equation, based on a treatment by Pethica (1955), to apply to the concentration of dye colour ions in a condensed polymer monolayer (Giles and McIver, 1977). Thus the surface excess Γ_D of the dye colour ion in the film is given by

$$\Gamma_D = -\frac{mf}{RT}\frac{d(\gamma - \pi_D)}{d(mf)}\frac{1}{\phi} \tag{4}$$

where Γ_D is the surface excess of the dye colour ion in the monolayer,

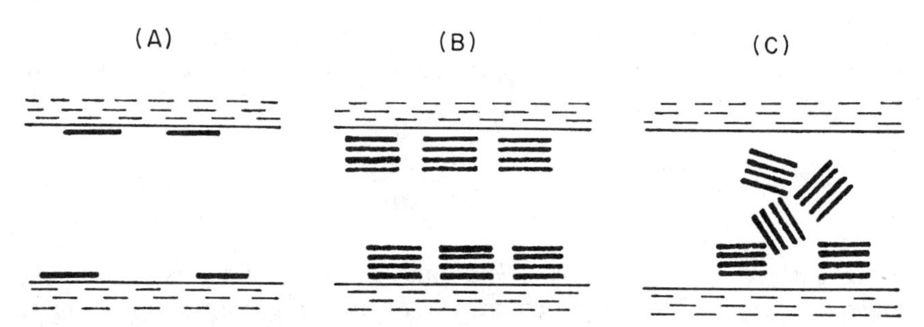

Figure 25 Sketch illustrating suggested successive stages in adsorption of a direct cotton dye from aqueous solution in a pore of cellulose. (A) At very weak depths of shade, a not fully condensed monolayer is formed, which develops into a multilayer (B) with increase in dye content, and finally breaks up to three-dimensional aggregates (C) at highest concentrations. (Giles and Haslam, 1978)

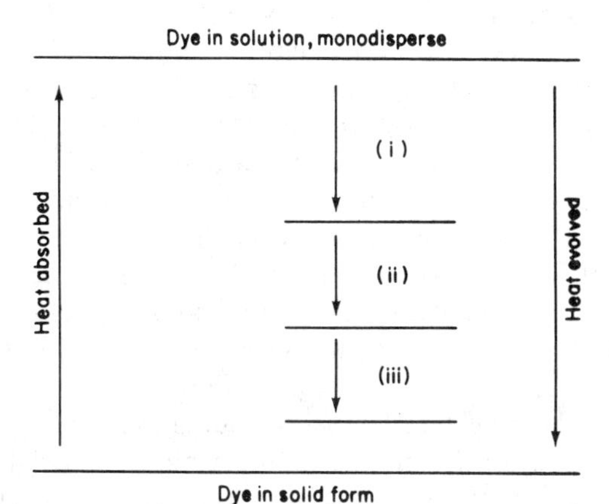

Figure 26 Energy-level diagram for adsorption of a dye by cellulose from aqueous solution. The stages in evolution of heat may be represented as: (i) extension of dye solution interface by contact with cellulose surface; (ii) replacement of water adjacent to cellulose surface by the first adsorbed dye monolayer; (iii) development of three-dimensional dye aggregates. (Giles and Haslam, 1978)

π_D is the measured film pressure on the trough, and ϕ is defined as $A_x/(A_x - \bar{A}_x)$, where A_x and $\bar{A}_x$ are respectively the area, per unit weight of the monolayer substance, of the polymer + dye and the pure polymer monolayer at the same film pressure, and m and f are the solution molarity and the molar activity coefficient respectively. From this expression and a calculated value for the activity coefficient (0.76) of the blue direct cotton dye C.I. 24410, the concentration of dye molecules under a cellulose monolayer was calculated to be $0.043 \times 10^{14}\ \mathrm{cm}^{-2}$, assuming they are oriented under the cellulose chains, an assumption based on the shape of the pressure/area curve. At the air/water interface the dye concentration calculated from the Gibbs equation,

$$\Gamma_D = \frac{mf}{RT}\frac{d\gamma}{d(mf)} \tag{5}$$

using surface tension values (Giles and Soutar, 1971), is 0.36×10^{14} molecules cm^{-2}. The lower concentration under cellulose is attributed to the mutual repulsion between dye anions and anionic groups in the cellulose. All these results were obtained at room temperature, with 0.4×10^{-3} mol dm^{-3} dye solutions.

In the bulk solution there are $0.3 \times 10^{-3} \times 6.1 \times 10^{23} \times 10^{-3} = 1.83 \times 10^{17}$ molecules cm^{-3}, and therefore if there were no surface excess and the dye concentration were the same in the surface layer as in the bulk solution, the surface concentration would be $(1.83 \times 10^{17})^{2/3} = 0.0032 \times 10^{14}$ molecules cm^{-2}. The dye is therefore $0.36/0.0032 = 112.5$ times as concentrated in the surface layer as in the solution. In the pores however a still greater increase in dye concentration can occur. In illustration of this, consider a pore of diameter 3 nm (Fig. 1), around the wall of which the dye monolayer is adsorbed. This layer is about 0.5 nm deep, and a simple calculation shows that it occupies 5/9 of the total volume of the pore. Therefore the total concentration of dye in the pore, allowing for the surface excess calculated above, increases considerably above that in the external solution, and this would cause aggregates of dye to form (see Pugh *et al.*, 1971). No data are available on the effect of rise in temperature on the interfacial tension, but associated dye is certainly present in wet substrates before the final drying, so association is not due simply to the drying-up of the internal solution. For example, the associated direct cotton dye particles measured by *p*-nitrophenol adsorption, in "never dried" cellulose film, were present in material that was maintained wet throughout (Giles and Haslam, 1977). Further evidence to the same effect is: (i) particle size measurement on a direct cotton dye, in cellulose dyed in the normal fashion and then dried, gave results for size distribution closely parallel to Daruwalla and D'Silva's (1963) heat of dyeing measurement changes, made of course during the actual dyeing operation (Giles and Haslam, 1978); (ii) the absorption spectrum of Methylene Blue in gelatine, showing the presence of associated dye, was found to be the same for films maintained wet as for those dried after incorporation of dye (Campbell and Giles, 1958).

The above argument applies to all protein and cellulose fibres, and also to a limited extent to the synthetic polymer fibres, which have a smaller volume of water-filled pores, in dyebaths, than have proteins and cellulose.

In these synthetic fibres however the (non-ionic) dyes penetrate deeper into the polymer structure, as already explained, and eventually appear to reach the voids in the structure and there develop aggregates. In inorganic porous solids there is some evidence that the above concentration effect comes into action with dyes of high molecular weight in very small pores (T. W. Smith, private communication, 1979).

Bach *et al.* (1963) established the effect of porosity in cellulose fibres on a firm basis by extended spectroscopic investigations of direct cotton and other anionic dyes under different conditions. Absorption spectra were determined, of these dyes in water and various solvents, in cellulose, in polyvinyl alcohol film, and solid in compressed KBr. Aggregation of the direct dyes in cellulose was clearly evident from a study of the spectra. These authors suggested that the long planar molecules of direct dyes are able readily to diffuse into the intermicellar spaces in the cellulose, where they lose their associated water and form aggregates. The dyes that have lower substantivity for cellulose do not so readily aggregate in the fibre and so cannot build up such heavy shades as those that do aggregate.

VI. Kinetics and Thermodynamics of Dyeing

These aspects of the present subject, while rather outside the scope of the treatment given above, are so intimately linked with this that brief consideration must be given to them.

A. KINETICS

There are at least four rate steps involved, viz. diffusion of dye (i) in external solution up to the fibre surface, (ii) across a liquid boundary layer at the surface, (iii) across an adsorbed layer on the surface, and (iv) into the pore or polymer structure of the fibre. In conditions where liquor circulation is sluggish, (ii) can become rate-determining, but normally (iv) is rate-determining. If the substrate is non-porous, also step (ii) is rate-determining, and the treatment of Boyd *et al.* (1947) and Kitchener (1959) then reveals a linear relation, with time, of the function $-\log[(1 - c_t/c_\infty)]$, where c_t and c_∞ are the amounts of dye present on a non-porous particle after time t and at equilibrium. Such linear curves have been obtained with sulphonated dyes on alumina powder (Cummings *et al.*, 1959; Giles, 1966a).

In porous fibres an empirical relation is found; the amount of dye adsorbed increases linearly with $(\text{time})^{\frac{1}{2}}$ over most of the period up to equilibrium. A law of this nature was originally derived for adsorption from an external solution of constant concentration, which does not apply in a normal (batch) dyeing operation. Barrer (1949) however has shown, by theory and experiment, that the law should still apply even if the external phase is not of constant composition. Figure 27 gives a typical $t^{\frac{1}{2}}$ rate curve.

Recent work (e.g. Peters *et al.*, 1961; Ratee and Breuer, 1974; cf. Vickerstaff, 1954) has been concerned with measurement of diffusion coefficients inside the fibre, which are changing with the progress of adsorption and also with distance into the fibre itself. The basis of the $t^{\frac{1}{2}}$ curve may be considered to be the expression

$$c_t/c_\infty = 2\sqrt{(Dt/\pi)} \qquad (6)$$

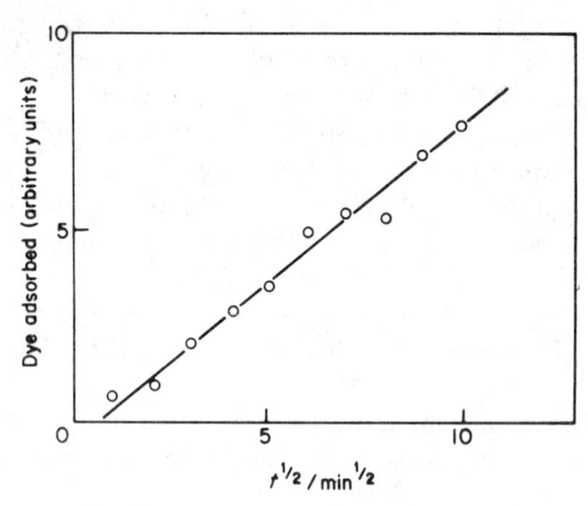

Figure 27 Rate-of-dyeing ($t^{\frac{1}{2}}$) curve for C.I. Direct Blue 1 on regenerated cellulose film at 50°C, in presence of 5 g dm^{-3} NaCl. Data from Peters *et al.*, 1961. (Giles, 1966a)

derived from Fick's law, where D is a diffusion coefficient (the "integral diffusion coefficient"). In spite of the difficulty caused by the variations just mentioned, this integral coefficient is most useful in comparing effects of e.g. the molecular structure of dyes or the morphology of fibres (Speakman and Smith, 1936; Vickerstaff, 1954; Giles *et al.*, 1962; Giles, 1966a). In practice, the $t^{\frac{1}{2}}$ curves may exhibit one or other of the three forms shown in Fig. 28. Type I curves are common with fibres. The initial induction period is caused by a physical barrier at the fibre surface, the nature of which varies with the substrate; with wool it is caused by the very thin epicuticle that covers the fibres, and when this is removed, by chemical (Lemin and Vickerstaff, 1946) or mechanical (Hampton and Rattee, 1979) means, the curve becomes type II or even type III.* The latter curve

* E.g. the sulphonated dye C.I. 15510 gives type I on untreated wool, but type III on wool that has had the epicuticle removed by alkali (Lemin and Vickerstaff, 1946).

represents a rapid build-up of a concentrated layer on the external surface, and this occurs with some highly surface-active dyes. Type II curves represent conditions where those causing both I and III are present and balance out.

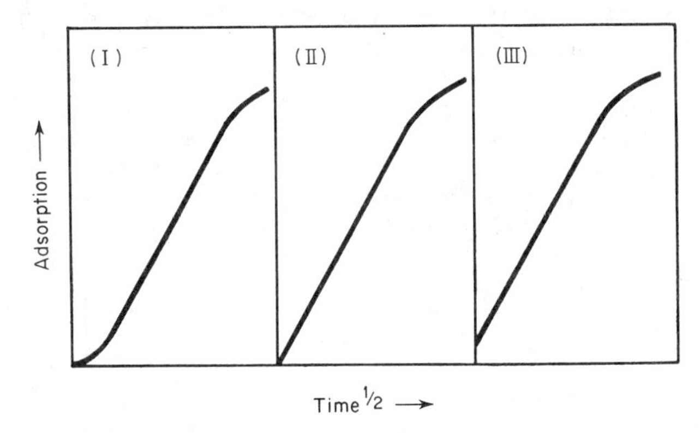

Figure 28 Classification of rate-of-dyeing ($t^{\frac{1}{2}}$) curves by position of origin. (Giles, 1966a)

Rate of dyeing can be related to the smallest cross-section of the dye molecule, i.e. its slimness. Merian (1966) demonstrated this for disperse dyes on cellulose acetate, using an equation experimentally derived by Park (1951) for diffusion of organic vapours into polystyrene:

$$\log D = \log K - AV_M - Bd \tag{7}$$

where D is the diffusion coefficient, V_M the molar volume of the vapour, d its minimum molecular diameter, and K, A, and B are constants. Merian (1966) and Husy *et al.* (1969) found a similar linear relation between log D for disperse dyes and a function of their molecular size and shape, based on the right-hand side of the above equation. Rate of dyeing measurements of anionic dyes, of progressively altered molecular structure, on wool, also shows a similar relation between rate and molecular structure (Fig. 29).

Nicholls (1956) found clear relations between rate of dyeing on the one hand, and rate of desorption on the other, of a series of wool azo dyes differing only in the number of their substituent sulphonate groups. The interpretation of these data is that rate of dyeing, which takes place in acid solution, is increasingly retarded by added $-SO_3^-$ ionic centres, which are attracted by the cationic ($-NH_3^+$) centres in the fibre; these exert a "drag" on the dye anions diffusing past them. In desorption however, in alkaline solution, the opposite effect operates, and the more $-SO_3^-$ centres there are in the dye molecule, the greater the electrostatic repulsion between the dye and the ionic centres in the fibre, which are now the $-COO^-$ ionic groups. Thus speed of dyeing is favoured by few sulphonate groups, and speed of desorption (and also of washing, which is carried out under neutral or mildly alkaline conditions) by the presence of many of these groups in the dye molecule (see Giles *et al.*, 1962).

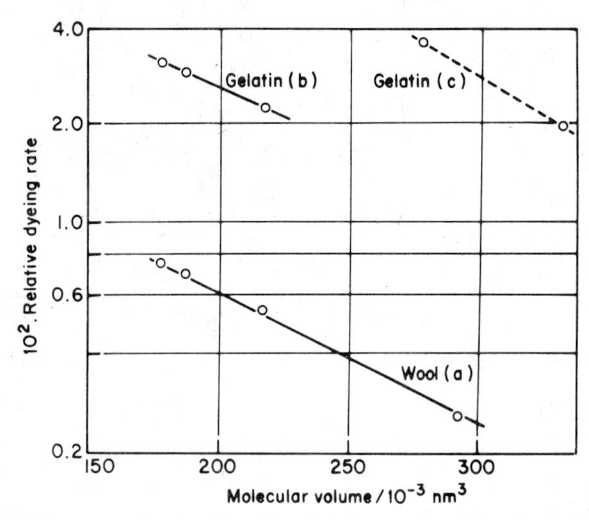

Figure 29 Influence of porosity of substrate, and molecular volume and ionic character of anionic dyes, on rate of dyeing of proteins, at 70°C.

R = H, CH_3, n-C_4H_9, n-$C_{12}H_{25}$ respectively: (a) wool; (b) gelatin. Line (c) is for a different type of dye, viz. two 2:1 dye–chromium complexes of similar structure to each other, with shielded anionic charge. (Giles *et al.*, 1962)

B. THERMODYNAMICS

Gilbert and Rideal's development of a treatment for determining the thermodynamic affinity of a dye for a protein fibre leads to the following expression for the difference in chemical potential $\Delta\mu^\circ$ of the dye D in the specified standard state, in the fibre F and the solution S:

$$-\Delta\mu^\circ = RT\ln([\mathrm{D}]_\mathrm{F}/[\mathrm{D}]_\mathrm{S}) \tag{8}$$

where the bracketed terms represent activities (Gilbert, 1944; Gilbert and Rideal, 1944). This expression however applies to unionized dyes, e.g. disperse dyes in non-ionogenic fibres. Complications arise when the dye is ionic and there is inorganic salt in the solution. In the dilute solutions used for dyeing, activity is assumed to be unity, but difficulty arises in deriving an expression for activity of dye in the fibre. The extent of the true volume of the fibre in which the dye is assumed to reside, and also the physical state of the adsorbed dye, should be taken into consideration. In practice the second of these factors is usually ignored and it is assumed that the dye in the fibre is monodisperse, though this is probably only true at very low dye concentrations, as we have discussed above. Three methods have been proposed for dealing with the effective dye concentration in the fibre.

(i) Use of the term $\theta/(1-\theta)$ for the activity of dye in fibre, θ being the proportion of available sites occupied. This was the term first used by Gilbert and Rideal (1944) for adsorption of simple mineral acids by wool, later by Vickerstaff (1954) for anionic dyes on wool, and more recently by Daruwalla and D'Silva (1963) for direct cotton dyes on cellulose. The treatment involves the assumption that the adsorbed ions do not interact, which is unlikely with many of the large aromatic colour ions of dyes.

(ii) Use of the hypothesis that the adsorbed solute is dissolved in a surface layer which is more closely associated with the substrate than is the main body of the solvent. The appropriate surface "volume" term is chosen empirically (Peters and Vickerstaff, 1948), e.g. the expression for the affinity of an ionized dye, $\mathrm{Na}_z\mathrm{D}$, diffusely adsorbed in the fibre, is thus defined as

$$-\Delta\mu^\circ = RT\ln\left(\frac{[\mathrm{D}]_\mathrm{F}[\mathrm{Na}]_\mathrm{F}^z/V^{z+1}}{[\mathrm{D}]_\mathrm{S}[\mathrm{Na}]_\mathrm{S}^z}\right) \tag{9}$$

where V is the above defined "volume" term.

(iii) Use of a measurement of the initial angle of the adsorption isotherm, whereby the required affinity value, $\Delta\mu^\circ = -RT\ln K$, is obtained, K being the equilibrium constant. The isotherm must be plotted with ordinate showing weight of solute per unit volume of an adsorbed layer, which again involves an estimate of the volume of this layer. Further, K is given by the slope of a line drawn through the origin, tangential to the initial slope of the isotherm. Considerable uncertainty in estimating the required slope is possible, and this is pointed out by the authors (Bartell *et al.*, 1951).

Both the kinetic and the thermodynamic approaches can have practical results, in indicating the types of dye molecular structure most conducive to rapid dyeing, fastness to washing, and efficient use (maximum adsorption) of dye.

VII. Conclusions

We may give a general picture of the sequence of events in the adsorption of dyes by solid substrates. Firstly the dye, of whatever type, is surface-active and this favours escape from the water solution to the fibre/water interface, where the dye molecules then concentrate. Electrostatic effects can however either aid or hinder the initial adsorption, if the dye is ionized. Most fibres have a negative charge in water and most dyes for cellulose or wool are anionic, so there is a repulsion between dye and fibre unless the charge is neutralized or reversed. On cellulose the charge is neutralized by inorganic salt (usually common salt) in the dyebath; on wool it is reversed by acidifying the bath, whereby the amino groups of the fibre protein acquire a positive charge. The repulsive effect of the fibre charge can however be overcome if the dye molecule has a high molecular weight, which increases the forces promoting the adsorption.

Secondly, the dye diffuses into the intermicellar spaces of the fibre, and becomes concentrated as a monolayer, by virtue of a combination of forces, including van der Waals forces and ion-exchange (with wool and to a minor extent with cellulose). As the dye concentration in the fibre rises, monolayers develop into multilayers, and these in turn break up into three-dimensional aggregates.

The size of the dye molecule and the attractive force between dye and fibre both influence the diffusion rate into the fibre, and also the degree of adsorption under given conditions (i.e. the affinity).

The above picture applies to the so-called hydrophilic fibres, i.e. those that have high water attraction and considerable volume of water-filled pores. In the hydrophobic fibres, i.e. man-made polymers (including cellulose acetates), the situation is different; there are insufficient water-filled spaces for ionic dyes to have as it were an adequate foothold. Instead non-ionic dyes of low water solubility are used. These diffuse into the polymer structure in regions beyond the water-filled pores, probably in monodisperse form, and attracted by—mainly—hydrogen bonds between the polar groups of dye and fibre. Eventually the dye molecules reach and concentrate in the voids in the polymer structure, where they associate, but into smaller particles than those forming in the hydrophilic fibres.

VIII. Covalent Bond Formation

The above discussion has been confined to the dyeing of textile fibres, which is by far the greatest use of dyes; in particular it has been concerned with the physical types of adsorption, i.e. those that are more or less readily reversible. This excludes the formation of covalent dye–fibre bonds which occurs in dyeing with the so-called reactive dyes (see typical formula above), introduced in the late 1950s. In their use, at first the normal adsorption processes take place, but later, usually by an induced change in pH of the dyebath, the covalent bond is formed. The process is applied to protein fibres and nylon, but its most important use is for cellulose fibres, for which it enables bright shades of very high wash fastness to be obtained. The high fastness is due to the difficulty of reversing the dye–fibre bond, and the brightness to the possibility of using dyes with relatively small molecules, rather than those with large molecules, which have duller shades, but which are necessary for the degree of substantivity required for satisfactory shade depths. A number of types of reactive group are used, substituted on normal dye molecules. Reactive dyeing, because of its technical importance, has been the subject of a considerable volume of research, which is outside the scope of the present treatment. Many accounts of its use and characteristics are available (e.g. Rattee, 1978).

References

Abrahart, E. N. (1977) *Dyes and their Intermediates,* 2nd edn, Edward Arnold, London.
Agnihotri, V. G. and Giles, C. H. (1972). *J. Chem. Soc. Perkin Trans. 2*, 2241.
Allen, R. L. M. (1971). *Colour Chemistry,* Thomas Nelson and Sons, London.
Allingham, M. M., Giles, C. H. and Neustädter, E. L. (1954). *Discuss. Faraday Soc.* **16**, 92.
Arshid, F. M., Connelly, R. F., Desai, J. N., Fulton, R. G., Giles, C. H. and Kefalas, J. D. (1954). *J. Soc. Dyers Colour.* **70**, 402.
Arshid, F. M., Giles, C. H., McLure, E. C., Ogilvie, A. and Rose, T. J. (1955). *J. Chem. Soc.* 67.
Arshid, F. M., Giles, C. H. and Jain, S. K. (1956). *J. Chem. Soc.* 559.
Aspland, J. R. and Bird, C. L. (1961). *J. Soc. Dyers Colour.* **77**, 9.
Astbury, W. T. and Dawson, J. A. T. (1938). *J. Soc. Dyers Colour.* **54**, 6.
Bach, H., Pfeil, E., Philippar, W. and Reich, M. (1963). *Angew. Chem.* **75**, 407.
Back, G. and Zollinger, H. (1958). *Helv. Chim. Acta* **41**, 2242.
Balmforth, D., Bowers, C. A. and Guion, T. H. (1964). *J. Soc. Dyers Colour.* **80**, 577.
Bamford, C. H. (1954). *Discuss. Faraday Soc.* **16**, 229.
Barrer, R. M. (1949). *Trans. Faraday Soc.* **45**, 358.
Bartell, F. E., Thomas, T. L. and Fu, Y. (1951). *J. Phys. Colloid Chem.* **55**, 1456.
Bird, C. L. and Firth, J. M. (1960). *J. Text. Inst.* **51**, T1342.

Bird, C. L. and Harris, P. (1957). *J. Soc. Dyers Colour.* **73**, 199.
Boyd, G. E., Schubert, J. and Adamson, A. W. (1947). *J. Am. Chem. Soc.* **69**, 2818.
Boyle, Robert (1664). *Experiments and Considerations Touching Colours; the Beginning of an Experimental History of Colours,* H. Herringman, London.
Burdett, B. C. (1975). In *The Theory of Coloration of Textiles* (C. L. Bird and W. S. Boston, eds), Society of Dyers and Colourists, Bradford.
Cameron, A. and Giles, C. H. (1957). *J. Chem. Soc.* 3140.
Cameron, A., Giles, C. H. and MacEwan, T. H. (1958). *J. Chem. Soc.* 1224.
Campbell, D. S. E. and Giles, C. H. (1958). *J. Soc. Dyers Colour.* **74**, 164.
Campbell, D. S. E., Cathcart, D. and Giles, C. H. (1957). *J. Soc. Dyers Colour.* **73**, 546.
Chandra, A. K. and Basu, S. (1960). *Trans. Faraday Soc.* **56**, 632.
Chipalkatti, H. R., Giles, C. H. and Vallance, D. G. M. (1954). *J. Chem. Soc.* 4375.
Chipalkatti, H. R., Chipalkatti, V. B. and Giles, C. H. (1955). *J. Soc. Dyers Colour.* **71**, 652.
Coates, E. (1967). *J. Soc. Dyers Colour.* **83**, 95.
Colour Index (1975). 3rd edn, Society of Dyers and Colourists, Bradford.
Coward, M. P. and Nursten, H. E. (1964). *J. Soc. Dyers Colour.* **80**, 405.
Cummings, T., Garven, H. C., Giles, C. H., Rahman, S. M. K., Sneddon, J. G. and Stewart, C. E. (1959). *J. Chem. Soc.* 535; in Fig. 4 and p. 542, line 13 from bottom, the expression should read "$-\log(1 - [Q_t/Q_\infty])$".
Daruwalla, E. H. (1974). *The Chemistry of Synthetic Dyes* (K. Venkataraman, ed.), Vol. 7, Ch. 3, Academic Press, New York.
Daruwalla, E. H. and D'Silva, A. P. (1963). *Text. Res. J.* **33**, 40.
Daruwalla, E. H., Rao, S. S. and Tilak, B. D. (1960). *J. Soc. Dyers Colour.* **76**, 418.
de Vries, M. J. and Smit, J. H. (1967). *J. S. Afr. Chem. Inst.* **20**, 11.
de Vries, M. J., Smit, J. H. and Raubenheimer, H. G. (1968). *J. S. Afr. Chem. Inst.* **21**, 47.
Dobozy, O. (1973). *Amer. Dyestuff Rep.* **62**, 36.
Ferguson, L. N. (1948). *Chem. Rev.* **43**. 385.
Flett, M. St. C. (1952). *J. Soc. Dyers Colour.* **68**, 59.
Gilbert, G. A. (1944). *Proc. Roy. Soc. A***183**, 167.
Gilbert, G. A. and Rideal, E. K. (1944). *Proc. Roy. Soc. A***182**, 335.
Giles, C. H. (1944). *J. Soc. Dyers Colour.* **60**, 303.
Giles, C. H. (1961). *Text. Res. J.* **31**, 141.
Giles, C. H. (1966a). *Chem. and Ind.* 92.
Giles, C. H. (1966b). *Chem. and Ind.* 137.
Giles, C. H. (1971). *Brit. Polymer J.* **3**, 279.
Giles, C. H. (1974). *Rev. Progr. Coloration* **5**, 49; this review includes references to detailed practical information.
Giles, C. H. (1975). In *The Theory of Coloration of Textiles* (C. L. Bird and W. S. Boston, eds), Society of Dyers and Colourists, Bradford.
Giles, C. H. (1978). *J. Soc. Dyers Colour.* **94**, 4.
Giles, C. H. (1979). *Trans. Inst. Metal Finsg.* **57**, 48.
Giles, C. H. and Datye, K. V. (1963). *Trans. Inst. Metal Finsg.* **40**, 113.
Giles, C. H. and D'Silva, A. P. (1969). *Trans. Faraday Soc.* **65**, 1943.
Giles, C. H. and Haslam, R. (1977). *Text. Res. J.* **47**, 347.
Giles, C. H. and Haslam, R. (1978). *Text. Res. J.* **48**, 490.

Giles, C. H. and Hassan, A. S. A. (1958). *J. Soc. Dyers Colour.* **74**, 846.
Giles, C. H. and MacEwan, T. H. (1959). *J. Chem. Soc.* 1791.
Giles, C. H. and McIntosh, A. (1973). *Text. Res. J.* **43**, 489.
Giles, C. H. and McIver, N. (1975). *J. Colloid Interface Sci.* **53**, 155.
Giles, C. H. and McIver, N. (1977). *J. Colloid Interface Sci.* **62**, 329.
Giles, C. H. and Soutar, A. H. (1971). *J. Soc. Dyers Colour.* **87**, 301.
Giles, C. H., McKay, R. B. and Good, W. (1961). *J. Chem. Soc.* 5434.
Giles, C. H., Montgomery, A. P. and Tolia, A. H. (1962). *Text. Res. J.* **32**, 99.
Giles, C. H., Elder, H. M. and Tolia, A. H. (1964). *Text. Res. J.* **34**, 839.
Giles, C. H., D'Silva, A. P. and Trivedi, A. S. (1969). In *Surface Area Determination*, p. 317, Butterworths, London.
Giles, C. H., Agnihotri, V. G. and Trivedi, A. S. (1970). *J. Soc. Dyers Colour.* **86**, 451.
Giles, C. H., Haslam, R., Hill, A. R. and Trivedi, A. S. (1971). *J. Appl. Chem. Biotechnol.* **21**, 5; the last three dyes are IX, VII, VIII, in that order and not as given.
Giles, C. H., Gallagher, J., McIntosh, A. and Nakhwa, S. N. (1972). *J. Soc. Dyers Colour.* **88**, 360.
Giles, C. H., MacEwan, T. H. and McIver, N. (1974a). *Text. Res. J.* **44**, 580.
Giles, C. H., Smith, D. and Huitson, A. (1974b). *J. Colloid Interface Sci.* **47**, 755.
Giles, C. H., Havard, D. C., McMillan, W., Smith, T. W. and Wilson, R. (1979). In *Characterisation of Porous Solids* (S. J. Gregg, K. S. W. Sing and H. F. Stoeckli, eds), p. 267, Society of Chemical Industry, London.
Greenland, D. J. (1965). *Soils Fert.* **28**, 415.
Greenland, D. J. and Quirk, J. P. (1962). Proc. Nat. Conf. Clays Clay Minerals, *Clays Clay Minerals*, **9**, 484.
Greenland, D. J., Laby, R. H. and Quirk, J. P. (1962). *Trans. Faraday Soc.* **58**, 829.
Griffiths, J. (1976). *Colour and Constitution of Organic Molecules*, Academic Press, London.
Hampton, G. M. and Rattee, I. D. (1979). *J. Soc. Dyers Colour.* **95**, 396.
Hartley, F. R. (1969). *J. Soc. Dyers Colour.* **85**, 66.
Herz. A. H. (1977). *Adv. Colloid Interface Sci.* **8**, 237.
Hodgson, H. H. (1933). *J. Soc. Dyers Colour.* **49**, 213.
Husy, H., Merian, E. and Schetty, G. (1969). *Text. Res. J.* **39**, 94.
Iyer, S. R. S. (1974). In *The Chemistry of Synthetic Dyes* (K. Venkataraman, ed.), Vol. 7, Ch. 4, Academic Press, New York.
Johnson, A., Peters, R. H. and Ramadan, A. S. (1964). *J. Soc. Dyers Colour.* **80**, 129.
Kitchener, J. A. (1959). In *Modern Aspects of Electrochemistry* (J. O'M. Bockris, ed.), Butterworths, London.
Kratky, O., Mittelbach, P. and Sekora, A. (1964). *Kolloid-Z.* **200**, 1.
Lamb, H. J. (1979). *Trans. Inst. Metal Finsg.* **57**, 53.
Langmuir, I. (1916). *J. Am. Chem. Soc.* **38**, 2221.
Langmuir, I. (1918). *J. Am. Chem. Soc.* **40**, 1361.
Lead, W. L. (1957). *J. Soc. Dyers Colour.* **73**, 464.
Lemin, D. R. and Vickerstaff, T. (1946). Society of Dyers and Colourists, Symposium on Fibrous Proteins, 129.
Lewis, G. N. (1945). *J. Am. Chem. Soc.* **67**, 770.
Meggy, A. B. (1950). *J. Soc. Dyers Colour.* **66**, 510.
Merian, E. (1966). *Text. Res. J.* **36**, 612.

Merry, E. W. (1936). *The Chrome Tanning Process,* A. Harvey, London.
Milicević, B. (1964). 4e Congr. Intern. de la Détergence, Sec. B, IV, 12.
Murrell, J. N. (1964). *Advancement of Science* **20**, 489.
Nicholls, C. H. (1956). *J. Soc. Dyers Colour.* **72**, 470.
O'Connor, D. J., Johansen, P. G. and Buchanan, A. S. (1956). *Trans. Faraday Soc.* **52**, 229.
Ohtsu, T., Nishida, K., Nagumo, K. and Tsuda, K. (1974). *Colloid Polym. Sci.* **252**, 377.
Park, G. S. (1951). *Trans. Faraday Soc.* **47**, 1007.
Peters, L. (1955). *J. Soc. Dyers Colour.* **71**, 725.
Peters, L. and Lister, G. H. (1954). *Discuss. Faraday Soc.* **16**, 24.
Peters, R. H. (1945). *J. Soc. Dyers Colour.* **61**, 95.
Peters, R. H. (1975). *Textile Chemistry*, Vol. 3, *The Physical Chemistry of Dyeing*, Elsevier, London.
Peters, R. H. and Sumner, H. H. (1955). *J. Soc. Dyers Colour.* **71**, 130.
Peters, R. H. and Sumner, H. H. (1956). *J. Soc. Dyers Colour.* **72**, 77.
Peters, R. H. and Vickerstaff, T. (1948). *Proc. Roy. Soc. A***192**, 292.
Peters, R. H., Petropolous, J. H. and McGregor, R. (1961). *J. Soc. Dyers Colour.* **77**, 704.
Pethica, B. A. (1955). *Trans. Faraday Soc.* **51**, 1402.
Preston, J. M. and Tawde, G. P. (1956). *J. Text. Inst.* **47**, T154.
Pugh, D., Giles, C. H. and Duff, D. G. (1971). *Trans. Faraday Soc.* **67**, 563.
Race, E., Rowe, F. M. and Speakman, J. B. (1946). *J. Soc. Dyers Colour.* **72**, 372.
Rattee, I. D. (1978). In *The Chemistry of Synthetic Dyes* (K. Venkataraman, ed.), Vol. 8, Ch. 1, Academic Press, London.
Rattee, I. D. and Breuer, M. M. (1974). *The Physical Chemistry of Dye Adsorption,* Academic Press, London.
Reynolds, W. B. (1943). *Amer. Dyestuff Rep.* **32**, 455.
Reynolds, W. B. (1944). *J. Soc. Dyers Colour.* **60**, 67.
Rollinson, C. L. (1956). In *The Chemistry of the Co-ordination Compounds* (J. C. Bailar, ed.), p. 448, Reinhold, New York.
Royer, G. L. and Maresh, C. (1943). *Amer. Dyestuff Rep.* **32**, 181.
Schuler, M. J. and Remington, W. R. (1954). *Discuss. Faraday Soc.* **16**, 201.
Skinner, B. G. and Vickerstaff, T. (1945). *J. Soc. Dyers Colour.* **61**, 193.
Smit, J. H., Smith, J. H. and de Vries, M. J. (1967). *J. S. Afr. Chem. Inst.* **20**, 144.
Speakman, J. B. and Smith, S. G. (1936). *J. Soc. Dyers Colour.* **52**, 121.
Stone, J. E. and Scallan, A. M. (1968). *Cellulose Chem. Tech.* **2**, 343.
Stone, J. E., Treiber, E. and Abrahamson, B. (1969). *TAPPI.* **52**, 108.
Taylor, J. B. and Rowlinson, J. S. (1955). *Trans. Faraday Soc.* **51**, 1183.
Vickerstaff, T. (1954). *The Physical Chemistry of Dyeing*, 2nd edn, Ch. 10, Oliver and Boyd, London.
Waters, E. (1950). *J. Soc. Dyers Colour.* **64**, 609.
Weissbein, L. and Coven, G. E. (1960). *Textile Res. J.* **30**, 58, 62.
Yoshida, Z., Ōsawa, E. and Oda, R. (1964). *J. Phys. Chem.* **68**, 2895.
Zollinger, H. (1954). *Discuss. Faraday Soc.* **16**, 123.
Zollinger, H. (1956). *Melliand Textilber.* **37**, 1316.
Zollinger, H. (1958). *Textil-Rundschau* **13**, 217.
Zollinger, H. (1959). *Textil-Rundschau* **14**, 113.
Zollinger, H. (1960). *Amer. Dyestuff. Rep.* **49**, 29.
Zollinger, H. (1965). *J. Soc. Dyers Colour.* **81**, 345.

8. Adsorption of Polyelectrolytes from Dilute Solution

F. TH. HESSELINK

I. Introduction

Polyelectrolytes are long-chain molecules carrying ionized groups. One type of polyelectrolyte is the flexible linear polymer molecule with ionized groups and having a structure such as $[-CH_2-CH(X)-]_y$, where X can be a strongly acidic group such as $-SO_3^-H^+$ (polyvinyl sulphonic acid), or a weakly acidic group such as –COOH (polyacrylic acid), or some cationic group. These molecules often adopt extended random coil type confirmations in solution.

Other types of polyelectrolyte are biological molecules such as proteins and nucleic acids which may adopt in solution very specific intramolecular structures, stabilized by hydrogen bonds and/or hydrophobic interactions. Under certain conditions these molecules may show a conformational transition from the ordered structure to a disordered, random-like conformation. The properties of polyelectrolyte solutions that are relevant to our discussion are dealt with further in Section II.

Adsorption of polyelectrolyte molecules at interfaces is a special case of polymer adsorption as discussed by Lyklema and Fleer in Chapter 4. For polyelectrolytes having a specific conformation in solution a major question is how much of this conformation is retained upon adsorption. This type of adsorption will be discussed in Section V.

The flexible polyelectrolyte coil goes through a conformational transition when it becomes adsorbed with a variable number of segments on the interface. Attachment of a segment increases the probability of a neighbouring segment becoming adsorbed. This cooperativity between adjacent segments causes the adsorption to occur in trains on the interface, alternated by loops dangling in solution while both ends of the polyelectrolyte chain may form free-dangling tails.

Whether the unadsorbed segments of the adsorbed polyelectrolyte occur predominantly in loops or in free-dangling tails depends on the chain length of the polymer. A loop has less conformational freedom than a tail of the same length which does not have to return to the adsorbing interface; from entropy considerations tails are favoured over loops. On the other hand, for a very long molecule the alternation of loops and trains produces more entropy than adsorption of a single train with one or two dangling tails. Therefore, for molecules that are not too long, the unadsorbed segments will be predominantly found in the tails, whereas for an infinitely long molecule loops should predominate.*

Adsorption of polyelectrolytes is more complex than the adsorption of nonionic polymers because of the electrostatics of the polymer–adsorbent interaction. In general, polyelectrolytes can become adsorbed on oppositely charged adsorbents (ion-exchange phenomena may then play a role) as well as on adsorbents carrying the same type of charge as the polyelectrolyte. For instance, polyacrylic acid (Joppien, 1978) and DNA (Chattoraj *et al.*, 1967) can become adsorbed on negatively charged silica. However, such adsorption of an anionic polyelectrolyte on a negatively charged particle will be hindered by the negative electrostatic potential and prevented at a sufficiently high potential. For instance, polyacrylic acid does not adsorb on silica above pH 7, where both adsorbent and polyelectrolyte become too negatively charged to allow adsorption (Joppien, 1978). The electrostatic repulsion is then stronger than the nonionic adsorption energy. Clearly, the adsorption energy as it figures in the adsorption of nonionic polymers has to be modified to include electrostatic interaction; this approach will be discussed in Section III.

* It should be noted that these considerations about loops, tails, and trains have been given before for the melting of double stranded helical DNA. Here, the double stranded helical portions correspond to adsorbed trains, and the internal loops and loose ends correspond to the dangling loops and tails discussed in polymer adsorption. Actually, one of the polymer adsorption theories (Hoeve *et al.*, 1965; see Chapter 5, Section III.B) has been formulated using this analogy.

Polyelectrolyte adsorption should resemble polymer adsorption at high ionic strength of the solution or at very low charge density on the polymer. Therefore, it is not surprising that the adsorption isotherm is often of the same high-affinity type (see Fig. 1) as usually observed for nonionic polymers.

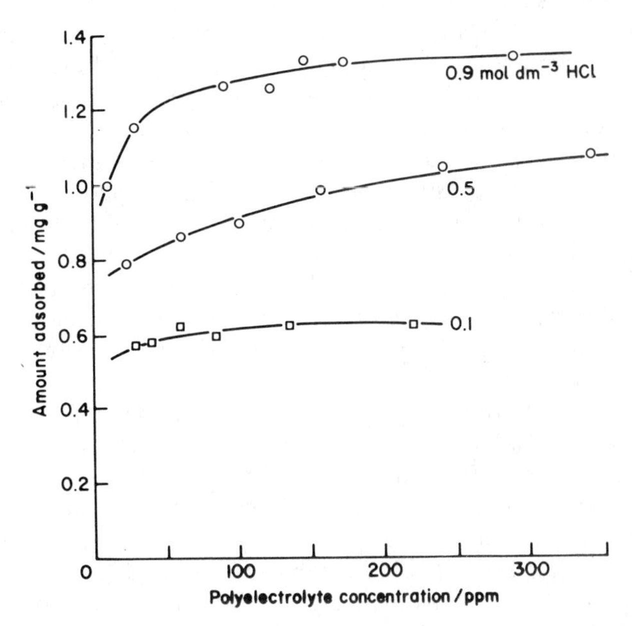

Figure 1 High-affinity type adsorption isotherms of ionized polyvinylpyridine on Vycor silicate glass from various aqueous HCl solutions. (Peyser and Ullmann, 1965)

A consequence of this high-affinity isotherm is the often mentioned "irreversibility" of adsorbed polyelectrolytes, i.e. washing away with solvent does not re-dissolve the polymer. This is to be expected, since a practically polymer-free solution is in equilibrium with a practically saturated adsorbent. Also, from a microscopic viewpoint instantaneous desorption is unlikely; a polymer attached with, say, 10–100 segments can only become desorbed if all adsorbed segments become loose simultaneously. For these reasons rearrangements of adsorbed polyelectrolyte molecules are very time-consuming and it is understandable that in an adsorption experiment it may take hours, or even longer, before the amount adsorbed has become constant.

Desorption, however, is often possible when conditions are drastically changed from those prevailing during the adsorption experiment. For instance, desorption can be caused by creating significant electrostatic repulsion between polyelectrolyte and adsorbent, either through a change in pH as in the case of polyacrylic acid adsorbed on silica (Joppien, 1978) or nylon (Cole and Howard, 1972), or through a change in applied potential as in the case of polyglutamic acid becoming desorbed from a mercury electrode with a highly negative potential (Pavlovic and Miller, 1971). Also, addition of a strongly adsorbing polymer or a drastic reduction in ionic strength (Greene, 1971) may cause desorption. A systematic analysis of the various parameters determining polyelectrolyte adsorption, such as charge density on the polymer and on the adsorbent, adsorption energy, ionic strength, etc., is given in Sections III and IV.

Polyelectrolyte adsorption plays a role in a wide variety of fields of science and technology, as may be illustrated by the following examples.

Flocculation. Polymeric flocculants are used extensively in effluent water clarification and other industrial solid–liquid separations. Very long and extended polyelectrolyte chains may become adsorbed on more than one particle in a suspension. These bridges between the particles then induce flocculation, producing strong and open flocs which are easily filtered (see Section IV).

Paper making. Polyelectrolytes are commonly used in the paper making process as wet-end additives; they increase the retention of fines and fillers in the fibre mat. Since the cellulose fibres have a negative surface charge, the polyelectrolytes chosen for this application are often cationic (see e.g. Lindstrom and Söremark, 1976).

Soil conditioning. Water-soluble polymers are used (Stewart, 1973) to prevent erosion in dry areas by fixing the silty soil particles on the sand; upon adsorption the polymer helps aggregation of the particles. The high affinity character of the adsorption is essential for this application; it prevents the polymer from being washed away from the soil by rain water.

Drilling. Partially hydrolysed polyacrylamides are added to drilling fluids since, by adsorbing on penetrated shale layers and clay cuttings, they prevent swelling and subsequent dispersion of the clay into the fluid system (Clark *et al.*, 1975).

Precipitation. Polyelectrolytes can drastically affect the precipitation of salts such as $CaSO_4$, $CaCO_3$, and $BaSO_4$ from supersaturated solutions by adsorbing on the growing crystals. The adsorbed polyelectrolytes ("scale inhibitors") act as immobile impurities on the crystal surface and inhibit crystal growth by reducing the rate of step movement along the crystal surface (Smith and Alexander, 1970). For instance, the presence of 14 ppm polyacrylate increased the induction period for $CaSO_4$ precipitation from 15 minutes to several hours, and some 30–40 days were needed for complete precipitation as opposed to $1\frac{1}{2}$ hours in the absence of polymer (Crawford and Smith, 1966). The total amount of precipitate is not much affected by the polyelectrolyte, as is to be expected since these minute amounts of polymer can hardly influence the solubility product, but the resulting crystals are much smaller and their occurrence is drastically retarded. This can be of great practical value, since precipitates of these salts can hamper the operation of steam boilers, pipelines, and water-injection facilities for oil recovery (see review by Cowan and Weintritt, 1976).

Drag reduction. Some investigators have speculated (e.g. Hand and Williams, 1973) on a possible role of polyelectrolyte adsorption in drag reduction, for instance via an adsorbed-entangled layer, but this mechanism has been rejected by others (Peyser and Little, 1974).

Oil recovery. A potentially enormous application of water-soluble polymers is in enhanced oil recovery operations. When water is injected into an oil-bearing formation to displace the oil towards the producing wells, this water may bypass a considerable part of the oil, especially if it is considerably more viscous than the water. Polyelectrolytes such as polysaccharides and partly hydrolysed polyacrylamide are added to the injection water to produce a viscous aqueous solution capable of displacing the oil from the reservoir with a better sweep efficiency than that produced by a plain water drive (e.g. Hirasaki and Pope, 1974). Polymer adsorption is a complicating factor in this process; it causes a loss in viscosity of the drive fluid which is detrimental to the process, but it may decrease the relative mobility for water in swept portions of the reservoir, which is a favourable factor. Adsorption of the polymer is determined on the basis of retention in a chromatographic sandstone column, but there is uncertainty as to how much polymer is actually adsorbed on the rock and how much is just trapped in the porous medium without being attached to an adsorbing interface.

Finally, two examples will be given of biosystems where understanding of polyelectrolyte adsorption may be advantageous.

Blood clotting. At interfaces clotting appears to be the result of the adhesion of thrombocyte platelets to the saccharide parts of adsorbed glycoproteins, fibrinogen and γ-globulin (see Bantjes, 1978, for references); adsorbed albumin inhibits platelet adhesion. Surfaces coated with the strongly anionic polyelectrolyte heparin do not adsorb fibrinogen, and platelets do not adhere to these surfaces. The molecular basis for this anticoagulant activity appears to be related to the negative surface potential induced by the polyelectrolyte adsorption (see, for instance, the streaming potential measurements by Ramasamy *et al.*, 1975), and also to specific chemical interactions involving the sulphamate groups of this polyelectrolyte.

Immunoadsorption. Antibody proteins can be purified from, and quantitatively determined in, a polyvalent serum by this technique (see Nezlin, 1979, for references). Antigen covalently coupled to an insoluble matrix serves as adsorbent capable of highly specific isolation of its antibody protein; the antibodies are subsequently desorbed by adding a low pH buffer or a concentrated urea solution.

In the remainder of this chapter we discuss the salient features of adsorption of polyelectrolytes from dilute aqueous solutions. As a basis for this discussion we first describe some of the properties of polyelectrolytes in solution. Then, some theoretical work on polyelectrolyte adsorption is discussed followed by a review of experimental findings. In Section V the adsorption of biological molecules that have a specific native structure in solution is discussed. Full coverage of the literature is not attempted; the selection of topics discussed is biased by the author's interest. For reviews of the literature on polymer and polyelectrolyte adsorption the reader is referred to papers by Fontana (1971) and by Vincent (1974), and especially to the recent book by Theng on adsorption on clays (1979).

II. Polyelectrolytes in Solution

A. FLEXIBLE POLYELECTROLYTES

Flexible polyelectrolyte molecules can have a wide variety of conformations, the polyelectrolyte character being superimposed on a linear polymer with random flight type coil conformations. In principle, the partition function Q for this set of conformations is given by

$$Q = \sum_{\text{all conformations}} \; \sum_{\text{all charge distributions}} \exp[-(g_e + g_0)/kT] \tag{1}$$

where g_e and g_0 are the electrical and non-electrical free energy of a particular conformation with a particular distribution of charged groups. The set of most frequently occurring conformations (or the average conformation) is clearly a compromise between the minimization of the non-ionic free energy and the electrical free energy.

For a relatively smooth multidimensional energy surface (energy as a function of conformation and charge distribution) not showing locally deep minima, it may be expected that slight changes in g_0 and g_e have a considerable effect on the average conformation of a polyelectrolyte macromolecule in solution. Such changes can be brought about for instance by changes in ionic strength or solvent quality. For nonionic polymers the effect of solvent on conformational properties such as the hydrodynamic volume or the radius of gyration is well documented.

Polyelectrolytes, however, are usually investigated in aqueous solutions, and here the electrical term g_e is usually varied via changes in ionic strength and degree of dissociation. In addition, because of the large number of charges in the polyelectrolyte, the properties of the system are much more sensitive to the nature of the counter-ions than of the co-ions.

To understand polyelectrolyte adsorption the size of the polyelectrolyte molecule in solution is of considerable importance. When the number of charges on the polyion increases, the molecule becomes more extended in solution, as a result of the mutual repulsion of charges on the chain. This is reflected in the well known drastic increase in viscosity of polyelectrolyte solutions with an increase in charge density of the polymer. Addition of neutral salts shields the ionic charges on the polyion and reduces the electrostatic extension of the chains and thus the viscosity of the solution. The effect of salt on viscosity can often be represented by $[\eta] = a + (b/C_s)^{1/2}$ (Pals and Hermans, 1952; Noda *et al.*, 1970). These expanded polyelectrolyte chains can still be regarded as hydrodynamically impermeable swollen polymer coils not much more asymmetric than in the absence of electrostatic interactions (Morawetz, 1975, p. 361). In concentrated salt solutions these long-range interactions between the ionized groups on the chain vanish and the polyelectrolyte assumes dimensions very similar to those of nonionic polymers (Eisenberg and King, 1977, p. 233).

As an example, Fig. 2 shows the effect of increasing NaCl concentration on the viscosity of a 0.3 wt % partially hydrolysed polyacrylamide solution. Note also the effect of shear rate on viscosity, especially for the low-salinity cases where the polyelectrolyte coils are extended far beyond their random flight dimensions.

Theories describing this extension of flexible polyions depend on the model adopted for the flexible polyion to calculate g_e in equation (1) and

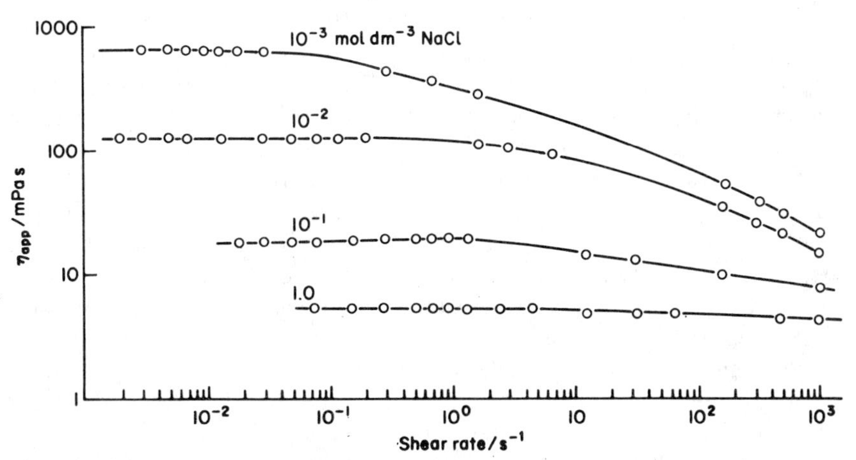

Figure 2 Apparent viscosity versus shear rate for a 3000 ppm aqueous solution of a 60% hydrolysed polyacrylamide at various NaCl concentrations.

the methods employed to evaluate the summations in equation (1) over all charge distributions and all conformations. Eisenberg and King (1977) listed nine equations that relate the expansion of the polyion to the charge density, unperturbed coil dimensions, and neutral salt concentration. The trends are often predicted fairly well but unfortunately the absolute value of the expansion is much less than most theories predict (Nagasawa, 1975), and Fixman and Skolnic (1978) concluded that those theories which do agree with experiment do so at the expense of *ad hoc* assumptions of uncertain merit. However, theoreticians are still active in this field; at present the favoured model appears to be a worm-like chain with a persistence length depending on charge density and ionic strength (Fixman and Skolnic, 1978; Odijk, 1978). Since the persistence length is taken as considerably larger than the Debye length, the potentials and the interactions with the counterions can be calculated using rod or cylindrical models.

The charge density on the polyion depends on the number of ionizable groups in the polymer and on the degree of dissociation. However, strong theoretical evidence (Manning, 1969, 1972, 1974) exists for a maximum charge density ξ, given for univalent counterions by

$$\xi = e^2/rDkT \leqslant 1 \tag{2}$$

where r is the distance between the charge groups along the chain, and D is the dielectric constant. If the number of ionized sites on the chain is greater than that corresponding to $\xi_{crit} = 1$, counterions will condense on the chain and reduce ξ to unity. In the presence of divalent ions, these ions condense preferentially on the chain until $\xi \leqslant \frac{1}{2}$.

Experimentally, a levelling off of the effective charge density on the chain, when the degree of dissociation α was still increasing, has been observed; for instance, Noda *et al.* (1970) report the excluded volume parameter, being a measure of the mutual repulsion of ionized chain segments, to increase drastically with increasing α up to $\alpha \approx 0.5$ but at higher values of α to be almost constant, practically independent of the salt concentration. This is in agreement with equation (2), which indicates that the charge density and the condensation of counterions does not depend on the concentration of polyelectrolyte or of added salt.

This theory may be somewhat simpler than real life, and it is still in full discussion, but it is elegant and it appears to be successful in predicting a large number of polyelectrolyte phenomena.

In principle, the concept of a critical charge density above which ion condensation occurs does not have to hold in cases of adsorption of polyions on an oppositely charged interface, since the interactions of the polyion with the charged adsorbing interface may thoroughly disturb the interactions of the polyion with condensing counterions. For instance, Norde and Lyklema (1978) observed ion-pair formation between cationic groups on the polyelectrolyte and anionic groups on the adsorbent.

The dissociation of a weak acid group on the polyion depends on the overall degree of dissociation, since dissociation of a proton from an already ionized polyacid is hampered by the negative potential of such a polyacid. For instance, Mandel (1970) finds a $pK_{apparent}$ for the first dissociating groups of polyacrylic acid around 5.2, whereas at $\alpha = 0.8$ he finds $pK_{app} = 7.5$. This effect is suppressed in the presence of salt; for instance, at a salt concentration of 0.35 mol dm^{-3} the pK_{app} increases only from 4.5 at $\alpha = 0.1$ to 5.7 at $\alpha = 0.9$. Thus, it can be expected that the electrical double layer of an adsorbing particle also has a significant effect on the degree of dissociation of a weak polyacid. However, this effect has hardly been discussed in the literature. Norde and Lyklema (1978) have observed such differences in degree of dissociation and ion binding for adsorbed and unadsorbed protein molecules (see Section V).

B. POLYELECTROLYTES WITH AN INTERNAL STRUCTURE

Complex biological molecules such as proteins and nucleic acids are also examples of polyelectrolytes. These molecules often have very specific conformations stabilized by a balance of hydrophobic, chemical, dipolar, and electrostatic interactions; in other words, their multidimensional potential energy surface has distinct local minima.

Conformations corresponding to such minima may be able to withstand minor changes in the environment, whereas upon major changes such a native conformation may break down rather abruptly. The molecule may then unfold and adopt a set of more random, often quite extended, conformations that are more or less similar to those of a flexible polyelectrolyte. Examples are the unfolding of proteins and the melting of DNA.

Conformational transitions of this type have been studied extensively and in depth for model compounds such as polyaminoacids and polynucleotides (see Poland and Scheraga, 1970). For our discussion charge-induced conformational transitions are particularly relevant, since adsorption of such polyelectrolytes on charged interfaces depends on the conformation in solution and may in itself give rise to a conformational transition. Well studied examples of charge-induced conformational transitions are the helix-to-coil transition observed on increase in charge density of the chain for poly-L-glutamic acid (PGA, an anionic polymer) and for poly-L-lysine (cationic polymer). For instance, PGA in aqueous solution at low pH exists in an α-helical conformation. At increasing pH the acid groups on the polymer dissociate, and at a particular degree of dissociation, depending on ionic strength, the mutual repulsion between the charged groups becomes strong enough to overcome the helix-stabilizing forces and the helix degenerates to an extended random coil conformation.

Theories to describe this charge-induced conformational transition require calculation of the difference in electrostatic free energy, g_e, between the helical and the coil conformation. Zimm and Rice (1960) have calculated g_e for a transition from an α-helix to a locally extended conformation, assuming a Debye–Hückel screened potential, via

$$g_e = \sum_{i>j} \frac{e^2}{DR_{ij}} \exp(-\kappa R_{ij}) \qquad (3)$$

where R_{ij} is the distance between the ionized sites and the summation is taken over all pairs of interacting charges; κ is the reciprocal Debye length where $\kappa^2 = 8\pi ne^2/DkT$. Hesselink *et al.* (1973) have refined this approach by taking into account the conformational freedom of the extended chain, and by allowing only partial dissociation of the ionizable groups (taking into account the entropy of mixing charged and uncharged groups). The conformational freedom of the extended coil is taken into account via Monte Carlo generation of a set of coil conformations, the statistical weight of each conformation depending on the electrical free energy of the particular conformation. Then, g_e for the extended coil is obtained by averaging over the whole set of conformations, each with its particular statistical weight calculated with equation (3). It should be noted that in this way also the extension of the polypeptide chain due to the mutual repulsion between the ionized groups has been calculated. For instance, this extension increases from a factor of 1.02 in 1 mol dm^{-3} NaCl to a value of 1.28 in 0.02 mol dm^{-3} NaCl.

Adsorption of a flexible polyelectrolyte on a charged interface can also be viewed as a particular case of a conformational transition. However, as will be discussed in the following section, the theory of polyelectrolyte adsorption has not been worked out to this level of sophistication.

The distinction between flexible polyelectrolytes adopting a wide variety of conformations and the class of complex biomolecules such as proteins and nucleic acids, which adopt a specific "native" conformation in solution, turns out not to be sharp. As indicated, the specific conformations of biomolecules can easily become disturbed (the function of these biomolecules often depends on their ability to change conformation when required), whereas replacing a hydrogen atom in the repeat unit of polyacrylic acid by an alkyl group results in polyelectrolytes which can show a conformational transition from a condensed to an expanded coil.

For instance, for a maleic acid copolymer with the repeat structure $[-CH(COOH)-CH(COOH)-CH_2-CH(OR)-]_y$, Dubin and Strauss (1975) investigated the effect of the alkyl R group on the solution properties. For the ethyl polyacid (R = ethyl) the viscosity data as a function of the degree of dissociation, α, showed no indication of a conformational transition of the condensed coil–expanded coil type. The data only indicated the usual coil expansion of a weak polyacid caused by the electrostatic repulsion between the ionized acid groups.

The butyl and hexyl polymers (R = butyl, hexyl), on the other hand, exist at low values of α in an abnormally compact state having a hydrodynamic volume smaller than that corresponding to θ conditions. In this compact state they can solubilize a water-insoluble lipophilic dye. Upon an increase in α, the butyl polymer shows a conformational transition to an expanded coil at $\alpha \approx 0.25$, while the hexyl polymer shows a similar transition at $\alpha \approx 0.5$. Beyond this transition the solubilizing ability of these polymers has vanished. The initially compact state, presumably stabilized by hydrophobic interactions, is destroyed by the electrostatic repulsions between the dissociated acid groups.

III. Theories for Polyelectrolyte Adsorption

A. THEORETICAL MODELS FOR RIGID MOLECULES

The theory of polyelectrolyte solutions is still in development and therefore it is not surprising that the theory of polyelectrolyte adsorption is still in its infancy. In the first theoretical paper on the subject, Frisch and Stillinger (1962) examined the effect that the charges on the polyelectrolyte have on the free energy of adsorption at the interface between two dielectrics. They took into account the work necessary to bring an ionized group to the surface against the potential of the image charge. The paper applies to adsorption of rigid rods from extremely dilute solution where the surface concentration $\bar{C}$ will be proportional to the macro-ion concentration in solution C_p according to

$$\bar{C} = K^+ C_p + \text{terms of order } C_p^2 \text{ and higher}$$

and a relation is derived for the constant K^+. In view of the high-affinity character of polyelectrolyte adsorption (see Fig. 1) this model is hardly applicable to experimental systems.

A simple isotherm for the adsorption of a cationic polyion on an anionic exchange column has been derived (Semenza, 1965) from the equilibrium

$$\text{polyion}^{\nu+} + \nu(\text{Na}^+)_{\text{adsorbed}} \rightleftarrows (\text{polyion}^{\nu+})_{\text{adsorbed}} + \nu\text{Na}^+$$

since the adsorption of a polyion carrying ν charges has to be accompanied by the desorption of ν cations. At equilibrium, the rate of adsorption $k_1 C_p(1-\theta)^\nu$ equals the rate of desorption $k_2[\text{Na}^+]^\nu\theta/\nu$, where θ is the fraction of the ionic adsorption sites occupied by polymer, $K = k_1/k_2$, and

$$KC_p = \theta[Na^+]^\nu/\nu(1-\theta)^\nu \tag{4}$$

At high NaCl concentration, $[Na^+]$ becomes constant and equation (4) reduces to

$$KC_p = \theta/\nu(1-\theta)^\nu \tag{5}$$

the adsorption isotherm derived by Frisch *et al.* (1959) for the adsorption of nonionic polymers, an equation that reduces for $\nu = 1$ to the classical Langmuir adsorption isotherm. For high values of ν, equation (4) shows the well known high-affinity adsorption isotherm and could fit the adsorption of polylysine in chromatography columns of weakly acidic carboxymethyl cellulose (Semenza, 1965).

Of course, equation (4) can only have limited applicability, since it denies the typical polymer character of polyelectrolytes, and the only mechanism of adsorption taken into account is the electrostatic attraction between a polyion and an oppositely charged adsorbent.

B. TRAIN/LOOP MODEL FOR ADSORBED POLYELECTROLYTES

The most comprehensive theoretical description of polyelectrolyte adsorption available appears to be that by Hesselink (1977). Therefore we shall discuss this work in some detail. The starting point of this work is the consideration that polyelectrolyte adsorption should resemble polymer adsorption at a high ionic strength of the solution or at a low charge density of the polyelectrolyte. The paper extends the treatment by Hoeve (1971) concerning the adsorption of uncharged polymers of very high molecular weight (see Chapter 5).

1. *Description of the Model*

The polyelectrolyte is adsorbed in series of trains of segments on the interface interconnected by loops dangling in solution (see Fig. 3); end effects of free-dangling tails are neglected.*

The electrical and non-electrical contributions to the free energy of the adsorption process are treated separately, a common, though not innocent, approximation in polyelectrolyte theory. Since we compare adsorbed polyelectrolytes with polyelectrolytes in solution, we mainly neglect the effect of different electrical fields on chain dimensions.

The partition function for N_a polyelectrolyte molecules adsorbed on

* A preliminary paper (Hesselink 1972) describes the adsorption of a polyelectrolyte with a single train on the interface and a long tail dangling in solution.

surface area A from a solution of volume V containing, after adsorption, N_f polyelectrolyte molecules is given by

$$Q = \sum \exp\left(\frac{\Delta G}{kT}\right) \frac{(A\delta)^{N_a}}{N_a!} \frac{V^{N_f}}{N_f!} \tag{6}$$

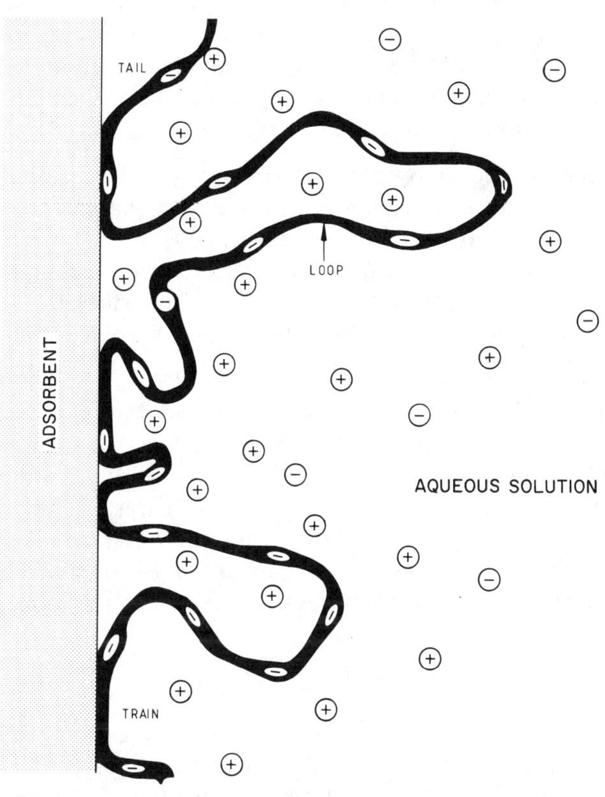

Figure 3 Model of an anionic polyelectrolyte chain adsorbed from an aqueous salt solution on a solid adsorbent. Note the conformational transition from an extended coil in solution into a conformation with adsorbed trains on the interface alternating with loops dangling in solution and terminating with loose-dangling tails.

where $(N_a + N_f)$ is constant. A molecule is assumed to be adsorbed if at least one of its segments is located within a distance δ (approximately equal to the thickness of a segment) from the interface. The term ΔG contains the following contributions to the free energy of adsorption:

(1) The free energy ΔG_1 of the configurational change upon adsorption of a random coil in solution into a sequence of trains on the interface

interconnected by loops dangling in solution; the interaction of the trains with the interface is also incorporated in this term.
(2) The interactions of the adsorbed trains with each other and with the solvent in the first layer of thickness δ on the interface, ΔG_2.
(3) The interactions between the dangling loops and solvent, ΔG_3.

These terms are incorporated into the theory of adsorption of nonionic polymers as developed by Hoeve (1971) and by Silberberg (1968).

The electrostatics of the adsorption of charged polyions on a charged interface requires two additional components to the free energy of adsorption, ΔG, to be taken into account:

(4) The change in free energy of the electrical double layer upon adsorption of charged trains on a charged interface with charged loops dangling in the solution part of the double layer, ΔG_4.
(5) The adsorption energy of uncharged segments on a charged interface or of a charged segment on an uncharged interface depends on the charge density because of dipole and/or polarizability effects, ΔG_5.

In order to formulate contributions ΔG_3 and ΔG_4, Hesselink assumes a step function of thickness s for the density distribution of the segments in the loop layer. In addition, it is assumed that the average segment density in the loop layer is the same as in a polyelectrolyte coil in solution. Although this assumption has been used before (Hirasaki and Pope, 1974), and some experimental evidence for it has been obtained (Chough *et al.*, 1974), it is bound to be of limited validity as will be discussed later in this chapter. The assumption of equal segment density in the loops and in the coil gives

$$s = (1 - p) v_0 N_a / A \tag{7}$$

where $(1 - p)$ is the fraction of the segments in the loops and v_0 is the volume of polyelectrolyte coil in solution. An advantage of equation (7) is that the expansion of the loop layer and of the coil in solution due to the intramolecular repulsions between ionized groups is taken into account in a consistent manner. The resulting values for s are liable to the same error as the theoretical value of v_0.

The non-electrical contributions to the free energy of adsorption are calculated following the approach by Hoeve (1971). The electrical free energy of adsorption is calculated using

$$\Delta G_4 = A(G^{el} - G_0^{el}) - N_a g^e$$

where G_0^{el} and G^{el} are the free energy of the electrical double layer before and after adsorption of N_a polyions, and g^e is the electrical free energy of a polyion in solution. The free energy of the electrical double layer, G^e, is affected by the adsorption of ionized segments and the occurrence of ionized polyelectrolyte loops in the solution part of the double layer. G^e is calculated using an imaginary discharging process (Verwey and Overbeek, 1948), neglecting Stern layer effects, differences in degree of dissociation between adsorbed and free polyelectrolytes, and effects of uncharged polymer on the double layer. Most results have been calculated using the Debye–Hückel approximation and assuming the polymer layer to be much thicker than the electrical double layer, thus restricting the theory to not too low salt concentrations. The polyelectrolyte charge density in the loop layer and for the coil in solution is regarded as smeared-out (Hermans and Overbeek, 1948); the radius of the coil in solution and its electrostatic extension is calculated following Pals and Hermans (1952).

Having thus evaluated ΔG the partition function equation (6) is maximized with respect to the fraction, p, of segments actually adsorbed, and with respect to the total number, N_a, of adsorbed molecules. This leads to the following adsorption isotherm:

$$\frac{N_a}{A\delta} = \frac{N_f}{V}\exp\{P[\varepsilon + \ln(l - \theta_0) + 2\chi\theta_0 - \tau\alpha\kappa\sigma/2ne - w\sigma(3\sigma - 2\sigma_0)]\} \quad (8)$$

where ε is the net nonionic adsorption free energy, which includes energetic segment/adsorbent interaction, the entropy loss for a segment passing from a free to a bound state, the possible desolvation of a segment upon adsorption, and the accompanying desorption of solvent molecules from the interface; P is the number of segments, of length l, per polymer chain; w is an empirical constant indicating the effect on ε of the difference in polarizability of adsorbate and solvent (see Hesselink, 1977); τ represents the nature of the electrolyte; σ and σ_0 are the surface charge density in the presence and in the absence of adsorbed polyelectrolyte; χ is the Flory–Huggins interaction parameter. The terms containing θ_0 (the fraction of surface in actual contact with polymer) represent the competition for space in the first layer on the surface of the adsorbent, whilst the terms containing σ originate from the electrostatics of adsorption (ΔG_4 and ΔG_5).

Equation (8) is the adsorption isotherm for polyelectrolyte adsorption. Numerical examples of this equation are shown in Figs 4 and 5. In the calculations it has been assumed that $\tau = -1$ (anionic polyelectrolyte), $T = 300$ K, $\delta = 0.5$ nm, $l = 2$ nm; the density of the polymer is equal to that of water; the volume of a polymer segment $= \pi\delta^2 l/4 = 0.39$ nm^3; then, the molecular weight of a segment $= 236$; $w = 0$. In these figures the amount of polymer adsorbed is plotted against the equilibrium concentration in solution.

When the term in square brackets in equation (8) has a not too small positive value, the multiplication by the number of segments, P, causes the exponent to be very large; at a finite number of adsorbed molecules

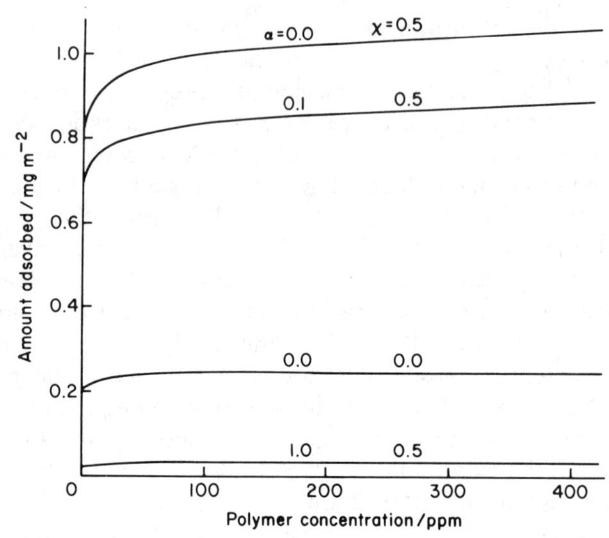

Figure 4 Theoretical adsorption isotherms [equation (8)]; the amount of polyelectrolyte adsorbed versus the equilibrium concentration of polymer in solution for $P = 2500$, $\varepsilon = 0.1$, $\sigma_0 = 0$, and $C_s = 0.1$ mol dm^{-3}. The curves for $\alpha = 0$ pertain to nonionic polymers; here the effect of solvent is shown, i.e. the amount adsorbed from a good solvent ($\chi = 0.0$) is much less than from a poor solvent ($\chi = 0.5$). The other curves illustrate the effect of increasing degree of dissociation; it decreases the polymer adsorption.

the concentration of polymer in solution will then be extremely small, i.e. a description of the well known experimental fact that polymer adsorption is effective from extremely dilute solutions. For nonionic polymer adsorption on an uncharged interface the last two terms in the exponent vanish and the result is similar to that obtained before by Silberberg (1968) and Hoeve (1971) for the adsorption of nonionic polymers.

This is illustrated in Fig. 4 where the adsorption isotherms for $\alpha = 0$ pertain to nonionic polymers. They show the well known result that, in the case of a small adsorption energy, adsorption from a good solvent ($\chi = 0$) is much smaller than that from a poor solvent ($\chi = \frac{1}{2}$). For high adsorption energies the effect of solvent is less important.

Polyelectrolyte adsorption is determined [see equation (8)] by: (a) properties of the adsorbent (surface charge density σ_0 and surface area A); (b) properties of the polyelectrolyte (degree of polymerization P and dissociation α); (c) interaction of polyelectrolyte and adsorbent (net adsorption energy ε); (d) properties of the solution (ionic strength or its effect on the Debye length κ^{-1}; amount of polymer present $(N_a + N_f)/V$).

The theory relates these parameters to the properties of the adsorbed layer: (a) amount of polyelectrolyte adsorbed per unit area, X; (b) fraction

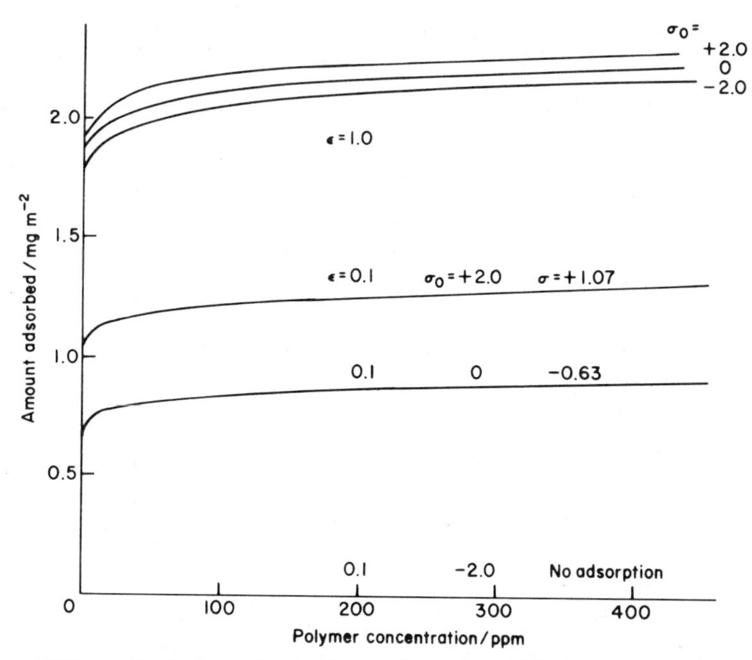

Figure 5 Theoretical adsorption isotherms [equation (8)]; the amount of polyelectrolyte adsorbed versus the equilibrium concentration of the polymer in solution for $P = 2500$, $\alpha = 0.1$, $\chi = 0.5$, and $C_s = 0.1$ mol dm^{-3}. The effect of variations in adsorption energy ε and surface charge density σ is shown. For $\varepsilon = 0.1$ and $\sigma_0 = -2\ \mu$C cm^{-2} no adsorption is found, in agreement with equation (9).

of the adsorbent interface covered, θ_0; (c) fraction of the segments of the polymer actually adsorbed on the interface, p; (d) the resulting surface charge density, σ; (e) the thickness of the adsorbed layer, s.

As an illustration, Table 1 shows some results of the analysis by Hesselink (1977) for the adsorption of a high molecular weight anionic polymer for various adsorbents, adsorption energies, and salt concentrations. For a high adsorption energy (ε = 1.0) and a low degree of dissociation (α = 0.1), the amount adsorbed is insensitive to ionic strength or charge density on the adsorbent. At a high degree of dissociation of the polymer (α = 1.0), however, the properties of the adsorbed layer are strongly dependent on adsorbent charge density and ionic strength, the adsorption is generally reduced, and no adsorption is predicted on a negatively charged particle. Note also the effects of ε and α on the fraction of the interface covered (θ_0), on the fraction of the segments actually adsorbed (p), and on the thickness of the adsorbed layer (s).

For a detailed analysis of how equation (8) predicts the effects of the adsorption-determining parameters on the properties of the adsorbed layer,

Table 1 Properties of the adsorbed layer as a function of adsorption energy, salt concentration, and surface charge density for $M = 2.36 \times 10^6$ ($P = 10^4$); $\alpha = 0.1$ (values in parentheses for $\alpha = 1.0$); $\chi = 0.5$; at 100 ppm anionic polyelectrolyte ($\tau = -1$) in solution. (Hesselink, 1977)

Input data			Results				
$\sigma_0/\mu C\ cm^{-2}$	$C/mol\ dm^{-3}$	ε/kT	adsorbed $X/mg\ m^{-2}$	ϕ_0	p	$\sigma/\mu C\ cm^{-2}$	$s/\mu m$
+2.0	0.01	0.1	2.5 (0.52)	0.55 (0.10)	0.11	+0.88	8.3
		1.0	3.8 (0.63)	0.85 (0.13)	0.11	+0.27	12.5
+2.0	0.1	0.1	2.1 (0.52)	0.46 (0.11)	0.11	+1.07	3.1
		1.0	3.8 (1.00)	0.84 (0.19)	0.11	+0.28	5.5
0.0	0.01	0.1	1.0 (0.02)	0.21 (0.003)	0.10	−0.43	3.3
		1.0	3.5 (0.15)	0.78 (0.03)	0.11	−1.6	11.7
0.0	0.1	0.1	1.5 (0.05)	0.32 (0.009)	0.11	−0.63	2.2
		1.0	3.7 (0.45)	0.82 (0.09)	0.11	−1.7	5.5
−2.0	0.01	0.1	no adsorption				
		1.0	3.0 (0.0)	0.67	0.11	−3.4	10.1
−2.0	0.1	0.1	no adsorption				
		1.0	3.5 (0.0)	0.80	0.11	−3.6	5.0

the reader is referred to the original paper (Hesselink, 1977). Here we discuss some of the salient points.

2. *Predictions on Polyelectrolyte Adsorption*

In the case of adsorption of a negative polymer on a negative interface (τ, $\sigma < 0$), the adsorption is caused by ε, whereas the other terms tend to reduce it. In general, if the expression in the exponent in equation (8) becomes negative, the concentration of polymer on the interface becomes smaller than in solution, i.e. we have no adsorption (negative adsorption). The terms containing θ_0 are then unimportant and, for not too highly charged interfaces, the condition for adsorption is

$$\varepsilon - \tau\alpha\sigma_0(2000\pi/DRT)C_s^{-1/2} > 0 \tag{9}$$

where R is the gas constant and C_s is the molar salt concentration. Thus, if τ and σ_0 have the same sign, ε requires a finite positive value for adsorption. On the other hand, if τ and σ_0 are of opposite sign, ε can be zero or negative and the adsorption is purely electrostatic in nature. An increase in salt concentration can cause desorption. It should be noted that the condition for adsorption, equation (9), is independent of polymer molecular weight and concentration.

An example of this condition for adsorption is shown in Fig. 5; for a high adsorption energy ($\varepsilon = 1.0$) the surface charge density has little effect on the adsorption, whereas for a low adsorption energy ($\varepsilon = 0.1$) significant adsorption is found for a positively charged interface ($\sigma_0 = 2\ \mu\mathrm{C\ m^{-2}}$), while no adsorption is observed on a negatively charged interface.

All calculated isotherms have the usual experimentally observed shape, the "high-affinity" isotherm. At extremely low polymer concentration in solution the amount adsorbed, X, rises very steeply, whereas at higher concentrations it levels off to a saturation value, although the curves never become completely horizontal.

This levelling off occurs when the fraction of the adsorbent interface covered, θ_0, has reached saturation. Further adsorption is accompanied by decrease in p, the fraction of the segments actually adsorbed on the interface. Then, the fraction of the segments in the loops increases dramatically, resulting in a transition of the adsorbed polymer from a relatively flat conformation with most of the segments on the interface into a "loopy" conformation with most of the segments in the loopy layer.

The maximum value for θ_0 increases with increasing values of the adsorption energy and decreasing values of the degree of dissociation of the polymer, while for an anionic polymer θ_0 increases with increasing σ_0; θ_0 is relatively insensitive to the molecular weight of the polymer.

The fraction p of the segments adsorbed appears to be surprisingly independent of ε, α, σ_0, and the electrolyte concentration; p decreases with increasing molecular weight and concentration of the polymer.

(a) *Effect of salt concentration, degree of dissociation, and chain length.* In general, adsorption increases with increasing salt concentration, since increase in salt concentration reduces the electrostatic interactions which usually oppose adsorption. Only when the polymer and the adsorbent carry opposite charges do the electrostatics promote adsorption. Then the adsorption decreases with increasing salt concentration, when the charge interaction is the main reason for adsorption (as for $\varepsilon = 0.1$ in Table 1). However, when the nonionic adsorption energy drives the adsorption (as for $\varepsilon = 1.0$ in Table 1), an increase in salt concentration tends to increase the adsorption, especially at a high degree of dissociation (see Table 1).

At salt concentrations above about 0.1 mol dm^{-3}, salting-out effects can become important. Then χ, and therefore also the amount of polymer adsorbed, will increase with increasing salt concentration. For very low salt concentrations the theoretical model is invalid. The predictions are estimated to be most reliable at monovalent salt concentrations between say 0.003 and 0.1 mol dm^{-3}.

The effect of multivalent ions is not investigated, since such ions may have drastic, often specific, effects on α and σ.

Most polyelectrolytes are susceptible to precipitation in the presence of multivalent counterions, but the interaction of the multivalent ions with the polymer may vary from straightforward electrical neutralization to the formation of intra- or inter-molecular bridges (Dubin, 1977). Such effects need to be considered separately from the polyelectrolyte adsorption problem.

The amount adsorbed usually decreases with increasing degree of dissociation α. Experimentalists have explained this influence of α by correlating the decreased amount of adsorbed polymer, X, with the increased hydrodynamic volume of the polyelectrolyte coil. Although this volume effect may play a role, it does not explain the influence of ε and σ_0 on the X versus α dependence. The data for θ_0 and σ for an initially uncharged adsorbent with $\varepsilon = 1.0$ suggest a different explanation. At $\alpha = 1.0$ only a very small fraction of the interface is covered by polymer; $0.003 < \theta_0 < 0.13$. Apparently, the highly negatively charged segments find it extremely difficult to adsorb on a negative interface. Thus, at $\alpha = 1.0$ the polymer adsorbs until $\theta_0 = 0.09$ and $\sigma = 1.9\ \mu C\ cm^{-2}$, and after that no more segments adsorb. The amount adsorbed, X, can still increase by a decrease in p (swelling of the loop layer) until $p \approx 0.1$ for this chain length and polymer concentration. Less negatively charged segments ($\alpha = 0.4$) adsorb until $\theta_0 = 0.48$, $\sigma = -3.9\ \mu C\ cm^{-2}$. Segments carrying only 0.1 charge can adsorb on even more negative interfaces. However, the surface charge becomes only $-1.7\ \mu C\ cm^{-2}$, less than for $\alpha = 0.4$. Apparently, further adsorption is not prevented by electrical forces but most probably by the competition for space in the first layer.

A special case arises when polymer and adsorbent are oppositely charged. Then the amount of polymer adsorbed may go through a maximum with increasing α. For low values of α the electrostatics promote adsorption, and saturation of adsorption is caused by competition for space in the first layer. At higher degrees of dissociation, the electrostatic forces promote adsorption until the surface charge is neutralized, after which they oppose further adsorption.

Polyelectrolyte molecular weight does not play a role in the condition for adsorption, equation (9). Also, the fraction of the interface covered, θ_0, and therefore also the number of segments actually adsorbed per unit area, are almost independent of molecular weight.

However, the amount of polymer adsorbed, X, depends markedly on the degree of polymerization, P, as shown in equation (8). A plot of $\ln X$ versus $\ln P$ shows straight lines according to

$$\ln X = a \ln P + b$$

where a and b are constants and $a \approx 0.4$.

Theoretically, one would expect X to continue to increase with P, since entropy considerations would always favour the adsorption of 1 molecule with $P = 10^5$ over 10 molecules with $P = 10^4$. The fact that this is not observed experimentally could be caused by irreversibility of adsorption. A molecule with $P = 10^4$, of which 10^3 segments are adsorbed, is extremely unlikely ever to become desorbed by local fluctuations, so it will not make way for a molecule with $P = 10^5$. Thus, theoretical predictions on molecular weight dependence of polymer adsorption become hazardous for very high molecular weight polymer (say $M > 10^6$) because the underlying assumption of equilibrium adsorption may become invalid.

(b) *Thickness of the adsorbed layer.* The predictions of the theory for the thickness, s, of the adsorbed layer of 5–10 μm (see Table 1) for a polymer with an extended length of 20 μm are clearly unrealistic. These too high values for s are a direct consequence of the assumption of equation (7). Three aspects of this equation cause the overestimate for s. First, ellipsometric studies indicate the loop layer to have a higher polymer density than a coil in solution (at least for nonionic polymers). Second, because of the charge excess in the electrical double layer compared with the bulk solution, equation (7) leads to some overestimate of the thickness of the adsorbed layer. Third, and probably most important, polyelectrolyte theories overestimate the expansion of the coil caused by the intramolecular electrostatic repulsions by a factor of about 2 (see discussion of polyelectrolytes in solution). This leads to an overestimate for v_0, and therefore for s, by a factor of 8. Reducing the data for s in Table 1 by a factor of 8 would probably bring them to the right order of magnitude.

3. *Shortcomings of the Theoretical Model*

Equation (8) has been derived for polyelectrolyte adsorption from solutions containing sufficient salt so that the thickness of the adsorbed polymer layer significantly exceeds the Debye screening length ($\kappa s \gg 1$). Silberberg (1978) has formally indicated how the isotherm should be modified to extend its validity to lower salt concentrations and solutions free from added salts. A new term in the equation accounts for the counterions accompanying the adsorbing polyelectrolyte and has the effect of considerably reducing polyelectrolyte adsorption from salt-free solutions. It vanishes for salt solutions where the fraction of counterions belonging to the polyion is small compared with unity.

The use of equation (8) amounts to assuming that the charge density of the polyelectrolyte and the "volume" of the molecule are not altered upon adsorption. Even apart from the inability of polyelectrolyte theory to predict the "volume" of the coil in solution, these assumptions are weak, as argued in Section II, for the following reasons:

(i) The degree of ion binding/condensation may be different for an adsorbed polyion as compared with a polyion in solution; especially when particle and polyion are oppositely charged, this may be expected to be an important effect drastically reducing the charge density and thus the dimensions of the polyelectrolyte coil.

(ii) When particle and polyion carry the same type of charge (both anionic or both cationic) the mutual repulsion between the charges on the polyion is now also screened by the counterions of the particle, leading to a smaller polyelectrolyte coil. On the other hand, the charges in the adsorbed portions of the polymer (loops, tails) are repelled by the particle, causing swelling of the loop layer. However, according to Silberberg (1978), the loop layer will dilute itself by expansion until the mean concentration in this layer equals the concentration in the polyelectrolyte coils in solution. In other words, Donnan effects, owing to differences in the two local counterion concentrations, are expected to be unimportant, an argument in support of equation (7).

(iii) The electrical double layer of an adsorbing particle will have a significant effect on the degree of dissociation of a weak polyacid or polybase. When adsorbed on a negative particle, the dissociation of a weak polyacid will be reduced and the dissociation of a weak polybase may increase because of the prevailing negative potential.

A sophisticated theory would give us insight into these phenomena, whereas the present model uses the *a priori* assumption that charge density and volume of the polyelectrolyte coil remain unchanged upon adsorption.

In addition, the present model uses the old smeared-out charge distribution model for a polyelectrolyte coil instead of the probably better model of a worm-like chain with a local cylindrical symmetry and a persistence length, depending on charge density and ionic strength (see Section II).

The absence of such a sophisticated theory, and of suitable data to test it, has led the author to believe that the polyelectrolyte theory is still in its infancy. Silberberg's (1978) considerations deserve to be worked out in more detail. For the present, Hesselink's (1977) work may provide useful, albeit qualitative, understanding of the adsorption of flexible polyelectrolytes.

IV. Experiments on Polyelectrolyte Adsorption

A. ADSORPTION BEHAVIOUR

The shape of the adsorption isotherm is usually that of the high-affinity isotherm as shown in Fig. 1. The first polyelectrolyte molecules coming into contact with the adsorbent become adsorbed with a large fraction p of the segments. If more molecules are present in solution, the first adsorbed molecules may desorb partly, so as to allow more molecules to become attached to the adsorbent. This is clearly illustrated in the work of Ueda and Harada (1968) and of Froehling *et al.* (1977).

Ueda and Harada (1968) studied the adsorption of a cationic polymer on a negatively charged bentonite clay (see Table 2). Adsorption of the polycation reduces the cation exchange capacity of the clay. As long as less than 30 meq/100 g polymer is adsorbed, the polymer–adsorbent complex has no anion exchange capacity, indicating that the polycation is adsorbed in a very flat conformation. When more polymer becomes adsorbed, the complex has a definite anion exchange capacity, indicating that part of the polymer is not adsorbed on the interface but is dangling in solution.

Froehling *et al.* (1977) studied the adsorption of an anionic heparinoid type polyelectrolyte on a PVC powder where cationic groups were introduced. In the absence of these cationic groups the polymer is not adsorbed, indicating that the adsorption is driven by the electrostatic attraction between polymer and adsorbent. This is confirmed by the decrease in adsorption observed with increases in the salt concentration; the salt screens the electrostatic interactions and therefore reduces the adsorption. The most interesting feature of this study, however, is the data reproduced in Table 3.

Upon adsorption of the anionic polymer the cationic adsorbent releases Cl^- ions and the amount of released Cl^- ions is taken as a measure of the number of anionic segments actually adsorbed on the adsorbent. Table 3 clearly shows that the number of adsorbed segments reaches a saturation value at very low amounts of polyelectrolyte adsorbed. Upon further supply of polyelectrolyte the adsorption still increases drastically but, since the number of segments actually adsorbed remains constant, this increase in adsorption is accompanied by a decrease in the fraction, p, of the segments actually adsorbed on the interface, i.e. a drastic swelling of the layer of segments dangling in solution.

Table 2 Adsorption of a cationic polyelectrolyte on bentonite. (Ueda and Harada, 1968)

Total adsorbed polymer, A	Cation-exchange capacity, B	Anion-exchange capacity, C	Bound polymer, $D = A - C$	Total cation-exchange sites, $B + D$
0	62.4	0	0	62.4
8.6	49.5	0	8.6	58.1
18.5	44.6	0	18.5	63.1
33.6	30.7	1.0	32.6	63.3
40.7	24.5	3.9	36.8	61.3
43.5	23.9	5.3	38.2	62.1
47.9	21.4	6.7	41.2	62.6
46.6	23.7	6.4	40.2	63.9
57.7	15.8	13.2	44.5	60.3
58.5	14.4	15.1	43.4	57.8
63.0	12.8	15.5	47.5	60.3
64.3	11.1	18.2	46.1	57.2

Amounts are in milliequivalents per 100 g bentonite.
Polyelectrolyte: diallyldimethylammonium chloride–SO_2 copolymer ($M = 167000$).

In addition, the number of segments adsorbed on the interface increases with increased cationic character of the adsorbent. All these observations, the reduction of adsorption with increasing ionic strength, the saturation of the number of adsorbed segments, the increase of this number with increasing surface charge and the swelling of the "loopy" layer, are in agreement with the predictions of the theoretical model described in the preceding section. This model would further predict an increase in the total amount of adsorbed polymer with increasing surface charge and a levelling off of the fraction of segments actually adsorbed towards the same value for the three adsorbents involved. This prediction has not been tested in this work; this would require experiments at higher polyelectrolyte concentrations.

Table 3 Adsorption of a heparinoid anionic polyelectrolyte ($M = 750000$) on cationic PVC powder. (Froehling *et al.*, 1977)

Adsorbent ion-exchange capacity/ μeq g^{-1}	Adsorbed polyelectrolyte/ μeq g^{-1}*	Released chloride/ μeq g^{-1}	Number of adsorbed segments per molecule
0.0515	3.9	0.034	34.4
	14.1	0.035	9.8
	24.0	0.036	5.9
	46.3	0.039	3.3
	62.7	0.039	2.5
0.2167	4.9	0.096	77.4
	14.4	0.096	26.3
	28.1	0.095	13.4
	36.4	0.096	10.4
	68.2	0.094	5.4
0.3051	3.5	0.156	176.0
	12.1	0.155	50.6
	28.6	0.153	21.1
	43.2	0.153	14.0
	78.4	0.153	7.7

* From adsorption experiments with labelled polyelectrolyte.

Adsorption of an ionic polymer on a negatively charged adsorbent may be possible; when, however, the electrostatic repulsion becomes too strong, adsorption becomes prohibited. For instance, partially hydrolysed polyacrylamides may adsorb on a negative clay such as montmorillonite, but at 100% hydrolysis of this polymer (yielding polyacrylates) no adsorption occurs. Other examples of electrostatic repulsion preventing adsorption are given in the introduction of this chapter.

Actually, polyelectrolyte adsorption data on clays should be interpreted very carefully, because of the possibility of adsorption on the positively charged edges of overall negatively charged clay particles, and because of interlayer penetration of not too long molecules, possibly accompanied by swelling and ultimately dispersion of the clay particles as numerous very thin platelets (see Theng, 1979).

The fraction p of the segments actually adsorbed on the interface was measured by Robb (1979) for spin-labelled polyacrylic acid adsorbed on calcium phosphate. An E.S.R. detection method was used to distinguish between spins with a long relaxation time (adsorbed segments) and spins with shorter relaxation times (segments in free-dangling loops/tails). At low surface coverage the polymer was found to be adsorbed in a flat conformation with more than 90% of its segments adsorbed (immobilized) on the interface. When a significant part of the Ca in the adsorbent was replaced by Na, considerably more segments were found in the loops. Robb also concluded that the osmotic repulsion between unadsorbed segments and a possible spatial fit between charged groups on the polymer and the adsorbing sites can dominate the adsorption behaviour.

The thickness of the adsorbed layer has been measured using ellipsometry (e.g. Peyser and Little, 1974), and various kinetic techniques such as the electrophoretic mobility, the sedimentation velocity, etc, where the thickness of the adsorbed layer is found from the increase of the radius of the particle caused by polymer adsorption. For instance, using such kinetic methods, Eirich and Kudish (1970) found thicknesses of 10–70 nm for adsorbed gelatin layers, and they conclude that the molecules retain their coil dimension to a large extent in the adsorbed state.

The swelling of the loopy layer after surface saturation as described above has been documented for the adsorption of polyvinylpyrrolidone and polyacrylamide on chromium surfaces; the swelling of the loopy layer is accompanied by a decrease in segment density in this layer (Killmann and Kuzenko, 1974). Ellipsometry has often been used for protein adsorption (Adams *et al.*, 1978; Morrisey and Stromberg, 1974) but comprehensive data on the thickness of the adsorbed layer for flexible polyelectrolytes appear not to be available.

B. EFFECT OF CHARGE DENSITY OF THE POLYELECTROLYTE AND IONIC STRENGTH OF THE SOLUTION

Adsorption of ionized polymer molecules creates a potential, hindering further adsorption of such polyions. Therefore it may be expected, and it follows from the theoretical model described in the preceding section, that adsorption is in general decreased when the degree of dissociation or the charge density of the polymer is increased. Only if adsorbent and polyelectrolyte are oppositely charged may one find the adsorption to increase with increasing degree of dissociation.

This expectation is confirmed in numerous experiments; for instance, Schmidt and Eirich (1962) already found that the adsorption of vinyl acetate–crotonic acid copolymers on titanium dioxide reduced from 5 mg g^{-1} at pH = 5.5 to less than 2 mg g^{-1} at pH = 7.5. Another example is the adsorption of ampholytic polyelectrolytes such as gelatin which is usually found to be maximal at the isoelectric point (i.e.p.) of the molecule (Fontana, 1971). In the presence of multivalent counterions the maximum shifts parallel to the shift of the i.e.p. (Eirich and Kudish, 1970). Collagen, on the other hand, has its maximum of adsorption on glass at pH = 3.5, one pH unit below its i.e.p.; apparently the molecules must have a small positive charge for highest adsorption (Bettelheim and Priel, 1979).

The desorption of polyacrylic acid from clay and silica at pH $\approx$ 7 mentioned in the introduction is also an example of reduced adsorption because of increased charge antagonism between polyelectrolyte and adsorbent. Similarly, Parfitt (1972) found polygalacturonic acid (a polysaccharide) to adsorb at low pH on sodium, calcium, and (most strongly) aluminium montmorillonite; however at pH = 5.9 no adsorption was found on these clays.

This decreased adsorption with increased ionization of the polymer is sometimes explained on the basis of the space requirements for the expanded coil of the highly ionized polymer. However, this space requirement theory can never explain the complete desorption of highly ionized polyelectrolytes.

Addition of salt can be expected to reduce the build-up of a repulsive potential and therefore will increase adsorption unless polyelectrolyte and adsorbent are oppositely charged as in Tables 2 and 3. For instance, Peyser and Ullmann (1965) reported drastic reductions (more than a factor of 5) in the amount of polyelectrolyte adsorbed with increasing α, especially at low salt concentrations (0.007 mol dm^{-3} NaCl). Increasing the salt concentration to values of 0.04 and especially 0.2 mol dm^{-3} NaCl increased the polymer adsorption by about a factor of 2. The data shown in Fig. 1 also illustrate this increase in adsorption with increasing ionic strength.

Böhm (1974) studied the adsorption of partly esterified polymethacrylic acid on polystyrene latex as a function of degree of dissociation, salt concentration, and surface charge density. His results are shown in Table 4. Again adsorption is clearly reduced with increasing α. At $\alpha = 0.9$ and 0.02 mol dm^{-3} NaCl no adsorption was observed for the latex with a high surface charge density. Using this observation and equation (9), Hesselink (1977) estimated the nonionic adsorption energy ε of 5.2 kcal mol^{-1}, and with this value and equation (8) the theoretical predictions were obtained for the amount of polymer adsorption given in Table 4.

Lipatov *et al.* (1978) studied the adsorption of polymethacrylic acid ($M = 10^4$) on positively charged zinc sulphide ($\zeta = +13$ mV) and on negatively charged aluminium oxide ($\zeta = -43$ mV) and quartz ($\zeta = -61$ mV).

As was to be expected, the adsorption, X, on aluminium oxide decreased with increasing degree of dissociation α (at $\alpha = 0$, $X = 2.3$ mg m^{-2}; at $\alpha = 0.26$, $X = 1.2$ mg m^{-2}; at $\alpha = 0.42$, $X = 0.6$ mg m^{-2}). On the positively charged ZnS the adsorption increases with α in various salt solutions (see Fig 6). Adding salt increases the adsorption values practically irrespective of the dissociation of the polymer.

Lipatov *et al.* (1978) also calculated the amount of adsorbed polymer that would correspond to a monolayer of adsorbed coils with a radius of

Table 4 Adsorption of PMA-pe on polystyrene latices. (Böhm, 1974)

	Surface charge $\sigma_0 = -8.2\ \mu C\ cm^{-2}$			Surface charge $\sigma_0 = -1.4\ \mu C\ cm^{-2}$		
α	0.02 mol dm^{-3} NaCl X/mg m^{-2}	0.20 mol dm^{-3} NaCl X/mg m^{-2}	0.02 N $CaCl_2$ X/mg m^{-2}	0.02 mol dm^{-3} NaCl X/mg m^{-2}	0.20 mol dm^{-3} NaCl X/mg m^{-2}	0.02 N $CaCl_2$ X/mg m^{-2}
0.90	0 (0)	0.58 (0.68)	0.50	0.24	0.80	0.74
0.70	0.40 (0.12)	1.25 (1.17)	1.72	1.08	1.74	1.94
0.50	0.64 (0.45)	1.44 (1.46)	1.66	1.34	1.84	1.94
0.30	0.70 (1.44)	1.50 (1.47)	1.84	1.22	1.74	2.14
0.10	0.95 (1.47)	2.18 (1.47)	1.97	1.50	2.54	2.38
0.0	1.87 (1.47)	— —	2.54	1.92	—	3.36

Results in parentheses are theoretical, calculated by Hesselink (1977).

It should be noted that, at $\alpha < 0.1$–0.2, a conformational transition causes the PMA coil to collapse.

Under these conditions the value for X may increase drastically above the saturation limit of about 1.5 mg m^{-2}, calculated for the flexible PMA coil.

the coil in solution. They found that the measured adsorption is about a factor of 10 higher than the adsorption calculated on the basis of the size of the coil in solution; thus the size of the coil in solution is hardly related to the amount adsorbed, in agreement with the conclusion reached in the preceding section that the size of the polyelectrolyte coil in solution is by no means the determining factor for the amount of polymer to be adsorbed.

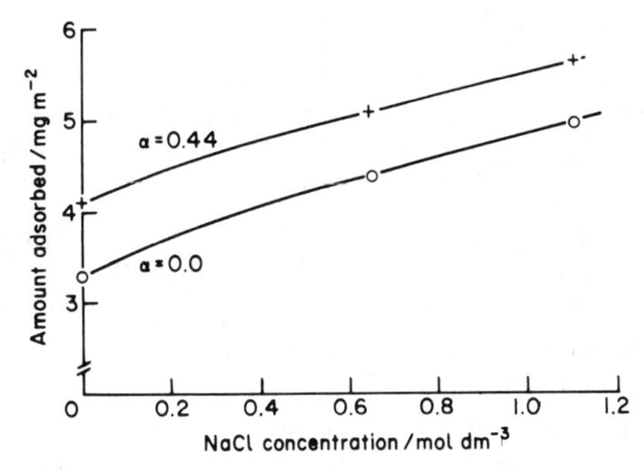

Figure 6 Adsorption of polymethacrylic acid on zinc sulphide ($\zeta = +14$ mV) at two degrees of dissociation and different NaCl concentrations. (Data from Lipatov *et al.*, 1978)

This same conclusion was also reached by Greene (1971) who found the adsorption of a strong polyacid (polysulphoethyl methacrylate) on polyethylene powder to depend much more strongly on the concentration of added NaCl than the radius of the polyelectrolyte coil in solution was affected by this added salt. Actually, the adsorption in deionized water was hardly measurable, whereas at 0.12 and 1.2 mol dm^{-3} NaCl the adsorption amounted to 0.6 and 5.0 meq/100g, respectively.

The coil size may, however, be an important parameter in porous media where it is often found that the adsorption decreases with increasing molecular weight (e.g. Cole and Howard, 1972). Similar observations have been made in concentrated suspensions.

C. CONFORMATIONAL TRANSITIONS FOR ADSORBED MOLECULES

Polyelectrolytes showing a conformational transition in solution may also do so when adsorbed at an interface. Actually, DiMarzio and Bishop (1974) have argued that such a transition would become sharper for an adsorbed polymer. A relatively well studied case is that of partially esterified polymethacrylic acid (PMA-pe, Van Vliet *et al.*, 1978, 1979). At a low degree of dissociation or in the presence of Ca^{2+} ions, the polyelectrolyte coil is contracted because of intramolecular interactions; in concentrated solutions this may lead to reversible gels. When adsorbed on emulsion droplets, addition of H^+ or Ca^{2+} activates these interactions between the free-dangling portions of the PMA-pe, leading to interparticle bridging and to emulsions with elastic properties.

D. POLYMERIC FLOCCULANTS

The mechanism of action of polymeric flocculants is particularly dependent on the mode of adsorption of these molecules, and therefore this application of polyelectrolyte adsorption is briefly discussed here. Both cationic and anionic polyelectrolytes are used in very low concentrations, since at higher concentrations these polymers would often stabilize the dispersion. Pioneering work in this field was done by La Mer and Healy (1963).

Since most dispersed particles are negatively charged, cationic flocculants often work via an electrostatic mechanism; the low molecular weight ($M \approx 10^4$) polyelectrolytes cause flocculation as they neutralize the surface charge. The high molecular weight cationics work via a combined electrostatic and bridging mechanism. Locally the polyelectrolyte becomes adsorbed, but part of the molecule protrudes into solution (tails, loops) and a second particle becomes attached on this positive spot on a negative particle. When an excess of cationic polyelectrolyte is added, the dispersion is stabilized and the particles show a positive ζ-potential.

Anionic polyelectrolytes work via a bridging mechanism; part of the molecule becomes adsorbed and part of it sticks out into solution, and the further it sticks out the more easily it catches a second particle. Thus, the electrostatic expansion of the polyelectrolyte coil improves the efficiency of the flocculant. Often some salt is required to bring the particles close enough for effective flocculation.

Kitchener (1972) formulates the salient features of polymeric flocculants: (a) polymer must be soluble in the medium, yet be able to become adsorbed, preferably with a long tail in solution, e.g. edges of clay particles; (b) optimum dosage is in principle at a surface coverage of 0.5 or somewhat less; (c) steric stabilization occurs at full surface coverage; (d) mode of mixing of polymer and adsorbent is important to ensure that at low surface coverage the polymer still has a long tail dangling in solution; (e) polymeric flocculants produce strong, somewhat elastic and open flocs, which can be filtered easily in comparison with electrolyte coagulation; (f) at sufficiently high shearing forces, flocs are irreversibly broken; (g) linear polyelectrolytes are more effective than branched ones.

An elegant application of our understanding of polyelectrolyte adsorption and flocculants is the development of selective flocculants; from a mixture of minerals the polymeric flocculant selectively flocculates one component. Read (1972) described how a strongly anionic polyacrylamide is able to flocculate, from a dispersion of haematite and silicate, a haematite-rich sediment; the haematite enrichment is improved when first NaF has been added to the dispersion. This fluoride increases the negative potential of both minerals, but the potential of the silicate becomes so negative that adsorption of the polyelectrolyte is prohibited and therefore the silicate is not flocculated by the polymer.

V. Adsorption of Polyelectrolytes with a Specific Conformation in Solution

Most biological polyelectrolytes such as proteins and nucleic acids are folded in a specific native conformation in solution. This conformation is often retained upon adsorption on a solid adsorbent; techniques such as immunoadsorption and the use of localized enzymes are based upon this feature. In this section we describe some work that addresses the possibility of a change in the native conformation (denaturation) upon adsorption. No large review is intended of the wide field of protein and nucleic acid adsorption.

Adsorption at liquid interfaces is outside the scope of this work but Miller's conclusion, that DNA preserves its double helical structure when adsorbed on negatively charged mercury but unfolds upon adsorption on a positively charged surface, is too beautiful not to mention here as an example of the effect of an adsorbing interface on the native conformation of a biological polyelectrolyte (Miller and Bach, 1973).

The adsorption of human plasma albumin (HPA) and ribonuclease (RNase) on sulphated polystyrene latex is probably one of the best studied cases of protein adsorption. Norde and Lyklema (1978, 1979) reported adsorption isotherms at various pH values, temperatures, and adsorbent charge densities (see Fig. 7); in addition to electrophoretic mobilities and heat effects upon adsorption they measured ion titration curves for adsorbed and free protein. Since their results are in general agreement with other studies, and provide a quite coherent body of information, we shall discuss this work in some detail.

The adsorption isotherms for HPA and RNase are usually of the high-affinity type* (see Fig. 7). Occasionally, definite steps in the isotherm are observed (see also Hlady and Füredi-Milhofer, 1979) indicating a reorientation of the adsorbed molecules (the enthalpy data disprove the possibility

* Silberberg (1973) has shown that also for globular polymers the adsorption isotherm can be expected to be of the high-affinity type.

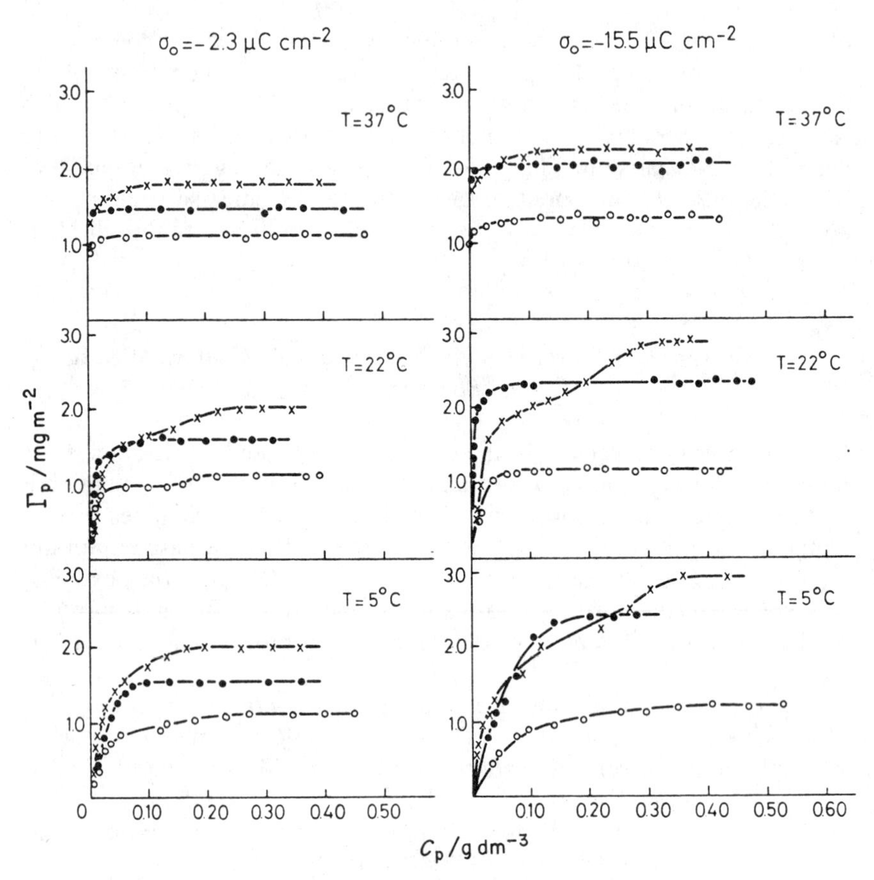

Figure 7 Adsorption isotherms for human plasma albumin (HPA) on sulphated polystyrene latex. (●) pH = 4.0, (×) pH = 4.7, (○) pH = 7.0; 0.05 mol dm^{-3} KNO_3. (Norde and Lyklema, 1978)

of multilayer adsorption). As in the case of flexible polyelectrolytes, adsorbed proteins are not desorbed by diluting the solution ("irreversible adsorption"); a change in pH may affect the adsorption.

Adsorption isotherms for RNase are less sensitive to variations in pH and σ_0 than those for HPA; for instance, for $\sigma_0 = -2.3\ \mu C\ cm^{-2}$ the plateau values for RNase are virtually independent of pH and temperature, whereas for HPA a variation with pH by a factor of 2 is observed. Only at a high surface charge density of the adsorbent ($\sigma_0 = -15.5\ \mu C\ cm^{-2}$) and a low ionic strength ($C_s = 10^{-2}$ mol dm^{-3}) does the adsorption depend markedly on pH.

The relative insensitivity to pH of RNase adsorption is related to the conformational stability of RNase in solution; it is usually adsorbed in its stable native conformation. The solution conformation of HPA, on the other hand, is sensitive to pH and so is its adsorption behaviour. HPA adsorption is maximal at its isoelectric point; then the adsorbed layer may correspond to a complete monolayer of more or less side-on adsorbed, conformationally unperturbed molecules. The reduction of the adsorption away from the i.e.p. is ascribed to conformational changes of adsorbed HPA molecules.

Increase in electrolyte concentration (0.01 to 0.05 mol dm^{-3}) appears to have little effect on adsorption; only for a negatively charged protein (high pH) on a highly negative particle did an increase in salt concentration increase the adsorption significantly. However, this may also be explained from the greater conformational stability of HPA at high ionic strength.

Particularly interesting are the titration results for adsorbed and free protein. Norde and Lyklema (1978) observe a change in pH upon adsorption of the protein molecules (see Fig. 8); apparently, upon adsorption the proteins absorb also some protons from the solution. Carboxyl groups of adsorbed HPA and RNase show a decreased degree of dissociation, an effect that becomes more pronounced at a highly negative adsorbent.

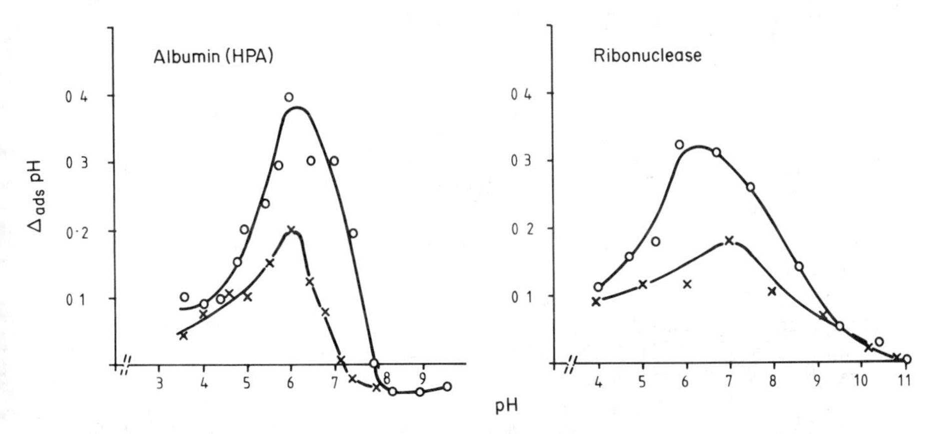

Figure 8 Variation in pH as a result of the adsorption of albumin and ribonuclease on polystyrene latex. (×) $\sigma_0 = -2.3\,\mu C\,cm^{-2}$, (○) $\sigma_0 = -15.5\,\mu C\,cm^{-2}$; 0.05 mol dm^{-3} KNO_3; 22°C. Upon adsorption the protein absorbs some extra protons at the expense of the solution. (Norde and Lyklema, 1978)

Apart from proteins, other ions (K^+, NO_3^-) also may become transferred from solution into the adsorbed layer. The trends are that at low pH a net uptake of anions occurs whereas at high pH, where both protein and adsorbent are negatively charged, a net amount of cations is incorporated in the adsorbed layer.

In addition, it was found that a fraction of the positively charged groups (ε-NH_3^+ groups) of the adsorbed protein is protected from deprotonation, probably because of ion-pair formation with the sulphate groups on the adsorbent.

Thus, differences between adsorbed and unadsorbed protein molecules in degree of dissociation and ion binding have been observed; in addition, indications are found for ion-pair formation between charged groups on the polyelectrolyte and the adsorbent. In Sections II and III the likelihood of these phenomena for adsorbed polyelectrolytes was mentioned, but Norde and Lyklema (1978, 1979) are among the first investigators who have observed them.

A number of optical techniques to study the conformation of adsorbed polyelectrolytes are being developed, such as ellipsometry (e.g. Adams *et al.*, 1978), circular dichroism (McMillin and Walton, 1974), quasielastic light scattering (Morrisey and Han, 1978), infrared spectroscopy (Morrisey and Stromberg, 1974), and fluorescence spectroscopy (Walton and Maepa, 1979). Studies using the full power of these methods in combination with the classical techniques should provide fascinating reading in the coming years. The elucidation of the conformational behaviour of polyelectrolytes near interfaces awaits such a combined approach.

References

Adams, A. L., Fisher, G. C. and Vroman, L. (1978). *J. Colloid Interface Sci.* **65**, 468.
Bantjes, A. (1978). *Br. Polym. J.* **10**, 267.
Bettelheim, F. A. and Priel, Z. (1979). *J. Colloid Interface Sci.*, **70**, 395.
Böhm, J. Th. C. (1974). "Adsorption of polyelectrolytes at the liquid-liquid interface", thesis, Wageningen.
Chattoraj, D. K., Chowrashi, P. and Chakravarti, K. (1967). *Biopolymers* **5**, 173.
Chough, E., Earl, C., Wadman, J. and Eirich, F. R. (1974). Prepr. 48th National Colloid Symp., Austin.
Clark, R. K., Scheuerman, R. F. and Rath, H. (1975). Soc. Petr. Eng. preprint 5514.
Cole, D. and Howard, G. J. (1972). *J. Polym. Sci. (A-2)* **10**, 993.
Cowan, J. C. and Weintritt, D. J. (1976). *Water Formed Scale Deposits*, p. 285, Gulf Publishing Company, Houston.
Crawford, J. E. and Smith, B. R. (1966). *J. Colloid Interface Sci.* **21**, 623.
DiMarzio, E. A. and Bishop, J. (1974). *Biopolymers* **13**, 2331.
Dubin, P. L. (1977). In *Structure-Solubility Relationships in Polymers* (F. W. Harris and B. R. Seymour, eds), Academic Press, New York.
Dubin, P. L. and Strauss, U. P. (1975). In *Polyelectrolytes and their Application*, Vol. 2 (A. Rembaum and E. Sélégny, eds), p. 3, Reidel, Dordrecht.

Eirich, F. R. and Kudish, A. (1970). Polym. prepr. ACS **11**, 1350.
Eisenberg, A. and King, A. (1977). *Ion-containing Polymers*, Academic Press, New York.
Fixman, M. and Skolnic, J. (1978). *Macromolecules* **22**, 863.
Fontana, B. J. (1971). In *The Chemistry of Biosurfaces* (M. L. Hair, ed.), Vol. I, Dekker, New York.
Frisch, H. L. and Stillinger, F. H. (1962). *J. Phys. Chem.* **66**, 823.
Frisch, H. L., Simka, R. and Eirich, F. R. (1959). *J. Polym. Sci.* **38**, 441.
Froehling, P. E., Bantjes, A. and Kolar, Z. (1977). *J. Colloid Interface Sci.* **62**, 35.
Greene, B. W. (1971). *J. Colloid Interface Sci.* **37**, 144.
Hand, J. H. and Williams, M. C. (1973). *Chem. Eng. Sci.* **28**, 63.
Hermans, J. J. and Overbeek, J. Th. G. (1948). *Rec. Trav. Chim. Pays-Bas* **67**, 761.
Hesselink, F. Th. (1972). *J. Electroanal. Chem.* **37**, 317.
Hesselink, F. Th. (1977). *J. Colloid Interface Sci.* **60**, 448.
Hesselink, F. Th., Ooi, T. and Scheraga, H. A. (1973). *Macromolecules* **6**, 542.
Hirasaki, G. J. and Pope, G. A. (1974). *Soc. Petr. Eng. J.* 337.
Hlady, V and Füredi-Milhofer, H. (1979). *J. Colloid Interface Sci.* **69**, 460.
Hoeve, C. A. J. (1971). *J. Polym. Sci.* (*C*) **34**, 1.
Hoeve, C. A. J., DiMarzio, E. A. and Peyser, P. (1965). *J. Chem. Phys.* **42**, 2558.
Joppien, G. R. (1978). *J. Phys. Chem.* **82**, 2210.
Killmann, E. and Kuzenko, M. (1974). *Angew. Makromol. Chem.* **35**, 39.
Kitchener, J. A. (1972). *Br. Polym. J.* **4**, 217.
LaMer, V. K. and Healy, T. W. (1963). *Rev. Pure Appl. Chem.* **13**, 112.
Lindstrom, T. and Söremark, C. (1976). *J. Colloid Interface Sci.* **55**, 305.
Lipatov, Yu. S., Fedorko, V. F., Zakordonskii, V. P. and Soltys, M. N. (1978). *Colloid J.* (U.S.S.R.) (English edn) **40**, 31.
McMillin, C. R. and Walton, A. G. (1974). *J. Colloid Interface Sci.* **48**, 345.
Mandel, M. (1970). *Eur. Polym. J.* **6**, 807.
Manning, G. S. (1969). *J. Chem. Phys.* **51**, 924, 3249.
Manning, G. S. (1972). *Ann. Rev. Phys. Chem.* **23**, 117.
Manning, G. S. (1974). In *Polyelectrolytes* (E. Sélégny, ed.), pp. 9–38, Reidel, Dordrecht.
Miller, I. R. and Bach, D. (1973). *Surf. Colloid Sci.* **6**, 185.
Morawetz, H. (1975). *Macromolecules in Solution*, J. Wiley, New York.
Morrisey, B. W. and Han, C. C. (1978). *J. Colloid Interface Sci* **65**, 423.
Morrisey, B. W. and Stromberg, R. R. (1974). *J. Colloid Interface Sci.* **46**, 152.
Nagasawa, M. (1975). *J. Polym. Sci.* (*C*) **49**, 1.
Nezlin, R. S. (1979). *Pure Appl. Chem.* **51**, 1421.
Noda, I., Tsuge, T. and Nagasawa, M. (1970). *J. Phys. Chem.* **74**, 710.
Norde, W. and Lyklema, J. (1978). *J. Colloid Interface Sci.* **66**, 257.
Norde, W. and Lyklema, J. (1979). *J. Colloid Interface Sci.* **71**, 350.
Odijk, Th. (1978). *Polymer* **19**, 989.
Pals, D. T. F. and Hermans, J. J. (1952). *Rec. Trav. Chim. Pays-Bas* **71**, 469.
Parfitt, R. L. (1972). *Soil Sci.* **113**, 417.
Pavlovic, O. and Miller, I. R. (1971). *J. Polym. Sci.* (*C*) **34**, 181.
Peyser, P. and Little, R. C. (1974). *J. Appl. Polym. Sci.* **12**, 2395.
Peyser, P. and Ullmann, R. (1965). *J. Polym. Sci.* (*A*) **3**, 3165.
Poland, D. and Scheraga, H. A. (1970). *Theory of Helix-Coil Transitions in Biopolymers*, Academic Press, New York.

Ramasamy, N., Weiss, B. R. and Sawyer, P. N. (1975). *J. Electroanal. Chem.* **62**, 179.
Read, A. D. (1972). *Br. Polym. J.* **4**, 253.
Robb, I. D. (1979). Personal Communication.
Schmidt, W. and Eirich, F. R. (1962). *J. Phys. Chem.* **66**, 1907.
Semenza, N. (1965). *J. Chromatogr.* **18**, 359.
Silberberg, A. (1968). *J. Chem. Phys.* **48**, 2835.
Silberberg, A. (1973). *Isr. J. Chem.* **11**, 153.
Silberberg, A. (1978). In *Ions in Macromolecular and Biological Systems* (D. H. Everett and B. Vincent, eds), Scientechnica, Bristol.
Smith, B. R. and Alexander, A. E. (1970). *J. Colloid Interface Sci.* **34**, 81.
Stewart, B. A. (ed.) (1973). *Soil Conditioners*, Soil Sci. Soc. Am., Madison.
Theng, B. K. G. (1979). *Formation and Properties of Clay-Polymer Complexes*, Elsevier, Amsterdam.
Ueda, T. and Harada, S. (1968). *J. Appl. Polym. Sci.* **12**, 2395.
Van Vliet, T., Lyklema, J. and van den Tempel, M. (1978). *J. Colloid Interface Sci.* **65**, 505.
Van Vliet, T. and Lyklema, J. (1979). *J. Colloid Interface Sci.* **69**, 332.
Verwey, E. J. W. and Overbeek, J. Th. G. (1948). *Theory of the Stability of Lyophobic Colloids*, p. 56, Elsevier, Amsterdam.
Vincent, B. (1974). *Adv. Colloid Interface Sci.* **4**, 193.
Walton, A. G. and Maepa, A. R. (1979). *J. Colloid Interface Sci.* **72**, 265.
Zimm, B. H. and Rice, S. A. (1960). *Mol. Phys.* **3**, 391.

Index

RETURN
LIBRARY
THE STANDARD OIL CO. (OHIO)
WARRENSVILLE LAB